D0321751

Principles *of* Vibration

SECOND EDITION

Benson H. Tongue

Department of Mechanical Engineering
University of California at Berkeley

New York   Oxford
OXFORD UNIVERSITY PRESS
2002

Oxford University Press

Oxford   New York
Athens Auckland Bangkok Bogotá Buenos Aires Cape Town
Chennai Dar es Salaam Delhi Florence Hong Kong Istanbul Karachi
Kolkata Kuala Lumpur Madrid Melbourne Mexico City Mumbai Nairobi
Paris São Paulo Shanghai Singapore Taipei Tokyo Toronto Warsaw
and associated companies in
Berlin Ibadan

Published by Oxford University Press, Inc.
198 Madison Avenue, New York, New York, 10016
http://www.oup-usa.org

Library of Congress Cataloging-in-Publication Data
Tongue, Benson H.
Principles of vibration / Benson H. Tongue.
p. cm.   Includes bibliographical references and index.
ISBN 0-19-514246-2
1. Vibration.   I. Title
TA355.T64   1996
620.2--dc20   95-45031

Printing (last digit): 9 8 7 6 5 4 3 2
Printed in the United States of America
on acid-free paper

To my wife, Claire, without whom this
book definitely would never have been written.

A subject is difficult only so long as you don't
understand it. Once you do, it becomes intuitively
obvious.

Benson Tongue (1957–present)

Contents

Preface

Well, several years have gone by and, happily, I've gotten requests for a second edition. So here it is. Since I now get another chance to wax eloquent, I'm going to make the most of it.

Those who're new to the book are possibly wondering why it exists, i.e., what makes it any different from "the competition." My claim to fame, if I have one, is that I write "to the students" as opposed to writing for the instructors using the book. These really are different audiences and I believe that the approach I've chosen makes a significant difference in the interest/motivation and ultimate learning that takes place. By "writing for the instructor" I mean writing a long and detailed text that strives for completeness and rigor, something that an expert in the field will appreciate and value. The problem is that at this point in their careers, students encountering the book are just beginning to learn vibrations. Just as a dance instructor would never overwhelm a beginning ballerina with complicated pirouettes and arabesques, there's nothing to be gained by piling vibrations esoterica onto someone at the start of his or her studies.

The approach I take is to introduce the topics of interest in a physically motivated way and to then show how one can use the material in a useful manner. Once that's done I go into some of the underlying details. The tone of the text is decidedly familiar. Many students have e-mailed and told me that it seemed that I was there with them, explaining the material to them in a one-on-one sort of way, as they read the book. That's exactly what I was aiming for. One student even went so far as to say the book makes good bedtime reading, but I think they might have been going a little overboard on that observation.

All of us that've gone through years of wading through textbooks know that many of them are boring to the point of tears. It's just not helpful, in my opinion, to have a fat book that's crammed with facts but is so inaccessible that the student doesn't get any motivation to read it. So I've chosen to limit the number of topics I present and to present them in as engaging a manner as I can manage. As an aside, it's been my experience that nobody can actually cover all the material in my book, let alone in some of the more densely packed chapters, in a typical semester or one- or two-quarter course. Thus, stuffing more material merely serves to make the book more difficult to lug around and more costly. I'd love to think that the majority of the students who use the book subsequently keep it close to their hearts, but the sad reality is that after a course is finished, most all textbooks find new life as used books in the school bookstore. Thus pumping the book full of facts so that it can serve as a lifetime reference becomes somewhat problematic.

I use MATLAB in the book and highly recommend that anyone pursuing a degree in engineering go out and buy a student copy of MATLAB. It'll help in doing the problems in this book and it'll help in most of the person's other courses. Or, check if your school has MATLAB already installed on school machines. When I was first learning vibrations I had to take a lot of the material on faith, since it was too much of a chore to verify some of the assertions through calculation. Having a convenient software package running on a PC or a larger machine makes it easy to check the book and also easy to try different cases—to poke around with the mathematics a bit.

It was a temptation to include a disk with presolved problems that students could fire up at will. But, as you can see, I didn't. There were a couple of reasons. First, some instructors have no interest in MATLAB and some schools don't have it resident on their machines. Thus, they'd be subsidizing the software for schools that do use it. More importantly, I feel that MATLAB is very user friendly and that the more time you put into it, the more you'll be inclined to use it and to learn more about it. None of the problems in the book really require more than a few lines of commands to run, and writing them is fairly quick and painfree. Thus I've indicated what commands will enable the reader to solve the problems in the book, and I've included some explicit MATLAB commands so that readers planning to use this package can get started.

The material in the book grew out of lectures to junior, seniors, and first-year graduate students at Georgia Tech and at Berkeley. My presumption is that the student using the text has encountered linear algebra as a sophomore (but has likely forgotten much of it), has had a course in deformable bodies, and has taken a first course in dynamics. The text starts off with the classical material (single-degree-of-freedom systems) and then moves into more modern material. A good deal of emphasis is given to multiple-degree-of-freedom systems, as this subject is probably the most important for the practicing engineer. Whether obtained through a modal approach or a lumped discretization, most practical vibrations problems ultimately lead to multiple-degree-of-freedom systems.

Many of the problems are meant to challenge the student to think deeply about a particular problem, to really analyze why a certain thing happens and what can be accomplished with a given technique. The instructor should definitely look at the solution book before assigning a problem, as some have very short solutions whereas others are much more involved. Many absolutely require the use of a computer and a good set of eigenanalysis algorithms.

One unique aspect of the book is Chapter 7, Seat-of-the-Pants Engineering. There are many ways to get a quick answer or to verify an analysis that are known to experienced engineers but are picked up only slowly, after years of working on vibration problems. And yet these approaches are useful for everyone to know right from the start. This chapter brings together a few of these approaches and applies them to all the systems that have been discussed. I felt that this material deserved the recognition of a separate chapter because it essentially carries the idea of approximate analyses (discussed in Chapter 6) to another level. Rather than being an approximate analysis that, with more effort, approaches the actual solution, these techniques are inherently approximate approaches that can't be made to emulate the real solution but can give the user a fast and accurate estimation of the real solution.

I hope that this book strikes the right balance between detail and accessibility and that it makes learning vibrations just a bit more fun than might otherwise be the case. I'd really enjoy hearing from you and finding out if the book achieved its aims. If you want to pass on your compliments or criticisms, I can be reached at bhtongue@newton.me.berkeley.edu. Errata, updates, and answers to selected problems can be found at

http://www.me.berkeley.edu/faculty/tongue/prinvib4.html.

Berkeley, California   B. H. T.

ACKNOWLEDGMENTS

This book wouldn't exist without contributions from many people. My views on the whole field of dynamics were molded by my Ph.D. advisor, Earl H. Dowell, a masterful researcher whose abilities span an impressive range and whose ability to see the simple idea that underlies a confusingly complex phenomenon is much admired (by me) and but poorly approximated.

Many students have given me the benefit of their reactions to this material, both in lecture notes and in book form, and their input has led to a better book. To everyone that pointed out something that could be improved—thanks.

Since the first edition of this book appeared, I've received feedback from numerous people, all of which was appreciated. One of these correspondents, Colonel Wayne E. Whiteman, of the US Military Academy, truly deserves special mention. Colonel Whiteman went over my book with a fine-tooth comb and e-mailed me on many, many occasions to point out areas that were unclear, ambiguous or, most embarrasingly, incorrect as stated. His contributions helped to markedly improve the book in its second printing. My fond hope is that he'll have nothing to write to me about in this second edition.

Thanks to my son Curtis for transcribing some of my sloppily written problems into electronic form.

Peter Gordon and Karen Shapiro, editors at Oxford University Press, were there to help guide the book along into finished form and their help was greatly appreciated.

And finally, I'd like to thank Brenda Griffing for her marvelously complete and truly excellent job of copy editing. She found typos that I certainly wouldn't have noticed and showed a deft hand at sentence management that kept my writing style intact while simultaneously increasing clarity.

1

Free Vibration of Systems with a Single Degree of Freedom

1.1 INTRODUCTION

In this chapter we'll examine the responses of systems with a single degree of freedom. This is one of the most important topics to master, since the more complicated cases (multi-degree-of-freedom and continuous systems) can often be treated as if they are simply collections of several, individual, single-degree-of-freedom systems. Once the responses of these simpler systems are well understood, branching out to the more complicated situations becomes relatively easy. Initially, we will ignore damping, since for many common structures the presence of damping changes the overall responses only slightly. Once we have a good grasp of how undamped problems behave, we'll look at the effects of damping.

Before diving in, it makes some sense to explain what "single degree of freedom" actually means. This idea is actually pretty simple—if an object is free to move in a particular direction, then it's got a degree of freedom in that direction. Thus, a bead that slides on a wire of an abacus has one degree of freedom. Figure 1.1 shows another example of a single-degree-of-freedom system, a mass in a channel. Since the mass is free to move only up or down the channel, it has one degree of freedom. We've used the coordinate x to describe the position of the mass. With this representation, it is clear

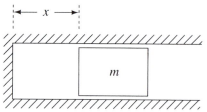

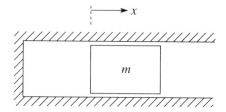

Figure 1.1   Single-degree-of-freedom system, single mass

Figure 1.2   Single-degree-of-freedom system, single mass

what x stands for; it gives us the displacement of the mass from the left end. The more common way that we'll illustrate coordinates is shown in Figure 1.2. It looks likes the x coordinate is attached to the mass, but it really isn't. x motions are still referenced to the fixed channel, and the starting point of the x coordinate is aligned with the mass because the initial (zero displacement) reference occurs exactly at the mass's initial position. Any variations of the mass from the position it's currently in will produce a finite x reading.

Also, you should ignore any physical dimensions of the mass unless they're explicitly mentioned. For the purposes of this chapter we'll consider all masses to act as *point masses*; i.e., we can neglect their dimensions. Thus, although the masses are drawn in Figures 1.1 and 1.2 as having finite dimensions, that's really done just as an aid in visualizing them.

What about more than a single degree of freedom? A point mass on a flat plane has two degrees of freedom (it can move both up or down, and horizontally, for instance, as shown in Figure 1.3). A penny on a flat plane has three degrees of freedom—two to define the position of the penny's center and a third (an angular rotation) to describe what angle the penny is oriented in. Rigid bodies in three-space have six degrees of freedom—three translational (telling the location of the center of mass) and three rotational (telling the angular orientation of the body). In this text, you'll sometimes see *degree of freedom* written out but most often it will be abbreviated as DOF. Thus 2 DOF should be read as *two degrees of freedom.*

Conceptually, here's how the book will develop. We'll start by looking at SDOF, vibratory systems with no externally applied forces (Chapter 1). After we've gotten a feel for their characteristics, we'll add forcing and see how the systems behave (Chapters 2 and 3). This will give us an excellent grounding in the basic physics that all vibrational systems will display. Logically, we'll then move

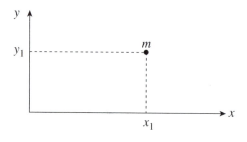

Figure 1.3   Two-degree-of-freedom system, single mass

on to multiple-DOF systems (Chapter 4). If we understand these systems, we've learned the most important part of the book, since essentially all modern vibrational analyses consider multiple-degree-of-freedom problems. As the number of degrees of freedom gets larger and larger, the system begins to look more like a continuum, rather than a collection of discrete pieces. Thus we'll spend some time looking at continuous systems and how they behave (Chapter 5). Following this, we'll develop some approximate solution methods that allow us to handle continuous problems that are just too difficult to solve exactly (Chapter 6). Finally, we'll show how some insight can help speed the solution procedure (Chapter 7) and touch on some of the aspects of modern modal analysis (Chapter 8).

1.2 TRANSLATIONAL VIBRATIONS—UNDAMPED

The first vibrational device that we'll discuss is a spring-mass system, something that you can make very easily. In fact, I strongly encourage you to take a minute right now and make one, since doing so will really help you in understanding several basic vibrations concepts. Although you can follow along without making an actual device, using one will greatly aid in developing your physical intuition.

In the good old days you'd have gotten a nifty spring included with the book. Sadly, too many people were ripping the springs out of the packaging and so the publisher (Oxford University Press) had to stop including them. Happily, they've agreed to supply the springs to the course instructor and, if all went well, he/she should already have them on hand. If not, ask him/her to give Oxford a buzz at PCG@OUP-USA.org and ask for them. If all else fails, you can try to use rubber bands instead of the spring.

Okay, assuming you've got a spring, the next thing to do is attach one end of it to a handy weight (a ceramic coffee cup will do nicely), and hold the free end of the spring in your hand. Make sure that the weight you've chosen isn't so heavy that the spring is stretched to near its limit, i.e., there must be a good bit of stretch remaining in the spring beyond its extension due to gravity. When you're done, you'll have a high-class vibrational test device like the one shown in this chapter's opener. (A cautionary note is needed here. If you decided to use rubber bands because you can't get hold of a spring, you should be aware that the rubber dissipates *much* more energy than a spring does, and so the vibrations will die away quickly and the forced tests to come in the next chapter will require more effort to make them work.)

These two elements, the mass and spring, are the most basic building blocks for vibrational analyses. Even though we know that our mass has some degree of internal flexibility, our model is considered to be infinitely rigid. Similarly, although the physical spring clearly has a mass of its own (it's not weightless, after all), our model will have none. Thus we are already approximating the real world. This idealization is necessary if we are to come up with reasonable mathematical models, but you should keep in mind that we are using approximations, which never exactly match reality. The fact that actual objects combine mass and stiffness (and damping) will become much more important when we look at the vibration of continuous systems, such as strings, rods, and beams. Happily, even though our system models will not *exactly* match reality, they can come very close to doing so, thus enabling us to use the approximations and feel confident that our results have physical meaning.

Having introduced masses and springs with words, it is appropriate to pin down exactly what we mean. Mass is a fundamental quantity, possessed by all macroscopic objects. This being a vibrations text, we won't get into the details of how physicists currently view matter, but instead use our well-known Newtonian view that mass times acceleration equals force and that the force generated by a mass in a gravitational field is called its weight.

Although we won't get any vibrations out of a mass by itself, it's useful to analyze the behavior of simple masses, especially since complicated objects such as airplanes and spacecraft can behave in a similar manner to a lumped mass under normal operating conditions. If we write out Newton's second law for a constant mass, we have

$$m\ddot{x} = f \tag{1.2.1}$$

Note that an overdot indicates differentiation with respect to time. If the applied force f is zero, then the solution to this equation is simply

$$x(t) = a + bt \tag{1.2.2}$$

where a and b are constants. Thus the mass can be stationary (if $b = 0$), or it can move with a constant speed b, in what we call rigid-body motion. In our case, there really isn't any body to speak of, since we're dealing with a mass particle. But more realistic extended bodies, such as aircraft, show the same behavior. If they aren't restrained, then they can remain in one spot or, depending on the initial conditions, translate and/or rotate at a constant speed. Those of you who've been in a jet when it hits a sudden downdraft can certainly attest that the whole plane can move sharply down without pitching or rolling to any great extent.

If we plot the relation between acceleration and force, we obtain Figure 1.4. The plot is linear and the slope varies with the mass m. Thus, the larger the mass, the less it accelerates for a given force.

Springs are a little more complicated than masses, but not by much. As we've already mentioned, to accelerate a given mass m at an acceleration a, we must apply a force f, where $f = ma$. In the case of a spring, an applied force doesn't produce an acceleration, but rather a displacement. Figure 1.5 shows the relationship between force and displacement for the illustrated spring: the graph

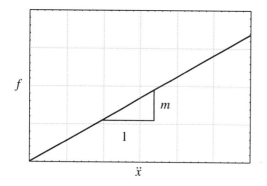

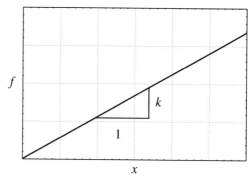

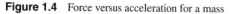

Figure 1.4   Force versus acceleration for a mass   **Figure 1.5**   Force versus displacement for a spring

is linear, indicating that an increase in the applied force will produce a proportional change in the displacement. The constant of proportionality is commonly denoted by k and is called the spring constant. Thus

$$f = kx \qquad (1.2.3)$$

When we increase k we're increasing the stiffness of the spring, making it harder to deflect.

Example 1.1

Problem   A force deflection test yields the data shown in Figure 1.6. Determine the spring constant of the tested element.

Solution   The slope of the data is equal to $\frac{40\,\text{N}}{5\,\text{mm}} = 8000\,\text{N/m}$. Thus the spring constant is equal to 8000 N/m.

Example 1.2

Problem   Find the local spring constant for motions about $x = 1$ m for the nonlinear spring characteristic shown in Figure 1.7. The f axis is given in newtons and the x axis is in meters.

Solution   Figure 1.7 shows a force versus deflection curve of a nonlinear spring for which

$$f(x) = x + x^3$$

To determine the linear spring characteristic, we have to examine the system for motions about a particular deflection, in this case $x = 1$ m. The local linear spring characteristic is given by

$$k = 1 + 3x^2 \Big|_{x=1} = 4\,\text{N/m}$$

If the system is in static equilibrium when $x = 1$ m, then we now know that, for small oscillations about this equilibrium, the local spring constant will be 4 N/m.

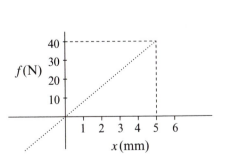

Figure 1.6   Experimental force/deflection data

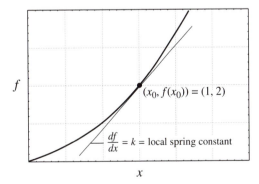

Figure 1.7   Local spring constant

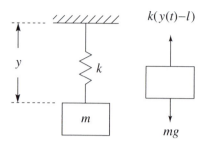

Figure 1.8   Free body diagram of a spring-mass system

Throughout the rest of the book we will use m_1, m_2, k_1, k_2, etc. to indicate both the particular object we're looking at and its associated physical quantity. Thus we might call a particular spring k_1 and also indicate its stiffness by k_1. Otherwise we'd have to say "consider the spring s_1, with spring constant k_1 and the mass n_1 with mass equal to m_1," which is long-winded and also uses up too many symbols. It'll be obvious from the sentence whether m_1 refers to the object itself or the mass of the object and whether k_1 refers to the spring or its spring constant.

The first experimental test we'll run will be a free vibration experiment. Keeping the free end of the spring fixed with the mass suspended beneath it, pull the mass down a bit and then release it. You'll note that the mass oscillates up and down at a fairly regular rate. Furthermore, it seems to oscillate with close to a constant amplitude. We can easily demonstrate that the mathematical equations of motion of this system exhibit these same properties. From the free body diagram of Figure 1.8 we see that the equation of motion for the system is

$$m\ddot{y}(t) = -k\big(y(t) - l\big) + mg \tag{1.2.4}$$

The spring's unstretched length is equal to l (thus the spring force is proportional to the stretch in the spring beyond this length) and g is the acceleration due to gravity. Although this representation is perfectly valid, it doesn't utilize an important fact, namely, that the oscillations will occur about the static equilibrium position of the mass due to gravity. This position is the one at which the restoring force due to the spring just balances the force due to gravity

$$k(y_0 - l) = mg \tag{1.2.5}$$

If we let $y(t) = y_0 + x(t)$, i.e., if x represents the motions about the equilibrium position, then we can rewrite (1.2.4) as follows:

$$m\ddot{x}(t) = -k(y_0 + x(t) - l) + mg \tag{1.2.6}$$

or, using our equilibrium equation (1.2.5) to eliminate mg,

$$m\ddot{x}(t) = -k(y_0 + x(t) - l) + k(y_0 - l) = -kx(t) \tag{1.2.7}$$

Moving all the dependent variables to the left side of the equation leaves us with

$$m\ddot{x}(t) + kx(t) = 0 \tag{1.2.8}$$

The difference between $y(t)$ and $x(t)$ is often a point of confusion, make sure that it is clear in your mind. Gravity simply serves to set up an equilibrium configuration for the system, and the system's vibrations occur about this equilibrium. You'll examine this more fully in some of the homework problems.

At this point, we'll stop showing the time dependency as (t) when it is clear from the context which of the variables are time dependent and which are just constants:

$$m\ddot{x} + kx = 0 \tag{1.2.9}$$

Equation (1.2.9) is an autonomous (no forcing), constant coefficient, linear differential equation. All constant coefficient, linear, ordinary differential equations have solutions of the form

$$x(t) = ae^{\lambda t} \tag{1.2.10}$$

That is, they exhibit exponential behavior. This is worth repeating. *ALL* autonomous, constant coefficient, linear differential equations have exponential solutions. Burn this into your brain; it underlies all the solution procedures we'll be meeting later in the book.

Using this assumed solution, (1.2.9) becomes:

$$(m\lambda^2 + k)ae^{\lambda t} = 0 \tag{1.2.11}$$

Since a can't be equal to zero (unless $x(t) = 0$ seems interesting to you) and $e^{\lambda t}$ isn't equal to zero for any finite time t, we're left with requiring that

$$m\lambda^2 + k = 0 \tag{1.2.12}$$

Therefore $\lambda^2 = -\frac{k}{m}$ or $\lambda_{1,2} = \pm i\sqrt{\frac{k}{m}}$ Thus the full solution can be written out as

$$x(t) = a_1 e^{it\sqrt{\frac{k}{m}}} + a_2 e^{-it\sqrt{\frac{k}{m}}} \tag{1.2.13}$$

Using the fact that $e^{\pm i\omega t} = \cos(\omega t) \pm i \sin(\omega t)$, we can just as easily express $x(t)$ as

$$x(t) = b_1 \cos\left(\sqrt{\frac{k}{m}}t\right) + b_2 \sin\left(\sqrt{\frac{k}{m}}t\right) \tag{1.2.14}$$

You can check that both (1.2.13) and (1.2.14) are valid solutions by using them in (1.2.9) and observing that they satisfy the equation of motion.

We see from (1.2.14) that $\sqrt{\frac{k}{m}}$ has the units of reciprocal seconds (since the argument of a sine or cosine is nondimensional) and thus has the dimensions of a frequency. We commonly refer to this particular frequency as the system's natural frequency, ω_n

$$\omega_n \equiv \sqrt{\frac{k}{m}} \tag{1.2.15}$$

We'll see when we start looking at forced vibrations that this frequency plays a very important role in a system's vibrational response.

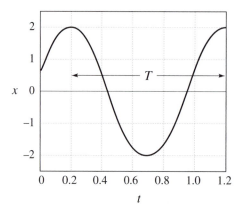

Figure 1.9 Sinusoidal waveform

If we plot (1.2.14) versus time for particular values of b_1, b_2, k, and m, we obtain Figure 1.9. You can see that the response is a periodic signal, meaning that the response repeats itself over and over. The time necessary for the signal to repeat (T) is the fundamental period of the oscillation, which is related to the natural frequency by

$$\omega_n = \frac{2\pi}{T} \tag{1.2.16}$$

In the absence of damping, the amplitude of the solution stays constant. Hopefully, when you performed the free vibration experiment with the spring and mass you saw the mass perform quite a few oscillations without much decay in the response. This was because the system didn't have much damping.

The details of what a particular response will look like depend upon the system's initial conditions. Given the right start, we could see a pure sine wave, a cosine, or a combination of the two. The next two examples will make this clear.

Example 1.3

Problem   Given a sinusoidal signal with frequency equal to 1.0 rad/s and initial conditions $x_0 \equiv x(0) = 2$ and $\dot{x}_0 \equiv \dot{x}_0 = 0$, find the form of $x(t)$.

Solution   From (1.2.14) we know

$$x(t) = b_1 \cos(t) + b_2 \sin(t)$$

Differentiating once yields

$$\dot{x}(t) = -b_1 \sin(t) + b_2 \cos(t)$$

Evaluating at $t = 0$ we have

$$x(0) = b_1 \cos(0) + b_2 \sin(0) = b_1 = 2$$
$$\dot{x}(0) = -b_1 \sin(0) + b_2 \cos(0) = b_2 = 0$$

$$x(0) = 2, \ \dot{x}(0) = 0$$

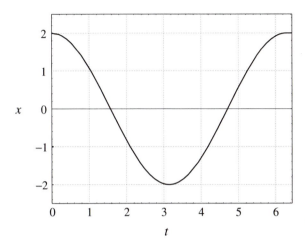

Figure 1.10   Free response of a spring-mass system

which gives us

$$x(t) = 2\cos(t) \tag{1.2.17}$$

This signal is shown in Figure 1.10.

Example 1.4

Problem   Given a sinusoidal signal with frequency equal to 2π rad/s and initial conditions

$$x_0 = 1$$
$$\dot{x}_0 = 5$$

find the form of $x(t)$.

Solution   From (1.2.14) we know

$$x(t) = b_1 \cos(2\pi t) + b_2 \sin(2\pi t)$$

and so, by differentiating, we have

$$\dot{x}(t) = 2\pi \left(-b_1 \sin(2\pi t) + b_2 \cos(2\pi t) \right)$$

Thus

$$x(0) = b_1 \cos(0) + b_2 \sin(0) = b_1 = 1$$
$$\dot{x}(0) = 2\pi b_2 = 5 \quad \Rightarrow \quad b_2 = \frac{5}{2\pi}$$

$$x(0) = 1, \quad \dot{x}(0) = 5$$

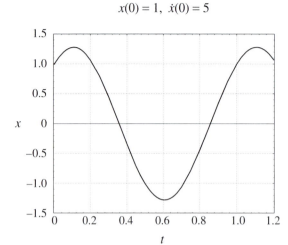

Figure 1.11   Free response of a spring-mass system

yielding

$$x(t) = \cos(2\pi t) + \frac{5}{2\pi} \sin(2\pi t) \tag{1.2.18}$$

This signal is shown in Figure 1.11.

Example 1.5

Problem   A mass is suspended from a nonlinear spring. Find the equilibrium position of the mass and the natural frequency of small oscillations about this position. $m = 1$ kg and the force necessary to deflect the spring is given by

$$f = 1.5x + .2x^2$$

where x has the units of millimeters and f has the units of newtons.

Solution   In an equilibrium condition, the force due to gravity is balanced by the force of the spring:

$$mg = 1.5x + .2x^2$$

Rewriting this (and using the known parameter values) yields

$$x^2 + 7.5x - 49.05 = 0$$

This quadratic equation has a positive root of $x = 4.194$ mm. k is the linearized spring constant of our nonlinear spring about the equilibrium position. Thus

$$k = \left.\frac{df}{dx}\right|_{x=4.194} = 1.5 + .4x\Big|_{x=4.194} = 3.178 \text{ N/mm} = 3178 \text{ N/m}$$

Although (1.2.17) and (1.2.18) are both perfectly valid ways to write the solutions, there are actually other forms that are just as correct. For instance, say we didn't like using cosines. Would it be possible to express (1.2.17) in terms of just a sine? The answer is "yes." There's no problem with doing this. All we need to do is include the appropriate phase shift. Our original solution was $x(t) = 2\cos(t)$. If we want to use a sine, then we'd say $x(t) = 2\sin(t + \frac{\pi}{2})$. Check it out. At every time t, both these expressions produce the same answer. The $\frac{\pi}{2}$ term is called a *phase shift*, and it does what the name implies—it shifts the phase of the sine wave by $\frac{\pi}{2}$. This can also be viewed as shifting the effective origin of the time axis by $t = \frac{\pi}{2}$.

Since the idea of phase shifts will be of fundamental importance to understanding vibrational responses, we'll spend some time on the concept now. You'll recall from trigonometry that sines (or cosines) of the sum of two quantities can be expanded in terms of sines and cosines of those quantities through the formulas

$$\sin(x \pm y) = \sin(x)\cos(y) \pm \cos(x)\sin(y) \qquad (1.2.19)$$

$$\cos(x \pm y) = \cos(x)\cos(y) \mp \sin(x)\sin(y) \qquad (1.2.20)$$

Thus, if we wished to express

$$a_1 \sin(\omega_n t) + a_2 \cos(\omega_n t) \qquad (1.2.21)$$

as a cosine with a phase shift ϕ we would say

$$d\cos(\omega_n t + \phi) = d\cos(\omega_n t)\cos(\phi) - d\sin(\omega_n t)\sin(\phi) \qquad (1.2.22)$$

and, comparing the coefficients of $\cos(\omega t)$ and $\sin(\omega t)$ in (1.2.21) and (1.2.22), see that

$$d\cos(\phi) = a_2 \qquad (1.2.23)$$

and

$$d\sin(\phi) = -a_1 \qquad (1.2.24)$$

Squaring both sides of (1.2.23) and (1.2.24), adding them together and then taking the square root yields

$$d = \sqrt{a_1^2 + a_2^2} \qquad (1.2.25)$$

and dividing $d\sin(\phi)$ by $d\cos(\phi)$ yields

$$\tan(\phi) = -\frac{a_1}{a_2} \qquad (1.2.26)$$

or

$$\phi = -\tan^{-1}\left(\frac{a_1}{a_2}\right) \qquad (1.2.27)$$

Thus we can always express the overall vibration as a cosine function with a phase shift and, in an exactly parallel manner, also express it as a sine function with a phase shift.

Example 1.6

Problem　Put $2\sin(3t) + 4\cos(3t)$ into the form $d\cos(3t + \phi)$.

Solution　For this problem, $a_1 = 2$ and $a_2 = 4$. Thus, using (1.2.25) we have

$$d = \sqrt{2^2 + 4^2} = \sqrt{20}$$

and (1.2.27) gives us

$$\phi = -\tan^{-1}\left(\frac{2}{4}\right) = -.464$$

Therefore

$$2\sin(3t) + 4\cos(3t) = \sqrt{20}\cos(3t - .464)$$

Figure 1.12 shows what the final waveform looks like and shows how the phase shift acts to move the cosine wave over to the right. The effective time origin of the cosine wave is at $t = \frac{.464}{3} = .155$.

　　Okay, now we know how to find the free response of a system that consists of one mass and one spring. But what if our system is more complex than this? Suppose it involves several springs. What do we do? Well, let's consider the problem.

　　Say we have a mass supported by two springs in *series*, as shown in Figure 1.13a. This is a very common situation. For instance, if you were modeling an automobile, you'd want to take account of the spring stiffness of the tire first (our k_1). Next, you'd run into the actual suspension springs of the vehicle (our k_2). Finally, you'd have the mass of the car body (our m).

　　How can we determine the equivalent k (Figure 1.13b) for this system? Well, the approach we'll use is one that you'll find helpful in the future when you need to determine the boundary conditions for continuous problems (and, of course, it's helpful here as well). What we'll do is create a new problem, shown in Figure 1.14a. For this problem, we've taken the two springs that are in series and

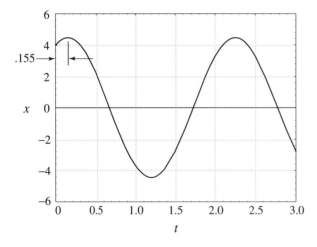

Figure 1.12　$4.47\cos(3t - .464)$

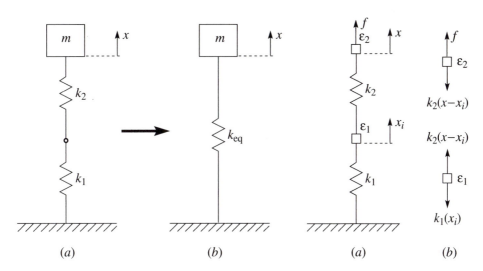

(a)   *(b)*   *(a)*   *(b)*

Figure 1.13   Representing two springs in series as a single equivalent spring

Figure 1.14   Free-body diagram with two springs in series

put a tiny mass ε_1 between them and also put another tiny mass ε_2 at the end of k_2. Finally, we've applied a force f to the top mass.

As indicated in Figure 1.13, what we eventually want is an expression like

$$f = k_{eq}x \tag{1.2.28}$$

where k_{eq} is our equivalent spring constant. The reason for introducing the tiny masses is to make it easier to apply Newton's laws to this problem in a way that you're familiar with. However, once we get the equations of motion, we'll set ε_1 and ε_2 to zero, thus reducing the problem to our original one. So the ε's are just a conceptual aid.

Going ahead and determining the equations of motion for this system (using the free body diagrams from Figure 1.14*b* gives us

$$\varepsilon_2 \ddot{x} = -k_2(x - x_i) + f \tag{1.2.29}$$

and

$$\varepsilon_1 \ddot{x}_i = k_2(x - x_i) - k_1 x_i \tag{1.2.30}$$

Note that no gravitational force (mg) appears in the free body diagram because, as mentioned previously, we're concerned with vibrations about the system's static equilibrium position.

We'll now set the ε's to zero, thus giving us the correct static equations

$$k_2(x - x_i) = f \tag{1.2.31}$$

$$k_2(x - x_i) - k_1 x_i = 0 \tag{1.2.32}$$

If we use (1.2.31) to eliminate $k_2(x - x_i)$ from (1.2.32), we'll obtain

$$f = k_1 x_i \tag{1.2.33}$$

or

$$x_i = \frac{f}{k_1} \tag{1.2.34}$$

Substituting this value for x_i into (1.2.32) gives us

$$k_2 \left(x - \frac{f}{k_1} \right) - f = 0 \tag{1.2.35}$$

which, when rearranged, yields

$$\left(\frac{k_1 k_2}{k_1 + k_2} \right) x = f \tag{1.2.36}$$

Comparing (1.2.36) with (1.2.28), we see that we've found our equivalent spring constant:

$$k_{\text{eq}} = \frac{k_1 k_2}{k_1 + k_2} \tag{1.2.37}$$

If you've taken a circuits course, this will remind you strongly of the formula for the equivalent resistance of two resistors in parallel, *not* two resistors in series. This is a point that sometimes causes confusion, since the schematic representation of two springs in series looks very much like two resistors in series (Figure 1.15). The relevant equations for this electrical example are:

$$v - v_i = r_2 i \tag{1.2.38}$$

and

$$v_i = r_1 i \tag{1.2.39}$$

Combining (1.2.38) and (1.2.39) produces

$$v = (r_1 + r_2)i \tag{1.2.40}$$

yielding an equivalent resistance of $r_{\text{eq}} = r_1 + r_2$.

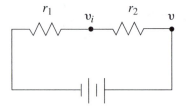

Figure 1.15   Electrical resistors in series

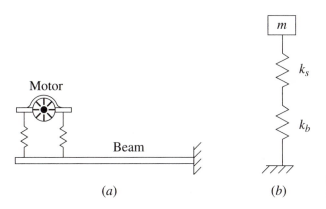

Figure 1.16   Motor on an elastic support

(a)   (b)

Example 1.7

Problem   The system shown in Figure 1.16a is a motor, m, with a suspension, k_s. The motor is attached to a flexible support beam, with stiffness k_b. The simplified model is shown in Figure 1.16b. Find the natural frequency of the combined system. $m = 20$ kg, $k_s = 1000$ N/m, and $k_b = 200$ N/m.

Solution   From (1.2.37) we have

$$k_{eq} = \frac{k_s k_b}{k_s + k_b} = 166.6 \text{ N/m}$$

$$\omega_n = \sqrt{\frac{166.6}{20}} = 2.887 \text{ rad/s}$$

Calculating the equivalent spring constant for springs in parallel is easier than the series arrangement. Figure 1.17 shows this case and the equivalent formulation we're trying for. Figure 1.18 illustrates the relatively straightforward free body diagram for this problem. Applying $f = ma$

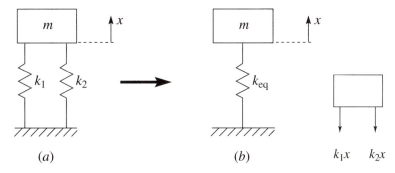

(a)   (b) $k_1 x$ $k_2 x$

Figure 1.17   Representing two springs in parallel as single equivalent spring

Figure 1.18   Free-body diagram for a mass with two springs in parallel

in the vertical direction gives us

$$m\ddot{x} = -k_1 x - k_2 x \tag{1.2.41}$$

or

$$m\ddot{x} + (k_1 + k_2)x = 0 \tag{1.2.42}$$

Thus the equivalent spring constant for two parallel springs is given by

$$k_{eq} = k_1 + k_2 \tag{1.2.43}$$

Note that this looks just like the electrical resistance formula we derived a moment ago. If you look at the equivalent resistance of resistors in parallel, you'll have the completed picture—that series arrangements of springs are like parallel arrangements of resistors and parallel arrangements of springs are like series arrangements of resistors. There actually is a logical way to look at both systems in a way that predicts this behavior, but we won't go into it here. The interested reader can check [Reference 8] for more information. Now that we've got (1.2.37) and (1.2.43), we can determine the equivalent spring constant for any particular arrangement of springs simply by breaking the more complicated problem down into smaller pieces.

Example 1.8

Problem What is k_{eq} for the system illustrated in Figure 1.19a?

Solution For k_{eq_1} (Figure 1.19b) we have

$$k_{eq_1} = \frac{k_2 k_2}{k_2 + k_2} = \frac{k_2}{2}$$

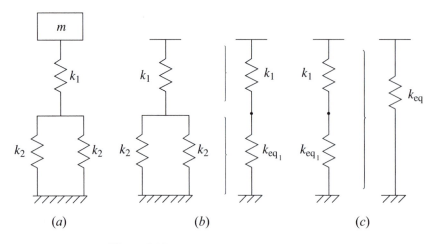

Figure 1.19 Mass on a multi-spring foundation

and for k_{eq} (Figure 1.19c) it's

$$k_{eq} = k_{eq_1} + k_1 = \frac{k_2}{2} + k_1$$

Even though we really never have pure springs and masses in practice, we often approximate our systems as if the lumped elements only exhibited a single characteristic. For instance, examine Figure 1.20, which shows an extensible bar with a lumped mass at the end. We know that the bar itself has mass, but if the lumped mass is much greater than the bar's mass, we might decide to model the bar as a pure spring. We'll revisit this problem in Chapter 7, to see how more exact solutions compare to our approximate ones.

If we're going to ignore the bar's own mass, then the system can be represented as a lumped mass m_1 at the end of a spring. All we need to do is determine the spring constant, and then we can solve for the system's natural frequency. The governing equation static for this case is

$$EA\xi_x = f \tag{1.2.44}$$

where f is a constant force applied at the end of the bar and EA is the bar's stiffness. Integrating this gives us

$$EA\xi(x) = fx + b \tag{1.2.45}$$

Our boundary condition of zero displacement at $x = 0$ causes b to equal zero and, evaluating this expression at $x = l$, we find

$$EA\xi(l) = fl \tag{1.2.46}$$

This can be rewritten as

$$f = \frac{EA}{l}\xi(l) \tag{1.2.47}$$

If we draw a parallel with our well-known spring force equation ($f = kx$), we can identify $\frac{EA}{l}$ as our equivalent spring constant k_b. Thus our natural frequency is given by

$$\omega_n = \sqrt{\frac{k_b}{m_1}} = \sqrt{\frac{EA}{lm_1}} \tag{1.2.48}$$

As we've mentioned, the accuracy of this approximation depends upon the end mass m_1 being substantially greater than the bar's mass. To see this, we can look at two cases. First, we'll consider the

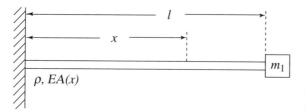

Figure 1.20   Schematic of a bar with a lumped mass

case for which the bar's density ρ is 1 kg/m. The other parameters will be $l = 1$ m, $m_1 = 10$ kg, and $EA = 10$ N. Thus the total mass of the bar is one-tenth that of the end mass. For these values we'll obtain $\omega_n = 1.00$ rad/s. If we were to solve this problem exactly (using the techniques of Chapter 5), we'd find that the exact answer is .9836 rad/s. Thus our error is only around 2%.

If we now increase the bar's density by a factor of 10, to 10 kg/m, our estimate doesn't change at all (since it ignores the bar's mass). However the mass certainly affects the actual answer, and lowers the natural frequency to $\omega_n = .8603$ rad/s, giving us an error of 16%. This isn't really too bad if you consider that the bar's mass is now equal to the end mass. It's actually possible to correct our approximation a good deal by using some engineering insight, and we'll go into this in Chapter 7.

As an aid to the homework problems, Appendix B contains lumped spring constants for a variety of common situations. In addition, Appendix C contains physical constants for several common structural materials.

1.3 ROTATIONAL VIBRATIONS AND LINEARIZATION

Although many systems involve only translational motion of the sort we've been studying, many also include rotational motions. As long as the rotation angles are small, these systems can be analyzed just like the systems we've already seen. Consider, for example, a simple pendulum (Figure 1.21). This consists of a point mass, suspended from a massless rod. The generalized coordinate is θ, the angle by which the rod deviates from the vertical position.

It's important to realize that there are actually two equilibrium positions for this problem. The obvious one is the one shown, for which θ is zero. A pendulum is quite happy to remain near this position, oscillating back and forth about it. However, there is another equilibrium, namely, when θ is 180 degrees. In this case the pendulum points directly upward. This is what we call an unstable equilibrium, since it is precarious at best. If we very, very carefully place the pendulum in this upright position, it should remain upright. At least that's what the mathematics will tell us. But the mathematics will also tell us that any disturbance, no matter how small, will cause the pendulum to fall away from this equilibrium (and end up moving about the stable equilibrium at $\theta = 0$). Thus in this and other examples, you'll want to be sure you're analyzing the system about the appropriate equilibrium. We'll mention how to do this after we've found the equations of motion.

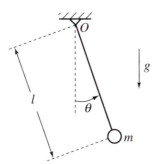

Figure 1.21   Simple pendulum

Constructing a free body diagram of this system and summing moments about the attachment point 0 leads to

$$\sum \text{moments}_0 = -mgl \sin(\theta) \qquad (1.3.1)$$

This is equal to the time rate of change of the angular momentum about O,

$$ml^2\ddot{\theta} = -mgl \sin(\theta) \qquad (1.3.2)$$

or

$$ml^2\ddot{\theta} + mgl \sin(\theta) = 0 \qquad (1.3.3)$$

Keep in mind that in this example the point O is stationary, and so the equation used is correct. If the attachment were moving, we'd have additional terms in our moment equation.

All right, at this point we'll look at equilibria. Simply put, an equilibrium position is one for which the system is happy to remain at rest. No motion. Not even a little. A straightforward way to find the equilibria is to take the equations of motion ((1.3.3) in our case) and just cross off the terms that imply motion. For us, this means crossing off the $\ddot{\theta}$ term. Obviously, if there's no motion, then all derivatives with respect to time will have to be zero. This leaves us with

$$mgl \sin(\theta) = 0 \qquad (1.3.4)$$

Since the mgl term isn't zero, we need to find the point at which $\sin(\theta)$ is zero. This occurs at $\theta = 0$ and at $\theta = \pi$. Thus our two equilibrium positions involve the pendulum hanging straight down or standing straight up, as mentioned a second ago.

We can't solve (1.3.3) in the same manner we used for the translational problems earlier because this equation is nonlinear (i.e., the equation doesn't just involve the first power of the dependent variable θ and its time derivatives). To solve this problem, we need to *linearize* the equation. Simply put, this means that we expand out all terms involving the dependent variables and their time derivatives and then throw away all that don't consist solely of the dependent variables (and their time derivatives) raised to the first power. Thus, any terms like $\theta\dot{\theta}$, θ^2, and $\theta^2\ddot{\theta}$ will be discarded, while terms like θ, $\dot{\theta}$, and $\ddot{\theta}$ will be kept. When equations involve only terms like θ, $\dot{\theta}$, and $\ddot{\theta}$ (dependent variables and their derivatives raised to the first power), we call them linear. If other terms, such as θ^2, $\dot{\theta}\theta$, and θ^3, are present, then the equation is nonlinear.

Since we know that $\sin(\theta)$ can be expanded in a power series:

$$\sin(\theta) = \theta - \frac{1}{3!}\theta^3 + \cdots \qquad (1.3.5)$$

we can discard everything except θ to obtain

$$\ddot{\theta} + \frac{g}{l}\theta = 0 \qquad (1.3.6)$$

We can safely discard the higher order terms because they become vanishingly small as θ is made small; i.e., if θ is small, θ^2 is *really* small—small enough to ignore.

Note that we've divided through by ml^2. Also note that when we expanded out, we were doing the expansion about a particular equilibrium position (in this case $\theta = 0$). This is always the case. You get your nonlinear equations of motion, find the equilibria, and then linearize about a particular equilibrium.

Equation (1.3.6) is linear and can be solved in the same manner as the translational problems. Since we've thrown away quite a few terms to get to this linearized form, we need to ask whether the resulting equation has any validity. Clearly, tossing out the terms as we did will be sensible only if, during the motion of the actual system, these terms are negligibly small. This condition is equivalent to saying that $\sin(\theta)$ is very close to equaling θ itself. We know from trigonometry that this is true only if θ is small. How small? You can easily verify with your calculator that, if θ is less than .24 radian, then the error in approximating $\sin(\theta)$ by θ is less than 1% and the error for angles up to .53 radian is under 5%. These are pretty fair-sized angles, and so we see that the approximation is a good one for realistic amplitudes of oscillation.

Example 1.9

Problem   How long must a simple pendulum (lumped mass on the end of a massless rod) be for the period of its oscillation to equal 1 second?

Solution   From (1.3.6) we have

$$\omega_n^2 = \frac{g}{l}$$

If the period is 1 second long, then

$$\omega_n = \frac{2\pi}{T} = 2\pi$$

Thus

$$(2\pi)^2 = \frac{9.81}{l} \quad \Rightarrow \quad l = .248\,\text{m}$$

Example 1.10

Problem   Find the natural frequency of the streetlight shown in Figure 1.22a, taking gravity into account. Compare the result to the case in which gravity is neglected. View the pole as being infinitely rigid. The mass of the light is the dominant mass and is represented by m (shown in Figure 1.22b). The rotational stiffness of the arm is represented by a lumped rotational spring, k_θ. The values of the various parameters are given by $m = 50$ kg, $g = 9.81$ m/s^2, $k_\theta = 29,393$ N·m, and $l = 3$ m.

Solution   When the arm is horizontal, the spring is unstretched. Gravity is going to work to push the end mass down while the torsional spring will resist this motion. Thus the equilibrium position will be somewhere between $\theta = 0$ and $\theta = \frac{\pi}{2}$. If we sum moments about 0, we'll find

$$ml^2\ddot{\theta} = -k_\theta\theta + mgl\cos(\theta)$$

Setting the time derivatives to zero gives us

$$-k_\theta\theta + mgl\cos(\theta) = 0$$

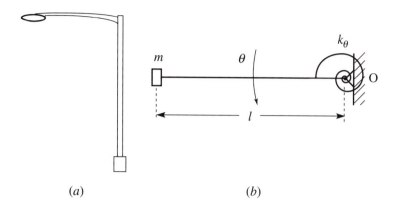

(a)   (b)

Figure 1.22   Streetlight and lumped approximation

or

$$k_\theta \theta = mgl \cos(\theta)$$

Using the given parameter values, this can be solved to give

$$\theta_{eq} = .05 \text{ rad}$$

If we expand about the equilibrium, θ_{eq}, by representing θ as

$$\theta = .05 + \eta$$

(where η is small), we'll have

$$ml^2 \frac{d^2}{dt^2}(.05 + \eta) + k_\theta(.05 + \eta) - mgl \cos(.05 + \eta) = 0$$

We now need to expand $\cos(.05 + \eta)$ into $\cos(.05) \cos(\eta) - \sin(.05) \sin(\eta)$. Since η is assumed to be small, this can be approximated by $\cos(.05) - \sin(.05)\eta$, (where we've used $\sin(\eta) \approx \eta$ and $\cos(\eta) \approx 1$).

Substituting this approximation into our equation of motion gives us

$$ml^2\ddot{\eta} + k_\theta(.05 + \eta) - mgl\Big(\cos(.05) - \sin(.05)\eta\Big) = 0$$

Expanded and simplified, this will finally yield

$$ml^2\ddot{\eta} + \Big(k_\theta + mgl \sin(.05)\Big)\eta = 0$$

If we go ahead and substitute in the appropriate values for m, l, and so on, we'll have

$$450\ddot{\eta} + 29,467\eta = 0$$

giving us a value of 8.09 rad/s as the system's natural frequency. Note that in the absence of gravity, the natural frequency would be $\sqrt{\frac{29,393}{450}} = 8.08$ rad/s. Thus the effect of the slight steady state rotation due to gravity had a negligible effect of the system's natural frequency.

Example 1.11

Problem   Find the equations of motion for the system shown in Figure 1.23. Neglect gravity.

Solution   The system we'll be examining contains both rotational and translational components. The horizontal bar can rotate about its hinge and the angle of rotation is given by θ. Rotations are resisted by the springs k_1 and k_2. As in the preceding examples, we'll assume that all rotations are small, and thus the displacement of the right end is given by $l_2\theta$. A mass is connected by a spring to the right end of the bar and an external vertical force $f(t)$ is applied to the same end.

    The first step in solving this problem is to construct a free body diagram that shows what forces are acting on the system (Figure 1.24). Note that no reaction forces are drawn for the left end of the bar because the bar is physically pivoting about that point and thus the reaction forces will apply no moments there.

    Summing moments about the pivot gives us

$$I_A\ddot{\theta} = -k_1(l_1\theta)l_1 - k_2(l_2\theta - x)l_2 + l_2 f(t)$$

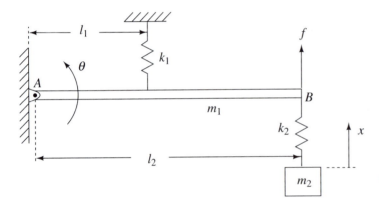

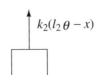

Figure 1.23   Spring-restrained/spring-mass system

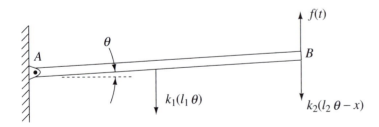

Figure 1.24   Free body diagram for spring-restrained/spring-mass system

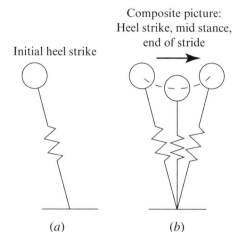

Composite picture:
Heel strike, mid stance,
end of stride

Initial heel strike

Figure 1.25   Spring-mass model of a
walking organism

(a)   (b)

while equating force and acceleration in the vertical direction for the lumped mass yields

$$m_2\ddot{x} = k_2(l_2\theta - x)$$

Putting all the dependent variables on the left-hand side then gives us the final equations of motion

$$I_A\ddot{\theta} + (k_1l_1^2 + k_2l_2^2)\theta - k_2l_2x = l_2 f(t)$$

$$m_2\ddot{x} + k_2x - k_2l_2\theta = 0$$

Finally, before moving on to the topic of damping, take a look at Figure 1.25. The illustrated model is used in state-of-the-art analyses of locomotion [4] and applies to creatures as diverse as crabs, cockroaches, and humans. As you can see, the model is a single-mass, single-spring system, exactly the same sort of system as we've been studying. The mass represents the creature's lumped mass and the spring represents the leg's stiffness. Figure 1.25a shows the system as the foot is just touching down and Figure 1.25b shows three snapshots that cover the full range of ground contact positions, from initial contact to the instant at which contact is lost.

The point of showing this example is to emphasize that even though you're not far into the book yet, you've already encountered models that are actually being used in the real world, and you have the knowledge to do something useful with these models.

1.4 VISCOUS DAMPING

The difficulty with the analyses so far is that our systems will never stop moving. They'll oscillate forever, something that clearly doesn't happen in the real world. When we pull down the coffee cup in our spring-mass experiment and then release it, the magnitude of the resulting oscillations becomes less and less as time goes on. The energy of the system is slowly dissipated away. Can we modify our equations so that they demonstrate this behavior? We've basically got two ways to go if we wish to add dissipation (damping) to our system model. First, we can try and accurately model the processes by which dissipation occurs in real life and add the appropriate terms to our equation of motion. In real life, damping is a very intricate and involved process, and to correctly model it we would need a complex set of equations that are very system dependent.

The other approach would be to ignore the "real" damping behavior and try and convince ourselves that a simple linear term is a good enough approximation. The advantages to this approach are that it's *much* simpler than doing it the "correct" way, and we can obtain closed-form solutions for our problems rather than having to resort to numerical (computer-based) solutions.

Given these two choices, the scientist in us would opt for the first approach, while our engineering half would probably vote for the simple solution. Since this is an engineering text, the second approach wins! Before those of you with a more scientific bent cry foul, you should know that the simple linear approach, although not exactly correct, is usually very close. When the level of dissipation is small, as it almost always is for engineering structures, it really doesn't make a great deal of difference in the final results if we use a linear damping approximation or a more realistic one. Only when the damping is large do we find sizable discrepancies creeping in between the different approaches. We'll see in a moment what is really meant by "large" and "small" values of damping.

Thus far, our formulations have involved terms for which the force is proportional to acceleration ($m\ddot{x}$) and displacement (kx). What we'll do now is add one that's proportional to velocity

$$f = c\dot{x} \qquad (1.4.1)$$

A damper that exhibits a linear characteristic like this is often referred to as a *viscous damper*. It makes physical sense to let our damping term scale with velocity, since we're very familiar with the fact that drag on cars and boats increases with velocity, and thus the damping force in our equations should increase with velocity. (By the way, the drag on cars actually doesn't increase linearly; it scales more closely with the square of the velocity.) Besides being called viscous, you'll also see this type of damping referred to as *linear damping*.

Putting this new term into our equation of motion yields

$$m\ddot{x} + c\dot{x} + kx = 0 \qquad (1.4.2)$$

Although we threw this damping term in, we could just as well have postulated the existence of a physical damper (shown in Figure 1.26) that is attached in parallel with the spring of our system. If we assume that the force generated is proportional to velocity, then a free body force balance would

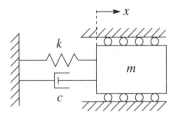

Figure 1.26 Spring-mass damper system

give us

$$m\ddot{x} = -c\dot{x} - kx \tag{1.4.3}$$

which is simply (1.4.2) arranged in a different order. The nice thing about visualizing the damper as a physical component is that it allows us to see why the damper will act to reduce the motions in the system. Since the force is directed toward the system's equilibrium position (just as the spring's force is) and is proportional to velocity, it generates a force that's directed toward the equilibrium point when the mass is moving away from that point. The faster the mass is moving, the greater the force acting on it.

Having added what we think will give us a damping effect, the next step is to solve the equations and see what we get. To start, we'll deal with the unforced case, just as we did for the undamped case. Since our equation is still linear, and remembering that all unforced, linear, constant coefficient, ordinary differential equations have exponential solutions, we'll solve the problem in exactly the same manner as we did for the undamped case, by assuming a solution of the form

$$x = ae^{\lambda t} \tag{1.4.4}$$

Using (1.4.4) in (1.4.2) we obtain

$$m\lambda^2 ae^{\lambda t} + c\lambda ae^{\lambda t} + kae^{\lambda t} = 0 \tag{1.4.5}$$

Grouping terms gives us

$$ae^{\lambda t}(m\lambda^2 + c\lambda + k) = 0 \tag{1.4.6}$$

Once again, $e^{\lambda t}$ is never zero for finite t and $a = 0$ is uninteresting, so we are left with

$$m\lambda^2 + c\lambda + k = 0 \tag{1.4.7}$$

Solving this gives us

$$\lambda_{1,2} = \frac{1}{2m}(-c \pm \sqrt{c^2 - 4mk}) \tag{1.4.8}$$

Although we could leave the solution in this form, we've already seen that dividing our undamped equation by m led us to the natural frequency ω_n and put the equation into a very neat form. If we carry through the division by m as in (1.4.8), we'll obtain

$$\lambda_{1,2} = -\frac{c}{2m} \pm \sqrt{\frac{1}{4}\left(\frac{c}{m}\right)^2 - \frac{k}{m}} = -\frac{c}{2m} \pm \sqrt{\frac{1}{4}\left(\frac{c}{m}\right)^2 - \omega_n^2} \tag{1.4.9}$$

Note that all we really need to do now is specify the ratio of c to m. The absolute values of these quantities don't greatly matter, which is one of the facts that motivated our division by m in the first place. Although it isn't clear why at the moment, it's convenient to replace $\frac{c}{m}$ with $2\zeta\omega_n$. If you'll suspend your disbelief for just a second, you'll see why.

Using this definition, our equation of motion becomes (after dividing by m)

$$\ddot{x} + 2\zeta\omega_n\dot{x} + \omega_n^2 x = 0 \tag{1.4.10}$$

Substituting $2\zeta\omega_n$ for $\frac{c}{m}$ in (1.4.9), we obtain

$$\lambda_{1,2} = -\zeta\omega_n \pm \omega_n\sqrt{\zeta^2 - 1} \tag{1.4.11}$$

This form is great as long as ζ is larger than one. But for $\zeta < 1$ we have the square root of a negative number. In this case we'll get an imaginary value, and the roots are most conveniently expressed as

$$\lambda_{1,2} = -\zeta\omega_n \pm i\omega_n\sqrt{1 - \zeta^2} \tag{1.4.12}$$

Note that we get two forms that depend upon ζ's value. For (1.4.11), both roots are real, negative numbers. This means that the overall solution to the free vibration problem will include two decaying exponentials. A typical response to an initial displacement is shown in Figure 1.27. For this example ω_n is 6.28 and ζ is 2. The mass starts off displaced from the origin and then smoothly approaches it as time increases.

This is different from the behavior we'll find for the kinds of root expressed by (1.4.12). For this case we no longer have two negative, real roots but, rather, complex roots. The real part of each root is the same, while the imaginary part differs in sign. Roots like these are called *complex conjugates* and are very common solutions for vibrations problems. To see what sort of overall solutions these roots support, we'll use the following trigonometric identity:

$$e^{ix} = \cos(x) + i\sin(x) \tag{1.4.13}$$

Thus, instead of solutions that look like

$$x(t) = a_1 e^{-\zeta\omega_n t} e^{i\omega_n t\sqrt{1-\zeta^2}} + a_2 e^{-\zeta\omega_n t} e^{-i\omega_n t\sqrt{1-\zeta^2}} \tag{1.4.14}$$

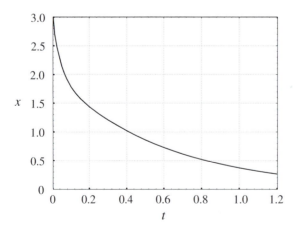

Figure 1.27   Overdamped response

we can express the solution as

$$x(t) = e^{-\zeta \omega_n t} \left[b_1 \cos(\omega_d t) + b_2 \sin(\omega_d t) \right] \tag{1.4.15}$$

where $\omega_d \equiv \omega_n \sqrt{1 - \zeta^2}$.

Thus we have a solution made up of a sinusoidal part multiplied by a decaying exponential. ω_d is known as the *damped natural frequency*. Since the exponential is equal to 1.0 for $t = 0$ and decays from there, the effect of multiplying the sine or cosine by it is to cause the oscillations to decay within an exponential envelope. This is shown graphically in Figure 1.28 for the response

$$x(t) = e^{-1.256t} \cos(6.153t) \tag{1.4.16}$$

which corresponds to $\omega_n = 6.28$ and $\zeta = .2$.

Note that the general form of this type of solution is $x(t) = e^{-c_1 t} \cos(c_2 t + \phi)$. If changes to our system cause c_1 to increase, then the speed at which the oscillations decay is increased. Also, increasing c_2 causes the frequency of the oscillations to increase. Since we can control either of these by varying ω_n and ζ, we have the tools necessary for modifying the decay characteristics of our system.

The critical parameter that divides the responses between oscillatory ones (1.4.12) and purely decaying ones (1.4.11) is ζ. For $\zeta > 1$ we'll have exponential decay (because the number under the square root of (1.4.11) is positive) and for $\zeta < 1$ we'll have decaying oscillations. This is the main reason for the introduction of ζ; it characterizes these two response regimes. Different books call ζ different things (damping coefficient, damping ratio, damping factor). In this book we'll refer to it as the damping factor. If our system has a ζ that's less than 1.0, we call the system *underdamped*. Systems with ζ's greater than 1.0 are *overdamped*, while systems having ζ's of exactly 1.0 are *critically damped*.

Looking at our definition of ω_d ($\omega_d = \omega_n \sqrt{1 - \zeta^2}$), we see that the frequency of our damped oscillations depends upon both ω_n and ζ. For small values of ζ (lightly damped) the oscillation rate is essentially ω_n. Increasing the damping factor serves to reduce the oscillation rate. This makes physical

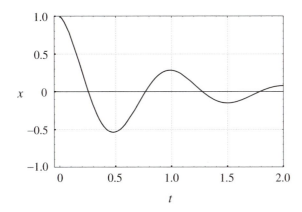

Figure 1.28   Underdamped response

sense, since the mass will have more difficulty in moving if the damping is raised, thus lowering the overall rate of oscillation.

Although ζ is the damping factor for our system, keep in mind that it is *not* the sole determiner of the response's decay rate. Equation (1.4.15) shows that the decay rate is determined by the product of ζ and ω_n. Thus the decay rate is increased by increasing the damping factor or by increasing the system's natural frequency. If we think physically about what's going on with our system, this makes sense. First of all, it certainly seems obvious that increasing the amount of damping in the system will increase the rate of decay, and ζ is linearly proportional to the damping physical damping coefficient, c. But it is also true that a larger ω_n means a higher rate of oscillation, leading to higher velocities and thus more dissipation (since the damping is proportional to velocity). Thus both the damping factor and natural frequency play a part in determining how fast an unforced oscillation will die away.

The general responses we've just examined are absolutely central to understanding and describing the responses of complex systems. The speed at which responses decay and the frequency at which they do so are determining factors in how well the systems will perform the tasks they were designed for.

Example 1.12

Problem   How large must c be made for the system described by the following equation to be critically damped?

$$5\ddot{x} + c\dot{x} + 2000x = 0$$

Solution

$$2\zeta\omega_n = \frac{c}{m} = \frac{c}{5}$$

$$\omega_n = \sqrt{\frac{k}{m}} = \sqrt{\frac{2000}{5}} = 20$$

$$\zeta = \frac{c}{(5)(2)\omega_n} = \frac{c}{200}$$

Since critical damping means $\zeta = 1$ we have

$$c = (200)(1) = 200 \, \text{N·s/m}$$

Example 1.13

Problem   Consider a system like that shown in earlier Figure 1.26. For this example $m = 10$ kg, $c = 100$ N·s/m, and $k = 30,000$ N/m. If the initial conditions are $x(0) = .01$ m, and $\dot{x}(0) = 0$, what will the response look like for $t > 0$?

Solution   Knowing m, c, and k, we can immediately form

$$\omega_n^2 = \frac{30,000}{10}$$

and so

$$\omega_n = 54.8 \, \text{rad/s}$$

Knowing that $\frac{c}{m} = 2\zeta\omega_n$ we have

$$\zeta = .0913 \quad \text{and} \quad \omega_d = 54.5 \, \text{rad/s}$$

Since

$$x(t) = e^{-\zeta\omega_n t} \left(b_1 \cos(\omega_d t) + b_2 \sin(\omega_d t) \right)$$

we can apply the initial conditions and obtain

$$.01 = b_1$$

and

$$0 = -\zeta\omega_n b_1 + \omega_d b_2$$

Solving both these equations simultaneously gives us

$$b_1 = .01 \quad \text{and}$$

$$b_2 = b_1 \frac{\zeta}{\sqrt{1 - \zeta^2}} = .0921$$

Thus

$$x(t) = (.01 \cos(54.5t) + .0921 \sin(54.5t)) \, e^{-5t}$$

1.5 LAGRANGE'S EQUATIONS

No vibrations book is complete anymore without a discussion of Lagrange's equations. All of you have already taken a dynamics class or two, which, in the vast majority of cases, has involved Newtonian mechanics, i.e., finding the equations of motion by applying Newton's laws. The fundamental distinguishing feature of Newtonian mechanics is that it treats the force interactions within a system as the fundamental quantities of interest. Once you've determined all the relevant forces, you simply use $f = ma$ (for constant mass), and you're done. Sounds simple. However, as you probably discovered in class, actually applying these laws is a bit trickier than one might at first think. Correctly analyzing a system using Newton's laws requires us to form free body diagrams and demands that we account for all interaction forces. For instance, let's consider the problem shown in Figure 1.29a, that of a mass-pendulum system. The pendulum's base is attached to the mass at A and is free to rotate.

There's more than one way to solve this problem with Newton's laws, and one particular approach to accounting for the applied and interaction forces is shown in Figure 1.29b. Note that we *must* solve for the interaction forces (n and t_p) between the two masses, even if we don't really care about them. As vibration engineers, we probably will be satisfied if we have the governing equations of motion, which don't explicitly involve these forces at all.

In the past, the need to spend time getting intermediate results that might be of no interest bothered several researchers, Lagrange being one of them. Lagrange wanted to find an alternate way of finding the equations of motion, one that didn't necessarily involve interaction forces. The notion

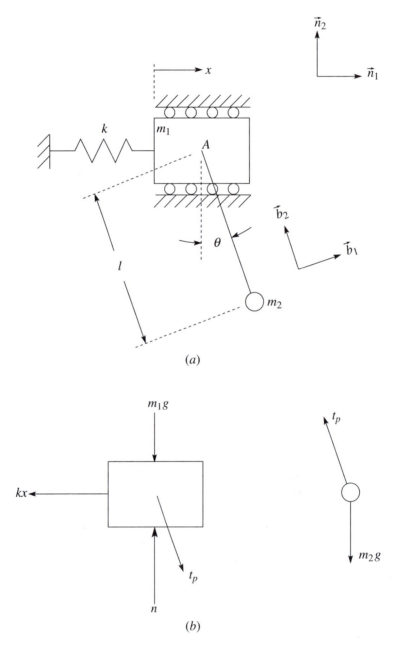

Figure 1.29 Mass/pendulum physical system (a) and free body diagram (b)

he came up with was to look, not at the forces in a system, but at the *energies*. This viewpoint is quite different from Newton's, and there is a huge body of literature detailing different energy-based approaches to dynamics. Since this isn't a dynamics course, we'll look at only one of these approaches, namely, Lagrange's equations.

It is important at the outset to tell you that we'll be indulging in some major handwaving in the next few pages. The "standard" derivation of Lagrange's equations involves the notion of virtual displacements, something that brings a shudder to some dynamicists, even today. Others couldn't care less. In any event, we'll present this approach, point out the places where one might question it, and leave the rest to you. If you're really interested in getting into the area, there are other, more involved approaches that some people find more appealing.

Okay, enough with the preliminaries. To apply Lagrange's equations, we first need to define what are called *generalized coordinates*. As you know, we can't really do much in the way of dynamics without some coordinates. At the very least, we want our coordinates to exactly pin down the physical configuration of our system, at any point in time. It is very possible for different people to come up with different, and equally valid, sets of generalized coordinates for a given problem. One example is shown in Figure 1.30. Here we have a bar that's supported by two springs. We could decide to use the displacement of the bar at each end as our coordinate set; $x_1(t)$ and $x_2(t)$ are all we need to know to accurately describe the bar's position (we're assuming small motions only). Alternatively, we could use $x(t)$ and $\theta(t)$ to describe the vertical motion of the bar's midpoint and the rotation about this point. This would also work just fine.

Actually, you can come up with any number of coordinate systems; all that matters is that they successfully describe the system's configuration. Most commonly, we refer to these coordinates by the letter q. Thus, for our second set of coordinates in Figure 1.30 we'd have $q_1 = x$ and $q_2 = \theta$.

If we have no externally applied forces and no dissipation, then Lagrange's equations are given by

$$\frac{d}{dt}\frac{\partial L}{\partial \dot{q}_i} - \frac{\partial L}{\partial q_i} = 0, \quad i = 1 \cdots n \tag{1.5.1}$$

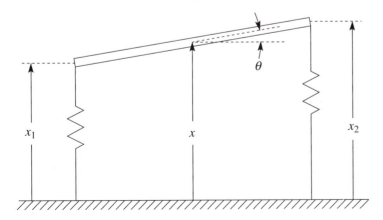

Figure 1.30 Two-coordinate systems

in which

$$L = KE - PE \tag{1.5.2}$$

where KE and PE are the system kinetic and potential energies, respectively, and L is called the Lagrangian.

Where did this result come from? I told you, we weren't going to go into details. You'll just have to believe that it's correct. Or, if you prefer, you can read more about it elsewhere [11]. To demonstrate that it actually works, check out the following example.

Example 1.14

Problem   Consider again the system of Figure 1.29. The system consists of a mass m_1 and an attached pendulum. m_1 can only move horizontally, and the pendulum (consisting of a massless rigid link and a lumped mass m_2) is attached to m_1 at the point A. The pendulum is free to rotate. There's no friction in the problem, and no external forces other than gravity. Because gravity will cause the pendulum to oscillate, and the spring k will induce oscillations in m_1, we should expect equations of motion that will involve periodic solutions, presumably ones in which both masses are moving. Find the system's equations of motion.

Solution   If we were going to approach this problem from a Newtonian standpoint, we'd first break the system up into two parts and then apply Newton's laws to them individually, as shown in Figure 1.29b. The unknown force t_p is the force of interaction between m_1 and m_2; it's what keeps the pendulum attached to m_1. We'll need to set up two force balance equations and one moment balance equation to solve this problem. Note that $\vec{n}_1, \vec{n}_2$ and $\vec{b}_1, \vec{b}_2$ are unit vectors that help define our coordinate systems. The $\vec{b}_1, \vec{b}_2$ frame is fixed to the pendulum while the $\vec{n}_1, \vec{n}_2$ set is fixed to the ground.

The two force balances in the x direction (for m_1 and m_2) are:

$$m_1 \ddot{x} = t_p \sin(\theta) - kx$$

and

$$m_2 \big(\ddot{x} + l\cos(\theta)\ddot{\theta} - \dot{\theta}^2 l \sin(\theta) \big) = -t_p \sin(\theta)$$

Using these equations together (and eliminating t_p) gives us

$$(m_1 + m_2)\ddot{x} + m_2 l\ddot{\theta}\cos(\theta) - m_2\dot{\theta}^2 l \sin(\theta) + kx = 0$$

Next, we have to sum moments about some convenient point. If we use the pendulum's point of attachment to m_1, our equation will be

$$-m_2 gl \sin(\theta) = m_2\big(l^2\ddot{\theta} + \ddot{x}l\cos(\theta)\big)$$

or, rearranging slightly and canceling out the m_2 term,

$$l^2\ddot{\theta} + \ddot{x}l\cos(\theta) + gl\sin(\theta) = 0$$

which gives us the second equation of motion.

You saw that the interaction force t_p showed up in the initial work and dropped out in the final equations, just as we expected, since we were using a Newtonian approach.

Now let's use Lagrange's equations. To find the kinetic energy, we need to know the velocities of the two masses. Finding m_1's velocity ($\vec{v}_1$) is the easiest, it's just $\dot{x}\vec{n}_1$. m_2 is a bit trickier, since it rotates about m_1. If we use the coordinate vectors shown in Figure 1.29b, we see that m_2's velocity is

$$\vec{v}_2 = \dot{x}\vec{n}_1 + l\dot{\theta}\vec{b}_1$$

To find the square of the velocities we need for the kinetic energy, we have to express the velocity in a consistent coordinate system. If we choose the fixed reference frame as the one to use, $\vec{v}_2$ becomes

$$\vec{v}_2 = \left(\dot{x} + l\dot{\theta}\cos(\theta)\right)\vec{n}_1 + l\dot{\theta}\sin(\theta)\vec{n}_2$$

The kinetic energy is simply

$$KE = \tfrac{1}{2}m_1 v_1^2 + \tfrac{1}{2}m_2 v_2^2$$

Using our knowledge of what $\vec{v}_1$ and $\vec{v}_2$ are equal to, we can reexpress this as

$$KE = \tfrac{1}{2}m_1\dot{x}^2 + \tfrac{1}{2}m_2\left((\dot{x} + l\dot{\theta}\cos(\theta))^2 + (l\dot{\theta}\sin(\theta))^2\right)$$

The potential energy comes from the translational spring and the gravitational potential that results when the pendulum swings up,

$$PE = \tfrac{1}{2}kx^2 + m_2 gl\left(1 - \cos(\theta)\right)$$

Using these expressions for KE and PE, we can form the Lagrangian ($L = KE - PE$) and apply Lagrange's equations. The two dependent variables (the q_i's) are x and θ. Applying Lagrange's equations with respect to x first, we find

$$\frac{d}{dt}(m_1\dot{x}) + m_2\frac{d}{dt}\left(\dot{x} + \dot{\theta}l\cos(\theta)\right) + kx = 0$$

or, performing the differentiations and grouping terms,

$$(m_1 + m_2)\ddot{x} + m_2 l\ddot{\theta}\cos(\theta) - m_2\dot{\theta}^2 l\sin(\theta) + kx = 0$$

If we next apply Lagrange's equations with respect to θ, we'll obtain

$$m_2\frac{d}{dt}(\dot{x}l\cos(\theta) + l^2\dot{\theta}) - m_2\dot{x}\dot{\theta}l\sin(\theta) + m_2 gl\sin(\theta) = 0$$

Note that the $-m_2\dot{x}\dot{\theta}l\sin(\theta)$ term came about when we differentiated the kinetic energy with respect to θ. Since the kinetic energy has a θ dependence, we obtained this additional term. This almost always happens when you're dealing with rotational systems and rarely occurs for translational ones, since a translational system's kinetic energy usually depends just on $\dot{x}$-type terms, not on terms proportional to x. It's easy to forget this fact if most of your examples are of the translational type, and you might begin to just differentiate the kinetic energy in the $\frac{d}{dt}(\frac{\partial}{\partial\dot{q}})$ term, forgetting about the $\frac{\partial}{\partial q}$ part of the equations. So now you've been warned—make sure to take account of all the relevant differentiations.

We can differentiate and group terms in this last equation to get

$$l^2\ddot{\theta} + \ddot{x}l\cos(\theta) + gl\sin(\theta) = 0$$

If you compare the two sets of equations of motion, you'll quickly notice that they are exactly the same. Thus it appears that Lagrange's equations do what was advertised: they come up with fully valid equations of motion.

Interestingly, you can get into some real battles with certain researchers over whether it is better to use Lagrange's equations or Newton's equations (or even other approaches). However, the simple fact of the matter is that you can almost always get your answer with either approach. In some problems a Newtonian approach might be less work and in others you might find Lagrange's equations to be preferable. It's not a bad idea to simply view the approaches as different tools with which to get the equations of motion and not get too caught up with which is "better." As a general guide, Lagrange's equations are "easier" in that you don't need to use any physical insight in modeling the force interactions. The method simply asks for the energies. Thus you need only find the relevant velocities with which to form the kinetic energy and account for any potential energies. Then you basically plug and chug in Lagrange's equations. The downside is that after you get the velocities, you have to square them and then differentiate. If you've got a bunch of reference frames because your system has several moving parts, then your velocity terms will be long and involved. Therefore you'll have a lot of terms to keep track of. For complicated systems, this can be a big downside, since errors can easily creep in when you're trying to keep track of so many terms. Also, computer simulations will tend to run more slowly because the equations are so complex. But for relatively simple problems, Lagrange's equations are just fine.

Newton's equations shine when you care about interaction forces, since you can't solve the problem without considering them. Often you will care about these interaction terms if you're evaluating the forces in a joint or a slider. The equations are also more physical, since they involve moments and forces, things we're comfortable dealing with as engineers. To apply Newton's equations, you need to find accelerations, something which is generally simpler than the manipulations needed for Lagrange's equations. Also, you need to think physically about the system to construct the necessary free body diagrams, something which isn't necessary with Lagrange's equations. Finally, you have to include all the interaction forces, even if you don't care about them, something that Lagrange's equations don't require.

Hopefully, this short discussion has shown that either approach will work. For the problems we'll be dealing with, most people find Lagrange's equations to be the more convenient. Thus you should probably become familiar with both approaches; it'll help make your life simpler in the long run.

All right, now we've got the unforced, undamped case under control. But what about damping? Can we add a viscous damper to our problem? Yes we can, but it is going to look like a Band-aid kind of fix. If our damper has a known force characteristic, what we need to do is integrate with respect to the time derivative of the generalized coordinates that apply to it. Thus, if we wanted to add a damper to our cart-pendulum problem, we'll get the system shown in Figure 1.31. From our earlier discussions, we recognize that the force of the damper is given by

$$f_d = c\dot{x} \tag{1.5.3}$$

and so, integrating with respect to $\dot{x}$ we obtain

$$\int f_d d(\dot{x}) = \frac{1}{2}c\dot{x}^2 \tag{1.5.4}$$

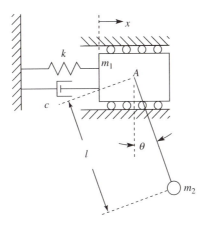

Figure 1.31   Damped mass/pendulum

We define this quantity as the Rayleigh dissipation function, RD. Thus

$$RD \equiv \tfrac{1}{2}c\dot{x}^2 \tag{1.5.5}$$

To use this in Rayleigh's equations, we need to modify the equation a bit, to

$$\frac{d}{dt}\frac{\partial L}{\partial \dot{q}_i} + \frac{\partial RD}{\partial \dot{q}_i} - \frac{\partial L}{\partial q_i} = 0, \quad i = 1 \cdots n \tag{1.5.6}$$

This modification will allow us to deal with any linear, unforced problem we might encounter.

At this point we have no place left to run. We have to deal with virtual quantities. And just what is a virtual quantity, you might ask? The simplest type is the virtual displacement. The first thing a virtual displacement must satisfy is that it *not* violate any of the physical constraints of our problem. This isn't too bad, in fact it sounds reasonable. If the displacement were allowed to violate physical constraints, like letting bodies pass through each other, it wouldn't stand much chance of being useful to us. Virtual displacements are also taken to be very small changes in the coordinates that occur in zero time. That's right, no time elapses during the virtual change of coordinates.

The following typical constraint equation

$$f(x_1,\ x_2,\ \ldots,t) = c \tag{1.5.7}$$

implies that there exists a constraint relation between the coordinates x_i. For instance, if our problem involved the motion of a mass on a table, and x_1 represented displacements to the right or left on the table, x_2 represented displacements forward or back on the table, and x_3 represented displacements above or below the table, our constraint equation would be

$$x_3 = 0 \tag{1.5.8}$$

which tells us that the particle stays on the surface of the table.

If we give our system a virtual displacement then we have

$$f(x_1 + \delta x_1,\ x_2 + \delta x_2,\ \ldots,t) = c \tag{1.5.9}$$

This tells us that, even though we've perturbed our system by the virtual displacements δx_1, etc., the constraints are still satisfied. Note that since the virtual displacements don't involve time, t wasn't varied.

We know that our usual idea of work is that force, multiplied by a displacement in the same direction as the force, gives us an associated work. A similar statement is made for virtual work. In this case, a force times a virtual displacement gives us a virtual work.

What we'll do is define the virtual work done on a mass particle as

$$\delta W_i \equiv Q_i \delta q_i \qquad (1.5.10)$$

where Q_i is the total force acting on the particle in the direction defined by q_i. This is exactly analogous to our usual definition of work—it's just force times displacement.

Having decided what virtual work is, we can now introduce what, for us, will be the final form of Lagrange's equations,

$$\frac{d}{dt}\frac{\partial L}{\partial \dot{q}_i} + \frac{\partial RD}{\partial \dot{q}_i} - \frac{\partial L}{\partial q_i} = Q_i, \quad i = 1 \cdots n \qquad (1.5.11)$$

where the Q_i are the virtual forces associated with external forces acting on our system. The use of this equation is best illustrated by an example.

Example 1.15

Problem　Figure 1.32 shows the system we've already examined (Figure 1.31) but with the addition of an external force $\vec{f}_{ext} = f_{ext}\vec{n}_1$. Determine the Q_i for this problem and form the overall equations of motion.

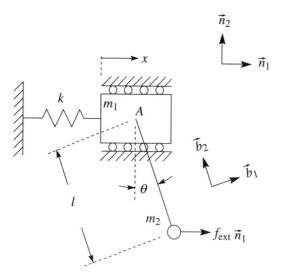

Figure 1.32　Mass/pendulum with external force

Solution The virtual work due to $\vec{f}_{\text{ext}}$ is given by $f_{\text{ext}}\vec{n}_1$ dotted into the virtual displacement of the tip mass. The quickest way to find this quantity is to form the tip mass's velocity, multiply by dt to clear t from the expression, and then switch the dq_i terms to δq_i,

$$\vec{v}_{m_2} = \frac{dx}{dt}\vec{n}_1 + l\frac{d\theta}{dt}\vec{b}_1$$

Multiplying by dt gives

$$d\vec{r}_{m_2} = dx\vec{n}_1 + ld\theta\vec{b}_1$$

and switching to δ yields

$$\delta\vec{r}_{m_2} = \delta x\vec{n}_1 + l\delta\theta\vec{b}_1$$

If we express this purely in terms of the $\vec{n}_i$ coordinate frame we'll see that the virtual displacement is equal to

$$\delta\vec{r}_{m_2} = \delta x\vec{n}_1 + l\delta\theta(\cos(\theta)\vec{n}_1 + \sin(\theta)\vec{n}_2)$$

or, by grouping terms,

$$\delta\vec{r}_{m_2} = (\delta x + l\delta\theta\cos(\theta))\vec{n}_1 + l\delta\theta\sin(\theta)\vec{n}_2$$

The total virtual work is given by

$$f_{\text{ext}}\vec{n}_1 \cdot \big((\delta x + l\delta\theta\cos(\theta))\vec{n}_1 + l\delta\theta\sin(\theta)\vec{n}_2\big) = f_{\text{ext}}\delta x + f_{\text{ext}}l\cos(\theta)\delta\theta$$

If we assign q_1 to be x and q_2 to be θ, then we have

$$Q_1 = f_{\text{ext}}$$
$$Q_2 = l\cos(\theta)f_{\text{ext}}$$
$$(m_1 + m_2)\ddot{x} + m_2l\ddot{\theta}\cos(\theta) - m_2\dot{\theta}^2l\sin(\theta) + kx = f_{\text{ext}}$$

and

$$l^2\ddot{\theta} + \ddot{x}l\cos(\theta) + gl\sin(\theta) = l\cos(\theta)f_{\text{ext}}$$

1.6 HOMEWORK PROBLEMS

Section 1.2

1.1. If $x(t) = a_1e^{i\omega t} + a_2e^{-i\omega t}$, $x(0) = 4$ and $\dot{x}(0) = 2$, what are a_1 and a_2 equal to?

1.2. Express $(1 + 2i)e^{i\omega t} + (1 - 2i)e^{-i\omega t}$ in terms of $\sin(\omega t)$ and $\cos(\omega t)$.

1.3. If $x(t) = b_1\cos(\omega t) + b_2\sin(\omega t)$, with the same initial conditions as in Problem 1.1, what are b_1 and b_2 equal to?

1.4. Express $2\cos(3t) + 4\sin(3t)$ in terms of e^{3it} and e^{-3it}.

1.5. Express $\cos^3(\omega t)$ in terms of $\cos(\omega t)$ and $\cos(3\omega t)$ by expressing $\cos(\omega t)$ as $\frac{1}{2}(e^{i\omega t} + e^{-i\omega t})$, cubing the expression, and then grouping the resulting terms.

1.6. Reexpress $\sin(\omega t)\cos^2(\omega t)$ as a sum of harmonic components by expressing $\sin(\omega t)$ and $\cos(\omega t)$ in complex exponential form, carrying through the multiplications and then grouping the terms into real components.

1.7. Express $\sin^2(\omega t)\cos^2(\omega t)$ in terms of $\sin(\omega t)$, $\cos(\omega t)$, $\sin(3\omega t)$, and $\cos(3\omega t)$ by putting the sinusoids in complex exponential form, carrying through the multiplications, and then grouping the resulting terms.

1.8. A sinusoid, $x(t) = a\cos(7t + \phi)$, has initial conditions $x(0) = 3$, $\dot{x}(0) = 20$. Determine a and ϕ.

1.9. Give the position, velocity, and acceleration at $t = 1.5$ if $x(t) = 2\cos(3t) - \sin(3t)$. $x(t)$ is given in meters.

1.10. What is the maximum value of $\dot{x}$ if $x(t) = -\cos(4t) + 2\sin(4t)$? $x(t)$ is given in meters.

1.11. Find the maximum value of $\dot{x}$ if $x(t) = 6e^{-2t} + 2e^{-t}$. $x(t)$ is given in meters.

1.12. A mass moves with the acceleration profile illustrated in Figure P1.12. Determine the position of the mass at $t = 8$ seconds. $x(0) = \dot{x}(0) = 0$.

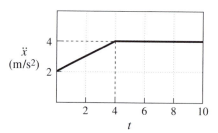

Figure P1.12

1.13. Plot the velocity and displacement vs time graphs for the acceleration profile illustrated in Figure P1.13. $x(0) = 4$ m/s and $x(0) = -2$ m/s.

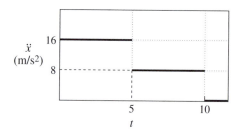

Figure P1.13

1.14. Plot $x(t)$, $\dot{x}(t)$, and $\ddot{x}(t)$, $(0 \le t \le 2\pi)$ for

$$x(t) = 2\cos(t) + .05\sin(10t)$$

What does this tell you about the advisability of differentiating a displacement signal to find the acceleration if the actual signal $(2\cos(t)$ in our case) is corrupted by sensor noise (represented by $.05\sin(10t)$)?

1.15. We know that the spring constant of two identical springs, used in series, is half the individual spring's spring constant. Show from the formula for the stiffness of a coil spring (Appendix B, Figure B.5) that doubling the number of coils (which is equivalent to putting two coil springs in series) will give the same result.

1.16. What is k_{eq} for a serial chains of n springs, each having a spring constant equal to k?

1.17. What is the equivalent stiffness for the set of springs illustrated in Figure P1.17?

n springs

Figure P1.17

1.18. Give the amplitude, frequency, and period of oscillation for the signal illustrated in Figure P1.18.

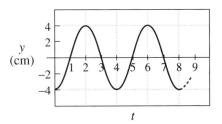

Figure P1.18

1.19. Plot a signal $x(t)$ having a frequency of 2 rad/s and an amplitude of 5 cm for $0 \le t \le 10$. Let $x(0) = \frac{5}{\sqrt{2}}$ cm and $\dot{x} = \frac{10}{\sqrt{2}}$ cm/s.

1.20. Determine the phase difference between the two signals illustrated in Figure P1.20. By how much is y lagging x?

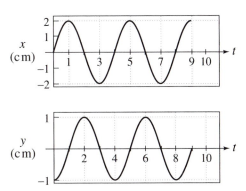

Figure P1.20

1.21. What is ω_n for the system illustrated in Figure P1.21 in terms of m, k_1, k_2, and k_3?

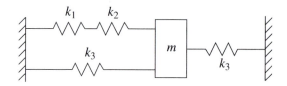

Figure P1.21

1.22. What is ω_n for the system illustrated in Figure P1.22?

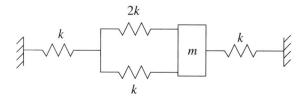

Figure P1.22

1.23. Find the equivalent spring constant for the two springs with widely differing spring constants shown in Figure P1.23. Comment on your result.

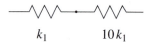

Figure P1.23

1.24. Find the equivalent spring constant for the two springs with widely differing spring constants shown in Figure P1.24. Comment on your result.

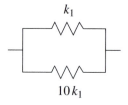

Figure P1.24

1.25. Figure P1.25 shows a Ferris wheel at a carnival. Assume that the sun is directly overhead and that the wheel has a radius of 15 m. It takes 3 minutes to complete a single revolution. Draw a plot of the shadow cast by a single car on the Ferris wheel as a function of time. Identify the amplitude, period, and frequency of the resulting waveform.

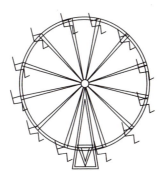

Figure P1.25

1.26. Consider a physical system governed by

$$m\ddot{x} + kx = 0$$

Given some m, k, and initial conditions, is it possible, for the free vibration solution to be given by the following equation?

$$x(t) = 3e^{-2it}$$

1.27. The platform illustrated in Figure P1.27 has a natural frequency of 50 Hz. It is supported by four identical springs. If the platform's mass is equal to 300 kg, determine the spring constant of a single spring. Assume that the platform's displacements are purely vertical.

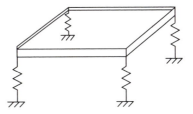

Figure P1.27

1.28. A mass is suspended from a nonlinear spring as in Figure P1.28. Determine the natural frequency of oscillation, ω_n, for small oscillations about the equilibrium position of the spring-mass system. Note that since the spring is not linear, you'll have to linearize about the system's equilibrium position to find its equivalent linear stiffness. The spring's force/displacement relation is

$$f_{\text{spring}} = 2x^3$$

where $f(x)$ is in newtons and x is in meters; $m_1 = 1$ kg and $g = 9.81$ m/s^2.

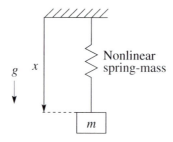

Figure P1.28

1.29. A spring is suspended from a nonlinear spring as in Figure P1.28. The force/displacement relation of the spring is given by

$$f_{spring} = 49(x + x^2)$$

where f_{spring} is in newtons and x is in centimeters. Find the linearized ω_n for the spring-mass system about its static equilibrium for $m = 10$ kg. Then find the new value of ω_n for $m = 20$ kg. Did the natural frequency drop by 29% (as linear theory would predict)?

1.30. A mass is suspended from a nonlinear spring as in Figure P1.28. The force/displacement relation of the spring is given by

$$f_{spring} = 10x - 4x^2 + x^3$$

x is measured in centimeters and f_{spring} is given in newtons. Find the linearized ω_n for the system illustrated about its static equilibrium using $m = .4209$ kg. Then find the new value of ω_n for $m = .8418$ kg. Did the natural frequency drop by 29% (as linear theory would predict)?

1.31. Consider the system illustrated in Figure P1.31, which has two linear springs. The unstretched length of k_1 is .5 m and the unstretched length of k_2 is .25 m. $k_1 = 1000$ N/m, $k_2 = 2000$ N/m, $m = 2$ kg, $l = .5$ m. Find the equilibrium position of the mass and determine the natural frequency of the system. Compare this natural frequency to that associated with $l = .75$ m (i.e., no precompression in the spring).

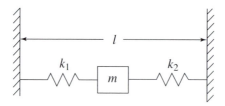

Figure P1.31

1.32. Consider the system in which the linear spring of Problem 1.31 is replaced with a nonlinear spring, as shown in Figure P1.32. The unstretched length of k_1 is .5 m and the unstretched length of the nonlinear spring is .25 m. $k_1 = 1000$ N/m, $f(x) = 24,000x^2$ N, $m = 2$ kg, $l = .5$ m. Find the equilibrium position of the mass and find the nonlinear spring's equivalent linear stiffness at that deflection. Calculate the natural frequency. Does the natural frequency change if l is set equal to .75 m? How does this situation differ from that of Problem 1.31?

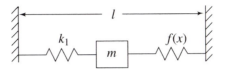

Figure P1.32

1.33. In an effort to add realism to cartoons, you've been asked to dynamically analyze a cartoon gag (assume that you're working during the summer as an animator). In this gag, Dopey Dog is falling from an airplane (see Figure P1.33) and, as he passes a ledge, he grabs an elastic rope he'd tied there earlier. His velocity at the instant he grabs the rope is 31.32 m/s. His mass is 35 kg, and the rope can be considered massless. The spring constant of the rope is 420 N/m. Calculate Dopey's position as a function after grabbing the rope and determine how long it will take for his velocity to reach zero. Assume he doesn't hit the canyon floor. (*Note*: Don't neglect the static offset he'd have due to gravity.) Calculate a trajectory for motions only for the period in which the rope is under tension.

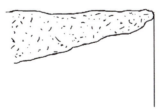

Figure P1.33

1.34. Determine the natural frequencies for the systems illustrated in Figure P1.34. What is the difference between the two cases? Which springs are most active and which are relatively uninvolved in the oscillations?

(*a*) $k_1 = 5\,\text{N/m}$, $k_2 = 5\,\text{N/m}$, $k_3 = 100\,\text{N/m}$, $m = 100\,\text{kg}$

(b) $k_1 = 50\,\text{N/m}$, $k_2 = 50\,\text{N/m}$, $k_3 = 10\,\text{N/m}$, $m = 100\,\text{kg}$

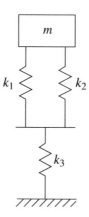

Figure P1.34

1.35. A static horizontal load of 80 N, applied to the lamp housing of the street lamp in Figure P1.35, caused a horizontal deflection of 5 cm. Assume that the motion of an earthquake can be modeled as a step change in position. Thus the initial condition is $x(0) = 18$ cm, $\dot{x}(0) = 0$. Determine the response of the lamp housing. Consider the lamp-pole system to be massless and the mass of the housing to be 25 kg.

Figure P1.35

1.36. The spring-mass system in Figure P1.36a has a natural frequency of ω_1 rad/s. By how much will the natural frequency change if the mass is attached as shown in Figure P1.36b? (The spring is cut into two equal pieces and reattached.)

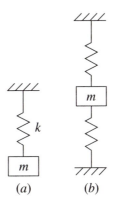

Figure P1.36

1.37. A 56.25 kg mass placed on top of an existing 100 kg mass (part of a spring-mass system) reduces the original natural frequency by 2 rad/s. What is the spring constant k?

1.38. Determine the free vibration of the system in Figure P1.38 (water in a U-shaped tube). The tube has an inner area of s, the water has density ρ, and the length of the water-filled section is l.

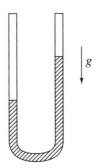

Figure P1.38

1.39. Is there any difference between the systems shown in Figure P1.39? If so, what is it?

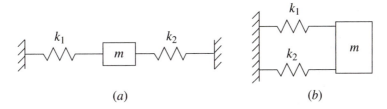

Figure P1.39

1.40. What is the natural frequency of oscillation for the system in Figure P1.40?

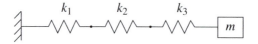

Figure P1.40

1.41. Determine how many natural frequencies you can come up with by combining the given springs in Figure P1.41 with a 1 kg mass.

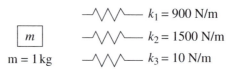

Some examples

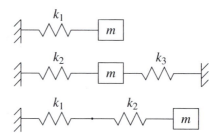

Figure P1.41

1.42. Consider two masses in Figure P1.42 joined by a spring and sliding on a frictionless surface. If the distance between them is given by z, determine the equation of motion of the system that involves only z, m_1, and m_2. (*Note*: The motion will be a "breathing" action in which the masses move together and apart.)

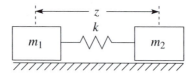

Figure P1.42

1.43. With use, the end conditions of the system in Figure P1.43 have gone from cantilevered-cantilevered to pinned-pinned. What is the percentage change in the system's natural frequency? Consider the beams to be massless. (Use Appendix B.)

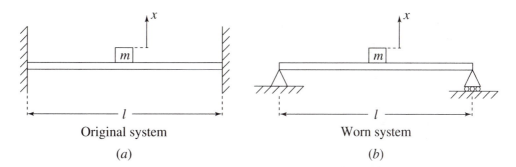

Figure P1.43

1.44. By what percentage will the natural frequency of the original system of Figure P1.43 change if the right support breaks, leaving the end unrestrained? Treat the beam as massless.

1.45. Figure P1.45a gives a simple model of a building consisting of a horizontal floor supported by two vertical beams. An even more simplified model would consist of a single beam and mass as shown in Figure P1.45b. If the stiffness of each original beam (for which both ends remain vertical) is k, determine what m' must equal in the single-mass, single-beam case if the natural frequencies are to agree. (For the single-mass, single-beam case, the beam end doesn't remain vertical.)

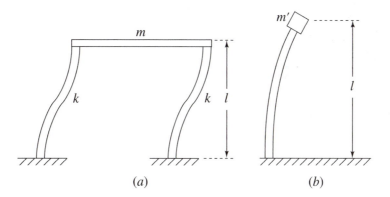

Figure P1.45

1.46. Demonstrate that the stiffness of the system in Figure P1.46a is equal to $\frac{192EI}{l^3}$. Use the knowledge that the stiffness of system in Figure P1.46b is $\frac{12EI}{l^3}$.

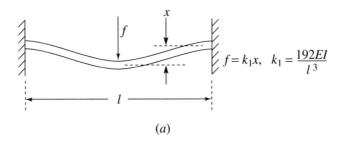

$$f = k_1 x, \quad k_1 = \frac{192EI}{l^3}$$

(a)

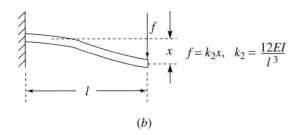

$$f = k_2 x, \quad k_2 = \frac{12EI}{l^3}$$

(b)

Figure P1.46

1.47. Demonstrate that the stiffness of the system in Figure P1.47a is equal to $\frac{48EI}{l^3}$. Use the knowledge that the stiffness of the system in Figure P1.47b is $\frac{3EI}{l^3}$.

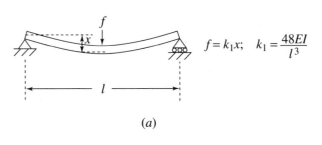

$$f = k_1 x; \quad k_1 = \frac{48EI}{l^3}$$

(a)

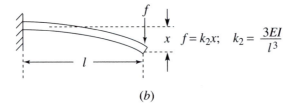

$$f = k_2 x; \quad k_2 = \frac{3EI}{l^3}$$

(b)

Figure P1.47

1.48. An SDOF spring-mass system has a natural frequency equal to 15 rad/s. How will this change if the suspension's spring constant is doubled?

1.49. When a 1500 kg mass is placed on a suspension, the springs deflect 1 cm. The spring force relation is linear. What is the natural frequency of oscillation for the system?

1.50. What is the natural frequency for the system illustrated in Figure P1.50? (Assume that it oscillates about its equilibrium position.) The radius of each pulley is .3 m, $k = 600$ N/m, and $m = 20$ kg. The pulleys are massless.

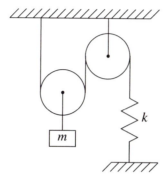

Figure P1.50

1.51. What is the natural frequency for the system illustrated in Figure P1.51? Keep in mind that we're looking at *small* oscillations about equilibrium, so you needn't worry about geometric effects from the springs due to the mass's motion. (*Hint*: Consider using polar coordinates.)

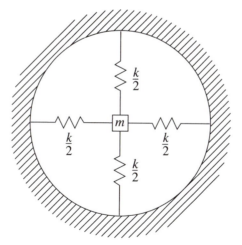

Figure P1.51

1.52. Consider a spring that has an unstretched length of .5 m and a spring constant equal to 500 N/m. Calculate the natural frequency if a 1 kg mass is suspended from the spring. Then calculate how the natural frequency will change if the spring is cut in half and the mass is suspended from just a half spring.

1.53. The natural frequency of the one-story building shown in Figure P1.53 can be calculated by determining the stiffness of the supporting beams and assuming that the roof moves horizontally. If the width of the building is l_1, the depth is l_2, the stiffness of each beam (4 total) is EI, and the length is l_3, calculate the natural frequency, assuming a roof mass of m. Treat the beams as massless.

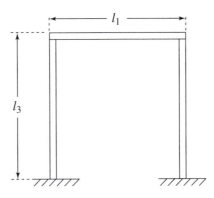

Figure P1.53

1.54. As a practical joke, you've put some fast-acting glue in your friend's slingshot as in Figure P1.54. Your friend puts a mass of .06 kg in the slingshot, pulls it back 1 m, and then releases it. If the mass travels out 1 m beyond the slingshot and then returns to strike his hand .5 second after the release, what must the spring constant be for the massless elastic band? Assume that the elastic band has an unstretched length equal to 0 m.

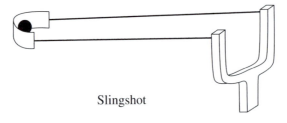

Slingshot

Figure P1.54

1.55. Given the spring force characteristics in Figure P1.55, determine the natural frequency of oscillation of the system. Assume the system executes small motions about the gravitationally induced static equilibrium. Take g to be 10 m/s^2 and $m = 2$ kg.

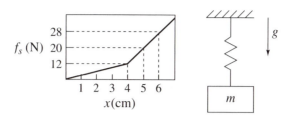

Figure P1.55

1.56. In an effort to add realism to cartoons, you've been asked to dynamically analyze a new episode. Dopey Dog, having attached himself by means of an elastic band to a rocky outcropping (Figure P1.56*a*), is dismayed when the outcropping becomes detached from the cliff (Figure P1.56*b*). After a drop of 49 m, the outcropping lands on a ledge with the part Dopey is attached to projecting outward from the ledge (Figure P1.56*c*). Thus Dopey continues to fall. The unstretched length of the elastic is 1 m. Just before the fall, Dopey was in static equilibrium and was 2 m from the outcropping (i.e., the total length of the unstretched elastic was 2 m). Assume that the separation between Dopey and the outcropping stayed constant during the fall.

Determine the value of linear spring constant k necessary to keep Dopey from hitting the ground (which is 3.1 m below his position at the instant the outcropping hits the second ledge). Considering the actual spring constant (as found from the static equilibrium conditions), will he hit the ground? Finally, which of the assumptions in the foregoing description doesn't make physical sense?

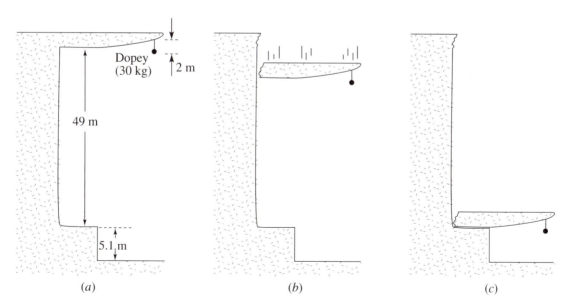

Figure P1.56

1.57. The coupling between the two railroad boxcars in Figure P1.57 can be viewed as a lumped spring. If the masses of the two cars are $m_1 = 2.0 \times 10^5$ kg and $m_2 = .9 \times 10^5$ kg and the spring constant $k = 8 \times 10^5$ N/m, determine the natural frequency of the system's oscillations.

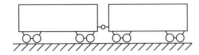

Figure P1.57

1.58. Determine the frequency of oscillation of a cube of wood seen in Figure P1.58 bobbing up and down in the water, in terms of the water's density, the wood's density, g, and l.

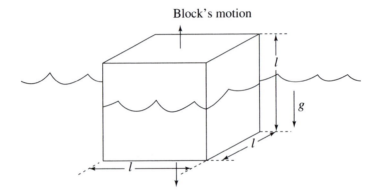

Figure P1.58

Section 1.3

1.59. Use the system illustrated in Figure P1.59 to determine the natural frequency of oscillation about $\theta = 0$. $l = 1.5$ m, $m = 2$ kg, $g = 9.81$ m/s^2.

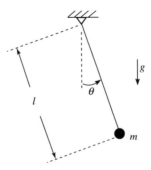

Figure P1.59

1.60. Find the natural frequency about $\theta = 0$ for the system in Figure P1.60. $m_1 = 5$ kg, $m_2 = 5.5$ kg, $l = 1$ m, $g = 9.81$ m/s^2.

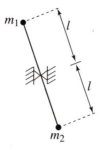

Figure P1.60

1.61. Find the natural frequency about $\theta = 0$ for the system in Figure P.161. $l_1 = 1$ m, $l_2 = 1.1$ m, $m_1 = m_2 = 5$, $g = 9.81$ m/s^2.

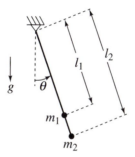

Figure P1.61

1.62. The pendulum illustrated in Figure P1.62 consists of a rigid bar of mass m_2 and length l. Attached to the end is a lumped mass m_1. Find the pendulum's natural frequency of oscillation.

Figure P1.62

1.63. Use the system illustrated in Figure P1.63 to determine the natural frequency of oscillation for the half-disk. Assume no-slip conditions.

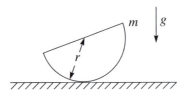

Figure P1.63

1.64. In Figure P1.64 Dopey Dog is rolling a large cylinder up a slope. The actual shape of the ground he's walking on is described by a circle, with a radius of $10r_1$. When $\theta = 30$ degrees, Dopey slips and the cylinder rolls over him. Assuming pure rolling, how long will it be before the cylinder returns to him? The cylinder's mass is m and it has a radius of r_1. Assume that linearization about $\theta = 0$ degrees is a valid approximation.

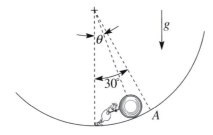

Figure P1.64

1.65. Determine how long it will take the cylinder of Problem 1.64 to return to Dopey if, instead of pure rolling, we have a frictionless interface between the cylinder and the slope.

1.66. For the system shown in Figure P1.66, the mass is constrained to ride in a tube that is rotating in a horizontal plane about the frictionless hinge O at the constant rate ω. The mass is connected to the spring k. Determine the response of the system about O for the following cases:

(a) $\omega^2 < \frac{k}{m}$

(b) $\omega^2 = \frac{k}{m}$

(c) $\omega^2 > \frac{k}{m}$

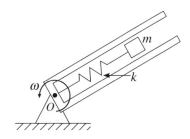

Figure P1.66

1.67. Find the linearized equation of motion about the equilibrium position of the system illustrated in Figure P1.67. You don't have to solve for the equilibrium position itself—just show what you'd need to do to obtain it. The torsional spring is uncompressed when $\theta = 0$. What is the linearized natural frequency?

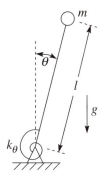

Figure P1.67

1.68. For the system shown in Figure P1.68, find the natural frequency of oscillation for the uniform bar of mass m in the xz plane. The supporting wires are massless and inextensible.

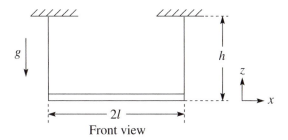

Front view

Figure P1.68

1.69. In Figure P1.69 shows a uniform bar (mass m, length $2l$) suspended by two inextensible wires (massless) of length h. At each end of the bar is a lumped mass m_1. What is the frequency of oscillation for rotary oscillations in the x-y plane?

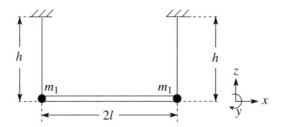

Figure P1.69

1.70. Find the equilibrium position for the system illustrated in Figure P1.70 and compute the natural frequency of oscillations about this equilibrium position. $m_1 = 1$ kg, $l_1 = 1.2$ m, $l_2 = .1$ m, $k = 6000$ N/m.

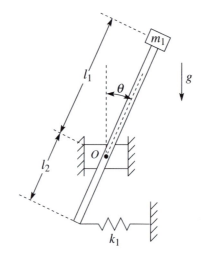

Figure P1.70

1.71. Use Figure P1.71 to determine the natural frequency of oscillation of a point mass sliding without friction on a circular track. (Use a small-angle approximation.)

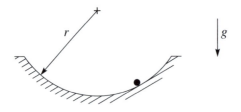

Figure P1.71

1.72. The system illustrated in Figure P1.72 rotates in a horizontal plane. What is the system's natural frequency?

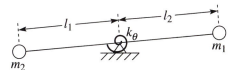

Figure P1.72

1.73. Find the natural frequency of oscillation for the system illustrated in Figure P1.73.

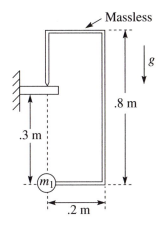

Figure P1.73

1.74. What are the equations of motion for the system shown in Figure P1.74? Consider motions about the vertical position.

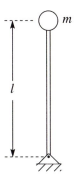

Figure P1.74

1.75. Use Figure P1.75 to find the solution for a mass particle sliding on a semicircle. Consider the system to be frictionless and the equilibrium position to be $\theta = 0$. Linearize the equations of motion (θ small), find the solutions, and explain what each solution physically represents. (Use $\theta(t) = ae^{\lambda t}$ as the assumed form for the solution.)

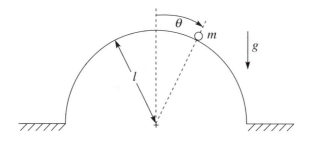

Figure P1.75

1.76. What is the natural frequency for the system shown in Figure P1.76?

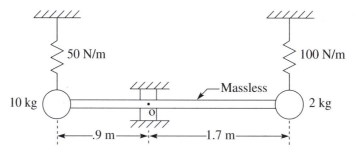

Figure P1.76

1.77. What is the natural frequency for the system shown in Figure P1.77?

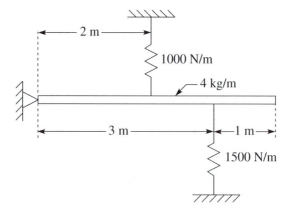

Figure P1.77

1.78. What is the natural frequency for the system shown in Figure P1.78? The hinge at O is frictionless; $k = 10,000$ N/m, $m_1 = 5$ kg, $m_2 = 7$ kg, $\rho = 1$ kg/m, $l_1 = 3$ m, and $l_2 = 5$ m.

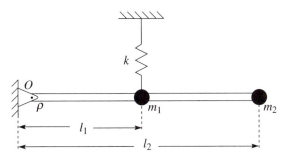

Figure P1.78

1.79. The system shown in Figure P1.79 consists of a mass m that slides without friction along a horizontal track and a tensioned string that connects the mass to the point 0. Determine the natural frequency of the system under the assumption that the tension doesn't change as the mass changes position. $m = 1$ kg, $\rho = .2$ kg/m, $T = 100$ N.

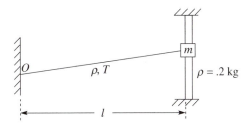

Figure P1.79

1.80. A band saw can be modeled as a continuous loop that runs around two pulleys, as shown in Figure P1.80. The steel has some flexibility, represented by the springs in the figure. Neglect the mass of the steel loop and determine the natural frequency of the pulley/band. Assume that the oscillation mode involves the pulleys moving in opposition to each other. The rotational inertia of each pulley about its shaft is equal to .001 kg·m², the spring stiffness is 10,000 N/m, and the radius of the pulleys is .4 cm.

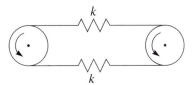

Figure P1.80

1.81. To analyze the motions of an automobile on its tires, we approximate the auto by a bar of uniform mass (1000 kg) with mass moment of inertia equal to $1,333.3$ kg·m². The bar is 4 m long. The tire-suspension assembly is approximated by a pair of springs, with spring constant equal to 11,697 N/m for each one. Use the system illustrated in Figure P1.81 to consider motions in which the car oscillates about its center of mass (no translation). What is the natural frequency of motion for the system?

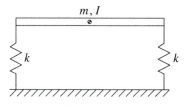

Figure P1.81

1.82. One way to determine a body's mass moment of inertia, $I_{\overline{AO}}$, is to attach the body to a torsional spring, observe the free vibrations, and deduce $I_{\overline{AO}}$ from the frequency of the oscillation. If the torsional stiffness of the vertical rod in Figure P1.82 is 81 N·m/rad and the free vibrations have a frequency of 150 rad/s, what must $I_{\overline{AO}}$ be equal to?

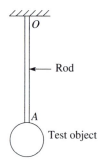

Figure P1.82

1.83. Determine the torsional stiffness of the vertical rod in the system illustrated in Figure P1.83. The rod is rigidly attached to the ceiling at 0 and attached to a bar of length 1 m and mass 3 kg (as shown). If the free vibrations occur at 5 rad/s, what is the rod's torsional stiffness?

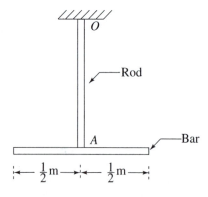

Figure P1.83

1.84. Figure P1.84 shows a lumped mass between two tensioned strings. Treat the strings as massless and neglect gravity. Determine the period of the free vibrations if the tension in the string is equal to 400 N, $l = 2$ m, and the mass is equal to .1 kg. (*Note*: Only small-amplitude motions of the mass take place.)

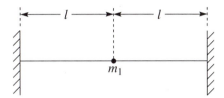

Figure P1.84

1.85. A circular disk is supported by three massless wires, as shown in Figure P1.85. The disk has a density ρ. Determine the frequency of oscillation if the disk is given an initial twist and released. Assume that the initial angular displacement of the disk is small.

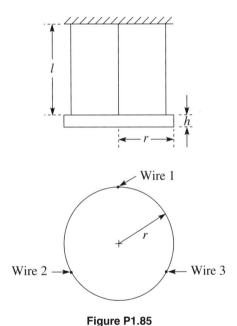

Figure P1.85

1.86. Find ω_n for the system illustrated in Figure P1.86. Consider the beams to be massless. The thin hoop of mass m and radius r can only rotate about O.

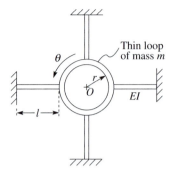

Figure P1.86

1.87. Find ω_n for rotational motions of the system illustrated in Figure P1.87 (a uniform disk constrained by four wires). Consider the tensioned restraining wires to be massless and to have uniform tensions T. All wires are of length l.

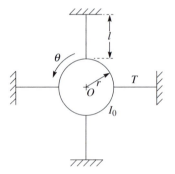

Figure P1.87

1.88. Determine the period of oscillation for the system illustrated in Figure P1.88. (Consider small oscillations, about $\theta = 0$.) $m_1 = 1$ kg, $m_2 = 5$ kg, $l_1 = 2.4$ m, $l_2 = 4$ m.

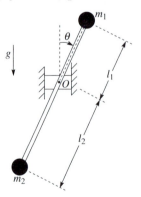

Figure P1.88

Section 1.4

1.89. As long as an SDOF spring-mass-damper is underdamped, its vibrational response will cross zero an infinite number of times, just as in the undamped case. By how much will the "period" of the oscillations be decreased from that of the undamped case if $\zeta = .05$? What if $\zeta = .8$? Comment on the results.

1.90. Does the addition of damping increase or decrease the frequency of the system oscillations?

1.91. If we wish to increase the decay rate for a viscously damped SDOF, is it more effective to alter ω_n or ζ?

1.92. What effect on the decay rate does m have for a viscously damped SDOF system?

1.93. Can the frequency of oscillation for a viscously damped spring-mass-damper system be altered without changing the oscillation decay rate by varying the system parameters?

1.94. If the mass of a spring-mass-damper system is doubled in value while the spring stiffness is halved, how will the decay rate and frequency of subsequent oscillations be affected?

1.95. How many oscillation cycles will elapse if you're waiting for the oscillation amplitude of a spring-mass-damper system to drop by 25% and $\zeta = .04$? Let $\omega_n = 100$. What if ζ is decreased to .0004?

1.96. Oscillation amplitudes are reduced by 50% in one second for a given spring-mass-damper system. Say you want to have the amplitude reduced to 25% of the original value in one second. How effective would it be to vary the value of k?

1.97. What does simultaneously doubling the mass and stiffness of a spring-mass-damper system do to the free vibration decay rate?

1.98. The mass illustrated in Figure P1.98 vibrates in a viscous flow. The force experienced by the mass as a function of velocity is plotted on the graph also shown in the Figure P1.98. Determine the damping factor for this system for $k = 1058$ N/m and $m = 2$ kg.

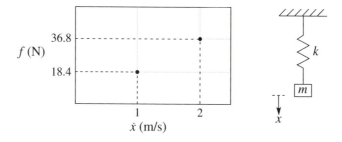

Figure P1.98

Section 1.5

(*Note*: Use Lagrange's equations to find the equations of motion for all the problems in this section.)

1.99. Find the equation of motion for the system illustrated in Figure P1.99.

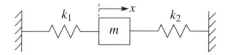

Figure P1.99

1.100. Find the equation of motion for the system illustrated in Figure P1.100.

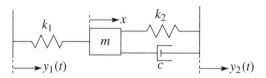

Figure P1.100

1.101. Find the equation of motion for the system illustrated in Figure P1.101.

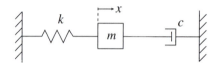

Figure P1.101

1.102. Find the equation of motion for the system illustrated in Figure P1.102.

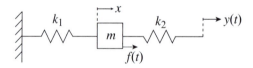

Figure P1.102

1.103. Find the kinetic and potential energies for the system illustrated in Figure P1.103 and determine the full equations of motion. After you've determined the equations of motion, linearize them (retaining terms like x, $\dot{x}$, $\dot{\theta}$, etc.).

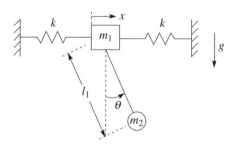

Figure P1.103

1.104. What are the equations of motion for the system illustrated in Figure P1.104?

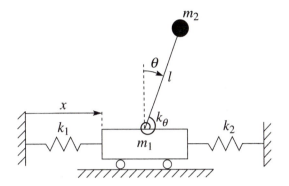

Figure P1.104

1.105. Find the equations of motion for the system illustrated in Figure P1.105.

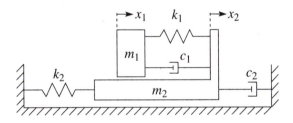

Figure P1.105

1.106. Find the equation of motion for the system illustrated in Figure P1.106 and linearize it about $\theta = 0$.

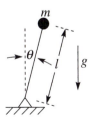

Figure P1.106

1.107. Find the equations of motion for the system illustrated in Figure P1.107.

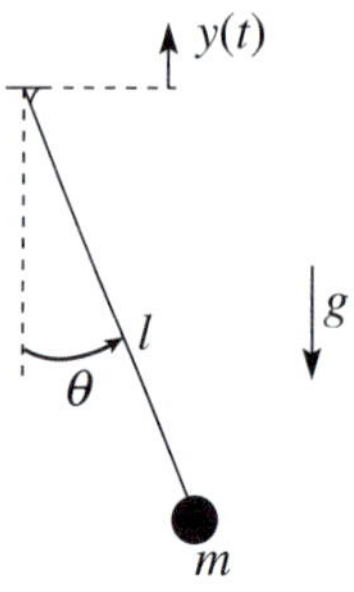

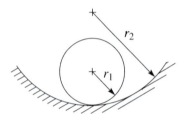

Figure P1.107

1.108. Find the linearized equation of motion for the system illustrated in Figure P1.108. Assume that the cylinder is of mass m and rolls without slipping on the circular path.

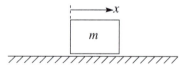

Figure P1.108

1.109. Find the equation of motion for the system illustrated in Figure P1.109. There is no interface friction.

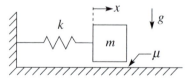

Figure P1.109

1.110. Find the equation of motion for the system illustrated in Figure P1.110. Assume that dry friction ($f = \mu W$) acts between the mass and the supporting surface.

Figure P1.110

1.111. Find the equation of motion for the system illustrated in Figure P1.111. There is no interface friction.

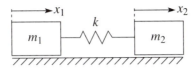

Figure P1.111

1.112. Find and linearize the equations of motion for the system illustrated in Figure P1.112. The bar has length l.

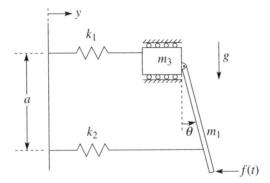

Figure P1.112

1.113. Find the equations of motion for the system illustrated in Figure P1.113. Assume small rotations and vertical deflections of the bar (no motion in the bar's longitudinal direction). Assume that the bar is uniform. As coordinates, use the vertical deflection of the bar's center of mass (y) and the rotation of the bar (θ).

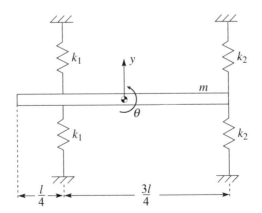

Figure P1.113

1.114. Find and linearize the equations of motion for the system illustrated in Figure P1.114. $m_1 = 1$ kg, $m_2 = 3$ kg, $l_1 = 2$ m, $l_2 = 1$ m, $g = 9.81$ m/s^2.

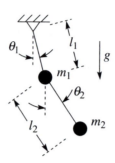

Figure P1.114

2

Forced Vibration of Systems with a Single Degree of Freedom

2.1 INTRODUCTION

In Chapter 1 we saw how to determine the equations of motion for systems that aren't experiencing any external forcing. Something must have started them into motion but, for the time under consideration, they were simply reacting to that initial condition, oscillating with or without some amplitude variation. Now we're going to ask what happens when we continually force the system.

We'll start by looking at step inputs, the simplest excitation, and one that is really a bridge between the preceding chapter and this one. Then we'll move on to the star of the chapter—sinusoidal excitations. Since our interest in this chapter is in steady state behavior, we're not going to concern ourselves about how the system reaches this steady state, but only about the specifics of the motion once all the transient effects have died down. Thus, this kind of an approach would tell us what the long-term vibrational forces are that act on a passenger in an automobile as a result of sinusoidal vibrations of the engine. It would also let us determine the forces transmitted to the ground because of an unbalanced shaft in a steam turbine. In fact, the number of examples you can come up with is almost unlimited. Any time there is a periodic disturbance you're going to get a periodic response.

Of course, the question of a system's transient response isn't an unimportant one, and thus we'll take on this particular subject in Chapter 3.

2.2 SEISMIC EXCITATION—STEP INPUTS

To introduce you to the behavior of forced, single-degree-of-freedom systems, we will again use our physical mass-spring experiment. As before, you need to hold the free end of the spring and let the mass hang down freely. There are two obvious ways to apply a force to a system of this sort. The first, and the one discussed at length in most texts, is to apply a force to the mass itself while keeping the free end of the spring fixed. The problem with this approach is that unless you have access to some noncontacting force generator, there is no way you can replicate this sort of forcing experimentally. Therefore we'll start with the second approach, that of moving your hand. This is called *base* or *seismic* excitation. Although our system hangs down, it is clear that we could just as well have a system supported by springs (as shown in Figure 2.1), and in this case movements of the spring support are obviously base motions. The term "seismic excitation" comes from the observation that this is precisely the sort of excitation acting on buildings during earthquakes. Since you are moving one end of the spring, its subsequent elongation or compression impresses on the mass a force that is proportional to the amount of stretch.

To perform the first experiment, let the mass hang down freely. Once it is hanging quietly, quickly raise your hand to a different height and keep it there. This will cause the mass to oscillate about a new equilibrium (one that's exactly the same distance up from the old one as the amount your hand moved). As time goes on, the oscillations die away and the steady state position of the mass will alter by the amount you moved your hand. Although this might not seem to be a big deal, it actually is pretty significant. For instance, say you're designing a positioning device, such as the one shown in Figure 2.2. We'll assume that a very powerful motor can move the left end to a specified position. The positioner arm is flexible, and at the right end there is a gripper that can hold onto a payload. If the arm is much lighter than the payload, we can neglect its mass and model it as a pure spring, with spring constant k. The mass of the payload and gripper is m. The simplified model for such a system would look like that shown in Figure 2.3, and the equations of motion for this system would

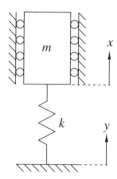

Figure 2.1 Seismically excited spring-mass system

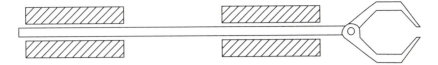

Figure 2.2   Positioner arm

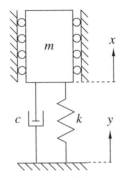

Figure 2.3   Seismically excited
spring-mass damper

be given by

$$m\ddot{x} + c\dot{x} + kx = ky + c\dot{y} \tag{2.2.1}$$

where y is the displacement of the controlled end and x is the displacement of the gripper. The damping $c\dot{x}$ is included to model energy losses in the overall system. If we divide (2.2.1) by m we can then reexpress it in terms of ω_n and ζ:

$$\ddot{x} + 2\zeta\omega_n\dot{x} + \omega_n^2 x = \omega_n^2 y + 2\zeta\omega_n\dot{y} \tag{2.2.2}$$

You'll note that this is exactly the same situation we had with our physical experiment: a mass-spring combination and a controlled end. We even had the damping, since it's certain you noticed that the oscillations didn't go on forever. Situations like this are extremely common in machine design. Very often we'll design a structure that's meant to translate or rotate. Since the structure has mass and flexibility, it will allow vibrations, vibrations that are often quite undesirable. To see this, let's analyze (2.2.2). Our input is illustrated in Figure 2.4 and is simply a step change in position. This is one of the simplest forced examples, since all the transient excitation occurs at $t = 0$; after that the forcing is constant.

Since the steady-state solution is just a constant displacement (as you saw from your experiment), we'll assume a solution of the form

$$x(t) = c + e^{-\zeta\omega_n t}(a\cos(\omega_d t) + b\sin(\omega_d t)) \tag{2.2.3}$$

We've explicitly assumed here that the system is underdamped (which is by far the most common situation). Note that we're using ω_d, which is equal to the damped frequency ($\omega_d = \omega_n\sqrt{1 - \zeta^2}$). The constant c is meant to account for the steady state displacement of the mass due to the base movement, and a and b are needed to account for initial conditions.

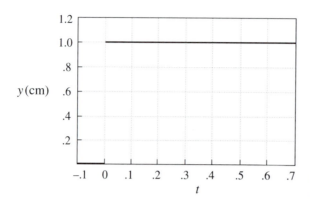

Figure 2.4   Unit step displacement input

If the system was initially at rest, then $x(0) = \dot{x}(0) = 0$. If the base displacement is 1 cm, then our forced solution must match our steady state conditions for large t.

$$a = 1 \text{ cm} \tag{2.2.4}$$

Thus our solution becomes

$$x(t) = 1 + e^{-\zeta \omega_n t}\left(a \cos(\omega_d t) + b \sin(\omega_d t)\right) \tag{2.2.5}$$

where $x(t)$ is measured in centimeters.

Finally, we account for the zero initial conditions:

$$x(0) = 1 + a = 0 \tag{2.2.6}$$

and

$$\dot{x}(0) = \zeta \omega_n + b \omega_d = 0 \tag{2.2.7}$$

The solutions to this are $a = -1$ and $b = -\dfrac{\zeta}{\sqrt{1-\zeta^2}}$. Thus

$$x(t) = 1 + e^{-\zeta \omega_n t}\left(-\cos(\omega_d t) - \frac{\zeta}{\sqrt{1 - \zeta^2}} \sin(\omega_d t)\right) \tag{2.2.8}$$

Now let's consider a couple of cases. First, let's assume that our system has very little damping, say a damping factor of .05. We'll let $m = 1$ kg and $k = 900$ N/m. The response for a step input to this system is shown in Figure 2.5a. You'll note that there's some good and some bad news associated with this response. If we continue with our analogy to a positioner, the point of the device is to position the end at the steady state position quickly and accurately. Although the payload is brought to the correct position (1 cm) quickly, it then overshoots and *rings* (or oscillates) for a long time. Only after many oscillation cycles does the final response start to look like a constant offset. Obviously, this ringing is not generally considered to be desirable. If our positioning arm is part of a robotic welder in an automotive plant, the robot will have to wait for a good while before making the weld, since if the oscillations are appreciable in magnitude, they will degrade the welding process.

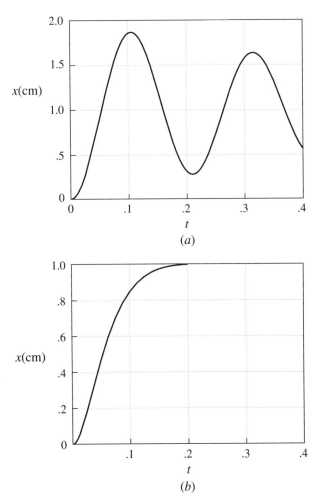

Figure 2.5   Effect of damping on system's time response: (a) $\omega_n = 30$ rad/s, $\zeta = 0.5$; (b) $\omega_n = 30$ rad/s; $\zeta = .8$

One way to reduce the amount of ringing is to increase the damping. If we increase ζ from .05 to .8, then our response will look that of Figure 2.5b. Note that now the ringing is greatly reduced, a good characteristic for our welder. However the response time has increased a great deal. Instead of reaching the .9 cm position in .05 second (as it did for the preceding case), it now requires almost .12 second. This is a not-so-good result. Whereas before we had lots of oscillations to deal with, now we have the problem of a relatively slow response.

This illustrates the sort of trade-off you'll be facing when in any design problem. Changing a parameter to make some aspect of the problem better invariably makes another aspect worse. Your job will be to find the optimum trade-off that gives you the behavior you need at an acceptable cost.

The general response shown in Figure 2.6 is a good one because it allows us to illustrate how transient responses are used to characterize a system. If a step input is our input, we'd like the response

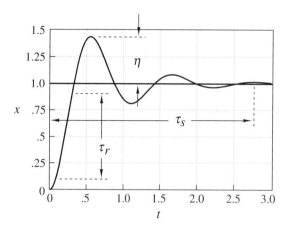

Figure 2.6 Step response for an underdamped system

to also be a step. Unfortunately, this is never the case, and thus we characterize the actual response by how much it varies from the ideal. One commonly used quantity is the *rise time*, shown in Figure 2.6 as τ_r. This can be defined in different ways; we've chosen to define it as the time it takes for the response to go from 10% of the steady state value to 90% of this value. We can also talk about *settling time*, shown in Figure 2.6 as τ_s. This is the time necessary for the oscillation to die away almost completely. The amount of an oscillation that is allowed is somewhat arbitrary; in Figure 2.6 we've set it equal to the time needed for the oscillations to die down to 2% of the steady state value. The last quantity we'll introduce is the overshoot, i.e., the amount by which the response exceeds the steady state response, indicated by η.

Example 2.1

Problem   A given spring-mass-damper system (like the one shown earlier in Figure 2.3) has an unacceptably slow settling time. The original system parameters are $m = 10$ kg, $k = 100,000$ N/m, and $c = 200$ N·s/m. What should c be changed to in order to halve the settling time?

Solution   The solution given by (2.2.8) shows an exponentially decaying envelope $(e^{-\zeta \omega t})$ with accompanying oscillations. For the given parameters $\omega_n = \sqrt{\frac{k}{m}} = 100$ rad/s, $\zeta = \frac{c}{2m\omega_n} = .1$, and $\omega_d = \omega_n \sqrt{1 - \zeta^2} = 99.5$ rad/s.

For the exponential envelope to decay down to 2% of its original extent would require

$$e^{-\zeta \omega_n t^*} = .02$$

Taking the natural log of both sides and using our known values of ζ and ω_n yields

$$10 t^* = 3.912$$

$$t^* = .39 \text{ second}$$

To halve this time we'll need to solve

$$e^{-100\zeta'(\frac{.39}{2})} = .02$$

or

$$19.5\zeta' = 3.912$$

$$\zeta' = .2$$

thus giving us

$$c' = 2m\omega_n\zeta' = 2(10)(100)(.2) = 400 \text{ N} \cdot \text{s/m}$$

The type of analysis we've been going over here is also quite central to controls work, and you'll be seeing the same material if you take a controls course. These transient measures of system response are very easy to visualize, and they play a large part in our intuition about how systems respond. Although we won't be going into any great detail in this book, these concepts also form the basis of several methods of system identification. When you first start to analyze a system, you may not know its mass, stiffness, or damping properties. In fact, you probably won't. Thus, a central concern is simply figuring out how to mathematically represent the system. This problem is the motivation behind system identification methods. Simply put, these methods allow you to run tests on a system and to then determine what the mass, stiffness, and damping properties are. You can intuitively see how this could be done just from our example. If you know how SDOF systems respond to a step input, you can apply a step to the system, record the output, and then calculate τ_s, τ_r, etc. By comparing these values to those you'd obtain from (2.2.1), you can figure out what the physical parameters of your system must be. We'll talk a bit more about identification in Section 2.10, but for now let's move on to more involved types of forcing input.

2.3 SEISMIC EXCITATION—SINUSOIDAL INPUT

To start our next experiment, hold the free end of the spring, begin to *slowly* move your hand up and down, and observe the mass. It moves up and down in the same direction as your hand and also seems to move the same amount. If your hand moves up one inch, the mass moves up around one inch as well. This is our first observation about vibrational systems. When you excite a vibrational system at a *low* frequency (which is what we were doing), the spring in the system acts like a rigid rod. The motion is the same as if the spring were actually replaced by a rigid rod. This is a limiting case response, a topic we'll look at more closely in Chapter 7.

Now start moving your hand up and down as quickly as you can. This is a *high* frequency excitation. You'll notice that the mass essentially stays fixed in space. Even though your hand is moving up and down, the mass doesn't move (at least not much). Observation number two! In the limiting case of high frequency vibrations, it is as if the spring disappears; almost no motion is transmitted through the spring to the mass.

At this point we're ready to move on to a more dramatic effect. If you experiment a very little bit, you'll find that there is a frequency of excitation at which, with almost no motion of your hand, the mass exhibits a very large response. In fact, you'll probably find that the motion keeps growing and that soon the mass is flying out of control. This phenomenon is called *resonance*.

Let's recap for a moment. At very high frequencies, the mass didn't move at all. At resonance it moved a tremendous amount, while at low frequencies the mass moved only as much as your hand moved. The remaining question is, What happens for other frequencies of excitation? Now that you know the resonant frequency, this will be easy to answer. Start oscillating your hand again, but at a frequency below the resonance frequency. You'll see that the mass again moves in the direction of your hand (up when your hand goes up, down when your hand goes down), but the amount of motion is now *greater* than your hand's motion. Next, try moving your hand at a frequency above the resonant frequency. This time we see something new. Depending on the precise frequency with which you move your hand, the mass's motion can be greater than or less than your hand's motion. And in either case, the motion of the mass is in opposition to that of your hand. When your hand goes up, the mass goes down, and vice versa. This is called *out-of-phase* motion. Similarly, for the case in which the mass moved in the same direction as your hand, we had *in-phase* motion. The situation will become a bit more complicated when damping is added, but for now, the mass can only be completely in phase or out of phase with the excitation. We'll explain why we use the word *phase* once we've examined the problem from a more mathematical viewpoint.

Now that we've gotten some physical feel for the system, we can analyze it mathematically and then correlate the results to our physical experiments. Looking again at Figure 2.1, we can see that the governing equation is

$$m\ddot{x} + kx = ky \tag{2.3.1}$$

We've neglected damping here, a reasonable assumption for responses of the kinds we're going to be looking at. Dividing by m, we obtain

$$\ddot{x} + \frac{k}{m}x = y\frac{k}{m} \tag{2.3.2}$$

or, using our knowledge that $\frac{k}{m} = \omega_n^2$,

$$\ddot{x} + \omega_n^2 x = \omega_n^2 y \tag{2.3.3}$$

We can model the excitation y as

$$y = \bar{y}\sin(\omega t) \tag{2.3.4}$$

i.e., a purely sinusoidal forcing. You'll recall from your course in ordinary differential equations that the general solution for such a system is

$$x = a\sin(\omega t) + b\cos(\omega t) \tag{2.3.5}$$

where ω is not equal to ω_n. If ω was equal to ω_n then our assumed solution would be the same as the forcing itself, in which case (2.3.5) would no longer be adequate [1].

Note that (2.3.5) is the solution for the steady state problem. We're ignoring the effect of initial conditions (for which we'd need to add the homogeneous solution $c\sin(\omega_n t) + d\cos(\omega_n t)$). Although we can obviously deal with both the homogeneous and nonhomogeneous solutions simultaneously, it is vastly more common in vibration analyses to be concerned only with the steady state responses of a system. Only if we cared about the transient response would we retain both solutions. Problems for which we'd care about the transient motion would include the response of an aircraft to a sudden wind gust or to a landing, the response of a building to an earthquake, and the motion of an offshore oil rig in response to a large wave.

Just this once, we'll work the problem out, keeping both the $\sin(\omega t)$ and $\cos(\omega t)$ terms to illustrate why only one is actually needed for the undamped case. Differentiating (2.3.5) twice to obtain an expression for $\ddot{x}$ and substituting for x, $\ddot{x}$, and y in (2.3.3), we obtain

$$a(\omega_n^2 - \omega^2)\sin(\omega t) + b(\omega_n^2 - \omega^2)\cos(\omega t) = \bar{y}\omega_n^2 \sin(\omega t) \tag{2.3.6}$$

Now $\sin(\omega t)$ and $\cos(\omega t)$ are linearly independent functions (i.e., you cannot combine multiples of them to get zero). Therefore, the only way that the left-hand side of the equation can equal the right-hand side is for all the terms multiplying $\sin(\omega t)$ on the left to equal the terms multiplying $\sin(\omega t)$ on the right (and the same for the $\cos(\omega t)$ terms).

Thus we must have

$$a(\omega_n^2 - \omega^2) = \omega_n^2 \bar{y} \tag{2.3.7}$$

and

$$b(\omega_n^2 - \omega^2) = 0 \tag{2.3.8}$$

It's pretty clear from (2.3.8) that unless $\omega = \omega_n$, the only way to satisfy the equation is for b to be zero. And we've already mentioned that ω can't be equal to ω_n. Thus we can drop the $\cos(\omega t)$ part of the solution.

What we've seen is a general result. When dealing with a sinusoidally forced oscillator that only has mass and spring terms, the assumed solution need only include the trigonometric terms found in the forcing. As we can see, this means that the response can be motion in the same direction as the forcing (in-phase motion) or motion in opposition to the forcing (out-of-phase). At this point we can clear up what we mean by the word "phase". Instead of calling the response $a \sin(\omega t)$, we can write it as $a \sin(\omega t - \phi)$. a represents the amplitude of the response, and it will not switch sign as the frequency goes above resonance. Simply the phase angle, ϕ, will change. When the mass moves in the same direction as the excitation, the phase angle is equal to zero. After resonance, the phase angle will abruptly switch to π radians (180 degrees), indicating that the response lags the input by π radians. This kind of representation will be essential once we've added damping to the problem, since we'll then have values of ϕ between 0 and 180 degrees.

Now that (2.3.8) is out of the way, it's easy to use (2.3.7) to solve for a,

$$a = \frac{\omega_n^2 \bar{y}}{\omega_n^2 - \omega^2} \tag{2.3.9}$$

Thus the final solution is

$$x(t) = \frac{\omega_n^2 \bar{y}}{\omega_n^2 - \omega^2} \sin(\omega t) \tag{2.3.10}$$

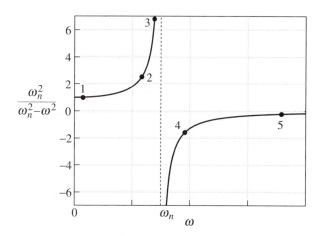

Figure 2.7 Displacement response of a seismically excited spring-mass system

This is a good time to compare our experimental findings with the mathematical result we've just obtained. Figure 2.7 shows a plot of $\frac{\omega_n^2}{\omega_n^2-\omega^2}$ versus ω. The dots labeled 1–5 correspond to the test cases of our physical experiments. At point 1, we're forcing the system at a low frequency and find that the response amplitude is almost equal to the forcing amplitude. Furthermore, the sign of a is the same as the sign of $\bar{y}$. We call this in-phase motion since, if we had used $a\sin(\omega t - \phi)$ as our guess for the response, ϕ would equal zero, and thus the input and output would be locked together. As we increase in frequency to point 2, the response stays in phase with the forcing but the amplitude grows. At point 3, near the resonance condition, the amplitude of the response is going to infinity. Increasing the forcing yet further (point 4) causes the response to go out of phase with the input (indicated by the negative values for the response). Finally, as the frequency becomes very large (point 5), the response becomes quite small.

Happily, all the mathematical predictions match the experimental results. Even at this early stage, we are able to do a good bit of engineering. We can determine the frequencies at which the response amplitude of the mass is less than the forcing amplitude and those for which it is greater. Furthermore, it's a simple matter to consider the accelerative loadings encountered by the mass (an item of interest when packaging fragile goods). All we need do is differentiate (2.3.10) twice to obtain $\ddot{x}$, the acceleration of the mass:

$$\ddot{x} = \frac{-\omega^2\omega_n^2\bar{y}}{\omega_n^2 - \omega^2}\sin(\omega t) \tag{2.3.11}$$

Figure 2.8 shows a plot of the normalized acceleration, $\frac{-\omega^2\omega_n^2}{\omega_n^2-\omega^2}$ versus the frequency of excitation. Although the overall results are somewhat similar to those shown in Figure 2.7, we see that the acceleration goes to zero at low frequencies (instead of going to a nonzero constant) and becomes constant at higher frequencies (instead of going to zero). Once again, the response switches phase at the system's resonant frequency.

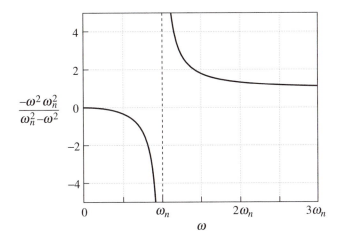

Figure 2.8 Acceleration response of a seismically excited spring-mass system

From the foregoing, we can see that the system's natural frequency is a very special frequency. With this in mind, we might well want to consider introducing the following, nondimensional, frequency:

$$\Omega \equiv \frac{\omega}{\omega_n} \tag{2.3.12}$$

In this case, resonance occurs when $\Omega = 1.0$. Dividing both numerator and denominator of (2.3.9) by ω_n^2 and using (2.3.12), we find

$$a = \frac{\bar{y}}{1 - \Omega^2} \tag{2.3.13}$$

This is certainly a clean and compact form for the amplitude response. It is also a very general solution. Dividing it by $\bar{y}$ makes it even more general by removing the particular value of excitation

$$\frac{a}{\bar{y}} = \frac{1}{1 - \Omega^2} \tag{2.3.14}$$

If we plot (2.3.14) versus Ω, as is shown in Figure 2.9, we have a response plot that is valid for any seismically excited, SDOF spring-mass system. To relate this figure to a particular system, we simply need to be given the actual excitation amplitude, mass, and stiffness values (to be able to convert the nondimensional Ω into an actual frequency and correctly scale the result).

Just as we could put (2.3.9) in a very clear form by using Ω, we can also reexpress (2.3.11) if we wish:

$$\ddot{x} = \frac{-\omega_n^2 \Omega^2 \bar{y}}{1 - \Omega^2} \sin(\omega t) \tag{2.3.15}$$

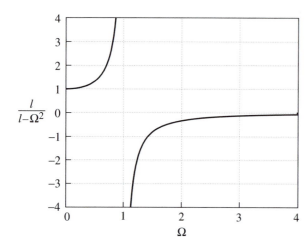

Figure 2.9 Response of seismically excited spring-mass system, nondimensional frequency

Example 2.2

Problem Consider again the system shown in Figure 2.1. Determine the absolute value of the response if the base is oscillated at 25 rad/s with an amplitude of 1 mm and if the system has a spring constant equal to 1000 N/m and a mass of 2 kg.

Solution The governing equation is given by

$$m\ddot{x} + kx = ky$$

Putting this into the form of (2.3.3) gives us

$$\ddot{x} + \omega_n^2 x = \omega_n^2 y$$

Since $\omega_n^2 = \frac{k}{m} = 500$, we have

$$\ddot{x} + 500\,x = 500\,y$$

Since the particular phase of the forcing isn't important, we'll assume a cosine for the forcing

$$\ddot{x} + 500\,x = (500).001\cos(25t)$$

where the particular amplitude and frequency of the base excitation have been used.
The solution amplitude is given by (2.3.9) as follows:

$$a = \frac{(500)(.001)}{500 - 25^2} = -.004$$

Thus the absolute value of the response is 4 mm.

Example 2.3

Problem The system shown in Figure 2.1 is excited at a frequency ω. The absolute value of the response is 3.6 mm; $m = .05$ kg, $k = 45$ N/m, and $y(t) = 2\sin(\omega t)$ mm. Can you determine ω?

Solution   Our equation of motion is given by

$$m\ddot{x} + kx = k\bar{y}\sin(\omega t)$$

or

$$.05\ddot{x} + 45x = 45\bar{y}\sin(\omega t) = 90\sin(\omega t)$$

Our solution is given by

$$x(t) = \frac{90}{45 - .05\,\omega^2}\sin(\omega t) = \bar{x}\sin(\omega t)$$

Since $|\bar{x}| = \left|\frac{90}{45 - .05\,\omega^2}\right|$, we have either

$$3.6 = \frac{90}{45 - .05\,\omega^2}$$

or

$$-3.6 = \frac{90}{45 - .05\,\omega^2}$$

depending upon whether ω is greater than or less than the system's natural frequency.
If $\omega < \omega_n$, then

$$3.6 = \frac{90}{45 - .05\omega^2} \quad \Rightarrow \quad 45 - .05\,\omega^2 = 25$$

$$\omega = 20 \text{ rad/s}$$

If $\omega > \omega_n$, then

$$-3.6 = \frac{90}{45 - .05\omega^2} \quad \Rightarrow \quad 45 - .05\,\omega^2 = -25$$

$$\omega = 37.42 \text{ rad/s}$$

Without knowing whether the response was in or out of phase with the input, we won't be able to determine which of these solutions for ω to choose.

2.4 DIRECT FORCE EXCITATION

Now that we've examined the case of unforced seismic excitation, we can move on to the problem for which the free end of the spring is fixed and a force is directly applied to the mass (Figure 2.10). Clearly, we won't be able to experimentally investigate this particular case as we did for seismic excitation but, as we'll soon see, the two problems are almost identical in their mathematics. Thus we can have some confidence in the analysis even though we don't perform a supporting experiment. Keep in mind that the two problems are almost identical only when there is no damping in the systems. The presence of damping will cause some complications and will be examined in Sections 2.6 and 2.8.

As you can easily determine from Figure 2.10, the equation of motion for this system is simply

$$m\ddot{x} = f(t) - kx \qquad (2.4.1)$$

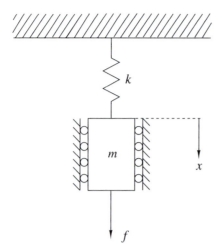

Figure 2.10 Direct force excited, spring-mass system

or, dividing by m and using the definitions already introduced

$$\ddot{x} + \omega_n^2 x = \frac{f(t)}{m}$$ (2.4.2)

Assuming that the excitation can be expressed as

$$f(t) = \bar{f} \sin(\omega t)$$ (2.4.3)

and using the general solution

$$x(t) = \bar{x} \sin(\omega t)$$ (2.4.4)

leads to

$$-\omega^2 \bar{x} \sin(\omega t) + \omega_n^2 \bar{x} \sin(\omega t) = \frac{\bar{f}}{m} \sin(\omega t)$$ (2.4.5)

Canceling out the $\sin(\omega t)$ factor and solving for $\bar{x}$, we find that

$$\bar{x} = \frac{\bar{f}}{m(\omega_n^2 - \omega^2)}$$ (2.4.6)

If we wish to express this in terms of the nondimensional frequency, Ω, we have

$$\bar{x} = \frac{\bar{f}}{m\omega_n^2(1 - \Omega^2)}$$ (2.4.7)

Some authors prefer to clean up this solution by letting $f(t)$ equal $f'k \sin(\omega t)$, which would lead to

$$\bar{x} = \frac{f'}{1 - \Omega^2}$$ (2.4.8)

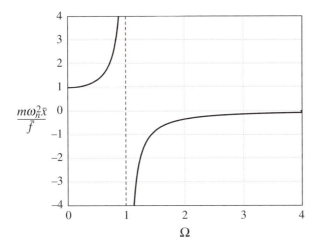

Figure 2.11   Displacement response of direct force excited, spring-mass system, nondimensional frequency

just the same form as we had in (2.3.13). However this often leads to confusion, and so we'll express the forcing as just $f \sin(\omega t)$.

The nondimensional response $\frac{m\omega_n^2 \bar{x}}{\bar{f}}$ (from (2.4.7)) is plotted versus Ω in Figure 2.11. Clearly, the forms of (2.3.13) and (2.4.7) are the same; the only difference is in their magnitude at a particular frequency.

Example 2.4

Problem   For the system shown earlier in Figure 2.10, determine the response if the input is a cosine excitation with a magnitude of 2 N. The mass of the system is 5 kg, the spring constant 20,000 N/m, and the frequency of excitation is 60 rad/s.

Solution   The solution for this problem will be equal to the system's transfer function multiplied by the input. The system's natural frequency is found from

$$\omega_n^2 = \frac{20,000}{5} = 4000$$

Using (2.4.6) we'll find

$$\bar{x} = \frac{2}{5(4000 - 60^2)} = .001$$

Thus $x(t) = .001 \cos(60\,t)$, where the units of displacement are meters.

2.5 TRANSFER FUNCTIONS

A very useful characteristic of linear systems is that they support a phenomenon known as superposition. That is, an input i_1 produces on output o_1 for a particular linear system, and another input i_2 produces the output o_2, then the output of the same system for the combined input $\alpha i_1 + \beta i_2$ (where

α and β are arbitrary constants) is simply $\alpha o_1 + \beta o_2$. Thus, if you already know the output of the system when the forcing is $\sin(\omega t)$, you don't have to recalculate the response if the input amplitude is changed to, say, $5 \sin(\omega t)$. It's simply equal to the original response multiplied by 5.

We've already seen, both experimentally and theoretically, that a seismically excited spring-mass system, when acted on by an input sinusoid, reacts by producing a sinusoidal motion of the mass. The only things that can change are the degree to which the mass responds and whether it is in or out of phase with the input. Thus we can view the system as a sort of amplifier, one that takes in sinusoidal signals and produces sinusoidal signals. The degree of amplification, or gain, is dependent upon frequency. Therefore, we really should be concerning ourselves not with the actual response for a particular sinusoidal input but with the gain characteristics of the system as a function of frequency. This information is contained in what we shall call the *transfer function* of the system.

A quick examination of (2.3.10) reveals that the response actually contains the entire input excitation $\left(\bar{y} \sin(\omega t)\right)$, along with other terms. We are now in a position to recognize these other terms as the transfer function of the system. Dividing (2.3.10) by $\bar{y} \sin(\omega t)$, we find that the transfer function $g(\omega)$ is given by

$$g(\omega) \equiv \frac{x(t)}{y(t)} = \frac{\omega_n^2}{\omega_n^2 - \omega^2} \tag{2.5.1}$$

or, in terms of the nondimensionalized frequency Ω,

$$g(\Omega) = \frac{1}{1 - \Omega^2} \tag{2.5.2}$$

found by dividing the numerator and denominator by ω_n^2. A plot of $g(\Omega)$ versus frequency is shown in Figure 2.12. Figures like this are called frequency-response plots, for the obvious reason that they show us the response amplitude of the system to an input sinusoid as a function of frequency.

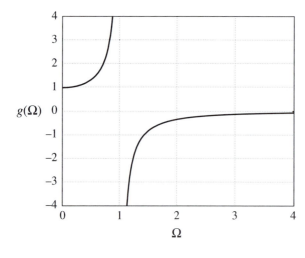

Figure 2.12   Transfer function of a seismically excited spring-mass system, nondimensional frequency

Although you will sometimes encounter frequency responses in the form shown, it is more common to present this information by expressing the transfer function in the form $|g|e^{i\phi}$, where $|g|$ is the absolute value of the transfer function and $e^{i\phi}$ represents any complex phase shifts. This is especially important for transfer function that is complex (as usually is the case when damping is present), since to adequately express the complete response, we'd then need to plot the real part versus frequency as well as the imaginary part versus frequency. Looking at the magnitude and phase is usually more physically meaningful. In addition, we won't be able to find the transfer function of damped systems in precisely this way (dividing the actual response by the forcing) but will instead use the complex exponential representation that'll be introduced in Section 2.7.

Since our transfer function is real (for now), ϕ will be either 0 or $-\pi$ radians. The first plot simply presents the absolute value of the amplitude response as a function of frequency ($|g(\Omega)|$) and is shown in Figure 2.13a. In this case, the resonant response at $\omega = \omega_n$ shows up as an infinitely tall spike. The other plot (Figure 2.13b) gives us the phase information that is lost when the absolute value of $g(\Omega)$ is taken.

The reason for choosing ϕ to be $-\pi$ rather than π is that the output will always lag the input. Thus we'd expect a negative phase shift. This will be much clearer in a few pages, when we introduce damping.

Example 2.5

Problem   Determine the transfer function for the system of Figure 2.1 as discussed in Example 2.2.

Solution   From (2.5.1) we have

$$g(\omega) = \frac{\omega_n^2}{\omega_n^2 - \omega^2}$$

Thus for our system we'll have

$$g(\omega) = \frac{500}{500 - \omega^2}$$

Example 2.6

Problem   Determine the transfer function for the system illustrated in Figure 2.14.

Solution   A free body diagram followed by a force balance gives us

$$m\ddot{x} = -k_3 x - (k_1 + k_2)x + \bar{f}\sin(\omega t)$$

or, putting this into a driven oscillator form

$$m\ddot{x} + (k_1 + k_2 + k_3)x = \bar{f}\sin(\omega t)$$

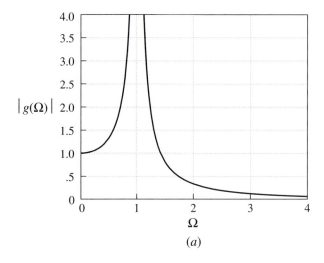

(a)

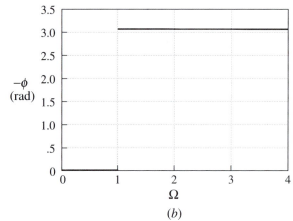

(b)

Figure 2.13 (a) Magnitude and (b) phase transfer function plots of a seismically excited, spring-mass system, nondimensional frequency

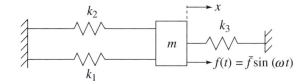

Figure 2.14 Three-spring system with $f(t) = \bar{f} \sin(\omega t)$

Assume that $x = \bar{x} \sin(\omega t)$. Substituting this into our equation of motion and canceling the $\sin(\omega t)$ terms yields

$$\bar{x}(k_1 + k_2 + k_3 - \omega^2 m) = \bar{f}$$

which means

$$g(\omega) = \frac{\bar{x}}{\bar{f}} = \frac{1}{k_1 + k_2 + k_3 - \omega^2 m}$$

2.6 VISCOUS DAMPING

All the work we've done so far has involved undamped systems. If no damping is present, then once a motion has been started in the system it will persist forever. Clearly, undamped systems are not highly accurate models of the real world—the unforced oscillations of real systems always decay away eventually. Thus, we're motivated to add damping to our system description. We've already described the most common type of damper, the viscous damper, in Chapter 1. Section 2.11 of this chapter will introduce other types of damping but, for the vast majority of vibrational analyses, viscous damping will be the damping model of choice.

If we add a viscous damper to our system (as shown in Figure 2.15) then the mass will have three forces acting on it: the applied external force $f(t)$, the force due to the spring $(-kx)$, and the new force due to the damper $(-c\dot{x})$. Our equation of motion becomes

$$m\ddot{x} + c\dot{x} + kx = f(t) \tag{2.6.1}$$

Dividing by m as we did for the undamped case allows us to write this as

$$\ddot{x} + \frac{c}{m}\dot{x} + \frac{k}{m}x = \frac{f(t)}{m} \tag{2.6.2}$$

If we go ahead and use the identities we derived in when looking at free vibrations, namely $\omega_n^2 = \frac{k}{m}$ and $\zeta = \frac{c}{2\sqrt{mk}}$, we'll obtain

$$\ddot{x} + 2\zeta\omega_n\dot{x} + \omega_n^2 x = \frac{f(t)}{m} \tag{2.6.3}$$

If we next assume that our forcing is sinusoidal (say a sine), then we'll have

$$\ddot{x} + 2\zeta\omega_n\dot{x} + \omega_n^2 x = \frac{\bar{f}}{m}\sin(\omega t) \tag{2.6.4}$$

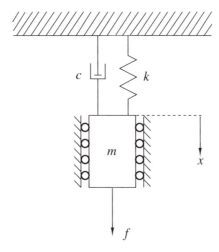

Figure 2.15 Direct force excited, spring-mass damper system

Now that we're including damping, our approach to solving the problem becomes important. The obvious method is to assume that the response is of the form

$$x(t) = a \cos(\omega t) + b \sin(\omega t) \tag{2.6.5}$$

just as we did with the undamped case. Although the additional harmonic wasn't needed in that case (the response was exactly in or exactly out of phase with the forcing), in this case it definitely *is* needed. The advantage to this assumed form is that it is quite physical; the values of a and b will indicate how much of the cosine and sine solutions make up the total response. The disadvantage, which will become clear in a moment, is that the solution is a bit messy. Since this approach is the most physically obvious, we'll start with it, completely characterize our solution, and then introduce a more compact way of getting the same answer.

If we use (2.6.5) in (2.6.4), we'll obtain

$$-\omega^2 \Big(a \cos(\omega t) + b \sin(\omega t) \Big) + 2\omega_n \zeta \Big(-\omega a \sin(\omega t) + \omega b \cos(\omega t) \Big) + \omega_n^2 \Big(a \cos(\omega t) + b \sin(\omega t) \Big)$$

$$= \frac{\bar{f}}{m} \sin(\omega t) \tag{2.6.6}$$

If we group the sines and cosines, we'll have

$$\left(-\omega^2 a + 2\zeta \omega_n \omega b + \omega_n^2 a \right) \cos(\omega t) + \left(-\omega^2 b - 2\zeta \omega_n \omega a + \omega_n^2 b \right) \sin(\omega t)$$

$$= \frac{\bar{f}}{m} \sin(\omega t) \tag{2.6.7}$$

We now need to realize that sines and cosines are linearly independent functions; there's no way to get one from the other. Therefore, the coefficient of the $\sin(\omega t)$ term on the left-hand side of the equation has to equal the coefficient of the $\sin(\omega t)$ term on the right, and the coefficients of the cosines must match also. Matching the coefficients of the cosine terms gives us

$$-\omega^2 a + 2\zeta \omega_n \omega b + \omega_n^2 a = 0 \tag{2.6.8}$$

and matching the sine coefficients yields

$$-\omega^2 b - 2\zeta \omega_n \omega a + \omega_n^2 b = \frac{\bar{f}}{m} \tag{2.6.9}$$

(2.6.8) and (2.6.9) are both equations in two unknowns. If we group the a and b terms we'll have

$$(\omega_n^2 - \omega^2)a + 2\zeta \omega_n \omega b = 0 \tag{2.6.10}$$

and

$$-2\zeta \omega_n \omega a + (\omega_n^2 - \omega^2)b = \frac{\bar{f}}{m} \tag{2.6.11}$$

If we go ahead and solve for a and b we'll obtain

$$a = -\frac{(2\zeta\omega\omega_n)\frac{\bar{f}}{m}}{(\omega_n^2 - \omega^2)^2 + (2\zeta\omega\omega_n)^2} \tag{2.6.12}$$

and

$$b = \frac{(\omega_n^2 - \omega^2)\frac{\bar{f}}{m}}{(\omega_n^2 - \omega^2)^2 + (2\zeta\omega\omega_n)^2} \tag{2.6.13}$$

Only b can ever equal zero, and this occurs only if the forcing frequency is equal to the system's natural frequency. At all frequencies a is nonzero, and thus we always will have some finite response. Therefore, the addition of damping has eliminated the infinite resonant response we saw in the undamped case when the system was forced at its natural frequency.

Our total response looks like this:

$$x(t) = -\frac{(2\zeta\omega\omega_n)\frac{\bar{f}}{m}}{(\omega_n^2 - \omega^2)^2 + (2\zeta\omega\omega_n)^2}\cos(\omega t) + \frac{(\omega_n^2 - \omega^2)\frac{\bar{f}}{m}}{(\omega_n^2 - \omega^2)^2 + (2\zeta\omega\omega_n)^2}\sin(\omega t) \tag{2.6.14}$$

The new behavior that's due to the damping is the inclusion of another harmonic component in the response (in addition to the one displayed by the forcing). When we had no damping, a sine input would give us a purely sine output. Now we've got both a cosine and a sine in the response. Therefore the question of the day is: What do we get when we add a sine and a cosine? Luckily, we've already answered this, way back in Chapter 1. If you check back, you'll see that a sine and cosine of the same frequency add together to yield either a sine with a phase shift or a cosine with a different phase shift.

Just to make sure all is clear, we'll work the problem out again, this time using the particular forms we've been dealing with. We're given

$$x(t) = a\cos(\omega t) + b\sin(\omega t) \tag{2.6.15}$$

Next, we'll write (2.6.15) as either

$$a\cos(\omega t) + b\sin(\omega t) = |x|\cos(\omega t - \phi) \tag{2.6.16}$$

or

$$a\cos(\omega t) + b\sin(\omega t) = |x|\sin(\omega t - \phi) \tag{2.6.17}$$

Both forms for the solutions are equally possible; the one we select will simply depend on which one is more convenient to use. Generally, the problem you're working on will suggest the appropriate form. For instance, if your forcing is a cosine, then it makes sense to assume that the solution is of the form $\cos(\omega t - \phi)$. In this case ϕ has a physical meaning; i.e., it's the amount by which the output lags the input. For illustration purposes, we'll choose (2.6.17), and you can derive the appropriate results for (2.6.16) for yourself, if you're interested in doing so.

The first step in showing that (2.6.17) is a reasonable equation is to expand $\sin(\omega t - \phi)$. To do this, we'll use the trigonometric identity

$$\sin(a - b) = \sin(a)\cos(b) - \cos(a)\sin(b) \qquad (2.6.18)$$

In the future you may also find it useful to use the associated cosine formula

$$\cos(a - b) = \cos(a)\cos(b) + \sin(a)\sin(b) \qquad (2.6.19)$$

If we apply (2.6.18) to (2.6.17), we'll have

$$a\cos(\omega t) + b\sin(\omega t) = |\bar{x}|\Big(\sin(\omega t)\cos(\phi) - \cos(\omega t)\sin(\phi)\Big) \qquad (2.6.20)$$

Just as we did when we solved the forced, undamped problem, we have to equate the coefficients of the cosine and sine terms separately to satisfy (2.6.18). This means that

$$a = -|\bar{x}|\sin(\phi) \qquad (2.6.21)$$

and

$$b = |\bar{x}|\cos(\phi) \qquad (2.6.22)$$

If we square both sides of (2.6.21) and (2.6.22) and add them, we obtain

$$|\bar{x}|^2 = a^2 + b^2 \qquad (2.6.23)$$

or

$$|\bar{x}| = \sqrt{a^2 + b^2} \qquad (2.6.24)$$

Dividing (2.6.21) by (2.6.22) gives us

$$-\tan(\phi) = \frac{a}{b} \qquad (2.6.25)$$

or

$$\phi = -\tan^{-1}\left(\frac{a}{b}\right) \qquad (2.6.26)$$

Note that the amplitude of the resultant wave is equal to the square root of the sum of the squares of the sine's and cosine's amplitudes. This is highly reminiscent of how one finds the length of a right triangle's hypotenuse: the length of the hypotenuse is equal to the square root of the sum of the squares of the other sides. This is also the way we find a vector's length. The length of a vector is equal to the square root of the squares of its projection onto the x and y axes. This geometric view is very much applicable to our problem. People will often think of a sinusoidal response as a vector that rotates about the origin. We'll get back to this in the next section when we start discussing the alternative way of finding the forced oscillator's response.

If we substitute in the actual values of a and b, our final results are

$$\phi = \tan^{-1}\left(\frac{2\zeta\omega\omega_n}{\omega_n^2 - \omega^2}\right) \tag{2.6.27}$$

and

$$|\bar{x}| = \frac{\bar{f}}{m}\frac{1}{\sqrt{(\omega_n^2 - \omega^2)^2 + (2\zeta\omega_n\omega)^2}} \tag{2.6.28}$$

We can find the magnitude of the system's transfer function by dividing (2.6.28) by $\bar{f}$

$$|g(\omega)| = \frac{1}{m\sqrt{(\omega_n^2 - \omega^2)^2 + (2\zeta\omega_n\omega)^2}} \tag{2.6.29}$$

Finally, we can reexpress the magnitude and phase in terms of the nondimensional frequency Ω, which you'll remember is equal to $\frac{\omega}{\omega_n}$

$$\phi = \tan^{-1}\left(\frac{2\zeta\Omega}{1 - \Omega^2}\right) \tag{2.6.30}$$

and

$$|g(\Omega)| = \frac{1}{k\sqrt{(1 - \Omega^2)^2 + (2\zeta\Omega)^2}} \tag{2.6.31}$$

These results are worth examining. Let's start by looking at what happens to the phase angle given by (2.6.27) as we increase ω from zero. Figure 2.16b shows a plot of ϕ versus $\frac{\omega}{\omega_n}$ that you can refer to as well. For very low values of ω we see that ϕ is close to zero. As ω increases, the denominator of (2.6.27) approaches zero (as ω approaches ω_n). Thus the overall fraction becomes very large. This tells us that the phase angle is approaching 90 degrees. In fact, when $\omega = \omega_n$, the phase angle is exactly 90 degrees. Since our assumed form for the solution subtracted ϕ from the sine output ($x(t) = |\bar{x}|\sin(\omega t - \phi)$), we realize that the solution is *lagging* the input force. This is always going to be the case. When you add damping to a system like the one we're examining, the output waveform will lag the input. Note also that the particular value of ζ doesn't affect the phase angle when $\omega = \omega_n$. All directly forced systems will have phase shifts of 90 degrees when forced at their natural frequencies. This fact can be used as an identification tool, and we'll look into this in Section 2.10.

As ω increases just a bit more, the magnitude of $\frac{2\zeta\omega\omega_n}{\omega_n^2-\omega^2}$ is still huge, but now it's negative. Therefore the phase angle continues to increase beyond 90 degrees. As ω continues to increase, $\omega_n^2 - \omega^2$ becomes large, thus driving $\frac{2\zeta\omega\omega_n}{\omega_n^2-\omega^2}$ to 0. Therefore ϕ approaches 180 degrees and the output lags the input by almost 180 degrees. The speed at which the phase angle goes from 0 to 180 degrees depends upon the damping factor ζ.

You can see from Figure 2.16b that for small values of ζ the phase response resembles that of an undamped oscillator. Below the natural frequency, the response is essentially in phase, and above

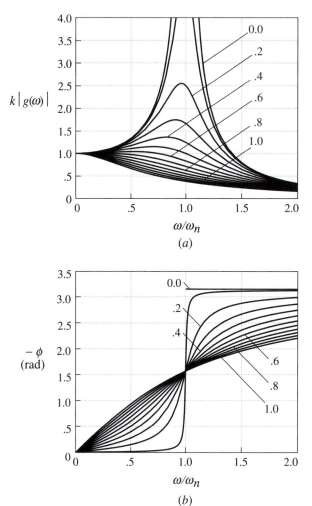

Figure 2.16 Amplitude and phase response of a direct force excited, spring-mass system damper (varying ζ values indicated by numbers on plot)

it the response is mostly out of phase. As ζ is increased, the change from in phase to out becomes much slower.

Now that we've looked at the phase response, it's time to look at the magnitude response. Figure 2.16a shows what $k|g(\omega)|$ looks like for a variety of ζ values ($k|g(\omega)|$ is more convenient to plot than $|g(\omega)|$ because it's nondimensional).

Just as was the case for the phase response, the amplitude response resembles that of the undamped oscillator for small values of ζ. The magnitude starts out at the static deflection value, increases slowly at first, grows to a large value (but not infinity), and then goes to zero. As already noted, the failure of the response to go to infinity, even at resonance, is one of the advantages that damping brings us.

You'll note that for low values of ζ, the peak response is greater than the zero frequency value and that it occurs around the system's natural frequency. As the damping is increased, the peak value decreases and the frequency at which the peak occurs is reduced. Finally, at $\zeta = \frac{1}{\sqrt{2}}$, we lose the peak entirely. As the damping is increased yet further, the response becomes smaller and smaller at all frequencies.

This basic kind of response is absolutely key to understanding a whole range of applications. For instance, those of you interested in hi-fi systems know that amplifiers are often referred to as having a flat response from 0 to 100 kHz (or some other upper limit). The response starts off pretty much flat and then, when the natural frequency is reached, the response begins to peak noticeably. Of course, an amplifier is a lot more complicated than a simple SDOF system. The essential features, however, are the same. Amplifiers have their own internal resonances and, when the excitation frequency reaches them, the response peaks. The same thing is seen in loudspeaker specifications; they're fine until you try to play too high a frequency.

The degree to which a system peaks at resonance is very much tied to the type of system being examined. For instance, if you were designing an oscillator circuit, you'd want the peak to be very large and you'd want to be operating within the peaked region. If you were designing a robotic manipulator, however, you'd want a much lower peak. As in all our work, engineering trade-offs have to be made, and there's no way to get everything you might want. For example, a manipulator should have a low peak response because you're using it to manipulate something, say a welder in an automobile factory. A peaky response at resonance implies that the arm won't quickly move to position and stay there; rather, it will oscillate a bit - which is bad for accuracy and speed. Decreasing the peak response (by increasing the damping) will fix this problem but means that the speed with which the arm can move is reduced. Of course, this is a transient behavior and can be evaluated using the descriptors shown earlier in Figure 2.6. Logically, we should be able to examine either the steady state forced response *or* the transient response to deduce what the relevant natural frequency, damping factor, and so on might be. Both these approaches are indeed possible, and the steady state approach is the second major avenue through which identification is done. We'll consider this approach in Section 2.10.

Example 2.7

Problem   Consider again the system shown in Figure 2.15. Let the mass be equal to 4 kg, the damping coefficient be 80 N·s/m, and the spring coefficient be 40,000 N/m. How much will the output lag the input if the forcing is at 80 rad/s?

Solution   The natural frequency for the system is

$$\omega_n^2 = \frac{40,000}{4} = 10,000$$

Since $2\zeta\omega_n = \frac{c}{m}$, we have

$$\zeta = \frac{c}{m}\frac{1}{2\omega_n} = \frac{80}{4}\frac{1}{200} = .10$$

and therefore, from (2.6.27), we have

$$\phi = \tan^{-1}\left(\frac{(2)(.1)(80)(100)}{(100^2 - 80^2)}\right) = .418 \, \text{rad}$$

Thus the output lags the input by .418 radian.

2.7 COMPLEX REPRESENTATIONS

Now that we've taken an in-depth look at how a directly forced system behaves, it's time to introduce a new way of looking at vibratory responses. Although representing a system's response as

$$x(t) = a\cos(\omega t) + b\sin(\omega t) \tag{2.7.1}$$

is physically attractive, it's a bit awkward mathematically, as you've just seen. Our solution to this problem will be to represent $x(t)$ in a new way, namely,

$$x(t) = \bar{x}e^{i\omega t} \tag{2.7.2}$$

where $\bar{x}$ can be complex.

With this representation, we're using a complex exponential description of the motion, rather than a sinusoidal one. To illustrate why this is a good idea, let's again examine a directly forced, damped system,

$$\ddot{x} + 2\zeta\omega_n\dot{x} + \omega_n^2 x = \frac{f(t)}{m} \tag{2.7.3}$$

Furthermore, let's represent the forcing as $\bar{f}e^{i\omega t}$, where $\bar{f}$ is real, and the response as $\bar{x}e^{i\omega t}$. This will give us

$$\left(-\omega^2\bar{x} + 2i\zeta\omega\omega_n\bar{x} + \omega_n^2\bar{x}\right)e^{i\omega t} = \frac{\bar{f}}{m}e^{i\omega t} \tag{2.7.4}$$

It's a simple matter to cancel the common factor of $e^{i\omega t}$ and solve for $\bar{x}$, leaving us with

$$\bar{x} = \frac{\bar{f}}{m}\frac{1}{\left(\omega_n^2 - \omega^2 + 2i\zeta\omega\omega_n\right)} \tag{2.7.5}$$

Recalling that the transfer function is just the output divided by the input, we can see from (2.7.5) that it's equal to

$$g(\omega) \equiv \frac{\bar{x}}{\bar{f}} = \frac{1}{m\left(\omega_n^2 - \omega^2 + 2i\zeta\omega\omega_n\right)} \tag{2.7.6}$$

Note that this is a complex quantity. If we express it in the form

$$g(\omega) = |g|e^{-i\phi} \tag{2.7.7}$$

i.e., with a magnitude and phase component, we'll have

$$|g| = \frac{1}{m\sqrt{(\omega_n^2 - \omega^2)^2 + (2\zeta\omega_n\omega)^2}} \tag{2.7.8}$$

and

$$\phi = \tan^{-1}\left(\frac{2\zeta\omega\omega_n}{\omega_n^2 - \omega^2}\right) \tag{2.7.9}$$

Compare how easily we obtained (2.7.8) and (2.7.9) with the work that was required to obtain (2.6.27) and (2.6.29). Instead of a few pages, it took only a few lines. That's the beauty of a complex representation—it's compact and quick to use.

Although a complex representation is certainly an efficient way of looking at our problem, there is a downside. The difficulty is that we're no longer dealing with real quantities; our variables are now explicitly complex. Let's take a moment to think about what this means. If we say that $x(t) = \bar{x}e^{i\omega t}$, then we can use our trigonometric identities to obtain

$$x(t) = \bar{x}\big(\cos(\omega t) + i\,\sin(\omega t)\big) \tag{2.7.10}$$

Your first thought might be to say "Hold on a second! $x(t)$ is a *physical* displacement. How can it have an imaginary component?"

The answer is—it can't. If $x(t)$ is a physical displacement, it must always be real. So must $f(t)$. However, what's stopping us from using $f(t) = \bar{f}e^{i\omega t}$ and remembering that we're going to use the only *real part* of it? If we take $\bar{f}$ to be real, then the real part of $\bar{f}e^{i\omega t}$ is just $\bar{f}\cos(\omega t)$.

This actually gives us some real flexibility. For instance, if we wanted our input to be $\sin(\omega t)$ then we simply have to use the imaginary part of $\bar{f}e^{i\omega t}$. Of course, once we've decided that we're going to pay attention to a particular part of the input (say the real part), then we also have to look only at the real part of the output. Is this difficult? Not at all. Our response is $\bar{x}e^{i\omega t}$. Since $\bar{x}$ is complex in general, we can write it as

$$\bar{x} = \bar{x}_r + i\bar{x}_i \tag{2.7.11}$$

Substituting (2.7.11) into (2.7.10) gives us

$$x(t) = (\bar{x}_r + i\bar{x}_i)\big(\cos(\omega t) + i\,\sin(\omega t)\big) \tag{2.7.12}$$

Expanding this out yields

$$x(t) = \bar{x}_r\cos(\omega t) - \bar{x}_i\sin(\omega t) + i\big(\bar{x}_r\sin(\omega t) + \bar{x}_i\cos(\omega t)\big) \tag{2.7.13}$$

Taking the real part leaves us with

$$x(t) = \bar{x}_r\cos(\omega t) - \bar{x}_i\sin(\omega t) \tag{2.7.14}$$

How do we find $\bar{x}_r$ and $\bar{x}_i$? Just use our recent results. From (2.7.5) we'll have

$$\bar{x} = \frac{\frac{\bar{f}}{m}}{\omega_n^2 - \omega^2 + 2i\zeta\omega\omega_n}$$

(2.7.15)

If $\bar{f}$ is real, then this can be rewritten as

$$\bar{x} = \frac{\frac{\bar{f}}{m}(\omega_n^2 - \omega^2 - 2i\zeta\omega\omega_n)}{(\omega_n^2 - \omega^2)^2 + (2\zeta\omega\omega_n)^2}$$

(2.7.16)

(The numerator and denominator were multiplied by $\omega_n^2 - \omega^2 - 2i\zeta\omega\omega_n$)
We can immediately identify $\bar{x}_r$ and $\bar{x}_i$ as

$$\bar{x}_r = \frac{\frac{\bar{f}}{m}(\omega_n^2 - \omega^2)}{(\omega_n^2 - \omega^2)^2 + (2\zeta\omega\omega_n)^2}$$

(2.7.17)

and

$$\bar{x}_i = \frac{-\frac{\bar{f}}{m}(2\zeta\omega\omega_n)}{(\omega_n^2 - \omega^2)^2 + (2\zeta\omega\omega_n)^2}$$

(2.7.18)

Example 2.8

Problem   Determine the response of

$$\ddot{x} + 2\dot{x} + 100x = 2\sin(6t)$$

Solution   An examination of the equation of motion shows that for this problem $\zeta = .1$ and $\omega_n = 10$ rad/s. Since the forcing is in the form of a sine wave, we'd want to represent the forcing by the imaginary part of $2e^{6it}$. Thus our solution will consist of the imaginary part of (2.7.13)

$$x(t) = \bar{x}_r \sin(\omega t) + \bar{x}_i \cos(\omega t)$$

Using the foregoing values, along with the expressions for $\bar{x}_r$ and $\bar{x}_i$ ((2.7.17) and (2.7.18)), we'll obtain

$$\bar{x}_r = \frac{2(10^2 - 6^2)}{(10^2 - 6^2)^2 + (2(.1)(6)(10))^2} = .0302$$

and

$$\bar{x}_i = \frac{-2(2)(.1)(6)(10)}{(10^2 - 6^2)^2 + (2(.1)(6)(10))^2} = -.00566$$

Thus

$$x(t) = .0302\sin(6t) - .00566\cos(6t)$$

A method for visualizing all this that some people find quite useful is to plot the force and displacement vectors on the complex plane. Consider for now just the forcing $f = \bar{f}e^{i\omega t}$ with $\bar{f}$

being real. Expanding this out in terms of the real and imaginary parts gives us

$$f = \bar{f}\big(\cos(\omega t) + i\sin(\omega t)\big) \tag{2.7.19}$$

Both real and imaginary components are sinusoidal, and they're exactly 90 degrees out of phase with each other. The overall magnitude is $\bar{f}$ and the phase angle on the complex plane is given by

$$\phi = \tan^{-1}\big(\tan(\omega t)\big) = \omega t \tag{2.7.20}$$

Thus the phase angle is increasing linearly with time. A series of snapshots of the forcing vector is shown in Figure 2.17 for the case of $\omega = \frac{\pi}{4}$. As you can see, the magnitude is a constant and the vector simply rotates around in a counterclockwise direction.

   If we look at the response vector (with components given by (2.7.17) and (2.7.18)) for a few chosen frequencies (fixing the time t), we see from Figure 2.18 that it simply lags the force vector (as previously discussed) and has a magnitude that varies with frequency.

   In Figure 2.18 the force vector is real and at $t = 0$ it lies along the real axis. For a low frequency input ($\omega = 2$ rad/s) the response is almost in phase with the force. Near resonance ($\omega = 10$ rad/s), the amplitude is larger than its low frequency magnitude and lags the force by a bit more than 90 degrees. For a forcing frequency well above resonance (30 rad/s), the response is almost completely out of phase with the force and has a small magnitude ($|\bar{x}| \to 0$ as $\omega \to \infty$).

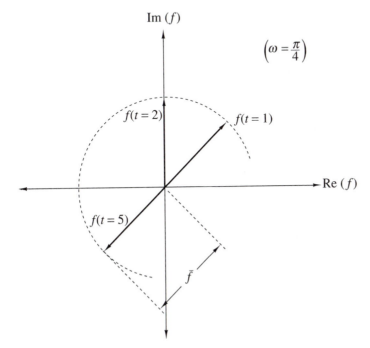

Figure 2.17 Complex representation of forcing vector

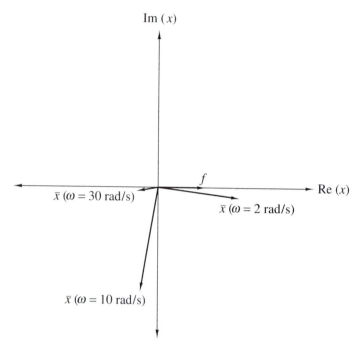

Figure 2.18 Complex representation of response vector

This sort of graphical viewpoint also can be used for visualizing the relationship between displacement, velocity, and acceleration. If we let the displacement for a particular coordinate be

$$x(t) = \bar{x}e^{i\omega t} \tag{2.7.21}$$

then we can differentiate with respect to t to find the velocity and acceleration:

$$\dot{x}(t) = i\omega\bar{x}e^{i\omega t} \tag{2.7.22}$$

$$\ddot{x}(t) = -\omega^2\bar{x}e^{i\omega t} \tag{2.7.23}$$

We can easily plot all three on the complex plane, as shown in Figure 2.19. The plotted results correspond to a value of ω that's greater than 1.0. We can see that the velocity leads the displacement by 90 degrees and, in turn, the acceleration leads the velocity by 90 degrees. Since the oscillation frequency is greater than 1.0, the magnitude of the velocity is *greater* than that of the displacement and the acceleration's magnitude is *greater* than that of the velocity. This means that, even if a part is vibrating with a small displacement, the accelerative loads might be significant if the oscillation frequency is substantially above 1.0.

Just to recap, you may well find it easier to analyze a system by using complex representations and then obtain the actual response by taking the real (or imaginary) part, rather than by explicitly using sines and cosines. This efficiency is not limited to SDOF systems but also applies to the multi-DOF vibration problems that we'll be encountering.

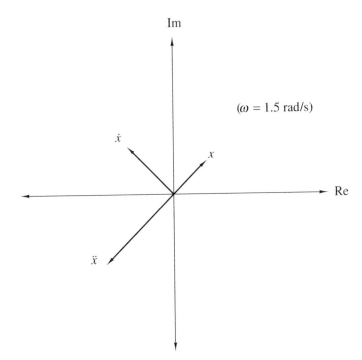

(ω = 1.5 rad/s)

Figure 2.19   Complex representation of displacement, velocity, and acceleration

Example 2.9

Problem   Find how much the response will lag the excitation for a direct force excited system having $m = 10$ kg, $c = 50$ N/m/s, $k = 6250$ N/m, and $f = 10\cos(20t)$ N. In addition, determine the magnitude of the response that's in phase with the disturbance.

Solution   The equation of motion is given by

$$10\ddot{x} + 50\dot{x} + 6250x = 10\cos(20t)$$

and

$$\omega_n = \sqrt{\frac{6250}{10}} = 25 \text{ rad/s}, \quad \zeta = .1$$

Equation (2.7.9) gives us

$$\phi = \tan^{-1}\left(\frac{2(.1)(20)(25)}{25^2 - 20^2}\right) = 24 \text{ degrees}$$

Since the forcing is given by $.01\cos(20t)$, we can use (2.7.17) to find the component of the response that's in phase (cosine component) with the disturbance,

$$\bar{x}_r = \frac{\frac{10}{10}(25^2 - 20^2)}{(25^2 - 20^2)^2 + (2(.1)(20)(25))^2} = .0037$$

Thus the in phase response has an amplitude of 3.7 mm.

2.8 DAMPED SEISMIC MOTION

Recall that there was no qualitative difference in the form of the equations of motion for an undamped seismically excited system and an undamped, direct force excited system. This fact allowed us to draw conclusions from our seismic system that also held for the force excited one. Unfortunately, this happy state of affairs ends when damping is added. The equations of motion for a seismically excited system with damping (illustrated in Figure 2.3) were presented earlier in (2.2.2) and are repeated here for convenience:

$$\ddot{x} + 2\zeta\omega_n\dot{x} + \omega_n^2 x = \omega_n^2 y + 2\zeta\omega_n\dot{y} \tag{2.8.1}$$

The part of this equation that qualitatively differs from that of the direct force excited system is in the $2\zeta\omega_n\dot{y}$ term. Since the excitation is transmitted from the ground motion to the mass by means of a damper, as well as a spring, this velocity-dependent forcing term arises. Luckily for us, our new grasp of complex representations means that solving this new problem will not be overly difficult. If we assume that $x = \bar{x}e^{i\omega t}$ and $y = \bar{y}e^{i\omega t}$, then (2.8.1) becomes

$$\bar{x}\left(\omega_n^2 - \omega^2 + 2i\zeta\omega\omega_n\right) = \bar{y}(\omega_n^2 + 2i\zeta\omega\omega_n) \tag{2.8.2}$$

The transfer function is simply

$$g(\omega) \equiv \frac{\bar{x}}{\bar{y}} = \frac{\omega_n^2 + 2i\zeta\omega\omega_n}{\omega_n^2 - \omega^2 + 2i\zeta\omega\omega_n} \tag{2.8.3}$$

If we compare (2.8.3) with (2.7.6) we see that the important difference is our gain of an additional phase shift from the numerator. (The difference in the overall magnitude isn't too important, since we had that in the undamped case also.) Going ahead and expressing the numerator and denominator of (2.8.3) in magnitude-phase form, we'll have

$$g(\omega) = \frac{\sqrt{\omega_n^4 + (2\zeta\omega\omega_n)^2}e^{i\phi_1}}{\sqrt{(\omega_n^2 - \omega^2)^2 + (2\zeta\omega\omega_n)^2}e^{i\phi_2}} \tag{2.8.4}$$

where $\phi_1 = \tan^{-1}\left(\frac{2\zeta\omega}{\omega_n}\right)$ and $\phi_2 = \tan^{-1}\left(\frac{2\zeta\omega\omega_n}{\omega_n^2 - \omega^2}\right)$.

Pulling the magnitude and phase angles together finally gives us

$$g(\omega) = \sqrt{\frac{\omega_n^4 + (2\zeta\omega\omega_n)^2}{(\omega_n^2 - \omega^2)^2 + (2\zeta\omega\omega_n)^2}}e^{i\phi}, \qquad \phi = \phi_1 - \phi_2 \tag{2.8.5}$$

Of course, we can also express this in terms of the nondimensional frequency Ω;

$$g(\Omega) = \sqrt{\frac{1 + (2\zeta\Omega)^2}{(1 - \Omega^2)^2 + (2\zeta\Omega)^2}}e^{i\phi}, \qquad \phi = \tan^{-1}\left(2\zeta\Omega\right) - \tan^{-1}\left(\frac{2\zeta\Omega}{1 - \Omega^2}\right) \tag{2.8.6}$$

What conclusions can we draw? Well, the first thing to note is that the magnitude response is similar to the direct force excited case. The characteristic polynomial $((1 - \Omega^2)^2 + (2\zeta\Omega)^2)$ is the same for both cases. This is reassuring because the characteristic polynomial depends only upon the system being excited, not the excitation itself. Since the system hasn't changed (it's a mass, spring, damper for both cases), the characteristic polynomial shouldn't alter. Also, we see that the numerator is almost the same as long as the damping is small. As the damping goes to zero, the two cases become more and more similar. When the damping increases, the numerators differ appreciably. In addition, the differences are magnified by the frequency of the excitation. Since the new term $2\zeta\Omega$ contains Ω explicitly, it will increase linearly as Ω increases. For the direct force excited case, the response was finite at small forcing frequencies and went to zero for very high frequency excitations like $\frac{1}{\Omega^2}$. For the seismically damped case, the zero frequency limit also yields a finite response. However, as the frequency goes to infinity, the amplitude scales as $\frac{1}{\Omega}$. It still goes to zero; however, it does so more slowly than the force excited amplitude does. This effect can be seen in Figure 2.20a. Finally, we see that increasing the damping will *decrease* the response if Ω is less than about 1.4 ($\sqrt{2}$ to be precise) but will actually *increase* the amplitude if Ω is above this point. This behavior is fundamentally different from the direct force excited case, for which increasing the damping will decrease the response amplitude for all values of forcing frequency. Thus, if you're trying to reduce the vibration levels for a seismically excited system, the frequency of the forcing becomes quite important. This is because increasing the damping will increase the vibration levels if you're in the wrong frequency regime.

As far as phase is concerned, we see that the seismic case actually has a positive phase effect due to the $2\zeta\Omega$ term. For small Ω this almost completely cancels the phase due to $\frac{2\zeta\Omega}{1-\Omega^2}$, thus slowing the initial buildup of phase lag. This can be seen in Figure 2.20b, for which the phase at small Ω is growing quite slowly as compared with that of Figure 2.16b. As you can see from Figure 2.20b, the phase doesn't approach $-\pi$ as the frequency approaches infinity. This is because the value of $\frac{2\zeta\Omega}{1-\Omega^2}$ goes to zero while $2\zeta\Omega$ goes to infinity. Thus the overall seismic response doesn't end up lagging the input by 180 degrees at high frequencies, but rather by 90 degrees. All the nonzero damping phase curves in Figure 2.20b therefore approach a phase of $-\frac{\pi}{2}$ radian at high frequencies, as seen more clearly in Figure 2.20c.

Example 2.10

Problem   Consider the system shown earlier in Figure 2.3. What is the difference between the amplitude and phase lag for the system with and without damping? Let $m = 2$ kg, $c = 80$ N·s/m, and $k = 20,000$ N/m. The seismic excitation has an amplitude of 5 mm and a frequency of 80 rad/s.

Solution   The natural frequency is found from

$$\omega_n = \sqrt{\frac{k}{m}} = 100 \text{ rad/s}$$

and the damping factor is given by

$$\zeta = \frac{c}{m}\frac{1}{2\omega_n} = .20$$

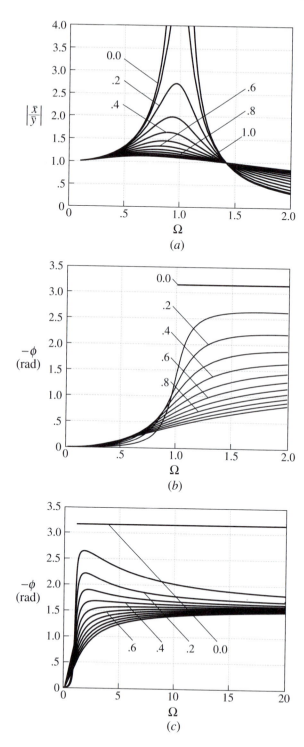

Figure 2.20 (a) Magnitude and (b) phase response of seismically excited, spring-mass-damper; (c) phase response of seismically excited, spring-mass-damper (varying ζ values indicated by numbers on plots)

From (2.8.6) we have

$$g(\Omega) = \sqrt{\frac{1 + (2\zeta\Omega)^2}{(1 - \Omega^2)^2 + (2\zeta\Omega)^2}} e^{i\phi}, \quad \phi = \tan^{-1}\left(2\zeta\Omega\right) - \tan^{-1}\left(\frac{2\zeta\Omega}{1 - \Omega^2}\right)$$

Thus for the damped problem we have

$$g(\Omega) = \sqrt{\frac{1 + (2\zeta\Omega)^2}{(1 - \Omega^2)^2 + (2\zeta\Omega)^2}}$$

which evaluated at $\Omega = \frac{80}{100} = .8$ gives us

$$|g(.8)| = 2.18$$

The phase angle is

$$\phi = -.417 \text{ rad}$$

Since the input amplitude is 5 mm, we have an output amplitude of $5(2.18) = 10.9$ mm. If ζ is set to zero we'll have

$$g(.8) = 2.78$$

and thus an output amplitude of 13.9 mm along with a phase lag of

$$\phi = 0 \text{ rad}$$

We see that the amplitude of the damped response has decreased by 28% over the undamped case, and the phase lag has increased from zero to 24 degrees. Thus the damping has had a very significant effect on the system's response.

2.9 ROTATING IMBALANCE

The final type of SDOF system we're going to look at, and perhaps the most important in a practical sense, is a system in which the excitation comes from a rotating (or reciprocating) mass. The number of problems that come under this heading are enormous. One common example is that of an automobile engine. In-line, six-cylinder engines (as well as opposed sixes) can be balanced through careful design. Most other cylinder arrangements, however, will tend to induce vibration due to the motions of the pistons and crankshaft. This vibration couples into the body through the engine mounts and causes a host of vibration problems for the car designer and for the driver. Another example is a clothes washer. Most all of us have seen what happens when the washer enters the spin cycle and the wet clothes have all gathered into a compact mass. In extreme cases the washing machine actually rocks its way across the floor because of the resulting vibrations. Even a rotating shaft can experience an inertially driven excitation if the shaft's center of mass isn't precisely centered. If it's off by a bit, this offset mass will excite vibrations in the shaft just as an independently rotating mass would. You can come up with any number of other examples. The basic example of a rotating excitation is shown in Figure 2.21. For this problem

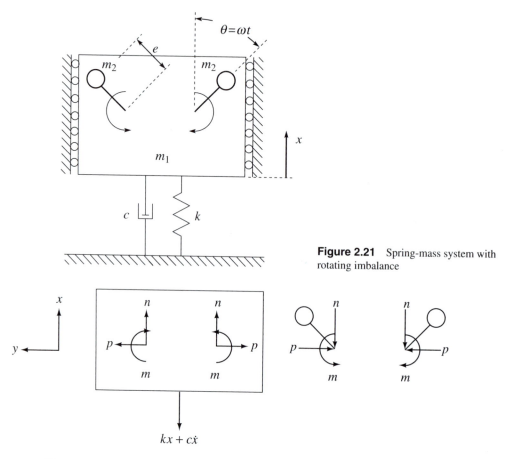

Figure 2.21   Spring-mass system with rotating imbalance

Figure 2.22   Free body diagram for a spring-mass system with rotating imbalances

we have a base mass (m_1) and two counterrotating masses (m_2). The rotation rate is constant and equal to ω.

To find the equations of motion for this system, we need to break it up into its component parts. The appropriate free body diagrams are shown in Figure 2.22. Note that since the two arms are counterrotating, we will have no excitations in the y direction, just in the x. Because the problem is symmetric, we know that the forces due to the arm on the left will be the same as those for the right arm, canceling each other in the y direction and adding together in the x.

Applying Newton's second law for the main mass gives us

$$m_1\ddot{x} = -kx - c\dot{x} + 2n \tag{2.9.1}$$

For completeness, the forces p and moments m are also shown in Figure 2.22 but, as just mentioned, they're equal and opposed and so will not affect the equation of motion for the main mass.

We need to look at either of the rotating arms to determine the unknown interaction force n. Taking the right arm, we have

$$m_2 a_x = -n \tag{2.9.2}$$

where a_x is the acceleration of the center of mass of the arm in the x direction. Since the arm is rotating at a constant rate with respect to the main mass, the only acceleration term is the centripetal acceleration, $l\omega^2$, which is directed toward the arm's attachment point to the body. Thus the total acceleration of the arm's center of mass is $\ddot{x} - l\omega^2 \cos(\omega t)$. Therefore

$$m_2 \left(\ddot{x} - e\omega^2 \cos(\omega t) \right) = -n \tag{2.9.3}$$

Combining both (2.9.1) and (2.9.3) yields the final equation

$$(m_1 + 2m_2)\ddot{x} + c\dot{x} + kx = 2m_2 e\omega^2 \cos(\omega t) \tag{2.9.4}$$

Although two counterrotating masses were used in this example, so that only forces in the x direction were present, we could have easily considered just a single rotating force as long as the body was constrained to allow motion in the x direction only. In this case we'd have had constraint forces between the moving body and the rigid walls. Thus, whether we really have two counterrotating masses or simply a single rotating mass with freedom to move in only one direction, the end result in terms of the moving mass will be the same. If we wish to consider a single rotating mass, then we could use m as the total rotating mass component (replacing $2m_2$). Making this substitution will give us

$$(m_1 + m)\ddot{x} + c\dot{x} + kx = me\omega^2 \cos(\omega t) \tag{2.9.5}$$

The unique aspect of this problem is that the rotational rate appears to the second power in the forcing term. This has a significant effect on the overall response, as we'll now see. If we let the forcing in (2.9.5) be given by $me\omega^2 e^{i\omega t}$ (using a complex representation and remembering that the actual excitation is the real part of $me\omega^2 e^{i\omega t}$) and assume a solution of the form $x(t) = \bar{x} e^{i\omega t}$, then after canceling the common exponential factor and solving for $\bar{x}$, we'll get

$$\bar{x} = \frac{me\omega^2}{-(m_1 + m)\omega^2 + k + i\omega c} \tag{2.9.6}$$

Both the direct force excited and seismically excited systems had a finite amplitude response at low frequencies and had responses that went to zero as the forcing frequency became very large. But in the present case we see that the amplitude response goes to zero as the forcing frequency becomes small and approaches a finite value ($\frac{-me}{m_1+m}$) as the forcing frequency becomes large.

We can also express our solution in a more general form by dividing (2.9.5) by $m_1 + m$ to obtain

$$\ddot{x} + 2\zeta\omega_n \dot{x} + \omega_n^2 x = e\beta\omega^2 \cos(\omega t) \tag{2.9.7}$$

where $\beta = \frac{m}{m_1+m}$, $\omega_n^2 = \frac{k}{m_1+m}$, and $2\zeta\omega_n = \frac{c}{m_1+m}$.

The solution is given by

$$\bar{x} = \frac{e\beta\omega^2}{\omega_n^2 - \omega^2 + 2i\zeta\omega\omega_n} \tag{2.9.8}$$

or, using $\Omega = \frac{\omega}{\omega_n}$

$$\bar{x} = \frac{e\beta\Omega^2}{1 - \Omega^2 + 2i\zeta\Omega} \tag{2.9.9}$$

Note that the real-valued solution can be found as we've done in the preceding sections, by utilizing a phase shift. The real-valued solution to (2.9.7) is given by

$$x(t) = \frac{e\beta\omega^2\cos(\omega t - \phi)}{\sqrt{(\omega_n^2 - \omega^2)^2 + (2\zeta\omega\omega_n)^2}}, \quad \phi = \tan^{-1}\left(\frac{2\zeta\omega\omega_n}{\omega_n^2 - \omega^2}\right) \tag{2.9.10}$$

The normalized magnitude response is given by $\frac{|\bar{x}|}{e\beta}$ and this is plotted versus Ω for a variety of damping values in Figure 2.23.

At this point it's of interest to draw a parallel between the response of a rotating imbalance problem and the response of our direct force excited system. You'll recall that the amplitude response of the two situations differs by a factor of ω^2 (and a constant multiplicative term). So it seems that no parallel exists. However, this conclusion is premature. What if we're not really interested in the amplitude response of the system but rather the acceleration? This would certainly be the case if we were concerned with limiting the accelerations that our system undergoes, something every engineer involved with packaging and shipping worries about. In this case we'll be looking at

$$a(t) = \ddot{x}(t) = \frac{d^2}{dt^2}(\bar{x}e^{i\omega t}) = -\omega^2\bar{x}e^{i\omega t} \tag{2.9.11}$$

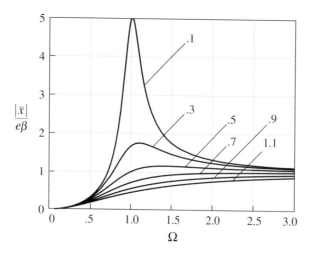

Figure 2.23 Normalized magnitude response for damped, rotating imbalance (varying ζ values indicated by numbers on plot)

We've found $\bar{x}$ earlier, in (2.7.15), and using this result can yield the magnitude of the acceleration ($\bar{a}$):

$$\bar{a} = \frac{-\frac{\bar{f}}{m}\omega^2}{\omega_n^2 - \omega^2 + 2i\zeta\omega\omega_n} \tag{2.9.12}$$

As you can see, (2.9.12) and (2.9.8) have the same form. Thus, the acceleration response of a direct force excited system behaves in the same manner as a system with a rotating imbalance. Actually, this shouldn't be all that surprising, since the force due to the rotating mass comes about *because* of the acceleration of that mass as it rotates.

Example 2.11

Problem   One of the most common examples of a rotating imbalance is associated with an automobile's tire–wheel assembly. If you own a car you probably know that all the tire stores will offer to balance your tires for you whenever you purchase new ones. The way in which the balance is effected is by adding little lead weights to the inside rim of the wheel. If this wasn't done, the system would behave exactly as predicted earlier, leading to a resonant response at certain rotation rates, which, for a car, means at certain forward velocities. Consider the system illustrated in Figure 2.24. The circle represents the tire-wheel system, k_2 represents the stiffness of the spring connecting the wheel to the car, k_1 represents the stiffness of the tire, and c represents any damping in the system. We want to determine the response of the tire as a function of frequency.

Solution   We'll look at the problem in two ways, since this kind of a problem often causes confusion. The first viewpoint is shown in Figure 2.24b. We'll consider the mass of the tire-wheel to be m_1 and let m be the mass of the rotating imbalance. This matches our earlier analyses. It may occur to you, however, that tire-wheel systems don't really have rotating imbalances—the problem is simply that because of manufacturing imperfections, the center of mass isn't exactly at the center of the wheel. Thus, the problem is more correctly one in which the mass of the system being excited is zero and

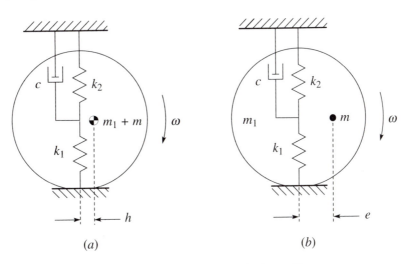

(a)   (b)

Figure 2.24   Rotating tire-wheel assembly

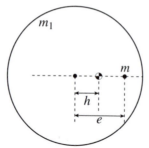

Figure 2.25 Center of mass for imbalanced wheel

the rotating imbalance is actually the mass of the entire system (shown in Figure 2.24a). As we'll see, both these viewpoints will lead to the same end result.

The key to understanding how both views are related is to ask where the center of mass of the *system* is for a perfectly balanced tire-wheel combination of mass m_1 with a rotating imbalance of mass m located e away from the center, as shown in Figure 2.25. The overall center of mass is located at h. From elementary statics we know that

$$hm_1 = (e - h)m$$

which gives us

$$h = \frac{em}{m_1 + m}$$

Thus, the location of the center of mass for Figure 2.24a is now known. Applying (2.9.6) to the second case (Figure 2.24b) gives

$$\bar{x} = \frac{me\omega^2}{-(m_1 + m)\omega^2 + k_1 + k_2 + c\omega i}$$

If we apply (2.9.6) for the case represented by Figure 2.24a (remembering what h is equal to), we'll get

$$\bar{x} = \frac{(m_1 + m)\frac{em}{m_1 + m}\omega^2}{-(0 + m_1 + m)\omega^2 + k_1 + k_2 + c\omega i}$$

which becomes

$$\bar{x} = \frac{me\omega^2}{-(m_1 + m)\omega^2 + k_1 + k_2 + c\omega i}$$

Thus we see that both views give us the same result. We've either got a small imbalance exciting the overall system with a large moment arm or a large imbalance exciting the system with a small moment arm. The final equations are identical.

Example 2.12

Problem   Having analyzed an imbalanced tire in Example 2.11, let's look at some particular parametric values. We'll consider the case for which the suspension stiffness (k_2) is 20,000 N/m, the tire stiffness is 180,000 N/m, the mass of the tire-wheel assembly is 15 kg, the mass of the stationary pieces (brakes, struts, etc.) is 20 kg, and the deviation of the center of mass from the center of the wheel is 2 mm. The distance from the ground to the center of the wheel's hub is .28 m. The damper for the car has been damaged and only gives us 265 N·s/m. We'll determine what the maximum vibration amplitude will be for this system and the vehicle speed at which it will occur.

Solution   Using (2.9.6), we write

$$\bar{x} = \frac{(.002)(15)\omega^2}{-(35)\omega^2 + 200,000 + 265\omega i}$$

(2.9.13)

Note that the mass in the numerator is 15 kg because this is the total rotating mass, whereas the mass in the denominator is 35 kg (the total mass, including rotating and nonrotating parts). To find the maximal response, we can differentiate $\bar{x}$ with respect to ω, set the resultant to zero, and solve for ω. An easier way to get the same result is to square $\bar{x}$ and differentiate with respect to ω^2. The peak response will occur at the same frequency, but the mathematical manipulations are simpler in the second approach.
   Carrying through will ultimately yield

$$\omega_{rp} = \frac{\omega_n}{\sqrt{1 - 2\zeta^2}}$$

where the subscript rp stands for rotating peak.
   For this problem $\omega_n^2 = \frac{200,000}{35}$ and so $\omega_n = 75.59$ rad/s. From $\frac{c}{m} = 2\zeta\omega_n$, we can find that $\zeta = .05$. Thus $\omega_{rp} = 75.78$ rad/s. Using this gives us a peak response magnitude of

$$|\bar{x}| = 8.57 \text{ mm}$$

If we can now find the corresponding velocity of the car, we're done. Since the distance from the ground to the hub of the wheel is .28 m, we can find the car's velocity from

$$v = \omega r = (75.78 \text{ rad/s})(.28 \text{ m}) = 21.2 \text{ m/s}$$

or 47 mph, for those of you who prefer English units.

2.10 IDENTIFICATION OF DAMPING AND NATURAL FREQUENCY

A topic of great practical importance is that of identification. In the real world, nobody is going to tell you the damping factor or natural frequency for the system you're examining. Thus, it's up to you to figure these quantities out for yourself. The field of system identification is a huge one and can easily fill a couple of courses. Although we can't go over all this material in this book, we'll look at a few useful techniques that can give us good approximations to the quantities we're looking for.
   The first response type we'll look at is the unforced system response. If the damping is relatively low, we can get a good estimate of both the natural frequency and the damping factor just by examining how the response decays after an initial excitation. If the damping is zero, we know that the natural frequency is simply the frequency of oscillation of the system. And if the damping is small, then the damped oscillation frequency ω_d ($\omega_d = \omega_n\sqrt{1 - \zeta^2}$) is very close to ω_n. We can, however, do much better than simply approximating our answers. Let's recall our solution to the unforced problem (1.4.15). If we include only the cosine term (implying an initial displacement with zero

initial velocity), then at $t = 0$ the response is equal to b_1. Next, look at the response one period later, i.e., at $t = t_p = \frac{2\pi}{\omega_d}$. Now the response is

$$x(t_p) = b_1 e^{\frac{-2\pi\zeta}{\sqrt{1-\zeta^2}}} \tag{2.10.1}$$

If we take the ratio of these two values, we obtain

$$\frac{x(0)}{x(t_p)} = e^{\frac{2\pi\zeta}{\sqrt{1-\zeta^2}}} \tag{2.10.2}$$

Taking the natural log of both sides yields

$$\ln\left(\frac{x(0)}{x(t_p)}\right) = \frac{2\pi\zeta}{\sqrt{1-\zeta^2}} \tag{2.10.3}$$

This result is known as the logarithmic decrement and is usually denoted by σ,

$$\sigma \equiv \ln\left(\frac{x(0)}{x(t_p)}\right) \tag{2.10.4}$$

Although we derived this from a signal that was maximal at $t = 0$, it can easily be found by taking the ratio of any two successive peaks. Once we have σ, it becomes a simple matter to solve for ζ,

$$\zeta = \frac{\sigma}{\sqrt{4\pi^2 + \sigma^2}} \tag{2.10.5}$$

This expression can be further simplified if the damping factor is small. In this case the denominator is closely approximated by 2π and our expression for ζ becomes

$$\zeta = \frac{\sigma}{2\pi} \tag{2.10.6}$$

If we wish to use more oscillations, the formula need be changed only slightly. As long as we're letting an integer number of oscillations elapse between samples, the cosine term is always equal to 1.0. Thus, the only change is that the response will decay due to the exponential envelope. If we wait for n periods, the response amplitude will be

$$x(nt_p) = b_1 e^{\frac{-2n\pi\zeta}{\sqrt{1-\zeta^2}}} \tag{2.10.7}$$

In this case, the ratio becomes

$$\frac{x(0)}{x(nt_p)} = e^{\frac{2n\pi\zeta}{\sqrt{1-\zeta^2}}} \tag{2.10.8}$$

Taking the natural log of this ratio leads to

$$\ln\left(\frac{x(0)}{x(nt_p)}\right) = \frac{2n\pi\zeta}{\sqrt{1-\zeta^2}} \tag{2.10.9}$$

This is equal to n times the log decrement as determined from two successive peaks. Thus

$$\sigma = \frac{1}{n} \ln\left(\frac{x(0)}{x(nt_p)} \right) \tag{2.10.10}$$

Once σ has been determined, we can find ζ using either (2.10.5) or (2.10.6). This way of determining ζ is also a very good means of determining whether the damping for your particular system is linear. If it *is* linear, then the damping factor found between peaks 1 and 2 of the response should be the same as that found between peaks 2 and 3. If the results differ, this implies that the damping isn't linear.

Example 2.13

Problem   Consider the response shown in Figure 2.26. This response corresponds to a spring-mass-damper that's been struck by an impact hammer, resulting in the vibrations shown. Determine the damping factor by means of the log decrement.

Solution   The first peak looks to be about .093 and the second is around .068. Thus $\sigma = \ln\left(\frac{.093}{.068} \right) =$ .313. Equation (2.10.5) lets us solve for ζ and leads to an estimate of $\zeta = .05$.

   If we try and estimate ζ from two successive peaks, we can use the first peak of .093 and the third peak of .05. This will produce a σ of .310, which will give a ζ of .049.

   Both these estimates are excellent ones: the equation of motion used to generate the illustrated response was

$$\ddot{x} + \dot{x} + 100x = 0$$

for which $\omega_n = 10$ rad/s and $\zeta = .05$.

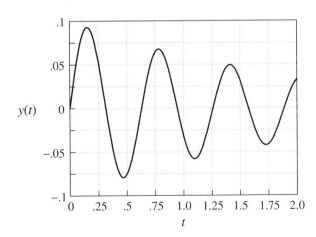

Figure 2.26   Free response of SDOF system

Example 2.14

Problem   You've been given the plot shown in Figure 2.27, which displays the displacement response of a vibrating microstructure. Estimate the damping factor from at least two data pairs and comment on whether the system damping is or is not linear.

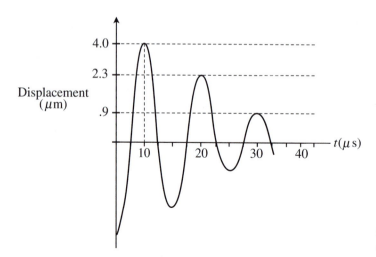

Figure 2.27 Displacement response of a vibrating microstructure

Solution   Using the first two peaks we have

$$\sigma = \ln\left(\frac{4}{2.3}\right) = .5534 \quad \text{from (2.10.4)}$$

and

$$\zeta_{1-2} = \frac{.5534}{\sqrt{4\pi^2 + (.5534)^2}} = .0877 \quad \text{from (2.10.5)}$$

Using the second and third peaks we'll obtain

$$\sigma = \ln\left(\frac{2.3}{.9}\right) = .9383$$

and

$$\zeta_{2-3} = \frac{.9383}{\sqrt{4\pi^2 + (.9383)^2}} = .1477$$

ζ_{1-2} and ζ_{2-3} differ appreciably, indicating that the actual damping mechanism is nonlinear.

What we've now seen is that a free vibration test will allow us to identify the damping factor in a single-degree-of-freedom system by means of the log decrement. Once the damping factor is known, we can find the natural frequency from a knowledge of the damped frequency of oscillation ($\omega_d = \omega_n\sqrt{1-\zeta^2}$). We're not limited, however, to free vibration tests. We can glean the same information from a forced test. In these tests, we're looking for the transfer functions of the system. We can obtain these transfer functions in a number of ways and we'll look at some of them in Chapter 8. For now we'll just assume that somehow we've obtained them and they're available for use in our identification tests.

If you look at Figure 2.28a you'll note that, for the direct force excited system, the phase angle goes through –90 degrees exactly at the system's natural frequency, regardless of the damping value.

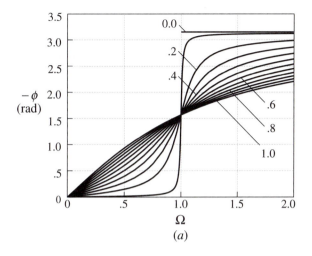

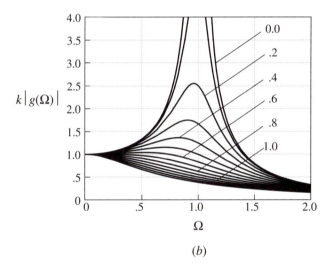

Figure 2.28   (a) Phase and (b) amplitude response of a direct force excited, spring-mass-damper system (varying ζ values indicated by numbers on plot)

Thus, all we need do is look at the phase, determine when it goes through -90 degrees, and read off the frequency at which this occurs. This will give us ω_n.

   Interestingly, there are even more indicators of ω_n and ζ. The first one is found by comparing the actual peak amplitude response of the system to the response at zero frequency. From (2.6.29), repeated here, we know that the transfer function of a direct force excited system is given by

$$|g(\omega)| = \frac{1}{m\sqrt{(\omega_n^2 - \omega^2)^2 + (2\zeta\omega_n\omega)^2}} \qquad (2.10.11)$$

   Thinking back to your first calculus class, you'll recall that the slope of a curve is zero at a local maximum. Thus, to determine the frequency at which our amplitude response curve is a maximum (which we'll call ω_p), we need only differentiate with respect to ω and set the result equal to zero. Actually, although this approach will give us the correct answer, we can simplify matters even further. Rather than differentiating with respect to ω, we can differentiate with respect to ω^2, since only even powers of ω show up in the equation. This will give us ω_p^2 instead of ω_p. We can easily take the square root to find ω_p. However we can simplify yet further. Taking the derivative of the reciprocal of the square root of a function is somewhat complicated. So instead, let's square $g(\omega)$. Since the function is positive everywhere anyway, this won't change our results. And finally, rather than looking at $g^2(\omega)$, let's differentiate $\frac{1}{g^2(\omega)}$. Thus, instead of looking for the maximum of $g(\omega)$ (shown in Figure 2.29), we'll be looking for the minimum of $h(\omega^2) \equiv \frac{1}{g^2(\omega)}$ (Figure 2.30). Thus we want

$$\frac{dh(\omega^2)}{d\omega^2}\bigg|_{\omega=\omega_p} = 0$$

or

$$2(\omega_n^2 - \omega_p^2)(-1) + (2\zeta\omega_n)^2 = 0$$

Solving for ω_p, we obtain

$$\omega_p = \omega_n\sqrt{1 - 2\zeta^2} \tag{2.10.12}$$

For small values of damping, ω_p is approximately equal to ω_n. So another way of estimating ω_n for lightly damped systems is to use the frequency at which the peak response occurs.

We can also use this result to verify what we can see from Figure 2.28b. We've already mentioned that, as ω increases from zero and approaches ω_n, the amplitude response increases if the damping is small but decreases if the damping is large. We can also see from Figure 2.26b that the response peak moves to the left as the damping increases.

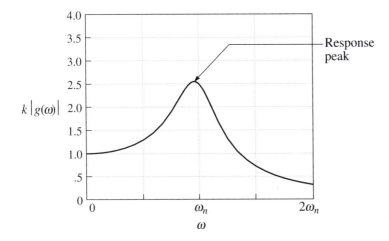

Figure 2.29 Typical underdamped response showing a peaked response

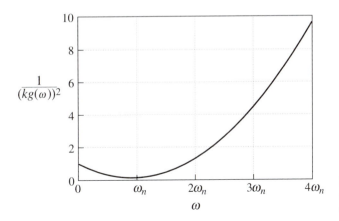

Figure 2.30 Graph with minimum corresponding to the maximum amplitude response

All these observations are reflected in (2.10.12). Clearly, as ζ is increased, ω_p drops. Furthermore, when $\zeta = \frac{1}{\sqrt{2}}$, $\omega_p = 0$. Thus the response peak occurs at $\omega = 0$. For $\zeta > \frac{1}{\sqrt{2}}$, no peak occurs and the response amplitude decays monotonically with increasing frequency.

Next, consider the ratio of the peak response amplitude to that at zero frequency $\frac{|g(\omega_p)|}{|g(0)|}$. Recalling that the response amplitude for a direct force excited system (2.6.28), repeated here, is

$$|x| = \frac{\bar{f}}{m}\frac{1}{\sqrt{(\omega_n^2 - \omega^2)^2 + (2\zeta\omega_n\omega)^2}} \tag{2.10.13}$$

we see that this ratio is equal to

$$\frac{\dfrac{\bar{f}}{m}\dfrac{1}{\sqrt{(\omega_n^2 - \omega_p^2)^2 + (2\zeta\omega_n\omega_p)^2}}}{\dfrac{\bar{f}}{m\omega_n^2}} \tag{2.10.14}$$

We've already found that $\omega_p = \omega_n\sqrt{1 - 2\zeta^2}$, and if we substitute this into (2.10.14) and simplify, we'll obtain

$$\frac{|g(\omega_p)|}{|g(0)|} = \frac{1}{2\zeta\sqrt{1 - \zeta^2}} \tag{2.10.15}$$

Since $\zeta^2 \ll 1$ for systems in which the response is appreciably peaked at ω_p, we can neglect the ζ^2 term and finally obtain

$$\frac{|g(\omega_p)|}{|g(0)|} = \frac{1}{2\zeta} \tag{2.10.16}$$

Now this is a nifty result. It tells us that if we've got the transfer function for our system, we can determine the damping factor just by finding the peak response and dividing by the zero frequency (or static) response.

The final way of finding the damping factor is to look at the *half-power points* of the response curve. By determining the frequencies at which the amplitude of the response drops by a factor of $\sqrt{2}$ from its peak value, we can determine the damping factor. It certainly isn't obvious why this should be so, but it is, as we'll now demonstrate.

We've already determined that the peak response of a mass forced system occurs at $\omega_p = \omega_n\sqrt{1 - 2\zeta^2}$. What we now want is to determine the frequencies, ω_1 and ω_2, at which the response amplitude is equal to $\frac{1}{\sqrt{2}}\left|g(\omega_p)\right|$. We'll call our unknown half-power frequencies ω_h. Using our previous results, we see that we'll need to solve

$$\frac{\bar{f}}{m}\frac{1}{\sqrt{(\omega_n^2 - \omega_h^2)^2 + (2\zeta\omega_n\omega_h)^2}} = \frac{\bar{f}}{m}\frac{1}{2\omega_n^2\zeta\sqrt{(2 - 2\zeta^2)}} \qquad (2.10.17)$$

The term on the left is the general expression for the amplitude at a given frequency (2.6.28) while the term on the right is equal to $\frac{1}{\sqrt{2}}$ times the peak response amplitude (from 2.10.15).

To solve this equation we can square both sides, cross-multiply (to get everything in the numerator), and end up with the equation

$$\omega_h^4 + \omega_h^2(4\zeta^2\omega_n^2 - 2\omega_n^2) + \omega_n^4 - 8\omega_n^4\zeta^2(1 - \zeta^2) = 0 \qquad (2.10.18)$$

If we solve this quadratic expression, we'll get two values for ω_h^2

$$\omega_{h_1}^2 = \omega_n^2(1 - 2\zeta^2) - 2\omega_n^2\zeta\sqrt{1 - \zeta^2} \qquad (2.10.19)$$

and

$$\omega_{h_2}^2 = \omega_n^2(1 - 2\zeta^2) + 2\omega_n^2\zeta\sqrt{1 - \zeta^2} \qquad (2.10.20)$$

Once we take their square roots, we can subtract ω_{h_1} from ω_{h_2}. If we also neglect terms that involve ζ^2 or higher powers (since they're negligible compared to ζ terms), we finally obtain

$$\omega_{h_2} - \omega_{h_1} = 2\omega_n\zeta \qquad (2.10.21)$$

Defining δ_h as

$$\delta_h \equiv \omega_{h_2} - \omega_{h_1} \qquad (2.10.22)$$

we can write

$$\zeta = \frac{\delta_h}{2\omega_n} \qquad (2.10.23)$$

Here is yet another way to determine ζ. For this identification, you need to find the frequency of the peak response, find the frequencies at which the response has dropped by a factor of $\sqrt{2}$, take their difference, and divide by $2\omega_n$ to obtain the damping factor. All the relevant quantities

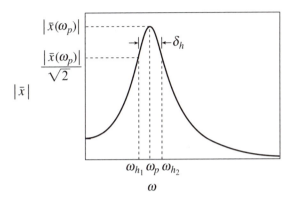

Figure 2.31   Half power points for a sinusoidally forced system

are shown in Figure 2.31. Note that this method is tougher to accomplish in actual practice than the peak response method because the slope of the response curve is quite steep at the half-power points when the damping is low. Thus it's going to be hard to find the half-power frequencies very accurately.

Example 2.15

Problem   In this example we'll try to determine the damping factor of a spring-mass-damper system under direct force excitation through use of the half-power points. Figure 2.32 shows the absolute value of the transfer function as a function of frequency. Calculate the desired parameter values.

Solution   We know that the magnitude of the transfer function for a direct force excited system is given by

$$|g(\omega)| = \frac{\bar{f}}{m} \frac{1}{\sqrt{(\omega_n^2 - \omega^2)^2 + (2\zeta\omega\omega_n)^2}}$$

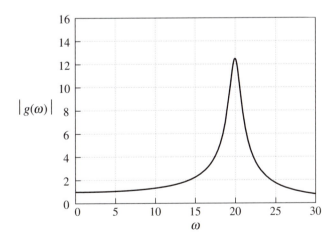

Figure 2.32   Absolute value of transfer function for underdamped spring-mass-damper system

We also know that the damping factor is approximated by $\frac{\delta_h}{2\omega_n}$ where δ_h is the difference in frequency of the two half-power points (review Figure 2.31).

From the graph we can estimate that $|g|_{max} = 12.5$ and occurs at 20 rad/s. The half-power points correspond to an amplitude of $\frac{12.5}{\sqrt{2}} = 8.84$. The frequencies corresponding to this are around 20.8 and 19.1. Thus we have $\delta_h = 20.8 - 19.1 = 1.7$ rad/s. We can then find ζ from the half-power formula:

$$\zeta = \frac{1.7}{(2)(20)} = .0425$$

We can validate this result by using the ratio of $|g_{max}|$ to $|g_0|$

$$\frac{1}{2\zeta} = \frac{|g_{max}|}{|g_0|} = \frac{12.5}{1} = 12.5$$

Solving for ζ gives us $\zeta = .04$, which is in close agreement with the damping factor value we found from the half-power analysis.

Just for interest, the values used to generate Figure 2.32 were $\bar{f} = 400$ N, $m = 1$ kg, $\omega_n = 20$ rad/s, and $\zeta = .04$.

2.11 OTHER TYPES OF DAMPING

It's already been mentioned that viscous damping, although convenient from a mathematical point of view, doesn't really model the damping found in actual structures very well. In this section we'll look other damping models and see how their inclusion affects the response of our systems.

The first type of damping we'll look at is dry friction (also called Coulomb damping). You've undoubtedly already seen this type of damping in your physics and dynamics classes. The simplest dry friction model assumes that two contacting bodies that are moving relative to each other will experience a constant friction force that always acts to oppose the motion. Figure 2.33 illustrates this behavior. The magnitude of the force is equal to μW, where W is the weight of the object that is sliding and μ is the coefficient of friction between the object and the surface it's sliding on. This force can be represented mathematically as

$$F_d = \mu W \text{sign}(\dot{x}) \tag{2.11.1}$$

where $\dot{x}$ is the relative velocity and sign( ) indicates the sign function (equal to 1.0 when the argument is greater than zero, zero at zero, and -1.0 when the argument is less than zero).

If we use this sort of damping with our spring-mass system, our equation of motion will be

$$m\ddot{x} + d \text{ sign}(\dot{x}) + kx = 0 \tag{2.11.2}$$

where for simplicity, we've used d to stand for μW.

The free vibration behavior of this system is rather interesting. Since the damping is constant (as long as the velocity's sign doesn't change), we can rewrite (2.11.2) as follows:

$$m\ddot{x} + kx = -d, \quad \dot{x} > 0$$

$$m\ddot{x} + kx = d, \quad \dot{x} \le 0 \tag{2.11.3}$$

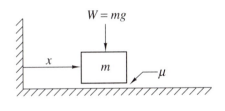

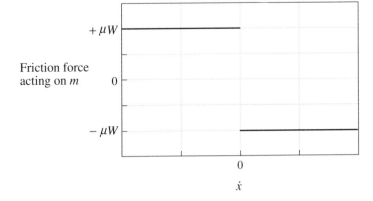

Figure 2.33 Force versus velocity relationship for dry friction

What this tells us is that we can consider the system to be a piecewise linear, forced system; i.e., the system looks like a linear spring-mass system being driven by a constant force. The only difference between this system and a completely linear one is that the forcing changes sign when the velocity changes sign. If we give this system an initial displacement and then release it, the envelope of the resulting oscillations is itself linear. This linear decay of amplitude is quite different from what we saw with linear damping, which produced an exponentially decaying envelope for the oscillations. A typical plot of a dry friction damped response is shown in Figure 2.34. Note that the oscillations stop in a finite time, whereas in the case of linear damping the oscillations continue forever (although they get very small). Also note that the response stops at a nonzero value. This is because the oscillations eventually become so small that the spring force is unable to overcome the dry friction force and the mass gets "stuck" at some deflected position. The precise position will depend upon the initial conditions.

A dry friction model is often used to approximate the damping that occurs when parts rub against each other. Thus if you were trying to estimate the damping that might exist in a large space structure, something made up of many truss elements, you might try to estimate the dry frictional losses due to relative motion between the truss elements and the joint assemblies that hold them together.

Example 2.16

Problem   From the response of a spring-mass system with a dry friction interface in Figure 2.34, determine the linear damping estimates that come from examining the change in the peak response between peaks 1 and 2, peaks 2 and 3, and peaks 1 and 3. Is the linear estimate found from two

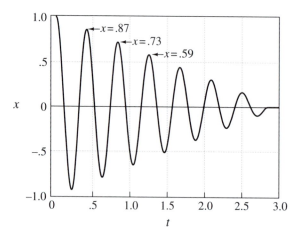

Figure 2.34   Time response of a dry friction damped system

oscillation cycles (peak 1 to peak 3) equal to the average of that found for the peaks 1-2 and peaks 2-3 estimates?

Solution   We saw in (2.10.5) that

$$\zeta = \frac{\sigma}{\sqrt{4\pi^2 + \sigma^2}}$$

and in (2.10.10) that

$$\sigma = \frac{1}{n}\ln\left(\frac{x(0)}{x(nt_p)}\right)$$

Using the data from Figure 2.34 we see that

$$\sigma_{1-2} = \ln\left(\frac{.87}{.73}\right) = .175$$

and therefore

$$\zeta_{1-2} = .028$$

Looking between peak 2 and peak 3 gives us

$$\sigma_{2-3} = \ln\left(\frac{.73}{.59}\right) = .213$$

Thus we have

$$\zeta_{2-3} = .034$$

Last, the log decrement between the first and third peaks is

$$\sigma_{1-3} = \ln\left(\frac{.87}{.59}\right) = .388$$

and the associated linear damping estimate is

$$\zeta_{1-3} = .062$$

We can immediately see that the damping couldn't have been linear in the original system. If the damping had been linear, all the linear damping estimates would have been identical (assuming no experimental error), as would all the log decrements. What these results show is that the log decrement over two cycles is equal to the sum of the log decrements from peak 1 to 2 and from peak 2 to 3. Furthermore, the linear damping estimate from peaks 1 to 3 is equal to the sum of those from peak 1 to 2 and from peak 2 to 3. Thus we now see what kind of trend should tip us off to the presence of dry friction. If the damping estimates increase as time increases (as ours did, going from .028 to .034), then we may have dry friction. If the overall damping estimate between three peaks is equal to the sum of those found from the two adjacent peaks, then we have even more reason to suspect dry friction. Of course, we might have a combination of linear and dry friction, and so we'd have to try to identify the amount of each particular component.

The other kind of damping we'll consider is called *structural* or *hysteretic* damping. This kind of damping does a reasonable job of approximating the kind of energy loss experienced by materials that are repeatedly loaded and unloaded. In testing for this type of damping, a specimen is cyclically loaded and the testing machine records the stress developed during the loading. If the system has internal damping, the response of such a test will look like curve shown in Figure 2.35. When the damping test is run, the curve is traversed in a clockwise manner. After one full cycle of testing, the material's state is back where it started. However the fact that there is a finite area enclosed by the curve implies that energy was lost during the test. If no damping had been present, the system would have unloaded along the same path that it took when loading was applied. When materials unload along a different path (as illustrated), the effect is called hysteresis, which tells us why this damping is also called hysteretic damping.

You can see that the area within the stress-strain curve is proportional to energy loss when you realize that stress is a force quantity and strain is a displacement quantity. Thus the integral represents a force times displacement quantity, i.e., a work term. The area within the curve is proportional to the work done on the specimen and indicates how much energy was dissipated by the specimen during the test.

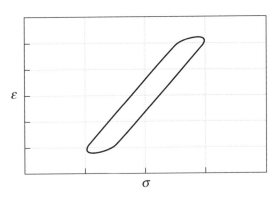

Figure 2.35 Stress-strain curve for hysteretically damped material

It turns out that the stress-strain curve doesn't depend upon frequency (at least within a reasonable range of frequencies). You can also see that doubling the amplitude of the strain will double the entire curve (both horizontally and vertically). The result of this is that the energy dissipated over one cycle will be proportional to the square of the amplitude of the motion. (The energy is proportional to the area enclosed in the curve and the area is proportional to the square of the amplitude.)

Next we need to find out how we can deal with this kind of damping. We'd like to come up with the equivalent viscous damping that will match the hysteretic damping in some sense. The most convenient approach to this question is to try to get the viscous damping to dissipate the same amount of energy over one cycle as the hysteretically damped material would. Therefore, our immediate task is to find out how much energy is dissipated over a cycle by a viscous damper.

As usual, we'll use complex notation to obtain our solution. We can assume that the forcing is actually equal to $\bar{f}\cos(\omega t)$ (with $\bar{f}$ real), and thus our solution is given by the real part of x that results from solving

$$m\ddot{x} + c\dot{x} + kx = \bar{f}e^{i\omega t} \tag{2.11.4}$$

We've solved this problem earlier in the chapter, and the complete solution is given by

$$x(t) = \bar{f}|g(\omega)|e^{i(\omega t - \phi)} \tag{2.11.5}$$

where $|g|$ and ϕ are defined in (2.7.8) and (2.7.9). The velocity is found by differentiating the displacement response, yielding

$$\dot{x}(t) = i\omega\bar{f}|g(\omega)|e^{i(\omega t - \phi)} \tag{2.11.6}$$

Now that we have expressions for the applied force and the velocity, we can determine the energy dissipated over one cycle. The work done over one cycle is given by

$$\Delta w = \int f\,dx \tag{2.11.7}$$

where the integral is taken over the entire displacement history. We can reexpress this in terms of time by realizing that $dx = \dot{x}\,dt$. Thus our work equation becomes

$$\Delta w = \int_0^{\frac{2\pi}{\omega}} f\dot{x}\,dt \tag{2.11.8}$$

Combining (2.11.6) and (2.11.8) gives us

$$\Delta w_v = \int_0^{2\pi/\omega} \text{Re}(\bar{f}e^{i\omega t}) \cdot \text{Re}(i\omega\bar{f}|g|e^{i(\omega t - \phi)})dt \tag{2.11.9}$$

$$= -\bar{f}^2|g|\omega\int_0^{2\pi/\omega}\cos(\omega t)\sin(\omega t - \phi)dt \tag{2.11.10}$$

$$= \bar{f}^2|g|\pi\sin(\phi) \tag{2.11.11}$$

where Re( ) indicates that the real part of the argument is being considered and w_v indicates that the energy dissipated is due to viscous damping. Recalling (2.7.9), we can find $\sin(\phi)$ to be

$$\sin(\phi) = \frac{2\zeta\omega\omega_n}{\sqrt{(\omega_n^2 - \omega^2)^2 + (2\zeta\omega\omega_n)^2}} \tag{2.11.12}$$

Comparing this with (2.7.8) and recalling that $2\zeta\omega_n = \frac{c}{m}$ gives us

$$\sin(\phi) = 2\zeta\omega\omega_n m|g| \tag{2.11.13}$$

Thus (2.11.11) becomes

$$\Delta w_v = (\bar{f}|g|)^2 \pi c\omega \tag{2.11.14}$$

Finally, $\bar{f}|g|$ is simply the magnitude of the resultant oscillations $|\bar{x}|$. Thus we have

$$\Delta w_v = |\bar{x}|^2 \pi c\omega \tag{2.11.15}$$

This result tells us that the energy absorbed because of viscous damping over one cycle of loading is proportional to the square of the amplitude of the response ($|\bar{x}|^2$), and it varies linearly with the damping coefficient c and with the frequency ω of the test. Thus, if we double the frequency, we'll double the energy absorbed. This makes sense because the force of a viscous damper is proportional to velocity, and doubling ω will double the velocity. It's also logical that the amount of damping will vary with the damping coefficient. You'd expect that doubling the damping coefficient should double the energy absorbed.

Now that we've seen how energy is consumed by a viscous damper, we can go ahead and determine what an equivalent viscous damper might be for hysteretic damping. We've already mentioned that the energy absorbed by a hysteretic damper is proportional to the square of the oscillation amplitude and isn't proportional to the forcing frequency, while we've just seen that the energy absorbed by a viscous damper has a frequency dependence. If the damping force generated by the viscous damper had been proportional to displacement instead of velocity, then we wouldn't have had a frequency dependence. However, a force that's simply proportional to displacement is a spring force and doesn't dissipate energy. It's necessary for the force to be out of phase with respect to the displacement for energy dissipation to occur. Realizing this, let's assume that hysteretic damping is proportional to displacement but is out of phase with it. We could do this by using complex exponentials as our solutions and assuming that the force developed is proportional to ix (with the spring force proportional to x and a viscous damper's force proportional to $i\omega x$). Of course, this would be tough to justify on physical grounds. However, let's run with this idea anyway and see where it leads.

If we take the damping force to be $ik\gamma x$, our equation of motion becomes

$$m\ddot{x} + kx + ik\gamma x = f \tag{2.11.16}$$

Since both spring and damper are proportional to displacement, we can combine them thus:

$$m\ddot{x} + (1 + i\gamma)kx = f \tag{2.11.17}$$

$k(1 + i\gamma)$ is called the *complex stiffness* and γ is the *structural damping factor*.

We can quickly go through the same sort of analysis we've just finished for the viscous case to find how much energy is dissipated by hysteretic damping. If we let $f = \bar{f}e^{i\omega t}$ (with $\bar{f}$ real) and $x = \bar{x}e^{i\omega t}$, we can solve (2.11.17) to find

$$\bar{x} = \frac{\bar{f}}{k - \omega^2 m + i\gamma k} \qquad (2.11.18)$$

Multiplying the real part of the force and the real part of the velocity, we can form the dissipated energy

$$\Delta w_s = \int_0^{2\pi/\omega} \bar{f}\cos(\omega t)\left(-\omega\bar{f}\right)\frac{(k - \omega^2 m)\sin(\omega t) - \gamma k\cos(\omega t)}{(k - \omega^2 m)^2 + (\gamma k)^2}\,dt \qquad (2.11.19)$$

Evaluating this gives us

$$\Delta w_s = \frac{\pi\gamma k\bar{f}^2}{(k - \omega^2 m)^2 + (\gamma k)^2} \qquad (2.11.20)$$

Finally, recognizing from (2.11.18) that $|\bar{x}|^2$ shows up in our work expression, we can form the final result

$$\Delta w_s = \pi\gamma k|\bar{x}|^2 \qquad (2.11.21)$$

This is exactly what we wanted, a dissipation that's proportional to the square of the amplitude and has no frequency dependence. It gives us a conceptual model for structural damping, i.e., a damping force that varies with displacement and is 90 degrees out of phase with it. If we compare the energy dissipated by viscous and structural damping

$$\Delta w_v = c\pi\omega|\bar{x}|^2$$

$$\Delta w_s = \gamma k\pi|\bar{x}|^2$$

we see that we can come up with an equivalent viscous damping coefficient c_{eq} by letting

$$c_{eq}\pi\omega|\bar{x}|^2 = \gamma k\pi|\bar{x}|^2 \qquad (2.11.22)$$

or

$$c_{eq} = \frac{\gamma k}{\omega} \qquad (2.11.23)$$

How would you go about actually determining c_{eq}? First you'd have to run a stress-strain test to determine what the energy absorption is for your material. Since you'll know the amplitude of your displacement during the test, you'll be able to use (2.11.21) to determine γ. Then you'd use (2.11.23) to calculate c_{eq}. You'll note that you also need ω. This implies that the c_{eq} will depend on the frequency at which your system will be oscillating. This is generally okay, since there's usually a primary vibration going on. It's important, however, to realize that your equivalent viscous coefficient is good only at this particular frequency. If the frequency changes, you have to recalculate c_{eq}.

A related observation is that these damping analyses are applicable only to a system that's vibrating in a sinusoidal way. It would be nice if we could also determine the response of the system to a nonsinusoidal input (a step or impulse), as we can with viscous damping. Unfortunately, that's not possible without changing our damping model drastically and increasing the complexity of the analysis by a very significant amount.

Example 2.17

Problem A sinusoidal stress-strain test has been run on a system and you're to determine the structural damping factor. The linear stiffness k was 100,000 N/m, the driving force had an amplitude of 350 N, and the amplitude of the response was .01 m. Furthermore, the output differed from the input by 90 degrees.

Solution An examination of (2.11.18) shows that the output and input differ by 90 degrees when $\omega = \sqrt{\frac{k}{m}}$. For this situation we have

$$|\bar{x}| = \frac{\bar{f}}{|\gamma k|}$$

Using the given data, this means that

$$.01 = \frac{350}{100,000\gamma}$$

which gives us $\gamma = .35$.

Example 2.18

Problem For the system discussed in Example 2.17, determine the energy dissipated during one cycle of the stress-strain test.

Solution The energy dissipated is given by

$$\Delta w_s = \pi \gamma k |\bar{x}|^2$$

Since we've already determined the value for γ and are given k and $|\bar{x}|$, we can immediately solve for the energy dissipated:

$$\Delta w_s = \pi(.35)(100,000)(.01)^2 = 11 \text{ N·m}$$

2.12 ACCELEROMETERS AND SEISMOMETERS

Transducer is the word given to describe the devices that allow us to measure the relevant variables in a vibrations experiment. One of the early decisions that must be made in the course of a vibrations test is the determination of which transducer is appropriate for the particular problem at hand. In this section, we'll discuss two kinds of transducer and give some insight into their uses and limitations.

Perhaps the most widely used vibrational transducer is the *accelerometer*. Although there's more than one way to build an accelerometer, the most common is the piezoelectric type, shown for two configurations in Figure 2.36. The basic mechanism at work for both is a mass that acts on a piezoelectric crystal. Because piezomaterials generate an electric charge that's proportional to their deformation, we can use them to determine the acceleration of the object on which the accelerometer is attached. In Figure 2.31*a* the piezoelectric material is sheared as the seismic mass moves relative to the surrounding casing, thus generating a charge. The accelerometer shown in Figure 2.31*b* induces a charge in the piezoelectric material by compressing it.

Both the configurations of Figure 2.36 can be viewed as simple SDOF systems (a mass attached to the accelerometer casing through a spring). Referring to Figure 2.37, we see that the internal movable mass within the accelerometer body is denoted by m, the spring that connects this mass to the accelerometer is denoted by k, the actual motion of the accelerometer body is given by y, and the relative motion of the mass m with respect to the accelerometer body is given by x. We've also included a damper c to account for frictional dissipation. It is very important to understand that x is the *relative* motion of the internal mass, not its actual motion with respect to an outside observer.

Remembering that the absolute motion of the seismic mass is given by $x + y$, we can perform a force balance to find

$$m(\ddot{x} + \ddot{y}) = -kx - c\dot{x} \tag{2.12.1}$$

Rearranging this yields

$$m\ddot{x} + c\dot{x} + kx = -m\ddot{y} \tag{2.12.2}$$

and dividing by m brings us to

$$\ddot{x} + 2\zeta\omega_n\dot{x} + \omega_n^2 x = -\ddot{y} \tag{2.12.3}$$

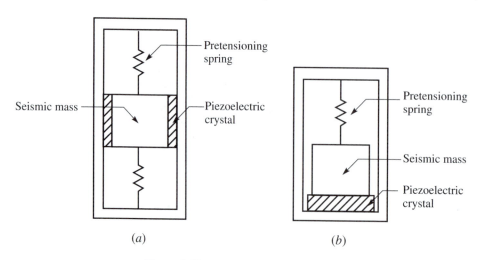

(a)   (b)

Figure 2.36 Accelerometer configurations

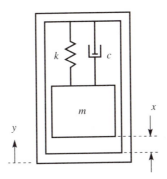

Figure 2.37   Simple accelerometer model

If we assume that the accelerometer body is oscillating sinusoidally, then using $y(t) = \bar{y}e^{i\omega t}$ and $x(t) = \bar{x}e^{i\omega t}$ will lead to

$$\left(\omega_n^2 - \omega^2 + 2i\zeta\omega_n\omega\right)\bar{x} = \omega^2\bar{y} \tag{2.12.4}$$

Therefore we have

$$x(t) = \frac{\omega^2\bar{y}e^{i\omega t}}{\omega_n^2 - \omega^2 + 2i\zeta\omega_n\omega} \tag{2.12.5}$$

This is the govening equation for our accelerometer. Although this doesn't look particularly simple, this situation changes when we look at a limiting case for which the excitation frequency is much smaller than the natural frequency of the accelerometer, i.e., $\omega \ll \omega_n$. In this case the velocity and acceleration terms become negligible in comparison to displacement ($|\bar{x}| \gg |i\omega\bar{x}| \gg |-\omega^2\bar{x}|$). Therefore (2.12.3) can be simplified to

$$x(t) = -\frac{1}{\omega_n^2}\ddot{y}(t) \tag{2.12.6}$$

Thus we have a direct measure of the object's acceleration: we need simply multiply $x(t)$ by $-\omega_n^2$ to obtain it.

This simple formula holds only as long as $\omega \ll \omega_n$, thus telling us the limitation on the usable frequency range.

In addition to being used as an accelerometer, the setup can be used to obtain a different kind of measurement, namely, the displacement of the mounting surface. Consider the case in which the natural frequency of our device is now much lower than the exciting frequency. This is the situation that corresponds to point 5 on our response graph of a seismically excited mass (as shown previously in Figure 2.7). You'll recall that in this case your hand moved up and down while the mass stayed fixed. Thus the relative motion between the mass and your hand was exactly the same as the actual motion of your hand, just 180 degrees out of phase. This is the result we should expect the math in this case to give us.

If $\omega \gg \omega_n$ in (2.12.5) we obtain

$$x(t) = -\bar{y}e^{i\omega t} \tag{2.12.7}$$

Thus, as expected, our transducer gives us a reading that's directly proportional to the mounting surface's motion. When used in this manner, the transducer is called a *seismometer*.

Physically, the constraint $\omega \gg \omega_n$ means that the natural frequency of the seismometer must be small, thus implying either a large mass or a weak spring. Consequently, seismometers are usually substantially larger than accelerometers.

Example 2.19

Problem   Assume a given accelerometer has a moving mass of 10 g and a spring constant of 3.95×10^7 N/m. We'll assume that there is no damping present. What is the upper usable frequency range for this accelerometer if the maximal error that we'll allow is 5%?

Solution   The general solution for an accelerometer was given by (2.12.5):

$$x(t) = \frac{\omega^2 \bar{y}e^{i\omega t}}{\omega_n^2 - \omega^2 + 2i\zeta\omega_n\omega}$$

We can recognize the numerator as simply $-1 \times$ the acceleration of the base. Thus the $-x$ response to acceleration ratio is given by

$$-\frac{x(t)}{\ddot{y}} = \frac{1}{\omega_n^2(1 - \frac{\omega^2}{\omega_n^2})}$$

As the frequency of excitation increases, the ratio of x to acceleration grows. Therefore a 5% error means we'll have

$$1.05 = \frac{1}{1 - \frac{\omega^2}{\omega_n^2}}$$

or

$$\frac{\omega}{\omega_n} = .2182$$

Since $\omega_n = 62,850$ rad/s (using the given values for k and m) this means $\omega = 13,700$ rad/s or 2180 Hz. Thus the maximum usable frequency range is about 20% of the system's natural frequency.

2.13 HOMEWORK PROBLEMS

Sections 2.2 and 2.3

2.1. An existing damped, SDOF system like the one shown in Figure P2.1 is too lightly damped. In the original system $m = 1$ kg, $k = 10,000$ N/m, and $c = 15.0$ N·s/m. To reduce the setting time by 70%, what must c be increased to?

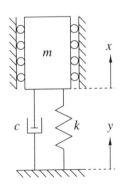

Figure P2.1

2.2. What will doubling the mass in the lightly damped SDOF system (of Figure P2.1) do to the settling time?

2.3. If k is doubled in the lightly damped SDOF system of Figure P2.1, what happens to the settling time?

2.4. We're going to analyze the motions of a unicyclist. The actual system is shown in Figure P2.4, as is our approximate model. If the unicyclist is traveling at 3 m/s and hits the illustrated bump, what will the cyclist's maximum deflection from x_{eq} be?

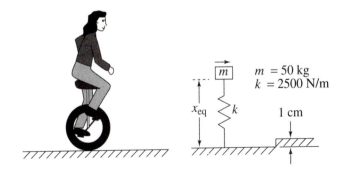

Figure P2.4

2.5. Consider the unicyclist of Problem 2.4. The road she's traveling on has a sinusoidal profile, as illustrated in Figure P2.5. What is the worst speed for her to travel at with regard to her vibrational response?

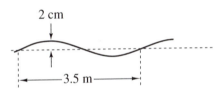

Figure P2.5

2.6. Dopey Dog, an animated film star, has just been handed a bone with an attached, stretched, elastic rubber band. After a momentary pause (inserted in violation of realism so the audience can see the amusing moment of shocked realization on Dopey's face) he begins to rocket to the right. If $m_{Dopey} = 45$ kg, $k_{elastic} = 300$ N/m, and the elastic is stretched 6 m when handed to Dopey, calculate Dopey's acceleration and velocity at $t = .3$ second (ignore the unstretched length of the rubber band, i.e., set it to zero).

2.7. A machine on its suspension deflects .6 cm under static conditions. What is the ratio of the mass's oscillation amplitude to the base oscillation amplitude if the base is oscillating at 40 Hz? Assume zero damping.

2.8. The spring-mass system shown in Figure P2.8 has a spring constant of 1600 N/m and a mass of 2 kg. What is the response amplitude to a ground input of

$$y(t) = .06 \sin(10 t) \text{ m}$$

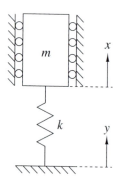

Figure P2.8

2.9. The spring-mass system shown in Figure P2.8 experiences a 2 mm in-phase oscillation when the base is oscillating at 300 rad/s with an amplitude of 1.1 mm. What is the static deflection of the mass?

2.10. The spring-mass system shown in Figure P2.8 experiences a 4 mm out-of- phase oscillation when the base is oscillated at 200 rad/s with an amplitude of 1.3 mm. What is the static deflection of the 5 kg mass?

2.11. Consider again the system shown in Figure P2.8. When excited by a ground motion of $.001\cos(20 t)$ m, the response has an magnitude of 1.25 mm and is in phase with the excitation. Find the natural frequency of the system.

2.12. Consider once more the system shown in Figure P2.8. This system is excited by a ground motion of $.001\sin(50 t)$ m. The response is out of phase and has a magnitude of 4 mm. What is the system's natural frequency?

2.13. An 18 kg mass is protected from vibration by being placed in a container and isolated from the walls by two springs. The force characteristic of the first spring is

$$f_1 = 5x \text{ (spring 1)}$$

and the other spring can be either of the following:

$$f_2 = x + x^3 \text{ (spring 2)}$$
$$f_3 = 31x \text{ (spring 3)}$$

In all three cases the displacement units are assumed to be meters and the forces are newtons. Initially the mass-spring unit is 12 m long. The available space inside the protective box is 8 m. Two possible configurations are being considered, shown in Figure P2.13. Which configuration will be more effective in isolating the mass from external disturbances, assuming that these disturbances occur around $\omega = 3$ rad/s and are of small amplitude ($y(t) = \varepsilon \sin(3.0t)$)? (*Note*: Best isolation here means minimum movement.)

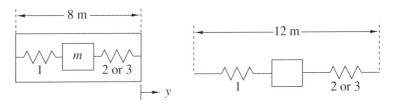

Figure P2.13

2.14. Consider a simple model of a car traveling along a road (the same model as in Problem 2.5). We'll look only at translational motions (vertical oscillations). The road profile is sinusoidal with amplitude .01 m and wavelength 8 m. The speed limit is 35 km/h. Would you experience a smaller oscillation amplitude if you reduced your speed to 25 km/h? $m = 1000$ kg and $k = 70, 224$ N/m.

2.15. Consider the system illustrated in Figure P2.15. A rigid ring with mass m is connected via four massless, inextensible wires to a central hub (radius r). Each wire has a tension T. If the motion of the hub is given by $\theta(t) = \theta_0 \cos \omega t$ (θ_0 small), determine the angular response of the ring.

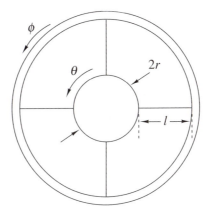

Figure P2.15

2.16. Figure P2.16 shows the steady state time response of an undamped SDOF system under seismic excitation (transients are ignored). Determine ω_n for this system. (*Note*: You *don't* need to try and back out the answer by plugging amplitudes and frequencies into the relevant dynamic equations.)

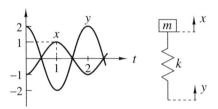

Figure P2.16

2.17. Figure P2.17 shows an enclosure containing a tensioned spring that presses on a mass. The initial compression of the spring is equal to h. The enclosure is subjected to a sinusoidally varying displacement $x(t) = a\cos(\omega t)$. Determine the frequency ω for which the mass will lose contact with the enclosure (consider m, k, h, and a to be fixed, and ignore gravity).

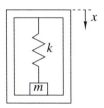

Figure P2.17

2.18. Consider a car that's moving along an undulating road as in Figure P2.18. The road's profile is given by

$$y(x) = .02\sin\left(\frac{2\pi x}{15}\right)$$

where both x and y are in meters. Treat the car as a point mass m. What is the normal force developed between the car and the road as a function of the car's speed $\dot{x}$? ($\dot{x}$ is constant and $m = 1300$ kg)

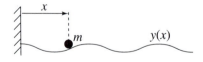

Figure P2.18

2.19. For the system of Problem 2.18, find the maximum and minimum normal force developed between the car and the road if $\dot{x} = 30$ m/s. What is the maximum acceleration that a passenger (m in Figure P2.18) would feel?

2.20. Consider the system of Problem 2.18 but now include the car's suspension, as shown in Figure P2.20. Let $k = 300,000$ N/m, and $m = 1300$ kg. What is the normal force developed between the spring and the road, and what is the maximal acceleration felt by the car? $\dot{x} = 25$ m/s.

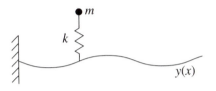

Figure P2.20

2.21. Consider a seismically excited spring-mass system. Find the range of oscillation frequencies ω for which the force transmitted to the mass m is less than or equal to 170 N, with $m = 1$ kg, $k = 25,000$ N/m, and $y(t) = \cos(\omega t)$ (y measured in centimeters).

2.22. Consider an undamped, direct force excited system (Figure P2.22). Let f be zero and let $x = .003 \sin(\omega t)$ (measured in meters). What is the force transmitted to the ground? $m = 10$ kg and $k = 5000$ N/m.

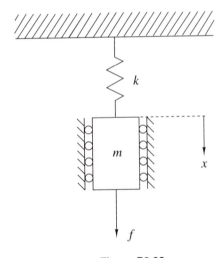

Figure P2.22

2.23. As shown in Figure P2.23, the spring k pushes the mass m against a moving surface. The motion of the surface is given by $x(t) = a(\cos \omega t)$. When $x(t) = 0$, the mass is being pressed into the surface with a force kh (i.e., the spring is precompressed on amount h). Determine the critical speed ω for which the mass will first lose contact with the surface. (*Note:* Identify the critical ω for $a > h$ and $a < h$.)

Figure P2.23

2.24. Given the displacement input $y(t) = \bar{y}\cos(\omega t)$, determine the equation of motion for the system illustrated in Figure P2.24 (neglect gravity). What is the natural frequency for the system?

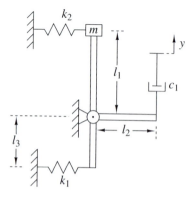

Figure P2.24

Section 2.4

2.25. Using the given data in Figure P2.25, determine the system's natural frequency.

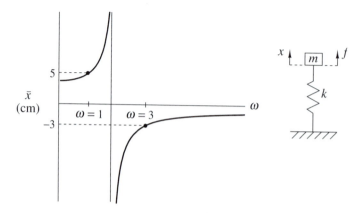

Figure P2.25

2.26. Consider again the undamped, direct force excited system from Figure P2.22. $f(t) = 10\cos(5t)\,\text{N}$, $m = 10$ kg, $k = 100$ N/m. By how much is the amplitude of the force between the spring and the floor reduced from that applied to the mass?

2.27. Determine the actual response of a spring-mass system when forced at its natural frequency. The relevant equation is

$$\ddot{x} + \omega_n^2 x = \overline{f}\cos(\omega_n t)$$

Assume a solution $x(t) = a_1\cos(\omega_n t) + a_2\sin(\omega_n t) + a_3 t\sin(\omega_n t)$.

2.28. Referring again to the system shown in Figure P2.22, determine the range of frequencies for which m's response amplitude will be below 1 mm if the forcing is equal to $1.1\cos(\omega t)$ N, $m = 10$ kg and $k = 10,000$ N/m.

2.29. Consider once more the system shown in Figure P2.22. Let $m = 5$ kg, $k = 20,000$ N/m, and $f(t) = 20\sin(\omega t)$ N. For what range of ω's will the amplitude of the response be less than 20 mm?

2.30. Consider the two-mass/two-spring system shown in Figure P2.30. m_1 is moved to the right until it contacts m_2. The left end of k_1 is then moved an additional distance $F_0\left(\frac{1}{k_1} + \frac{1}{k_2}\right)$ to the right, causing m_1 and m_2 to press tightly together. Further motions of k_1's left end will be labeled z, motion of m_1 will be labeled x, and motion of m_2 will be labeled y. As long as m_1 and m_2 remain in contact, $x = y$. Derive an expression for the force between the two masses as a function of frequency and show that for an appropriately chosen excitation frequency the masses will separate. Let $z = \frac{F_0}{2k_2}\cos(\omega t)$.

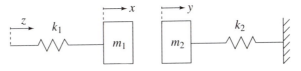

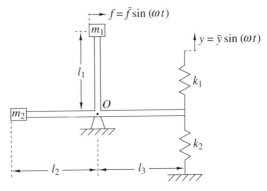

Figure P2.30

2.31. Find the equations of motion and natural frequency for system shown in Figure P2.31 (neglect gravity). What would f need to be for there to be no overall excitation of the system? ($\overline{f}$: newtons, $\overline{y}$: meters). The system is freely hinged at O.

Figure P2.31

2.32. Derive the equations of motion for the system shown in Figure P2.32 for a displacement input given by $y(t) = \bar{y} \cos(\omega t)$ (neglect gravity).

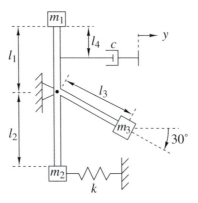

Figure P2.32

2.33. Figure P2.33 is a model of the valves in an engine. m is the mass of a valve, k is the restraining spring, and the rotating body is the cam that drives the valves. Consider the center of this cam to be fixed in space. As the cam rotates, the mass is forced to follow the cam's profile, inducing an x deflection equal to $x_0 + a \cos(\omega t)$. At what rotational velocity do the inertial forces of the piston become so great that the piston loses contact with the cam? When the cam is in the indicated position, the total spring deflection is equal to h_1.

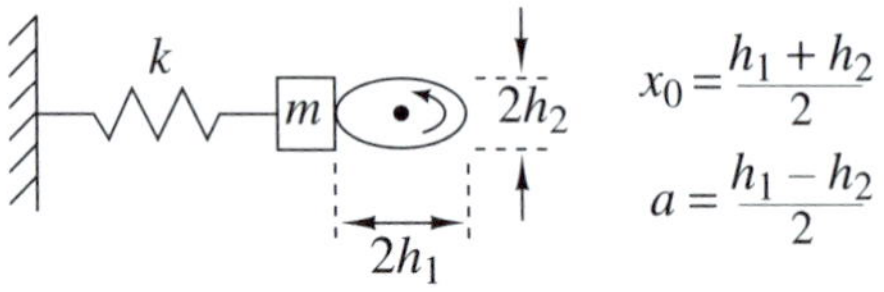

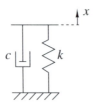

$$x_0 = \frac{h_1 + h_2}{2}$$

$$a = \frac{h_1 - h_2}{2}$$

Figure P2.33

2.34. In Figure P2.34, does the force transmitted to the floor reach a maximum as ω is increased or does it increase indefinitely? At high ω what is more important to the transmitted force, k or c? ($x = x_0 \sin(\omega t)$.)

Figure P2.34

2.35. Given the force input $f = \overline{f}\sin(\omega t)$, determine the equation of motion for the system illustrated in Figure P2.35. What is the system's natural frequency? The pivot at O is frictionless. Neglect gravity.

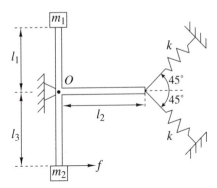

Figure P2.35

Section 2.5

2.36. Determine the transfer function for the system illustrated in Figure P2.36 that relates input force at the mass to output force felt at the floor. ($f = \overline{f}\sin(\omega t)$)

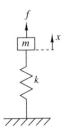

Figure P2.36

2.37. Find the transfer function between m's acceleration and the applied force f for the system shown in Figure P2.36.

2.38. Find the transfer function of acceleration $\ddot{x}$ to position x for the system of Figure P2.36.

2.39. Find the transfer function of support excitation y to response angle θ for the pendular system shown in Figure P2.38. Make sure to linearize your equations. The pendulum is of length l and the freely pivoted upper end of the pendulum is moved horizontally according to

$$y(t) = a\sin(\omega t)$$

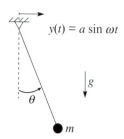

Figure P2.39

2.40. Find the transfer function between the displacement input y and the displacement output x for the system shown in Figure P2.40. $y = \bar{y}\sin(\omega t)$. The rigid bar pivots freely at O.

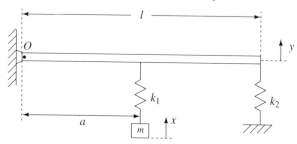

Figure P2.40

2.41. Find the transfer function between the input displacement y and the output force against the wall for the system shown in Figure P2.41. $y = \bar{y}\sin(\omega t)$

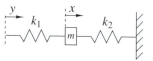

Figure P2.41

2.42. Consider the pendular system illustrated in Figure P2.42 (lumped mass m on the end of a rigid rod of length l). A small motor at O produces a sinusoidally varying torque ($M = \bar{M}\sin(\omega t)$) that acts on the pendulum. Find the transfer function from input torque to response angle $\theta(t)$. Linearize your system equations about $\theta = 0$.

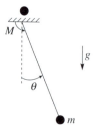

Figure P2.42

Section 2.6

2.43. Consider the spring-mass-damper system illustrated in Figure P2.43. You need to determine the actual values for k and c but have only a limited amount of information. Given the data shown, determine k and c.

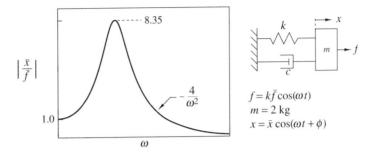

$$f = k\bar{f}\cos(\omega t)$$
$$m = 2 \text{ kg}$$
$$x = \bar{x}\cos(\omega t + \phi)$$

Figure P2.43

2.44. Is the plot illustrated in Figure P2.44 (actual displacement vs. frequency for a direct force excited, damped oscillator) a reasonable one?

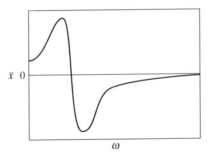

Figure P2.44

2.45. Figure P2.45 shows the steady state response (transients are ignored) of a direct force excited, SDOF system. The mass is equal to 2 kg. What other system parameters can you determine?

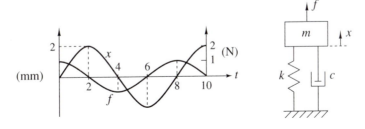

Figure P2.45

2.46. Let $l_1 = 1$ m, $l_2 = .5$ m, $l_3 = 2$ m, $m_1 = 1.1$ kg, $k = 100$ N/m, and $c = 3$ N·s/m. Determine the equation of motion for the system illustrated in Figure P2.46 in terms of the parameters ω_n and ζ. Neglect gravity and assume that the maximal response occurs when the system is forced at $\omega = \omega_n\sqrt{1 - 2\zeta^2}$.

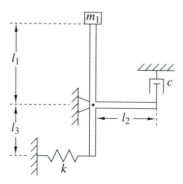

Figure P2.46

Section 2.7

2.47. Put $x(t) = 4\cos(3t) - 2\sin(3t)$ into complex form so that the real part of the complex number gives us $x(t)$.

2.48. Can $x(t) = 3\cos(4t) - 3\sin(2t)$ be put into complex form so that $x(t)$ is expressed as the real part of a single complex term?

2.49. Express $x(t) = \cos(t) - \sin(t)$ in purely complex form, i.e., so that there is no need to take the real or imaginary part of the complex variables.

2.50. Referring again to the system shown in Figure P2.1, find the complex transfer function between the input displacement y and the output force against the ground.

2.51. Find the complex transfer function between the velocity input $\dot{y}$ and the displacement output x for the system illustrated in Figure P2.51.

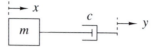

Figure P2.51

2.52. Find the transfer function from input force f to output velocity $\dot{x}$ for the system shown earlier in Figure P2.22. Let $f = \bar{f}e^{i\omega t}$ and $x = \bar{x}e^{i\omega t}$.

2.53. Find the transfer function from the input force f to output acceleration $\ddot{x}$ for the system shown in Figure P2.22. Let $f = \bar{f}e^{i\omega t}$ and $x = \bar{x}e^{i\omega t}$. Express the system equation in terms of ω_n and ζ and use the nondimensionalized frequency Ω.

2.54. What is the real response of $5\ddot{x} + 2\dot{x} + 300x = 4\sin\omega t$?

2.55. Find the real response of $\ddot{x} + 2\zeta\omega_n\dot{x} + \omega_n^2 x = \sin\left(\omega t + \frac{\pi}{4}\right)$

2.56. Solve for the velocity response $\dot{x}$ of the system illustrated in Figure P2.56. $y = .02\sin(150\,t)$, $k_1 = 10,000$ N/m, $k_2 = 5000$ N/m, $m = .5$ kg, $c = 8.66$ N·s/m.

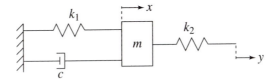

Figure P2.56

2.57. Find the velocity $\dot{x}$ response for the system illustrated in Figure P2.57. $y = .01\sin(100\,t)$, $k = 8000$ N/m, $c_1 = 4$ N·s/m, $c_2 = 2$ N·s/m, $m = .25$ kg.

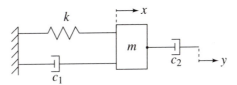

Figure P2.57

Section 2.8

2.58. Add a linear damper in parallel with the spring of the unicycle of Figure P2.5. Let $c = 132$ N·s/m. Under steady state conditions, what will the maximum amplitude response be, and at what velocity will it occur?

2.59. Find the transfer function $\frac{\bar{x}}{\bar{y}}$ for the system illustrated in Figure P2.59. Express the transfer function in both complex and in magnitude/phase form. $m = 10$ kg, $k_1 = 400$ N/m, $c = 14.14$ N·s/m, and $k_2 = 100$ N/m.

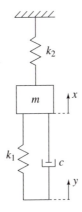

Figure P2.59

2.60. Assume that a seismically excited system is accurately modeled by as shown in Figure P2.1. You need to determine whether the weld connecting the spring to the mass will hold. The maximum stress that the weld can withstand is $1.0 \times 10^6 \, \text{N/m}^2$. The mass is 10 kg and the spring constant is 4000 N/m. The damping constant is 40 N·s/m. Is it reasonable to allow the excitation frequency to reach 10 rad/s? The attachment area of the spring is 1 mm × 1 mm and the base amplitude is 1 mm.

2.61. If $y_1 = 5 \sin(10t)$, determine what y_2 would have to be for there to be no unbalanced net force acting on m, for the system of Figure P2.61.

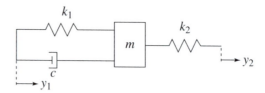

Figure P2.61

2.62. Consider again the seismically excited, damped SDOF system Figure P2.1. What is the magnitude of m's response to a ground excitation of $y(t) = .01 \cos(1.25t)$ for $k = 12,000$ N/m, $m = 20$ kg, and $\zeta = .001$? This should give you a general feel for how very lightly damped structures would respond in an earthquake.

2.63. What is the magnitude of the acceleration response for the system shown in Figure P2.1 for $m = 1$ kg, $c = 10$ N·s/m, $k = 100$ N/m, and $y = .005 \cos(50t)$?

2.64. Repeat Problem 2.63 and then re-solve for two different cases. In case 1, let $k = 0$ and in case 2, let $c = 0$. Which is dominant, c or k, in determining the actual acceleration response magnitude?

2.65. Repeat Problem 2.62 but use $\zeta = .01$. Does this improve the situation much?

2.66. When the damping is small, increasing the damping by a factor of α will reduce the peak response by the same factor. Does this hold true at large values of damping? Why or why not?

2.67. Show that all the amplitude response plots for varying ζ go through the point $\left(\frac{\bar{x}}{y}\right) = 1.0$ at $\Omega = \sqrt{2}$ for a seismically excited, damped SDOF system.

Section 2.9

2.68. The plot in Figure P2.68a shows amplitude response of the system illustrated in Figure P2.68b a centrifuge as a function of frequency *with* a specimen, m_2, in the machine. The mass of the centrifuge is m_1 and it's restrained by a spring of stiffness k. Given that the radius of the spinning chamber is 0.1 m and that the mass m_1 is 100 kg, find the mass of the specimen as well as the spring constant k.

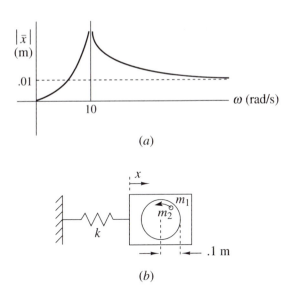

(a)

(b)

Figure P2.68

2.69. A massless rod (with end masses m_1 and m_2) rotates at an angular speed ω with respect to the massless collar A, which rides along the guide BC, restrained by the two springs, k_1 and k_2. Find the response amplitude of the collar in terms of $\omega, l_1, l_2, m_1, m_2, k_1$ and k_2 in Figure P2.69.

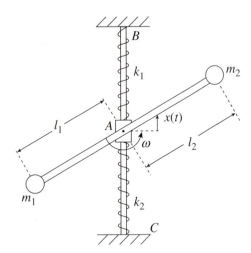

Figure P2.69

2.70. The overhead fan shown in Figure P2.70 isn't dynamically balanced. Because of this, the center of mass of the rotating fan-blade assembly lies off from the center of rotation by a distance l. The rotating mass has a magnitude m_R while the nonrotating motor housing has a mass m_N. Treat the support rod as massless. The radius of gyration of the blade assembly about the point of rotation is equal to k_R. Determine the fan's response when the blades are rotating at a fixed speed ω. (*Hint:* The spring stiffness in the problem comes from gravity. Look at the fan as a spring-mass system being excited by a rotating mass. Since the angle is small ($\phi \ll 1$), you can look at just the x, y motions.)

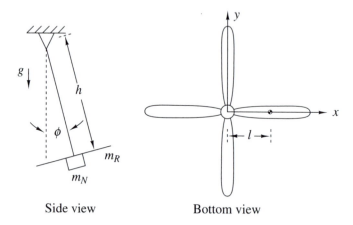

Side view   Bottom view

Figure P2.70

2.71. Assume that most of the wet clothes in a dryer (Figure P2.71) are uniformly distributed around the drum (mass = 10 kg). In addition to these clothes, a single .8 kg lump of clothing also lies against the drum's surface. How will these mass terms enter the equation of motion (2.9.5) for the system?

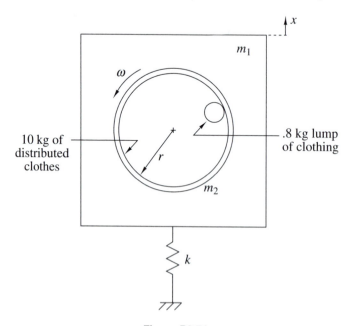

Figure P2.71

2.72. Figure P2.72 illustrates a top view of a top-loading washing machine. The rotational speed of the drum is $\omega = 35$ rad/s and the radius of the drum is $r = .3$ m. Assume that the wet clothes have all gathered into a ball of mass 7 kg (approximated by a point mass). Determine the lateral forces generated by the washing machine. Assume that the mass of the drum is 4 kg.

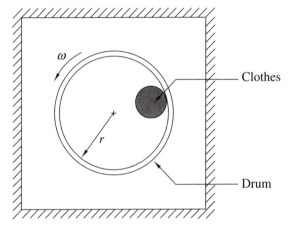

Figure P2.72

2.73. Consider the system of Problem 2.72 but refine the approximation as in Figure P2.73. The drum rotates within a nonrotating sleeve and the total assembly has a mass equal to 4 kg. The outer assembly is held in place by a spring, as shown in Figure P2.73. Note that motion is allowed only in the x direction. The spring constant k is 500 N/m. The ball of clothes is still 7 kg, and $\omega = 35$ rad/s. Determine the force applied to the ground.

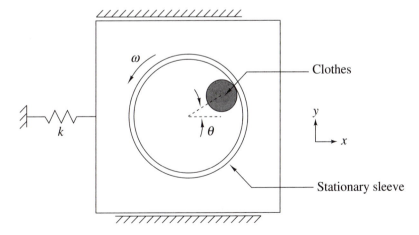

Figure P2.73

2.74. A simplified model of a rotating shaft within its bearings is shown in Figure P2.74. If the shaft is imbalanced, it will experience a time-varying force due to the rotating mass. The operating frequency is 60 rad/s, the support stiffness k_1 is 170,000 N/m, and the total rotating mass m is 100 kg. Determine the amplitude of vibration, given that at high frequencies ($\omega \to \infty$) the oscillation amplitude is equal to .001 m. You'll have to figure out l from the given data.

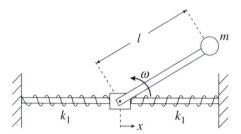

Figure P2.74

Section 2.10

2.75. All you have left from a vibration study is the scrap of paper illustrated in Figure P2.75. Can you use the data to reconstruct the peak response amplitude for the system? An answer that's within 5% is good enough. The system was a direct force excited, spring-mass system.

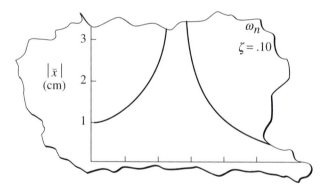

Figure P2.75

2.76. Assume that your car's shock absorber has different values of viscous damping on the extension stroke and on the compression stroke. If the value of the damping factor is $\zeta = .2$ during compression, what must it be on the extension stroke for the overall damping factor to be .4? (Ignore changes to ω_d due to varying damping rates; they'll be small for the values of ζ you're considering.)

2.77. Show that for small damping, the frequency at which the magnitude of the response of a seismically excited damped SDOF system (Figure P2.1) is maximized is the same as that of a direct force excited system.

2.78. Estimate the size of the peak amplitude response for

$$3\ddot{x} + 17.5\dot{x} + 4000x = 15\sin(\omega t)$$

without solving the equation exactly.

2.79. The peak amplitude response $|g(\omega_p)|$ is equal to 5 cm for a direct force excited (Figure P2.79), damped SDOF system. If you know that $m = .03$ kg, $c = .048$ N·s/m, and $k = 12$ N/m, can you determine $|g(0)|$?

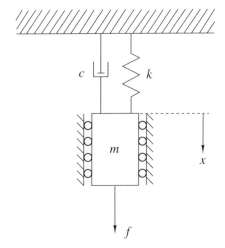

Figure P2.79

2.80. Can you deduce the natural frequency and damping factor of the system shown in Figure P2.80 from the following data? The damping factor is small ($\zeta \ll 1$) and the governing equation is

$$\ddot{x} + 2\zeta\omega_n\dot{x} + \omega_n^2 x = \omega_n^2 \bar{f}\sin(\omega t) = f(t)$$

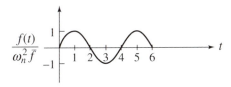

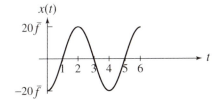

Figure P2.80

2.81. Ω_n is close to the average of Ω_{h_1} and Ω_{h_2} when ζ is small. Let's consider the case when ζ is relatively large. If $\zeta = .2$, what is the error in identifying Ω_n as equal to $\frac{\Omega_{h_1}+\Omega_{h_2}}{2}$?

2.82. Can you determine the value of ζ for the system and responses illustrated in Figure P2.82? The responses shown are two samples from a large number of tests. The governing equation is

$$m\ddot{x} + c\dot{x} + kx = f(t) = \bar{f}\sin(\omega t)$$

The test that produced the maximal amplitude response is the second one illustrated.

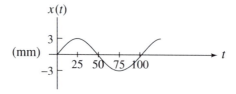

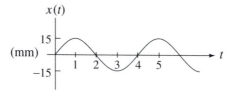

Figure P2.82

2.83. What is the error in using the half-power approximation $\zeta \approx \frac{\delta_h}{2\omega_n}$ if the actual damping factor is .3?

2.84. Why can't you use the half-power approach if $\zeta = .5$? Will the peak response approach, namely, $\frac{1}{2\zeta} = \frac{|g(\omega_p)|}{|g(0)|}$, work? How well?

2.85. Why will neither the peak response approach nor the half-power method work to determine ζ if the system's actual ζ is .8?

2.86. My dad bought a new car a few years back. Unfortunately, a vibration was noticeable when traveling at around 52 mph. The technicians took the car out for a test and gathered acceleration data from an accelerometer. Back at the shop they looked at the data and saw that a strong peak in the vibrational energy was present at 12 Hz when the car had been traveling at 52 mph, just as my father had claimed. I suggested to my dad that the vibration must be due to a resonance in the car that was excited because of the rotation of the tires. Why did I think this? (*Note:* The distance from the center of the wheel to ground for my dad's car was 1 ft.)

2.87. It's not necessary to have an entire cycle of damped response to determine the linear damping through the log decrement approach. Show how you can modify the approach detailed in Section 2.10 to produce an estimate for ζ based on the response information illustrated in Figure P2.87.

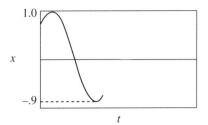

Figure P2.87

2.88. Was the system that produced the response illustrated in Figure P2.88 a linear one?

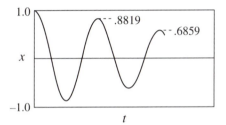

Figure P2.88

2.89. The dampers on automobiles are supposed to work in the same manner during compression and extension. You are examining one that owing to a manufacturing defect, is operative only during one half of the cycle (the compression stroke). If you know that the damping is viscous during compression and you also know that the amplitude drops from 1.0 to .15 over one cycle of free vibration, calculate the value of ζ that would be produced by a properly operating damper. The time elapsed between the two amplitude measurements was .5 second.

Section 2.11

2.90. In this exercise you'll determine the amount of energy dissipated over one cycle of forcing for Coulomb friction. Assume the response is given by $x(t) = a \sin(\omega t)$ and thus $\dot{x}(t) = a\omega \cos(\omega t)$. The equation of motion is

$$m\ddot{x} + \mu mg \, \text{sign}(\dot{x}) + kx = f_s \sin(\omega t) + f_c \cos(\omega t)$$

The forcing has been broken into distinct sine and cosine components so that our assumption of a purely sinusoidal response for x will hold. Calculate the response and retain only those terms corresponding to sinusoidal oscillations at the frequency ω. This information, along with (2.11.8), will allow you to compute the energy dissipated over one cycle of motion.

2.91. Using the same approximation to the sign function as in Problem 2.90, show that the steady state amplitude of oscillation for a direct force excited mass with Coulomb friction is given by

$$a = \frac{\sqrt{\bar{f}^2 - \left(\frac{4\mu mg}{\pi}\right)^2}}{k - m\omega^2}$$

where $\bar{f}$ is the magnitude of the forcing and a is the response magnitude.

2.92. Using the response magnitude relation of Problem 2.91, determine what the magnitude of forcing must be for finite oscillations to exist. What if the forcing is below this limit?

2.93. Consider a direct force excited, spring-mass system with Coulomb friction; $k = 20,000$ N/m, $m = 6$ kg, $f = 100$ N, $\mu = .1$, $\omega = 12$ rad/s. Use the information in Problem 2.91 to determine the magnitude of the steady state response.

2.94. How much will the magnitude of oscillation be reduced if the system of Problem 2.93 has its coefficient of friction increased from .1 to .5?

2.95. How much will the magnitude of oscillation be altered if the system of Problem 2.93 has the forcing frequency increased to 30 rad/s?

2.96. Consider the static problem of a spring-mass system with Coulomb friction (Figure P2.96). If $k = 12,000$ N/m, $m = 10$ kg, and $\mu = .3$, what is the maximal initial deflection of the mass that will not result in subsequent motion of the mass?

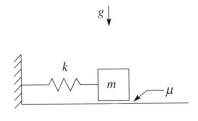

Figure P2.96

2.97. Consider a spring-mass system with Coulomb damping and direct force excitation. Let $f = 50$ N, $m = 1$ kg, $k = 1000$ N/m, and $\omega = 10$ rad/s. You've run two tests. You observe an oscillation magnitude of 55.5 mm for the first test and 55.6 mm for the second. Both readings seem to be within a reasonable experimental error. Calculate the value of μ that would correspond to these measurements and comment on whether determining μ from such a test is a reliable approach. Use the result from Problem 2.91.

2.98. Consider a spring-mass system with Coulomb damping and direct-force excitation. For $f = 25$ N, $\mu = .4$, and $m = 6$ kg, the system remains in static equilibrium. What should μ be reduced to if it is necessary to permit oscillations in the system? Use the result from Problem 2.91.

2.99. Consider a spring-mass system under direct force excitation with Coulomb damping. As in Problem 2.97, you've run two tests and recorded the oscillation magnitudes. For the first run you obtain a magnitude of 41.5 mm and for the second you obtain 42 mm. Use this information to calculate μ and comment on whether, for these parameters, this is a reliable means of determining μ. $k = 1000$ N/m, $m = 2$ kg, $\omega = 23$ rad/s, $f = 5.1$ N.

Section 2.12

2.100. Consider the accelerometer described in Example 2.19. Assume that we've included a damper in the accelerometer, with $\zeta = .2$. With all other parameters the same, what will the percentage error be for the accelerometer's prediction of the magnitude of the acceleration? What phase error will this damping induce?

2.101. Consider the accelerometer described in Example 2.19. Assume that a viscous damper ($\zeta = .15$) is included in parallel with the spring. At what frequency will the phase difference between the actual acceleration and the accelerometer's measurement differ by 15 degrees?

2.102. If an accelerometer is being approximated as an undamped spring-mass system, what percentage of ω_n is the maximum frequency for which the sensor error is less than or equal to .5 percent?

2.103. Consider a seismometer for which $m = .25$ kg and $k = 250$ N/m. What is the lowest frequency of vibration at which the seismometer can be used if we require accuracy of at least 15%? Neglect any damping for this problem.

2.104. Consider Problem 2.103. If the mass of the seismometer is doubled, how much lower will the lowest allowable frequency be? Again neglect damping and allow a maximum of 15% error.

3

Nonsinusoidal
Excitations
and Impact

HEY ANDY!

3.1 INTRODUCTION

This chapter addresses the problem of how to find a system's response when the input excitation is not a simple sinusoid. If you think about it, it seems pretty obvious that almost no real system will have a purely sinusoidal input. Many, many of the inputs will be periodic; i.e., the input will be made up of an endlessly repeating waveform. But the chances that the waveform is a pure sine wave are pretty low. For instance, the repeated firings inside the cylinder of an internal combustion engine will produce a periodic, but not sinusoidal, loading on the piston. Thus we're motivated to figure out how we can determine a system's steady state response for these cases.

Furthermore, the input to our system might not even be periodic. Oftentimes the loading will be very transient in nature. An earthquake is a good example of a transient loading. The ground will shake for a few seconds and then stop. Our problem then consists of determining a structure's response to a loading that exists only for a finite amount of time. If the earthquake lasts for a long time, then we might actually view the problem in a steady state way. But if the earthquake is of short duration, then a transient viewpoint is more appropriate. A similar problem will occur when an airplane encounters a momentary updraft. The updraft will load the wings for only a short time, and the wings will then bend (and the airplane will displace), in response to the input.

The loading might even be *very* transient. This would be the case for impact, such as the one shown in the chapter's opening cartoon. Impacts and explosions will create high levels of force that last for only a very short time. Dropping an object onto the ground is one way to create a severe impact, and we'll definitely want to know how to handle such cases if we want our designs to withstand normal usage. In industry, this type of analysis affects packaging and shipping. The manufacturer wants the product to withstand the normal operating environment and also to withstand the loads that may occur prior to installation. Interestingly, some of the most severe loadings applied to automobiles happen, not during normal driving but during shipment to dealers. Automobile manufacturers, who spend a good deal of effort to determine the nonsinusoidal loads vehicles will experience during shipment, work hard to ensure that they survive.

The final type of signal we'll consider, the random signal, falls between the first two types. Periodic signals have a repeating pattern that doesn't die down. Transient signals have no periodicity and a finite lifetime. Random signals are not periodic but don't die down. Random excitations are very common in real life. The force of wind on a building will be random. There's always some wind, but the direction and force vary randomly. The vibration of a floor in a manufacturing facility will be random: the inputs due to people walking about, parts being dropped or moved occasionally, doors being open and closed, etc. If an earthquake goes on for an extended time, its motions can be viewed as random.

We'll start the chapter by looking at the most immediately accessible extension to what we already know, the response of a vibratory system to a periodic, but not sinusoidal, input. Following this, we'll consider what needs to be done to deal with transient inputs. Finally, we'll look at the problem of random loading.

3.2 FOURIER SERIES ANALYSIS

This section will cover how to analyze systems for which the excitation is periodic but not made up of a single sinusoid. We've mentioned that superposition lets us easily determine the output of a system to an input $y_1 + y_2$ if we already know the corresponding outputs x_1 and x_2 for the individual inputs (it's simply equal to $x_1 + x_2$). Just to clarify this idea a bit more, let's work out a particular example. Say we want to know the response of a seismically excited system for the input

$$y(t) = \frac{8}{\pi^2} \sin(\omega t) - \frac{32}{36\pi^2} \sin(3\omega t) \tag{3.2.1}$$

A plot of $y(t)$ versus time is shown in Figure 3.1. Note that the addition of the $\sin(3\omega t)$ term has served to sharpen up the base sinusoid, making it more triangular in shape. This result isn't a coincidence, and it's due to the particular choice of amplitudes used. We'll look at this more in a few pages.

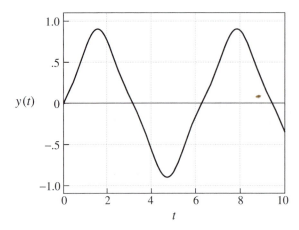

Figure 3.1   Sum of two sinusoids

The governing equation of motion (2.3.3) is repeated here for convenience:

$$\ddot{x} + \omega_n^2 x = \omega_n^2 y \tag{3.2.2}$$

Since the input consists of two, linearly independent excitations, and the dynamical system has no damping, we can assume a response

$$x(t) = b_1 \sin(\omega t) + b_2 \sin(3\omega t) \tag{3.2.3}$$

which, when differentiated twice and substituted into (3.2.2), leads to

$$b_1(\omega_n^2 - \omega^2)\sin(\omega t) + b_2(\omega_n^2 - 9\omega^2)\sin(3\omega t) = \frac{8\omega_n^2}{\pi^2}\sin(\omega t) - \frac{32\omega_n^2}{36\pi^2}\sin(3\omega t) \tag{3.2.4}$$

Since $\sin(3\omega t)$ and $\sin(\omega t)$ are linearly independent, (3.2.4) can be satisfied only if the coefficients of each sinusoid on the right and left sides of the equation are equal to each other, or

$$b_1(\omega_n^2 - \omega^2) = \frac{8\omega_n^2}{\pi^2} \tag{3.2.5}$$

and

$$b_2(\omega_n^2 - 9\omega^2) = -\frac{32\omega_n^2}{36\pi^2} \tag{3.2.6}$$

Thus we can easily solve for b_1 and b_2:

$$b_1 = \frac{8}{\pi^2}\frac{\omega_n^2}{\omega_n^2 - \omega^2} \tag{3.2.7}$$

$$b_2 = -\frac{32}{36\pi^2}\frac{\omega_n^2}{\omega_n^2 - 9\omega^2} \tag{3.2.8}$$

Therefore

$$x(t) = \frac{8}{\pi^2} \frac{\omega_n^2}{\omega_n^2 - \omega^2} \sin(\omega t) - \frac{32}{36\pi^2} \frac{\omega_n^2}{\omega_n^2 - 9\omega^2} \sin(3\omega t) \tag{3.2.9}$$

As advertised, this is exactly the same result as if we had solved

$$\ddot{x} + \omega_n^2 x = \frac{8\omega_n^2}{\pi^2} \sin(\omega t) \tag{3.2.10}$$

to obtain

$$x(t) = \frac{8}{\pi^2} \frac{\omega_n^2}{\omega_n^2 - \omega^2} \sin(\omega t) \tag{3.2.11}$$

and had then solved

$$\ddot{x} + \omega_n^2 x = \frac{32}{36\pi^2} \sin(3\omega t) \tag{3.2.12}$$

to obtain

$$x(t) = -\frac{32}{36\pi^2} \frac{\omega_n^2}{\omega_n^2 - 9\omega^2} \sin(3\omega t) \tag{3.2.13}$$

and added the two solutions together. Note that the solutions have been written in the form

$$\text{(amplitude of forcing)(transfer function)(time variation of forcing)} \tag{3.2.14}$$

This is to make it doubly clear that the response of a linear system is just the transfer function multiplied by the excitation.

Now we can begin to see why superposition is such a powerful concept. It implies that we can easily find the response of dynamical systems as long as we can consider the inputs to be made up of individual sinusoids. Thus the obvious question is: when can we view an input to a system as being made up of individual sinusoidal components? Clearly $y(t) = e^{-2t}$ isn't a likely candidate, since it is patently nonsinusoidal. But what about a triangular input, like that illustrated in Figure 3.2? Is this made up of sinusoids? We might be tempted to say yes, since this figure looks a good bit like Figure 3.1, which we've just seen is composed of $\sin(\omega t)$ and $\sin(3\omega t)$. In fact, Jean Baptiste Joseph Fourier, in the early 1800s, showed that one can approximate any well-behaved, periodic signal as the infinite sum of a series of pure sinusoids. Note the *periodic*. At this point we will consider only inputs that are periodic; i.e., if $f(t)$ is the function under consideration, then $f(t + T) = f(t)$, where T is the period of the function. Later in the chapter we'll consider what can be done when the functions of interest are not periodic.

According to Fourier, we should be able to represent a periodic signal, $f(t)$, as

$$f(t) = \sum_{n=0}^{\infty} a_n \cos(n\omega_0 t) + \sum_{n=1}^{\infty} b_n \sin(n\omega_0 t) \tag{3.2.15}$$

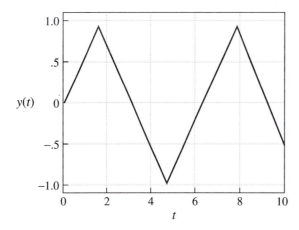

Figure 3.2 Triangle wave

The ω_0 component is the lowest frequency (longest wavelength) present in the signal $f(t)$ and is related to the signal's period T by

$$\omega_0 = \frac{2\pi}{T} \tag{3.2.16}$$

You'll note that the summations in the preceding equation go to infinity. This means that if we *really* wanted to precisely model an arbitrary periodic signal, we might need an infinite number of sinusoids. Luckily, we will never need an infinite number of terms for our analyses, for reasons that will become clear momentarily.

Assuming that we can indeed approximate $f(t)$ by a series of sinusoids, the problem still remains of how to determine a_n and b_n, the coefficients that multiply the sinuosoids. To make the general procedure clear, let's first work out a specific example. The function that we are going to analyze is

$$f_1(t) = 2\sin(\omega_0 t) - 3\cos(2\omega_0 t) \tag{3.2.17}$$

Thus we have

$$\sum_{n=0}^{\infty} a_n \cos(n\omega_0 t) + \sum_{n=1}^{\infty} b_n \sin(n\omega_0 t) = f_1(t) = 2\sin(\omega_0 t) - 3\cos(2\omega_0 t) \tag{3.2.18}$$

Of course, it's obvious that we can just read off the correct answers ($a_2 = -3$, $b_1 = 2$, and all other a_n and $b_n = 0$), but for the purpose of this discussion we'll assume that we don't know the answers in this way, that we'll have to calculate them from $f(t)$. This procedure is what we'd actually have to do in a real Fourier series analysis, one in which we didn't know the sinusoidal contributions beforehand.

Here's how we'll do it. First, multiply both sides of (3.2.18) by $\sin(m\omega_0 t)$. This is the sinusoid associated with b_m. For our particular example we'll expect answers only for $m = 1$, since that's the only sine component in $f(t)$. This gives us

$$\sum_{n=0}^{\infty} a_n \cos(n\omega_0 t) \sin(m\omega_0 t) + \sum_{n=1}^{\infty} b_n \sin(n\omega_0 t) \sin(m\omega_0 t)$$

$$= 2 \sin(\omega_0 t) \sin(m\omega_0 t) - 3 \cos(2\omega_0 t) \sin(m\omega_0 t) \tag{3.2.19}$$

We can simplify matters a bit by reexpressing these products of sines and cosines. Recall that

$$\sin(m\omega_0 t) = \frac{1}{2i}(e^{im\omega_0 t} - e^{-im\omega_0 t}) \tag{3.2.20}$$

and

$$\cos(m\omega_0 t) = \frac{1}{2}(e^{im\omega_0 t} + e^{-im\omega_0 t}) \tag{3.2.21}$$

Using (3.2.20) and (3.2.21), we can expand (3.2.19) out into its exponential parts and then combine these exponentials back into sines and cosines:

$$\sum_{n=0}^{\infty} a_n \left(\frac{1}{2}\right) \left[\sin\big((n+m)\omega_0 t\big) - \sin\big((n-m)\omega_0 t\big)\right]$$

$$+ \sum_{n=1}^{\infty} b_n \left(\frac{1}{2}\right) \left[-\cos\big((n+m)\omega_0 t\big) + \cos\big((n-m)\omega_0 t\big)\right]$$

$$= 2 \left(\frac{1}{2}\right) \left[-\cos\big((1+m)\omega_0 t\big) + \cos\big((1-m)\omega_0 t\big)\right]$$

$$- 3 \left(\frac{1}{2}\right) \left[\sin\big((2+m)\omega_0 t\big) - \sin\big((2-m)\omega_0 t\big)\right] \tag{3.2.22}$$

Next we're going to exploit a property called *orthogonality*, something we haven't talked about yet, but it will be easier to understand once you see it at work. To make use of orthogonality, we will integrate both sides of (3.2.22) from $t = 0$ to $t = T$, i.e. over one period. So we'll have

$$\sum_{n=0}^{\infty} a_n \left(\frac{1}{2}\right) \left[\int_0^{\frac{2\pi}{\omega_0}} \sin\big((n+m)\omega_0 t\big) \, dt - \int_0^{\frac{2\pi}{\omega_0}} \sin\big((n-m)\omega_0 t\big) dt\right]$$

$$+ \sum_{n=1}^{\infty} b_n \left(\frac{1}{2}\right) \left[-\int_0^{\frac{2\pi}{\omega_0}} \cos\big((n+m)\omega_0 t\big) \, dt + \int_0^{\frac{2\pi}{\omega_0}} \cos\big((n-m)\omega_0 t\big) dt\right]$$

$$= 2 \left(\frac{1}{2}\right) \left[-\int_0^{\frac{2\pi}{\omega_0}} \cos\big((1+m)\omega_0 t\big) \, dt + \int_0^{\frac{2\pi}{\omega_0}} \cos\big((1-m)\omega_0 t\big) \, dt\right]$$

$$- 3 \left(\frac{1}{2}\right) \left[\int_0^{\frac{2\pi}{\omega_0}} \sin\big((2+m)\omega_0 t\big) \, dt + \int_0^{\frac{2\pi}{\omega_0}} \sin\big((2-m)\omega_0 t\big) \, dt\right] \tag{3.2.23}$$

Taken all together, this looks *really* messy. However, if we look at each term individually we notice a remarkable thing. It seems that almost every term is equal to zero since, for each term, all we have is an integration over one or more periods of a pure sinusoid. And we know that integrating a cosine or sine over one or more periods yields zero as a result. For instance, if $n = 1$ and $m = 5$, then the first integral of the equation is

$$\int_0^{\frac{2\pi}{\omega_0}} \sin(6\omega_0 t)\, dt = -\frac{\cos(6\omega_0 t)\Big|_0^{\frac{2\pi}{\omega_0}}}{6\omega_0} = 0 \tag{3.2.24}$$

This will be true for any combination of n and m except those for which $n = m$. If $n = m$, then the $\sin((n - m)\omega_0 t)$ term equals zero. Thus it isn't a finite sinusoid anymore. However it's equal to zero and so gives us no trouble. However $\cos((n - m)\omega_0 t)$ goes to 1. Therefore the fourth integral on the left-hand side of (3.2.23) becomes

$$\int_0^{\frac{2\pi}{\omega_0}} dt = \frac{2\pi}{\omega_0} \tag{3.2.25}$$

Let's take a second to recall what we've found. We already know that only two harmonics show up in our signal, namely, $2\sin(\omega_0 t)$ and $-3\cos(2\omega_0 t)$. We formed a general Fourier series, set it equal to our signal, multiplied by $\sin(m\omega_0 t)$, and integrated over a cycle. We then found that almost every term went to zero. If m isn't equal to 1, then *all* the terms on the right-hand side of (3.2.23) are equal to zero. Only if $m = 1$ do we obtain a finite result.

If we choose $m = 1$, then (3.2.23) becomes

$$b_1 \frac{\pi}{\omega_0} = \frac{2\pi}{\omega_0} \tag{3.2.26}$$

or

$$b_1 = 2 \tag{3.2.27}$$

This is the correct result! Of course, we knew when we started that the coefficient of the Fourier series for $\sin(\omega_0 t)$ had to be 2 because that's how much of that particular harmonic was in the signal originally. The point is that we usually *don't* know ahead of time how much of a particular harmonic there will be, and the procedure we've just gone over will give us the answer. We could go ahead and find the cosine harmonics in the same way; simply multiply by $\cos(m\omega_0 t)$, integrate over a period, and vary m until we get a finite answer. The general procedure can be written as

$$a_n = \frac{\omega_0}{\pi} \int_0^{\frac{2\pi}{\omega_0}} f(t) \cos(n\omega_0 t)\, dt \tag{3.2.28}$$

and

$$b_n = \frac{\omega_0}{\pi} \int_0^{\frac{2\pi}{\omega_0}} f(t) \sin(n\omega_0 t)\, dt \tag{3.2.29}$$

The only other twist you'll need to remember is that the signal you're analyzing may not average out to zero. In such a case we say it has a bias or a constant offset. Since constants are even functions, we can view this as a cosine with zero frequency and so we'll have

$$a_0 = \frac{\omega_0}{2\pi} \int_0^{\frac{2\pi}{\omega_0}} f(t)dt \tag{3.2.30}$$

These coefficients are then used in

$$f(t) = \sum_{n=0}^{\infty} a_n \cos(n\omega_0 t) + \sum_{n=1}^{\infty} b_n \sin(n\omega_0 t) \tag{3.2.31}$$

to define our Fourier series.

You're probably thinking that this looks a bit tedious to do by hand. And it is. But it's a snap to do on a computer. You just need a short subroutine that takes in a signal and then decomposes it into its individual harmonics, using pretty much the same approach that we've just gone over.

Please don't dismiss all this lightly. The ideas behind the Fourier series will pop up again and again in the future. What we're doing is using (3.2.28)–(3.2.30) as *filters*. The indicated integration operations really do sift through $f(t)$ and filter out everything except the harmonic we're looking for. If that harmonic is in the signal, then these calculations will find it. This is a very powerful idea and finds uses not only in linear vibrations but in nonlinear vibrations as well.

By determining the Fourier series representation of the signal, we have done a very significant thing: we have moved from a time domain description into a frequency domain. We now can determine how much of a certain frequency is present in a given signal. This kind of a viewpoint is *very* prevalent in a wide range of applications. For instance, a great deal of control work is done almost entirely in the frequency domain. The control designer specifically looks for control strategies that will reduce the amount of motion in a system at specific frequencies and is willing to let the motion increase at other frequencies to achieve this. A frequency viewpoint also is useful in connection with the Tacoma-Narrows bridge disaster. Most of you saw film clips of the Tacoma-Narrows collapse in high school. For those who didn't, this spectacular bridge failure, several decades ago, occurred as a result of its own excessive motions. (Really excessive!) The Tacoma-Narrows bridge failed because the designers hadn't considered what sort of dynamic responses the bridge might have under different wind conditions. If they had known what to look for, they would have realized that a particular type of motion, occurring at a particular frequency, could be excited by certain wind conditions and that the wind would pump energy into the bridge's motion until the structure was destroyed (which is what occurred). The fundamental behavior is one that is most easily viewed from a frequency standpoint.

Finally, almost all studies of structural systems are concerned with the frequency content of the system. The reason for this is that periodic excitations close to the system's natural response frequencies can cause extremely large deflections of the system, deflections that can threaten the structural integrity of the system.

We'll examine spectral representations more fully later in the chapter. For now we will have to be content with knowing that any periodic forcing of our system can be broken down into several sinusoidal inputs and that we can therefore determine the response of our system by first determining the individual responses to all the separate harmonic components of the forcing and then adding them together, a process that is graphically shown in Figure 3.3. At the top we see a complex periodic input broken up into its constituent harmonic parts by means of a Fourier series analysis. Next, each harmonic is fed through the system being analyzed. As a result, the amplitude and phase of each component are altered. Finally, the resultant outputs are summed into the final response. Note that the input looks like a single harmonic with some superimposed higher harmonic "wiggles", while the output is much more irregular. This change occurred because resonance (due to the system's low damping) greatly amplified the harmonic near the system's natural frequency. Thus we see that a low magnitude component of the input signal can become significant or even dominant in the output.

It's worth reemphasizing that this Fourier series analysis technique is one of the powerful results that comes of having a linear system. Only when the governing differential equation is linear (all dependent variables and their derivatives appear only to the first power) can we break up a signal, calculate the individual responses, add them up, and end up with the actual solution. In a nonlinear system this would *not* work.

Example 3.1

Problem Determine the response of the system shown in Figure 3.4 when the base motion is in the form of the triangle wave illustrated.

Solution Although we could go ahead and just "plug and chug" using (3.2.28)–(3.2.31), we'll spend some time thinking about the problem and see how doing so can sometimes reduce our work level. The first thing to notice about the input forcing is that it is an odd function. Therefore we know immediately that all the a_n coefficients in (3.2.31) are going to equal zero. (If the function is odd, then it can't contain any even components.) Next, we can realize that only the odd harmonics ($n = 1, 3, 5, \ldots$) will show up because of the function's symmetry. Figure 3.5 shows why this is so. If an even harmonic ($n = 2, 4, \ldots$) was present, it would skew the function, making it look lopsided. Odd harmonics won't disturb the function's symmetry. Thus we need concern ourselves only with the odd harmonics. Furthermore, when we multiply the function by $\sin(n\omega_0 t)$ and integrate, we'll get an equal contribution from the section going from 0 to $\frac{\pi}{\omega_0}$ as we do from the section between $\frac{\pi}{\omega_0}$ to $\frac{2\pi}{\omega_0}$. Thus we can calculate the integral over just one interval and double the answer. And finally, we can notice that the same argument holds for our remaining interval. The integral from 0 to $\frac{\pi}{2\omega_0}$ is the same as that from $\frac{\pi}{2\omega_0}$ to $\frac{\pi}{\omega_0}$. Therefore we need only calculate the integral from 0 to $\frac{\pi}{2\omega_0}$ and multiply the answer by 4. Figure 3.6 shows this. The sectors labeled A all contain the same area.

Based on this analysis, our problem has simplified down to solving

$$b_n = 4\frac{\omega_0}{\pi} \int_0^{\frac{\pi}{2\omega_0}} \sin(n\omega_0 t)\left(\frac{2a\omega_0 t}{\pi}\right) dt, \quad n = 1, 3, 5, \ldots$$

This isn't actually all that bad to solve. If we integrate by parts, we'll have

$$b_n = \frac{8a\omega_0^2}{\pi^2}\left(-\frac{t\cos(n\omega_0 t)}{n\omega_0}\bigg|_0^{\frac{\pi}{2\omega_0}} + \int_0^{\frac{\pi}{2\omega_0}} \frac{\cos(n\omega_0 t)}{n\omega_0} dt\right), \quad n = 1, 3, 5, \ldots.$$

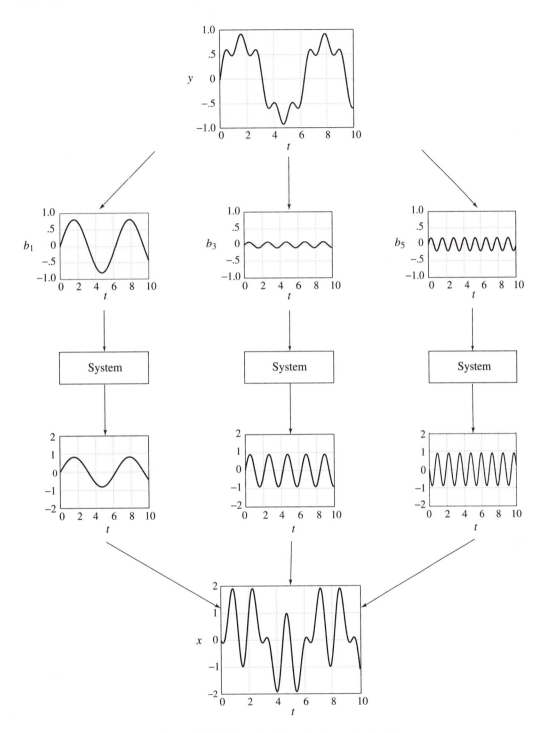

Figure 3.3   Schematic of how Fourier analysis works

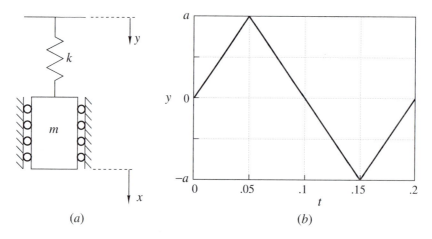

Figure 3.4   Seismically excited system, triangle wave input

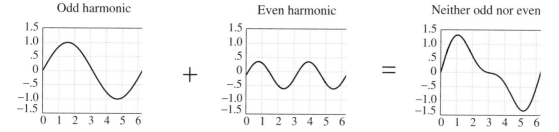

Figure 3.5   Adding odd- and even-order harmonics destroys symmetry

The first term is always zero for odd values of n, and the second term integrates out to

$$\frac{8a\omega_0^2}{\pi^2}\left(\left.\frac{\sin(n\omega_0 t)}{(n\omega_0)^2}\right|_0^{\frac{\pi}{2\omega_0}}\right), \quad n = 1, 3, 5, \ldots$$

The sine can only yield 1.0 or -1.0, depending on the value of n, and so the final solution becomes

$$b_{2n-1} = \frac{8a}{\pi^2}\frac{(-1)^{n+1}}{(2n-1)^2}, \quad n = 1, 2, \ldots$$

Now that we've got the b_n coefficients, we can put them together to obtain our Fourier series representation of the triangle wave:

$$y(t) = \frac{8a}{\pi^2}\sum_{n=1}^{\infty}\frac{(-1)^{n+1}}{(2n-1)^2}\sin\left((2n-1)\omega_0 t\right), \quad n = 1, 2, \ldots$$

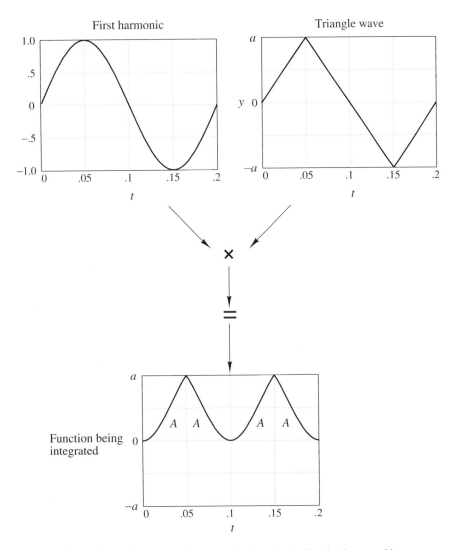

Figure 3.6   Symmetry in the integrals of a particular Fourier decomposition

The rest is reasonably straightforward. We know that the transfer function for the illustrated system is

$$g(\omega) = \frac{\omega_n^2}{\omega_n^2 - \omega^2}$$

Therefore we can immediately write the complete response by combining the two preceding equations to obtain

$$x(t) = \frac{8a}{\pi^2} \sum_{n=1}^{\infty} \left(\frac{(-1)^{n+1}}{(2n-1)^2} \right) \left(\frac{\omega_n^2}{\omega_n^2 - (2n-1)^2 \omega_0^2} \right) \sin\left((2n-1)\omega_0 t\right), \quad n = 1, 2, \ldots$$

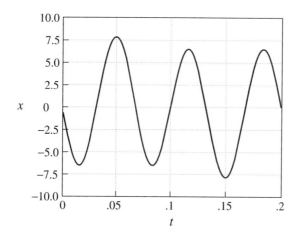

Figure 3.7 Steady state response of spring-mass system to triangle wave

Figure 3.7 shows what the response will look like if $k = 82,810$ N/m, $m = 10$ kg, and $\omega_0 = 31.4$ rad/s. For this k and m, the system's natural frequency is 91 rad/s, and thus the forcing's third harmonic ($3\omega_0 = 94.2$ rad/s) will be strongly excited owing to its proximity to the system's natural frequency. We'd therefore expect to see a large third harmonic response, which is exactly what's shown in Figure 3.7.

We've seen that the response harmonics in an input can occur near a system's natural frequency, causing a strong amplification of that harmonic. To make matters more complicated, MDOF systems and continuous systems have more than one natural frequency, any of which could potentially be excited by a harmonic of the input. Thus structural analysts must be aware of all the harmonics running around the system and how they might interact with the system's natural frequencies. If an excessive response due to resonance is found, then either the input or the system itself must be modified to move the excitation harmonic away from the natural frequency that's causing the problem.

3.3 FORCED RESPONSE VIA THE CONVOLUTION INTEGRAL

We've been concentrating on periodic excitations up until now because these are among the most common kinds of system input you'll be dealing with. There are, however, many situations in which the excitation isn't periodic and thus the approaches we've looked at aren't applicable. You know that the general solution to a forced differential equation involves both a transient and a steady state solution. So far, we've been consistently neglecting the transient motions, claiming that they'll die out as a result of dissipation, leaving us with the purely steady state response. Thus we were able to assume the existence of a steady state response (having an unknown amplitude and phase) and directly solve for this response. This section takes a different point of view. We're not going to presuppose a steady state response that will qualitatively match the excitation. Instead, we'll consider what the precise response is for *any* input and then determine the solution.

Just so you know where we're aiming, we're first going to find out how our systems respond to an impulsive force. Roughly speaking, an impulsive force is a force that occurs over a very short span of time, like that due to a quick tap on the mass of the system. Once we've got this response, we're going to claim that *any* forcing we might run into can be approximated as a string of impulses. Thus we'll just figure out the response due to all the individual impulses and sum them together. It's similar in a way to how we used a Fourier series approach to calculate the response of the system to a periodic but nonsinusoidal input. We figured out what particular sinusoids made up the periodic signal, determined the system's response to each individually, and then summed up all the responses.

Our first question should be, What precisely is meant by an impulse? Simply (and confusingly) stated, an impulse is an input that takes place over zero time and has infinite magnitude. Clearly, such an entity can't really exist. Infinite magnitudes over zero time intervals aren't the sort of thing you can produce in the real world. This is an important point—impulses are mathematical abstractions. All we can do in the real world is approximate them. There are many ways of approximating them, and the one we'll use is probably the most easily grasped. We'll start by looking at a forcing function that's equal to $\frac{1}{\varepsilon}$ and lasts only from $t = 0$ to $t = \varepsilon$ (Figure 3.8):

$$f_\varepsilon = \frac{1}{\varepsilon}, \quad 0 < t \leq \varepsilon$$

$$f_\varepsilon = 0, \quad t > \varepsilon \tag{3.3.1}$$

ε is small to start, and we'll make it smaller as we go. You can see that as ε gets smaller, the magnitude of f_ε grows and the length of time it takes on nonzero values shrinks. Thus, as ε approaches zero, the magnitude goes to infinity and the input takes place over zero time. We call such a function a Dirac delta function, $\delta(t)$, and we represent it graphically by a vertical arrow, as shown in Figure 3.9. Although we can't actually attain such an input in real life, we can come close. The easiest way to approximate such an input is to take a hammer and sharply strike the mass in a system. The hammer will hit the mass and almost immediately bounce back, imparting a large force to the mass which lasts only for a short time. As we'll see later, this approach is a very common tool in vibrational analysis.

You'll note that the area under the f_ε function is always equal to 1.0, regardless of how small ε becomes. Do we have any way of knowing what such a forcing will do when applied to a dynamical

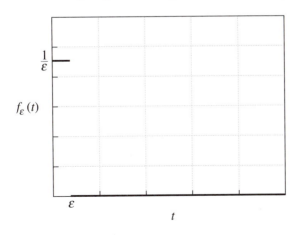

Figure 3.8   First approximation to an impulse function

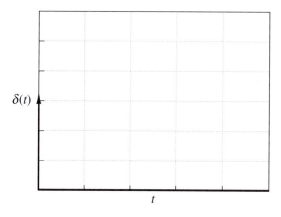

Figure 3.9   Dirac delta function

system? Actually, you've probably already learned the answer in a dynamics course. Any course in two-dimensional dynamics will cover the topic of momentum for a system and how it changes. What's taught is that the change in momentum of a mass m is equal to the *applied linear impulse*, where the applied linear impulse is defined as $\int f(t)dt$. Since this is simply the area under the f_ε curve, we see that the linear momentum must be altered as a result of the force. Thus, applying the force shown in Figure 3.8 will mean that

$$m(\dot{x}(\varepsilon) - \dot{x}(0)) = \int_0^\varepsilon \frac{1}{\varepsilon}dt = 1 \tag{3.3.2}$$

If the initial velocity is zero, this implies that the velocity at $t = \varepsilon$ is equal to

$$\dot{x}(\varepsilon) = \frac{1}{m} \tag{3.3.3}$$

As we take ε smaller and smaller (going to a limit of zero), we approach a simple change of initial conditions. As ε approaches zero, $\dot{x}(\varepsilon)$ approaches $\dot{x}(0)$. What about the displacement x, though? Is it possible that this might change also? To answer this, let's assume that it does change in response to the impulse and goes to a value x_0. We know that the velocity at time t is found from $\frac{x(t+\varepsilon)-x(t)}{\varepsilon}$. Thus, in the limit, as ε goes to zero, we'd have

$$\dot{x}(0) = \lim_{\varepsilon \to 0} \frac{x_0}{\varepsilon} \to \infty \tag{3.3.4}$$

Since we know the velocity can't go to infinity, we conclude that the displacement must actually still be zero after the impulse has been applied.

You can see the same results in a slightly different way by explicitly solving

$$m\ddot{x} = f_\varepsilon(t) \tag{3.3.5}$$

where $f_\varepsilon(t)$ is shown in Figure 3.8. Assume initial conditions $x(0) = 0$, $\dot{x}(0) = 0$ and let $m = 1$. We can integrate both sides to obtain

$$\dot{x}(t) = \frac{1}{\varepsilon}t \tag{3.3.6}$$

and

$$x(t) = \frac{1}{\varepsilon}\frac{t^2}{2} \tag{3.3.7}$$

for t between 0 and ε. (Note that the initial conditions were used in the solution to eliminate any constants of integration.) Let's assume that ε is small, say .01. Then $\dot{x}(.01) = 1.0$ and $x(.01) = .005$. It's pretty clear that the velocity is finite and the change in position is essentially zero. As ε goes to zero, this effect becomes even more pronounced, leading finally to a finite change in initial velocity and no change in displacement as ε hits zero.

What about using a Dirac delta function as the input for a more complicated system? To see, let's determine what will happen to a general spring-mass-damper system when an impulse is applied. In this case the equation of motion would be given by

$$m\ddot{x} + c\dot{x} + kx = \delta(t) \tag{3.3.8}$$

To aid in our physical understanding, let's go ahead and use the f_ε approximation for the delta function. Integrating once, from $t = 0$ to $t = \varepsilon$, gives us

$$m\left(\dot{x}(\varepsilon) - \dot{x}(0)\right) + c\left(x(\varepsilon) - x(0)\right) + k\left(\int_0^\varepsilon x\,dt\right) = 1 \tag{3.3.9}$$

Can $\int_0^\varepsilon x\,dt$ be finite (and thus part of the left-hand-side terms that add to 1.0)? If so, then $x(\varepsilon)$ must go to infinity like $\frac{1}{\varepsilon}$, since otherwise the integral would vanish. Since we already know the mass isn't going to go off to infinity in zero time, we know that the integral must therefore equal zero. Next we can ask if the displacement can be finite, letting the $c\left(x(\varepsilon) - x(0)\right)$ term contribute toward making up the 1.0 on the right-hand side. We've already seen that this would imply an infinite velocity and is therefore not possible. Thus we're left with the same result as before, namely,

$$m\left(\dot{x}(\varepsilon) - \dot{x}(0)\right) = 1 \tag{3.3.10}$$

Therefore the velocity undergoes a step change. In the limit of $\varepsilon \to 0$, this again produces the result that the effect of an applied impulsive force is to alter the initial velocity and leave the displacement undisturbed.

You can now see why using a hammer to apply an approximate Dirac delta function of force to a test object is a nice idea. If a Dirac delta is applied, the only result is to change the initial conditions of the system. Thus we'll end up with an *unforced* problem, one for which the displacement is zero and the velocity is finite. We can then observe the free response and, from it, deduce the damping, natural frequency, etc. Many techniques of modal analysis rely on just such an approach.

The other way in which the Dirac delta function is useful is in providing us with a method for determining the forced response of our system to *any* input. Figure 3.10 illustrates the first step in this approach, and Figure 3.10a shows the original, continuous, forcing function. The next step is to discretize the time line into segments Δ seconds long, shown in Figure 3.10b. If Δ is small,

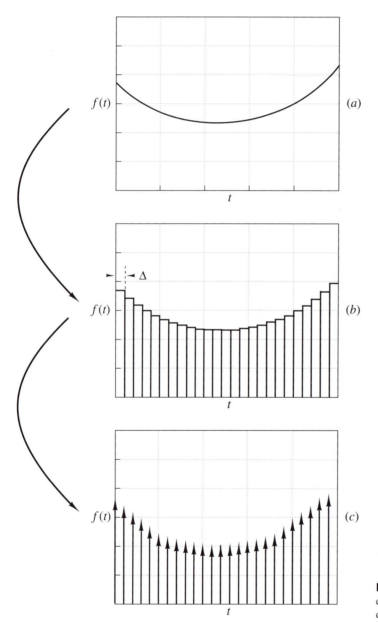

Figure 3.10   Approximation of a continuous function $f(t)$ by a train of delta functions

then the area within each four-sided box is given by the value of the function times the time interval; i.e., the first box has area $f(0)\Delta$, the second has area $f(\Delta)\Delta$, etc. The final step is to replace the function by a train of delta functions, as shown in Figure 3.10c. Each unit delta function is multiplied by the appropriate factor to make it match the actual function values. Thus the first delta function is multiplied by $f(0)\Delta$, the second by $f(\Delta)\Delta$, etc.

The idea here is to hit our system with this train of delta functions, rather than using the original, continuous, function. The reason for this is that we've already figured out how the system will respond to a delta function: it will have a step change in velocity and then undergo free vibration. So what we'll have is a succession of free vibration responses. The system will undergo free vibrations for Δ seconds, have a change in velocity that is due to the next impulse, and then freely vibrate for another Δ seconds. We can therefore sum up all these free vibrational responses to find out how the overall system responds.

Before doing this, let's pin down exactly the free vibration response of a spring-mass-damper system. Say our system is acted on by a Dirac delta function, with magnitude $\bar{f}$. Our equation of motion is

$$m\ddot{x} + c\dot{x} + kx = \bar{f}\delta(t) \tag{3.3.11}$$

We've seen that this is equivalent to

$$m\ddot{x} + c\dot{x} + kx = 0 \tag{3.3.12}$$

with initial conditions $x(0) = 0$ and $\dot{x}(0) = \frac{\bar{f}}{m}$.

The general solution to an unforced spring-mass-damper system is given by

$$x(t) = e^{-\zeta\omega t}\left(a_1 \cos(\omega_d t) + a_2 \sin(\omega_d t)\right) \tag{3.3.13}$$

where $\omega_n^2 = \frac{k}{m}$ and $\omega_d = \omega_n\sqrt{1 - \zeta^2}$. Applying the $x(0) = 0$ initial condition sets a_1 to zero and applying the $\dot{x}(0) = \frac{1}{m}$ initial condition gives us $a_2 = \frac{1}{m\omega_d}$. Thus the total solution is given by

$$x(t) = \frac{1}{m\omega_d}e^{-\zeta\omega t}\sin(\omega_d t) \tag{3.3.14}$$

We'll use the symbol $h(t)$ to represent this response and remember that $h(t)$ is the *impulse response* of any spring-mass-damper system:

$$h(t) \equiv \frac{e^{-\zeta\omega_n t}}{m\omega_d}\sin(\omega_d t) \tag{3.3.15}$$

Now let's look at our original problem again. We've already approximated our input function by a string of Dirac delta functions. Therefore we know that the response from $t = 0$ to $t = \Delta$ is given by

$$x(t) = \frac{f(0)\Delta e^{-\zeta\omega_n t}}{m\omega_d}\sin(\omega_d t) = f(0)h(t)\Delta \tag{3.3.16}$$

From $t = \Delta$ to $t = 2\Delta$, the response is given by

$$x(t) = f(0)\Delta h(t) + \frac{f(\Delta)\Delta e^{-\zeta \omega_n (t-\Delta)}}{m\omega_d} \sin(\omega_d (t - \Delta))$$

$$= f(0)h(t)\Delta + f(\Delta)h(t - \Delta)\Delta \qquad (3.3.17)$$

You can see the pattern shaping up here. We're going to get a longer and longer string of responses. Jumping to the response at the nth interval, we've got

$$x(t) = \sum_{i=0}^{n-1} f(i\Delta)h(t - i\Delta)\Delta \qquad (3.3.18)$$

which holds from $t = (n-1)\Delta$ to $t = n\Delta$. Look at what happens if we now let the interval Δ get smaller and smaller. The summation becomes an integral, the Δ becomes a differential, and we're left with

$$x(t) = \int_0^t f(\tau)h(t - \tau)d\tau \qquad (3.3.19)$$

For our problem this becomes

$$x(t) = \int_0^t f(\tau)\frac{e^{-\zeta \omega_n (t-\tau)}}{m\omega_d} \sin(\omega_d (t - \tau))d\tau \qquad (3.3.20)$$

Equation (3.3.19) is called a convolution integral and, not surprisingly, this approach is called the convolution method. This may look pretty heavy-duty, but if you remember where it came from, it should seem a bit more manageable. All we're doing is trying to determine the *impulse* response at any particular point in time, and then letting this response remain. It's even easier to see what's going on if we use a particular time t_1 rather than t. In this case the response at t_1 is given by

$$x(t_1) = \int_0^{t_1} f(\tau)h(t_1 - \tau)d\tau \qquad (3.3.21)$$

If we think about it in this way we see that the response at t_1 depends upon $f(\tau)$, with $f(\tau)$ going from $f(0)$ to $f(t_1)$ and upon $h(t_1 - \tau)$, with $h(t_1 - \tau)$ going from $h(t_1)$ to $h(0)$. For instance, take the first part of the integral, near $\tau = 0$. This is approximately equal to $f(0)h(t_1)\Delta$, where we're just taking a Δ-wide slice of time. This is just the response due to an initial impulse. Since we're looking at the response at time t_1, we let the system continue to vibrate for that long (i.e., we use $h(t_1)$).

Now look at the end of the integral. In this case τ is almost equal to t_1, say $t_1 - \Delta$, and the chunk of the integral near the end is equal to $f(t_1 - \Delta)h(\Delta)\Delta$. This indicates that some of the response is due to the latest excitation. This excitation occurred at $t_1 - \Delta$ and is given by $f(t_1 - \Delta)\Delta$. The free vibration lasts for only Δ seconds in this case; thus we use $h(\Delta)$.

If you look at the discrete way we first approached the integral, and think about what we've just gone over, you'll see the same result is indicated by both approaches. Taken in total, this integral operation is saying to take the response due to the early forcing and remember it. Then add to this the response due to the force that occurred a little bit later. Then add the part due to even later forcings. Continue doing this until you're at the time you care about.

Although (3.3.19) is totally valid, you may find it easier to express it in a slightly different way. To derive the new form, we'll define a new variable γ, where we'll say that $\gamma = t - \tau$. Remember that t is fixed in the convolution. Therefore if we take a differential of γ we'll get

$$d\gamma = -d\tau \tag{3.3.22}$$

Eliminating τ in favor of γ in (3.3.19) gives us

$$x(t) = -\int_t^0 f(t - \gamma)h(\gamma)d\gamma \tag{3.3.23}$$

We found the new limits of integration by substituting what τ is equal to at these limits (0 and t) into our expression for γ. We can flip the limits of integration (which simply generates a -1 factor), cancel this -1 with the existing -1 in front of the integral, and get

$$x(t) = \int_0^t f(t - \gamma)h(\gamma)d\gamma \tag{3.3.24}$$

Since γ is just a dummy variable (it doesn't show up in the final result), we can rename it τ so that the equation looks more like (3.3.19):

$$x(t) = \int_0^t f(t - \tau)h(\tau)d\tau \tag{3.3.25}$$

This is identical to (3.3.19) except that we're putting the time shift into the forcing function f, rather than into the impulse response h. You should use whichever form is more convenient for your particular problem.

What we've now derived is a general way of determining the response of a system to any input *if* you've already found out the response of the system to a delta input. Let's work out an example to pin it down a little more.

Example 3.2

> **Problem**   Consider the damped system and excitation shown in Figure 3.11. The excitation is a relatively simple one, namely, a step input. This is the sort of situation that would occur if you could have a mass suspended from a spring in the absence of gravity and, at $t = t_0$, suddenly switch gravity on. For clarity, we'll let $t_0 = 0$. Letting it be any other value doesn't affect the solution at all except to shift the time that the mass starts to move from $t = 0$ to $t = t_0$. Let's find the response to this input.
>
> **Solution**   To solve this problem by the method outlined in the preceding section, we'll need the response of the system to a unit delta function, i.e., the solution to
>
> $$m\ddot{x} + c\dot{x} + kx = \delta(t)$$

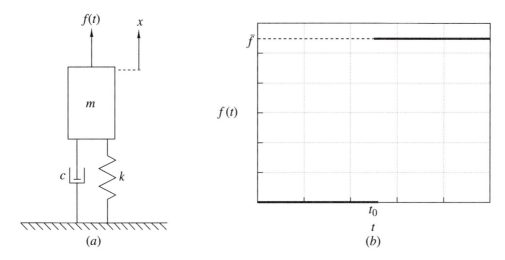

Figure 3.11   Damped system subjected to a step input force

We already know from (3.3.15) that the solution to this problem is given by

$$h(t) = \frac{e^{-\zeta\omega_n t}}{m\omega_d} \sin(\omega_d t)$$

Knowing $h(t)$, we can go ahead and use the convolution integral (3.3.25) to find the response to the illustrated step input

$$x(t) = \int_0^t h(\tau)f(t-\tau)d\tau = \frac{\bar{f}}{m\omega_d} \int_0^t e^{-\zeta\omega_n\tau} \sin(\omega_d\tau)\,d\tau \qquad (3.3.26)$$

At this point we could simply use a table of integrals to evaluate (3.3.26). What we'll do instead is show how to calculate the answer. That way you'll be able to carry on even if you don't have an integral table handy.

   If you try integration by parts, you'll still end up with an integral involving the product of a sine (or cosine) and an exponential. The trick to getting the answer is to integrate by parts *twice*. In this way we'll end up with an equation having the integral we want to evaluate on the left and also on the right. By combining the two, we'll be able to solve the problem. It goes like this. For the first integration by parts ($\int u\,dv = uv - \int v\,du$) we identify dv as $\sin(\omega_d\tau)d\tau$ and u as $e^{-\zeta\omega_n\tau}$. This gives us

$$\frac{\bar{f}}{m\omega_d} \int_0^t e^{-\zeta\omega_n\tau} \sin(\omega_d\tau)\,d\tau$$

$$= \frac{\bar{f}}{m\omega_d^2} \left(1 - e^{-\zeta\omega_n t} \cos(\omega_d t) - \zeta\omega_n \int_0^t e^{-\zeta\omega_n\tau} \cos(\omega_d\tau)d\tau \right)$$

When integrating the $e^{-\zeta \omega_n \tau} \cos(\omega_d \tau) d\tau$ term, we'll use $u = e^{-\zeta \omega_n \tau}$ and $dv = \cos(\omega_d \tau)$. This will give us

$$\frac{\bar{f}}{m\omega_d} \int_0^t e^{-\zeta \omega \tau} \sin\left(\omega_d(\tau)\right) d\tau$$

$$= \frac{\bar{f}}{m\omega_d^2} \left(1 - e^{-\zeta \omega_n t} \cos(\omega_d t)\right) - \frac{\zeta \omega_n \bar{f}}{m\omega_d^3} e^{-\zeta \omega_n t} \sin(\omega_d t)$$

$$- \frac{\bar{f}(\zeta \omega_n)^2}{m\omega_d^3} \int_0^t e^{-\zeta \omega_n \tau} \sin(\omega_d \tau) d\tau$$

You can see that we've got the same integrals on the left and on the right. Combining them on the left-hand side of the equation gives us

$$\frac{\bar{f}}{m\omega_d(1 - \zeta^2)} \int_0^t e^{-\zeta \omega_n \tau} \sin\left(\omega_d \tau\right) d\tau$$

$$= \frac{\bar{f}}{k(1 - \zeta^2)} \left(1 - e^{-\zeta \omega_n t} \cos(\omega_d t)\right) - \frac{\zeta \bar{f}}{k(1 - \zeta^2)^{\frac{3}{2}}} e^{-\zeta \omega_n t} \sin(\omega_d t)$$

Some of the ω_d's were reexpressed in terms of ω_n and ζ, and $m\omega_n^2$ was re-expressed as k in the preceding expression. Multiplying by $(1 - \zeta^2)$ gives us our final expression for $x(t)$

$$x(t) = \frac{\bar{f}}{k} \left(1 - e^{-\zeta \omega_n t} \cos(\omega_d t)\right) - \frac{\zeta \bar{f}}{k\sqrt{1 - \zeta^2}} e^{-\zeta \omega_n t} \sin(\omega_d t)$$

This solution is illustrated in Figure 3.12. Note that the response starts at zero, grows to a maximum, and then dies down as it oscillates about $\frac{\bar{f}}{k}$. Of course, this steady state response was predictable from the original differential equation by setting all the time-varying parts ($\ddot{x}$ and $\dot{x}$) to zero and solving the resulting equation. Varying the damping will serve to reduce or increase the rate of decay of the oscillations.

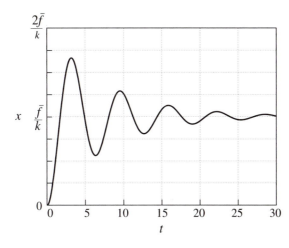

Figure 3.12 Displacement response for a damped system subjected to a step force input

Example 3.3

Problem Our example will be interesting for two reasons. The first is that we'll see what the response looks like for a finite ramp input force. All right, that really isn't all that interesting, is it? But the second reason is better. After we've found the response to a finite ramp, we'll ensure that the area within the ramp is always equal to 1.0 and we'll reduce the duration of the ramp to zero. This means that the ramp will look a lot like the Dirac delta function, with an infinite value, zero duration, and unit area. The interesting question will be whether the system responds to this input as it did to the force profile shown earlier in Figure 3.8.

The force profile is shown in Figure 3.13b. As you can see, the force ramps up linearly from zero to a maximum of $\bar{f}$ (at $t = t_0$). Our equation of motion is given by

$$m\ddot{x} + kx = f(t)$$

or

$$\ddot{x} + \omega_n^2 x = \frac{f(t)}{m}$$

We can find $h(t)$ either by solving the free vibration problem subjected to a unit velocity or, more simply, just using (3.3.15) and setting $\omega_d = \omega_n$ and $\zeta = 0$. This results in

$$h(t) = \frac{1}{m\omega_n} \sin(\omega_n t)$$

Using $h(t)$ in our convolution integral gives us

$$x(t) = \frac{1}{m\omega_n} \int_0^t f(t - \tau) \sin(\omega_n \tau) d\tau \tag{3.3.27}$$

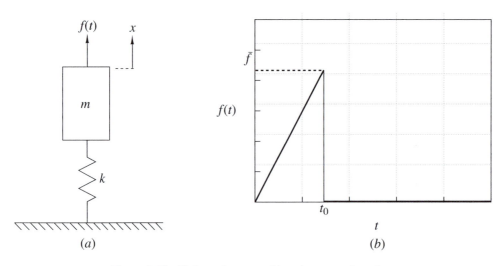

(a)   *(b)*

Figure 3.13 Undamped system subjected to a ramp input force

If t is less than t_0, then we can simply use this expression for $x(t)$. Since the forcing function is zero for $t > t_0$, we'd have to solve for $x(t_0)$ and $\dot{x}(t_0)$ and then solve for the system's free vibrational response for all $t > t_0$, using $x(t_0)$ and $\dot{x}(t_0)$ as initial conditions.

To solve (3.3.27), we'll need to reexpress $f(t)$ more explicitly. Since $f(t)$ is linearly increasing from 0 to $\bar{f}$, it's given by

$$f(t) = \frac{\bar{f}}{t_0}t, \qquad 0 < t \le t_0$$

Using this form for $f(t)$, we find that $x(t)$ becomes

$$x(t) = \frac{\bar{f}}{m\omega_n t_0} \int_o^t \sin(\omega_n \tau)(t - \tau)d\tau$$

Expanding this yields

$$x(t) = \frac{\bar{f}}{m\omega_n t_0} \left(t \int_o^t \sin(\omega_n \tau)d\tau - \int_o^t \tau \sin(\omega_n \tau)d\tau \right)$$

We were able to pull the t out of the first integral because it is considered to be a constant during the integration with respect to τ. Integrating these (using integration by parts for the second integral) yields

$$x(t) = \frac{\bar{f}}{m\omega_n^2 t_0} \left(t - \frac{\sin(\omega_n t)}{\omega_n} \right)$$

If we cared to, we could now use this formula to find $x(t_0)$ and $\dot{x}(t_0)$ and then apply these as initial conditions to determine the free vibrations of the subsequent motion. Instead, what we'll do is let $\bar{f} = \frac{2}{t_0}$. For this value of $\bar{f}$ you can see that the area under the $f(t)$ curve is equal to 1.0. The solution for $x(t)$ is now given by

$$x(t) = \frac{2}{m\omega_n^2 t_0^2} \left(t - \frac{\sin(\omega_n t)}{\omega_n} \right) \qquad (3.3.28)$$

The next step is to ask what $x(t_0)$ and $\dot{x}(t_0)$ are as t_0 goes to zero. For t_0 small, (3.3.28) becomes

$$x(t_0) \approx \frac{2}{m\omega_n^2 t_0} - \frac{2}{m\omega_n^3}\frac{\omega_n t_0}{t_0^2} = 0$$

Differentiating (3.3.28) gives us

$$\dot{x}(t) = \frac{2}{m\omega_n^2 t_0^2}(1 - \cos(\omega_n t))$$

Letting $t = t_0$ and expanding out $\cos(\omega_n t)$ to include the first two terms gives us

$$\dot{x}(t_0) = \frac{2}{m\omega_n^2 t_0^2} \left(1 - \left(1 - \frac{(\omega_n t_0)^2}{2} \right) \right)$$

which leads to

$$\dot{x}(t_0) = \frac{2}{m\omega_n^2 t_0^2}\frac{\omega_n^2 t_0^2}{2} = \frac{1}{m}$$

There we have it! We've found that as t_0 goes to zero, the application of the ramp input has served to alter the system initial conditions from $x(0) = 0$ and $\dot{x}(0) = 0$ to $x(0) = 0$ and $\dot{x}(0) = \frac{1}{m}$, exactly what happened when we applied the delta function that came from f_ε. Thus a delta function that's found from letting the duration of a step input go to zero, while holding the area fixed at 1.0 has the same effect as a delta function formed from a ramp input in which the duration of the ramp goes to zero and the integral is held at 1.0. As you now might guess, just about any function you can come up with that has a vanishingly small duration and an area equal to 1.0 will act in the same way. This fact is the reason for much of the success in impact testing in modal analyses. For these tests the structure under consideration must be struck with an instrumented hammer. The hope is that the force input of the hammer acts like a Dirac delta function. Sophisticated software can then capture the output of the sensors attached to the structure and determine the system properties. Obviously the hammer hit can't really input a Dirac delta function. Our results here show that this doesn't matter. As long as the hammer strike is quick, it is well approximated by a Dirac delta function.

Before leaving this section, we'll discuss a couple of additional properties of the Dirac delta function because they'll be useful to know later on. We've already seen that the Dirac delta function is defined as a function that's zero everywhere except at zero (where it's equal to infinity). The integral over the delta function is always equal to 1.0. Thus

$$\int_{-\infty}^{\infty} \delta(t)dt = 1 \tag{3.3.29}$$

We can be a little more general and say that the delta occurs not at $t = 0$ but at any time t_1. Such a function is shown in Figure 3.14. Since all we've done is displace the nonzero part of the delta function, the integral over t will still be equal to 1.0.

A helpful property associated with delta functions is that

$$\int_{-\infty}^{\infty} f(t)\delta(t - t_1)dt = f(t_1) \tag{3.3.30}$$

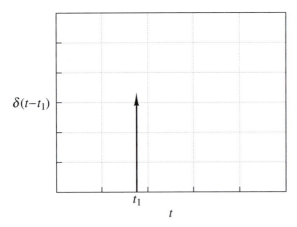

Figure 3.14   Delta function displaced in time

If we recall that the delta function can be viewed as the limiting case of the f_ε function shown in Figure 3.8 (and letting the nonzero part occur at $t = t_1$), this property is pretty easy to see. Since the delta function is zero almost everywhere, most of the integral in (3.3.30) is just equal to zero. The only nonzero contribution occurs right where the delta becomes nonzero. If ε is small, $f(t)$ is approximately equal to $f(t_1)$ over the entire interval. Thus the integral is approximated by

$$\int_{-\infty}^{\infty} f(t)\delta(t - t_1)dt \approx f(t_1)\frac{1}{\varepsilon}\varepsilon \tag{3.3.31}$$

In the limit of $\varepsilon \to 0$, this becomes an equality. Thus we see that the ε's cancel out, leaving us with

$$\int_{-\infty}^{\infty} f(t)\delta(t - t_1)dt = f(t_1) \tag{3.3.32}$$

Example 3.4

Problem   Express the following force input as a Fourier series expansion:

$$f(t) = \sum_{n=-\infty}^{\infty} \delta(t - n)$$

Solution

$$f(t) = \sum_{m=0}^{\infty} a_m \cos(m\omega_0 t) + \sum_{m=1}^{\infty} b_m \sin(m\omega_0 t)$$

$$\omega_0 = \frac{2\pi}{1} = 2\pi$$

The function in Figure 3.15 is even and thus $b_m = 0$. The a_i's are given by

$$a_0 = \int_0^1 f(t)dt, \quad a_m = 2\int_0^1 f(t)\cos(2\pi mt)dt$$

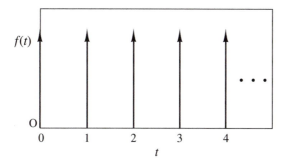

Figure 3.15   Function for Example 3.4

For simplicity, shift the integration interval to $-\frac{1}{2}$ to $\frac{1}{2}$. In this case we'll have

$$a_0 = \int_{-\frac{1}{2}}^{\frac{1}{2}} \delta(t)dt = 1$$

$$a_m = 2\int_{-\frac{1}{2}}^{\frac{1}{2}} \delta(t)\cos(2\pi mt)dt = 2\cos(0) = 2$$

$$f(t) = 1 + 2\sum_{m=1}^{\infty} \cos(2\pi mt)$$

3.4 SHOCK RESPONSE

A special class of transient loadings are those involving shock. As mentioned earlier, shock loadings can come about in a variety of ways. Explosions always generate shocks. When an object is suddenly brought to rest after falling, the result is a shock. When a car hits a pothole, a shock is produced that then affects the car and anything in it. Obviously, when one has a sharp loading of this sort, concern over an object's structural integrity becomes acute. The acceleration levels produced by shock loading are high and can easily lead to breakage within the system.

Some common shock profiles are shown in Figure 3.16a–c. In Figure 3.16a, the force input rises quickly to some maximum value and then stays constant at that value. This might be the case for the loading within a hydraulic system. Once the system is pressurized, the force on an interior surface will be higher than it was before pressurization. If the rise in pressure is rapid, the result is a shock load. In this case the shock is caused by the rapidly increase in force, not by the steady state force that exists after t_1.

Figure 3.16b shows a shock load that consists of a constant force with a short duration t_0. This sort of shock can occur as a result of a blast wave, for which the object in the path of the blast will feel a high level of force for a short time.

Finally, Figure 3.16c shows a common approximation to shock loadings, called a half-sine loading. Just as you'd expect from the name, the load has the profile of half a sine wave and then remains at zero for later times. If t_1 is small and the amplitude of the half-sine is large, the result is a shock load. Engineering experience and insight will often be needed to decide whether the loading shown in Figure 3.16b or c is the more appropriate for the actual shock being considered. If the time duration of the shock is short enough, it might be sufficient to simply view the input as a Dirac delta function.

Once we know the profile of the shock load, it's straightforward to apply the convolution approach to determine the response. Alternatively, considering the speed of modern computers, we might simply numerically integrate the equations of motion, using the shock profile as the input excitation. In either case, once we've determined the response, we can then characterize the effect of the shock. Commonly, we're concerned with the *peak* acceleration levels that the shock will induce

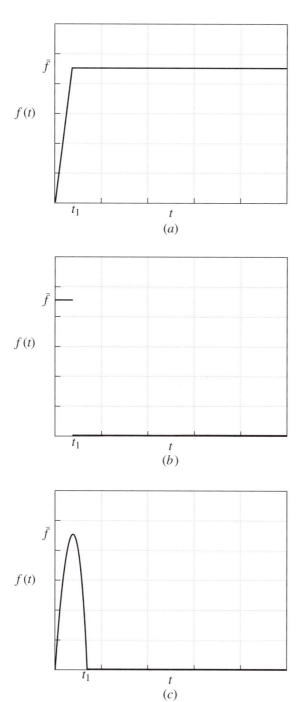

Figure 3.16 (*a*) Abrupt change in force level, (*b*) impulsive force, and (*c*) half-sine input force

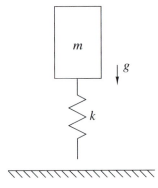

Figure 3.17   Spring-mass system subjected to shock due to impact

in the system. Thus we can plot peak acceleration versus some shock parameter (say shock duration) and get a feel for how the system will be affected by different shocks.

Let's consider the case for which a test object has been dropped and contacts the ground. We'll determine what the acceleration levels are for the mass and see how they vary as a function of contact velocity. Finally, we'll relate contact velocity to the drop height. This will allow us to set safe limits with respect to how high the structure should be carried during movement, since the height at which the object is dropped is directly related to the impact velocity. The system to be examined, shown in Figure 3.17, consists of a spring and a mass. When dropped, it will accelerate owing to gravity and eventually contact the floor with a velocity v. At the instant it strikes the floor, the spring will be uncompressed. We'll measure x from the inital position of the mass upon just contacting the floor. We'll also orient x downward so that the initial velocity upon contact is positive. The relevant equation of motion is given by

$$m\ddot{x} + kx = mg \tag{3.4.1}$$

with initial conditions $x(0) = 0$ and $\dot{x}(0) = v$.

Dividing by m gives us

$$\ddot{x} + \omega_n^2 x = g \tag{3.4.2}$$

The general solution for this problem is given by

$$x(t) = a_1 \cos(\omega_n t) + a_2 \sin(\omega_n t) + \frac{g}{\omega_n^2} \tag{3.4.3}$$

where the sine and cosine are the homogeneous solution and $\frac{g}{\omega_n^2}$ is the particular solution, i.e., the one due to the gravity loading. This problem therefore falls into the category of Figure 3.16a, one in which a constant loading is suddenly applied to the system.

Applying the initial condition $x(0) = 0$ tells us that $a_1 = -\frac{g}{\omega_n^2}$, and subsequent application of $\dot{x}(0) = v$ gives us $a_2 = \frac{v}{\omega_n}$. Thus our complete solution is given by

$$x(t) = \frac{g}{\omega_n^2}(1 - \cos(\omega_n t)) + \frac{v}{\omega_n}\sin(\omega_n t) \tag{3.4.4}$$

To find the accelerative response, we need to differentiate twice, yielding

$$\ddot{x}(t) = g\cos(\omega_n t) - v\omega_n \sin(\omega_n t) \tag{3.4.5}$$

We know from our work with harmonic vibrations that this response can be expressed as $a\cos(\omega_n t + \phi)$, where a is given by

$$a = \sqrt{g^2 + (v\omega_n)^2} \tag{3.4.6}$$

Therefore the maximum acceleration magnitude is given by a.

It's always nice to normalize our plots, and a convenient way to normalize the acceleration level is to divide by gravity. Thus our normalized maximum acceleration is given by $\frac{a}{g}$ or, expressed slightly differently,

$$\left|\frac{\ddot{x}}{g}\right|_{\max} = \sqrt{1 + \left(\frac{v\omega_n}{g}\right)^2} \tag{3.4.7}$$

An equation you've undoubtedly seen in your dynamics course is

$$\ddot{x}\,dx = \dot{x}\,d(\dot{x}) \tag{3.4.8}$$

a result that allows us to immediately determine the relationship between the drop height x_h and the impact velocity v. Using a constant acceleration (equal to g) to integrate this equation gives us

$$gx_h = \frac{v^2}{2} \tag{3.4.9}$$

Solving for v gives us

$$v = \sqrt{2gx_h} \tag{3.4.10}$$

Thus our normalized maximum acceleration versus drop height becomes

$$\left|\frac{\ddot{x}}{g}\right|_{\max} = \sqrt{1 + \frac{2x_h\omega_n^2}{g}} \tag{3.4.11}$$

Finally, look again at (3.4.2). As we've discussed, the steady state deflection of the mass due to gravity is found by setting the $\ddot{x}$ term to zero, giving us

$$\omega_n^2 x_{eq} = g \tag{3.4.12}$$

or

$$x_{eq} = \frac{g}{\omega_n^2} \tag{3.4.13}$$

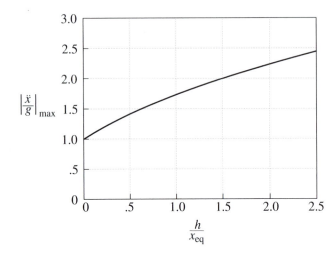

Figure 3.18   Normalized maximum acceleration versus normalized distance fallen

Interestingly, you'll see that this term pops up in (3.4.11). Substituting x_{eq} for $\frac{g}{\omega_n^2}$ in (3.4.11) gives us our final expression:

$$\left|\frac{\ddot{x}}{g}\right|_{max} = \sqrt{1 + \frac{2x_h}{x_{eq}}} \tag{3.4.14}$$

Figure 3.18 shows a plot of this maximal acceleration versus the normalized drop height $\frac{x_h}{x_{eq}}$. You can see that the maximal acceleration increases as we move along the horizontal axis. This certainly makes sense; as the item is dropped from greater heights, the maximal acceleration levels should increase. We can get some more insight by recalling that $\frac{x_h}{x_{eq}}$ is the same as $\frac{1}{2}(\frac{v\omega_n}{g})^2$. Since increases in v move us along the horizontal axis, we see that increasing impact velocities increase our maximal accelerations. This fact is directly related to the preceding observation. A new observation, however, is that increasing ω_n also moves us to the right on the horizontal axis, thus increasing maximal acceleration levels. This gives us some useful design insight. We now know that increasing the natural frequency of a system is bad, at least as far as the impact loads are concerned if the system is dropped. We might want a higher natural frequency for other reasons (such as strength), but this tells us that a trade-off we'll then have to live with is an increased susceptibility to certain shock loadings. Often one will find that structures that deal with shock very well are not particularly good with regard to steady state excitation and vice versa. Once again this is a classic case of how one must carefully consider the different factors of a design, weigh their importance, and then come up with the best overall design that produces acceptable performance in all the relevant categories.

Example 3.5

Problem   Determine how one can drop a mass of 1 kg without the maximal acceleration levels exceeding $10g$. Assume that the floor is massless and has an effective stiffness coefficient of 10,000 N/m. $g = 9.81$ m/s^2.

Solution   From (3.4.11) we have

$$|\ddot{x}|_{max} = g\sqrt{1 + \frac{2x_h\omega_n^2}{g}}$$

ω_n for this example is given by $\sqrt{\frac{10,000}{10}} = 31.6$ m/s^2 Thus we have to solve

$$10g = g\sqrt{1 + \frac{2x_h\omega_n^2}{g}}$$

for x_h. Rearranging this gives us

$$x_h = \frac{99g}{2\omega_n^2} = .486 \text{ m}$$

As a further illustration of how we can examine shock responses, we'll again look at a falling vibratory system, but this time we'll include damping. Figure 3.19 shows the system configuration. In this case, the analysis is a bit more complex, since we'll have to account for the exponential decay of the vibrations.

In this case we won't be able to simply find the overall amplitude of the sinusoidally varying acceleration and say it is equal to the maximum acceleration, as we did in Example 3.5. For this system we'll have to actually find the time at which the maximal response occurs and then evaluate its magnitude. Figure 3.20 shows a typical amplitude and acceleration response ($\zeta = .2$, $\omega_n = 1$ rad/s, $v = 10$ m/s) for such a system, while Figure 3.21 shows how the acceleration profile changes if the natural frequency of the system is increased to 100 rad/s. Clearly, increasing the natural frequency has increased the overall magnitudes of the accelerations.

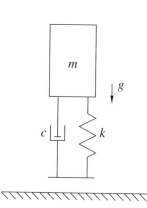

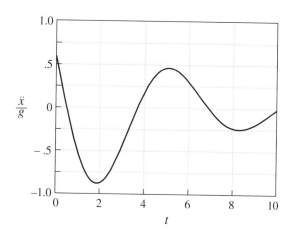

Figure 3.19 Spring-mass damper subjected to shock due to impact

Figure 3.20 Normalized acceleration for a damped system impacting a rigid floor

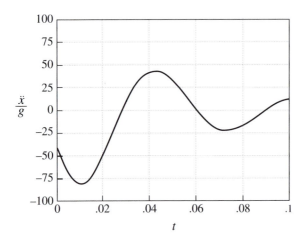

Figure 3.21   Normalized acceleration for a damped system impacting a rigid floor ($\omega_n = 100$ rad/s)

The equation of motion is given by

$$m\ddot{x} + c\dot{x} + kx = mg \tag{3.4.15}$$

or

$$\ddot{x} + 2\zeta\omega_n\dot{x} + \omega_n^2 x = g \tag{3.4.16}$$

with $x(0) = 0$ and $\dot{x}(0) = v$. The general solution for this problem is given by

$$x(t) = \frac{g}{\omega_n^2} + e^{-\zeta\omega_n t}(a_1 \cos(\omega_d t) + a_2 \sin(\omega_d t)) \tag{3.4.17}$$

where ω_d is the damped frequency of oscillation. Applying the initial conditions gives us

$$a_1 = -\frac{g}{\omega_n^2} \tag{3.4.18}$$

and

$$a_2 = \frac{v - \frac{\zeta g}{\omega_n}}{\omega_d} \tag{3.4.19}$$

Differentiating twice gives us the acceleration:

$$\ddot{x}(t) = e^{-\zeta\omega_n t}\left((g - 2\zeta\omega_n v)\cos(\omega_d t) + \left(\frac{v\omega_n(2\zeta^2 - 1) - g\zeta}{\sqrt{1 - \zeta^2}}\right)\sin(\omega_d t)\right) \tag{3.4.20}$$

As you can see, this is quite a bit more complicated than the undamped case was. If we were just going to check numerically for when the absolute value of the acceleration was a maximum, we wouldn't need to go any further. If we want to determine the peak response analytically, however, we'll have to differentiate once again. Doing so, setting the result to zero (to find the maximum value), and rearranging slightly will give us

$$\tan(\omega_d t_{\max}) = \frac{\omega_d b_2 - \omega_n \zeta b_1}{\omega_d b_1 + \omega_n \zeta b_2} \tag{3.4.21}$$

where $b_1 = g - 2\zeta \omega_n v$ and $b_2 = \frac{v\omega_n(2\zeta^2 - 1) - g\zeta}{\sqrt{1-\zeta^2}}$

To solve for $t_{\max}$, we've got to take the inverse tangent

$$\omega_d t_{\max} = \tan^{-1}\left(\frac{\omega_d b_2 - \omega_n \zeta b_1}{\omega_d b_1 + \omega_n \zeta b_2}\right) \tag{3.4.22}$$

This will then allow us to solve for the time at which the maximum response occurs.

Unfortunately, the nice way we plotted normalized acceleration versus normalized drop height won't really work here. The variables are too complex to allow a simple normalization such as we used before. It's more convenient in this case to plot the acceleration-to-gravity ratio versus the impact velocity. We'd then need to make a secondary calculation to relate impact velocity to drop height.

The results of following through on the procedure just outlined are shown in Figure 3.22. The normalized, maximum acceleration is plotted versus impact velocity for ζ ranging from 0 to .4. The natural frequency was taken to be 1.0 rad/s for this plot. There are a few observations that can be made from these results. First, you'll note that the maximal acceleration ratio is sometimes less than 1. This might seem odd, since the object was in a $1g$ free-fall before impact, but what we're looking at is the maximum acceleration ratio after impact has occurred. Once the object has struck the ground, the acceleration is simply a function of the initial velocity and the damping and stiffness characteristics. As you might guess, impacts typically result in accelerative loadings much higher than g, and thus this observation is mainly of academic interest.

The second observation is that the acceleration ratio is growing linearly for large values of impact velocity. You can see why this happens from looking at (3.4.20). As v increases, it will come to dominate the coefficients of both the cosine and sine terms. Since v appears linearly, we'd expect

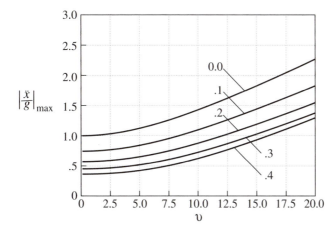

Figure 3.22 Normalized maximum acceleration versus impact velocity for damped system (damping factors indicated on graph)

a linear increase in our plot. This fact simply tells us that for large values of v, as the impact velocity increases, so does the maximal acceleration.

Finally, it's clear for this range of parameters that increasing the damping serves to reduce the maximal acceleration levels. Thus damping is another useful design tool to use in deciding how to build systems to withstand shock loading.

3.5 HOMEWORK PROBLEMS

Section 3.2

3.1. Find the response of the system illustrated in Figure P3.1 to the input force shown.

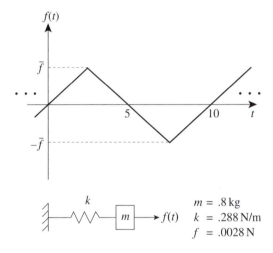

Figure P3.1

3.2. How many terms have to considered in a Fourier expansion of the signal shown in Figure P3.2 before the individual harmonic components are less than 5% of the magnitude of the fundamental component (the one at ω_0)?

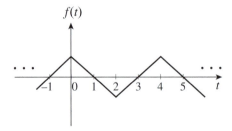

Figure P3.2

3.3. Calculate and plot several Fourier approximations (numerically is fine) for the function shown in Figure P3.3. How many terms are needed before the fit looks reasonable?

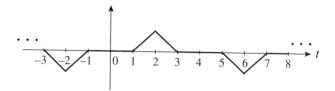

Figure P3.3

3.4. Plot the response of the system illustrated in Figure P3.4 to the illustrated seismic input.

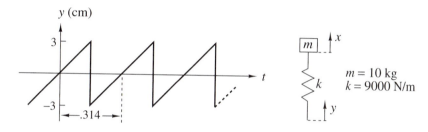

Figure P3.4

3.5. Given the Fourier series solution for $f(t)$ in Figure P3.5, construct the solution for $g(t)$ also shown in terms of the a_i and b_i

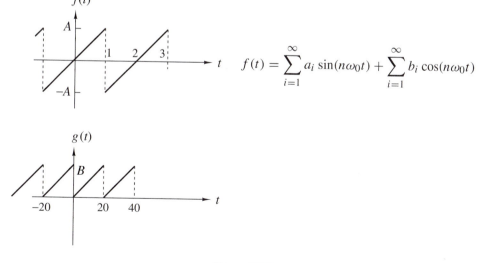

$$f(t) = \sum_{i=1}^{\infty} a_i \sin(n\omega_0 t) + \sum_{i=1}^{\infty} b_i \cos(n\omega_0 t)$$

Figure P3.5

3.6. How do the coefficients of the Fourier representation change if, instead of the signal $f(t)$, shown in Figure P3.5, your signal looks like that in Figure P3.6?

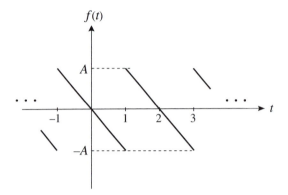

Figure P3.6

3.7. How do the coefficients of the Fourier representation change if, instead of the signal g(t), shown in Figure P3.5, your signal looks like that in Figure P3.7?

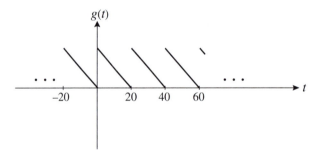

Figure P3.7

3.8. For the system shown in Figure P3.8, $m = 20$ kg and $k = 400$ N/m. The base of the system is given a vertical displacement $y(t)$ in the form of a sawtooth function with amplitude $A = 3$ cm and period $T = 2$ seconds. Compute the response of the system and identify the dominant response harmonic.

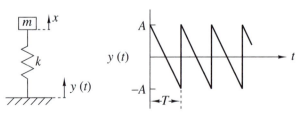

Figure P3.8

3.9. Compute and plot the response for the system illustrated in Figure P3.9 for the input force also shown. You'll note that the second harmonic ($2\omega_0 = 1.57$) is very close to the system's natural frequency. Since the damping is equal to zero, why isn't this harmonic dominating the response?

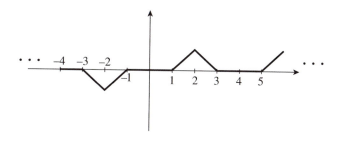

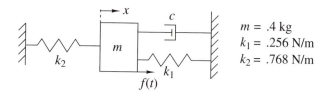

Figure P3.9

3.10. For Figure P3.10, determine whether the coefficients b_m and a_m are zero or nonzero if the function is approximated by

$$f(t) = \sum_{m=0}^{n} a_m \cos(m\omega_0 t) + \sum_{m=1}^{n} b_m \sin(m\omega_0 t)$$

What is the actual value of ω_0 and a_0?

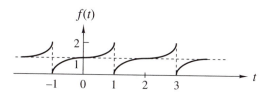

Figure P3.10

3.11. Why don't we retain an infinite number of terms in our Fourier series analyses? Give two reasons. What is a reasonable criterion for deciding how many terms to include our series?

3.12. Use a Fourier series analysis to plot out approximations to the function shown in Figure P3.12. Show how the approximation changes as you go from one included harmonic to several.

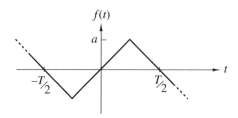

Figure P3.12

3.13. Determine the response of a direct force excited, spring-mass system to the given input. Two cases need to be examined: $T = 3$ seconds and $T = 9.24$ seconds. Look at the harmonic content of the input and the harmonic content of the output and use this information to explain why you obtained qualitatively different responses for the two cases. $m = 5$ kg and $k = 20$ N/m. $f(t)$ given in Figure P3.12.

3.14. Express the following function in terms of a Fourier series representation (illustrated in Figure P3.14). It might model the force input of a pair of woodpeckers located on opposite sides of a tree—sharp, regularly spaced impulses. What difficulties will the Fourier series analysis have with such a function?

$$f(t) = \sum_{n=-\infty}^{\infty} (-1)^{n+1}\delta(t - (2n - 1)T)$$

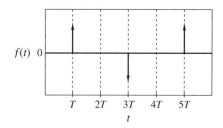

Figure P3.14

Section 3.3

3.15. Determine $x(2.)$ and $\dot{x}(2)$ for a direct force excited, spring-mass system for the input given in Figure P3.15 via a convolution approach. $x(0) = \dot{x}(0) = 0$, $m = 1$ kg, $k = 9$ N/m.

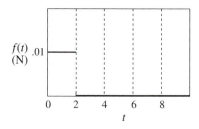

Figure P3.15

3.16. Consider a direct force excited, spring-mass system. Show via a convolution approach that the displacement at $t = .5T$ will be twice the static equilibrium displacement if the force input is constant from $t = 0$ to $t = .5T$.

3.17. Use a convolution approach to determine the response of a seismically excited spring-mass system to a step seismic input.

3.18. Will the force input illustrated in Figure P3.18 lead to a stable oscillation (no growth in amplitude) for a direct force excited, spring-mass system for which $m = 100$ kg and $k = 2500\pi^2$ N/m? Use a convolution approach.

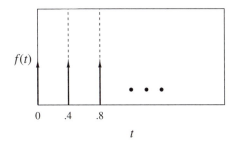

Figure P3.18

Section 3.4

3.19. Determine the maximal acceleration that the mass will experience if dropped onto the spring illustrated in Figure P3.19. $m = 20$ kg, $k = 20,000$ N/m, $h = 1$ m, $g = 9.81$ m/s^2.

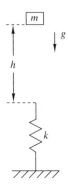

Figure P3.19

3.20. What is the accelerative loading on a mass that's uniformly decelerated from 35 mph to zero in .1 second?

3.21. By adding a crush zone to a car, the time for the occupants' velocity to uniformly decelerate from 60 mph to zero is raised from .07 second to .28 second. By how much does this reduce the deceleration felt by the driver?

3.22. By how much will the maximal acceleration of Problem 3.19 change if the gravitational acceleration is reduced by a factor of 6 (as would happen if one ran the experiment on Earth and then on the moon)?

3.23. In an egg drop contest, the goal is to drop an egg into a container from the greatest height possible without having the egg break. Assume that the egg doesn't break if dropped from a 2 m height in a container that has a stiffness of 400 N/m. The mass of the egg is .05 kg. How must the supporting stiffness be changed to double the height from which the egg can successfully be dropped?

3.24. What is the maximum acceleration felt by a lumped mass that falls onto an arbitrarily stiff surface?

3.25. Will reducing an item's mass decrease the maximal acceleration it will experience if dropped onto an elastic surface?

3.26. A shipping company has bought new trucks for which the height of the cargo bed has been reduced from 1.1 m to .9 m. Unfortunately for the consumer, this company unloads its cargo by dropping it off the cargo bed onto the ground. Model the package as a spring-mass system like the one in Figure P3.26. If $k = 1000$ N/m and $m = 10$ kg, determine how much the maximal accelerative loading will decrease because of the new cargo bed height.

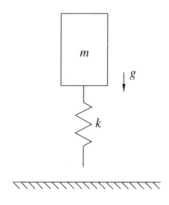

Figure P3.26

3.27. If the drop height of a spring-mass system like the one shown in Figure P3.16 is increased by 20%, by how much must the system's natural frequency be changed for there to be no overall change in the maximum accelerative loading?

4

Vibrations Involving More Than One Degree of Freedom

4.1 INTRODUCTION

Although some systems can be analyzed from a single-degree-of-freedom (SDOF) standpoint, there are a great many situations for which this viewpoint is simply insufficient. In one example, to be studied in some detail in this chapter, a designer purposely puts an additional degree of freedom in an existing system to modify the original system's vibrational response. The devices that do this are known as vibration absorbers. Automobile designers, when trying to obtain good handling characteristics for their cars, routinely have to include more than just one degree of freedom to capture the various motions the car exhibits, such as yaw, heave, pitch, and roll. Even the simplest model of a steam turbine should include at least two degrees of freedom, while reasonable models for large space structures routinely involve hundreds.

The fundamental difference between a multi-degree-of-freedom (MDOF) system and one with a single degree of freedom is that in the first case we've got additional masses to worry about. The important consequence of this fact is that a sinusoidal response of the MDOF system will automatically imply a relationship between the motions of the various masses in the structure. For instance, the motion of one mass in a 2 DOF system might be described by a cosine and the motion of the other by a cosine having twice the amplitude of the first. Thus, to describe the motion, we should concern ourselves with the *relative* motion between the two masses, since the fundamental characteristic (cosine oscillations) is the same for both masses. In the case just mentioned, this would mean noting that one mass moves twice as much as the other. This idea of relative motion is a crucial one, and since it underlies most of the remaining chapters, we'll study it in some detail.

4.2 FREE RESPONSE—UNDAMPED SYSTEM

We'll start by analyzing the system shown in Figure 4.1. External forcings and damping have been excluded for clarity and will be examined later in the chapter.

Since 2 DOF problems are easily visualized, can be analytically solved, and can be experimentally demonstrated by using a simple spring with some masses, we'll focus on them when introducing and discussing multi-degree-of-freedom characteristics. The findings, however, will be equally applicable to problems with any desired number of degrees of freedom.

As we can see from the free body diagram of Figure 4.1b, the equations of motion for this system can be written as

$$m_1\ddot{x}_1 + (k_1 + k_2)x_1 - k_2x_2 = 0 \tag{4.2.1}$$

and

$$m_2\ddot{x}_2 + (k_2 + k_3)x_2 - k_2x_1 = 0 \tag{4.2.2}$$

At this point we'll make the most important transition of the book. Up until now, everything we've been dealing with has involved scalar quantities. But we now have the opportunity to move beyond this stage and into the vector/matrix world, because the foregoing equations can just as easily be

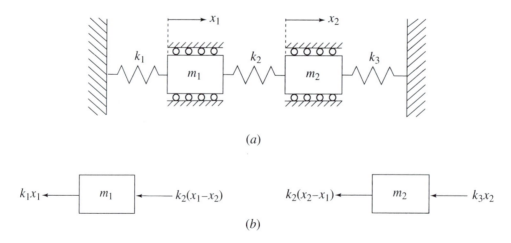

(a)

(b)

Figure 4.1   (a) 2-DOF system and (b) free body diagram

reexpressed as

$$\begin{bmatrix} m_1 & 0 \\ 0 & m_2 \end{bmatrix} \begin{Bmatrix} \ddot{x}_1 \\ \ddot{x}_2 \end{Bmatrix} + \begin{bmatrix} k_1 + k_2 & -k_2 \\ -k_2 & k_2 + k_3 \end{bmatrix} \begin{Bmatrix} x_1 \\ x_2 \end{Bmatrix} = \begin{Bmatrix} 0 \\ 0 \end{Bmatrix} \tag{4.2.3}$$

If we define

$$[M] \equiv \begin{bmatrix} m_1 & 0 \\ 0 & m_2 \end{bmatrix}, \ [K] \equiv \begin{bmatrix} k_1 + k_2 & -k_2 \\ -k_2 & k_2 + k_3 \end{bmatrix}, \ X \equiv \begin{Bmatrix} x_1 \\ x_2 \end{Bmatrix}, \ \text{and} \ O \equiv \begin{Bmatrix} 0 \\ 0 \end{Bmatrix} \tag{4.2.4}$$

then we can express (4.2.3) as

$$[M]\ddot{X} + [K]X = O \tag{4.2.5}$$

Thus we have a form that is, on the surface, identical to that of our SDOF oscillator. The difference now is that our "masses" and "springs" are matrices and our dependent variable is a vector, not a scalar.

The definitions just given follow the convention that will be used throughout the rest of the book. Lowercase letters will be used to represent scalar quantities. Thus, when you see something like b, you'll know it is a scalar. Vectors are denoted by an uppercase letter, such as X or Y. Finally, matrices are indicated by an uppercase letter surrounded by square brackets, such as $[A]$. When referring to the entries in a vector or matrix, the letter of the vector/matrix in lowercase form will be used, along with an added subscript. Thus $a_{1,2}$ is the entry in the first row and second columns of $[A]$. As a further example, the first row, first column entry in $[K]$ (defined in 4.2.4) is $k_{1,1}$ and

$$k_{1,1} = k_1 + k_2 \tag{4.2.6}$$

Even though the actual entries are denoted by k, you can tell the difference between them and the symbol indicating a particular matrix entry because the matrix entry will show the row and column number, by means of two subscripts, separated by a comma. The only exceptions to this convention are certain parameters that are traditionally written in uppercase letters (such as Young's modulus, E). It will be clear from the context that these symbols represent parameters, not vectors.

Now that we've entered the matrix/vector world, we're not going to leave it. For the rest of the book, we'll use matrix equations in one form or another. More specifically, we'll use (4.2.5) quite a bit, and thus all the analyses and conclusions we draw in this chapter will be very applicable to the problems we'll encounter in the future.

Before analyzing (4.2.5) any further, you should see what you can recall from your class in linear algebra. Almost certainly, part of what you learned included an analysis of the matrix equation

$$[A(\lambda)]X = O \tag{4.2.7}$$

It's also possible that you dealt with a different form of this equation, namely

$$[A]X = \lambda X \tag{4.2.8}$$

In either case, you'll recall that these were called *eigenvalue problems* and that the solution involved finding both the eigenvalues and eigenvectors for the system. What's important is that (4.2.5) will lead to just such an eigenvalue problem.

To see this, let's presume a solution to (4.2.5) in the form

$$X(t) = \bar{X}e^{i\omega t} = \begin{Bmatrix} \bar{x}_1 \\ \bar{x}_2 \end{Bmatrix} e^{i\omega t} \qquad (4.2.9)$$

We know that this is an appropriate form because, as we've stated before, we should always expect linear, constant coefficient, ordinary differential equations to support exponential solutions. We've got three unknowns to deal with, $\bar{x}_1$, $\bar{x}_2$, and ω. We chose $e^{i\omega t}$ instead of $e^{\lambda t}$ because of our earlier observation that a spring-mass system supports sinusoidal responses rather than real exponential solutions. However, you should realize that we could use $e^{\lambda t}$ just as easily; it won't affect the final results. You'll notice that we've got three unknowns and only two equations. Thus we know that we won't get unique numerical solutions for $\bar{x}_1$, $\bar{x}_2$, and ω. This is fine, as we'll see shortly.

To solve (4.2.5), we'll use our assumed solution, (4.2.9), differentiate it twice (for the $\ddot{X}$ term), substitute it in, and find (factoring out the $e^{i\omega t}$ term)

$$-\omega^2[M]\bar{X} + [K]\bar{X} = O \qquad (4.2.10)$$

We can now factor out the $\bar{X}$ term to obtain

$$\left[[K] - \omega^2[M] \right] \bar{X} = O \qquad (4.2.11)$$

This looks exactly like (4.2.7), (in which $[A]$ is equal to $([K] - \omega^2[M])$. Thus we have an eigenvalue problem. Although $\bar{X} = O$ solves this equation, we recognize this as the trivial solution—trivial in that $\bar{X} = O$ implies no motion at all. Consequently we can ignore this solution as being of no interest. We're left with asking when an equation of the form

$$[A]\bar{X} = O \qquad (4.2.12)$$

has a solution for finite $\bar{X}$. The answer is, when $[A]$ is not invertible. And why is this? Well, if $[A]$ **was** invertible, then we could compute its inverse and premultiply (4.2.12) by $[A]^{-1}$ to obtain

$$[A]^{-1}[A]\bar{X} = [A]^{-1}O \qquad (4.2.13)$$

or

$$\bar{X} = O \qquad (4.2.14)$$

Thus, if $[A]$ was invertible, we'd be left with the trivial solution. Therefore, if we want a more interesting solution, $[A]$ must not be invertible. And for a matrix to not be invertible, its determinant must be equal to zero. This means that our requirement for a solution is given by

$$\det\left([K] - \omega^2[M]\right) = 0 \qquad (4.2.15)$$

Thus

$$\det\left(\begin{bmatrix} k_1 + k_2 - \omega^2 m_1 & -k_2 \\ -k_2 & k_2 + k_3 - \omega^2 m_2 \end{bmatrix}\right) = 0 \qquad (4.2.16)$$

If we evaluate this determinant, we'll obtain:

$$(k_1 + k_2 - \omega^2 m_1)(k_2 + k_3 - \omega^2 m_2) - k_2^2 = 0 \qquad (4.2.17)$$

Expanding this out and grouping the powers of ω leads to

$$m_1 m_2 \omega^4 - \omega^2((k_2 + k_3)m_1 + (k_1 + k_2)m_2) + k_1 k_2 + k_1 k_3 + k_2 k_3 = 0 \qquad (4.2.18)$$

To provide further insight, let's assign specific values to the springs and masses. We'll let m_1 and m_2 each have a mass of 1 kg and let the k_i all be equal to 4 N/m. Under these conditions, (4.2.18) becomes

$$\omega^4 - 16\omega^2 + 48 = 0 \qquad (4.2.19)$$

This polynomial is called the *characteristic polynomial*, and, when set equal to zero, gives us the *characteristic equation* for the system. Although we'll refer to the roots of this polynomial as our natural frequencies (or damped frequencies of oscillation for the case with damping), we'd call them the "poles" of the system if we were looking at it from a controls perspective. So for those of you who have studied (or plan to study) control theory, remember that the poles of a system and an undamped system's natural frequencies are the same thing.

Recognizing that this equation isn't really a full fledged, fourth-order equation but is rather quadratic in ω^2, we can solve for ω^2 straightforwardly:

$$\omega_{1,2}^2 = \frac{1}{2}(16 \pm \sqrt{16^2 - (4)(48)}) \qquad (4.2.20)$$

or

$$\omega_1^2 = 4 \ (\text{rad/s})^2 \qquad (4.2.21)$$

and

$$\omega_2^2 = 12 \ (\text{rad/s})^2 \qquad (4.2.22)$$

Labeling the natural frequencies in ascending order ($\omega_1 \leq \omega_2 \leq \omega_3 \cdots$) is the typical convention and the one we'll follow. Just as in the SDOF case, we don't need to worry overly much about obtaining four values for ω when we take the square roots of ω_1^2 and ω_2^2, since these will combine to give us sine and cosine solutions just as they did before. These two values for ω^2 are eigenvalues for our problem, and their square roots are equal to the two possible frequencies of oscillation for the system. Note that for the single-degree-of-freedom case the only possible free vibration frequency was ω_n. Now we have two frequencies.

Once we've found the two solutions for ω^2, we're left with determining $\bar{x}_1$ and $\bar{x}_2$. To go further with our solution, we'll revisit (4.2.11). With ω^2 now known, we simply need to specify which of the two solutions we want to use. To start, we'll look at the lower frequency solution, that is, $\omega_1^2 = 4$ (rad/s)^2. Substituting this value for ω^2 and inserting the appropriate values for m_i and k_i gives us

$$-4\left[\begin{bmatrix} 1 & 0 \\ 0 & 1 \end{bmatrix} + \begin{bmatrix} 8 & -4 \\ -4 & 8 \end{bmatrix}\right]\begin{Bmatrix} \bar{x}_1 \\ \bar{x}_2 \end{Bmatrix} = \begin{Bmatrix} 0 \\ 0 \end{Bmatrix} \tag{4.2.23}$$

or more simply

$$\begin{bmatrix} 4 & -4 \\ -4 & 4 \end{bmatrix}\begin{Bmatrix} \bar{x}_1 \\ \bar{x}_2 \end{Bmatrix} = \begin{Bmatrix} 0 \\ 0 \end{Bmatrix} \tag{4.2.24}$$

Both equations in (4.2.24) indicate the same thing, namely, that $\bar{x}_1$ must be equal to $\bar{x}_2$. Getting the same equation makes sense: it implies that the two rows of the matrix are not independent, something that has to be true if the matrix is to be noninvertible.

We're now in a position to write down one of the solutions to our problem. Based on the results so far, we know that one possible solution is

$$X(t) = a_1 \begin{Bmatrix} 1 \\ 1 \end{Bmatrix} e^{2it} + a_2 \begin{Bmatrix} 1 \\ 1 \end{Bmatrix} e^{-2it} \tag{4.2.25}$$

where a_1 and a_2 are as yet undetermined constants we could solve for if we were given specific initial conditions.

A more accessible (and equivalent) form is to use a trigonometric representation instead of the exponential form shown in (4.2.25):

$$X(t) = b_1 \begin{Bmatrix} 1 \\ 1 \end{Bmatrix} \cos(2t - \phi_1) \tag{4.2.26}$$

Since we've included a phase shift, this representation can include pure sines and cosines as well as a mix of the two, as discussed earlier.

Equation (4.2.26) represents one of the fundamental responses of the system. Physically, both masses are locked together in their motions: when one moves to the right, so does the other, and to the same degree. You'll also note that the frequency of this oscillation, 2 rad/s, is the same as the natural frequency we'd find from analyzing one of the springs attached to one of the masses. The reason for this is not too difficult to see. Imagine that the middle spring has been removed and you move both masses the same amount to the left and then release them. Since they're now simply two independent, single-degree-of-freedom systems, they'll oscillate at their natural frequencies. And since the masses and springs are identical, the natural frequency of oscillation will be the same for both. Thus the masses will move in synchronism with each other and the distance between them will stay fixed. Therefore it doesn't matter whether there is a spring between them or not; since the relative distance between them stays constant, the middle spring isn't stretched or compressed anyway. This is a consequence of both the left and right spring-mass systems having the same natural frequencies. Whenever you have this situation, the first natural frequency of the 2 DOF system will be the same as the individual SDOF systems.

We call the vector $(1 \ 1)^T$ shown in (4.2.25) and (4.2.26) an *eigenvector* of our system. You'll also see it called modal vector, eigenmode, natural mode, and sometimes simply mode. It indicates the relative displacements (and later on will include velocities) of the different masses in the system. Since the system has no damping, the oscillations go on forever and, if the system is initially displaced into the first eigenmode, it remains there. Note that the amplitude of the eigenvector is arbitrary, just as the amplitude was arbitrary in the SDOF case. So the really important thing about this eigenmode is that both entries are the same.

We can now solve for the second mode of the system exactly as we solved for the first. In this case ω^2 is 12 rad/s^2 and our equation governing the second mode is given by

$$\begin{bmatrix} -4 & -4 \\ -4 & -4 \end{bmatrix} \begin{Bmatrix} \bar{x}_1 \\ \bar{x}_2 \end{Bmatrix} = \begin{Bmatrix} 0 \\ 0 \end{Bmatrix} \tag{4.2.27}$$

and our solution is

$$X(t) = a_3 \begin{Bmatrix} 1 \\ -1 \end{Bmatrix} e^{2\sqrt{3}it} + a_4 \begin{Bmatrix} 1 \\ -1 \end{Bmatrix} e^{-2\sqrt{3}it} \tag{4.2.28}$$

or, more familiarly

$$X(t) = b_2 \begin{Bmatrix} 1 \\ -1 \end{Bmatrix} \cos(2\sqrt{3}t - \phi_2). \tag{4.2.29}$$

Using X_i to identify our eigenvectors

$$X_1 \equiv \begin{Bmatrix} 1 \\ 1 \end{Bmatrix}, \quad X_2 \equiv \begin{Bmatrix} 1 \\ -1 \end{Bmatrix} \tag{4.2.30}$$

we can write our total solution for this unforced, 2 DOF oscillator as

$$X(t) = b_1 X_1 \cos(\omega_1 t - \phi_1) + b_2 X_2 \cos(\omega_2 t - \phi_2) \tag{4.2.31}$$

where ω_1 is the first natural frequency of the system (and is equal to 2 rad/s) and ω_2 is the second natural frequency (equaling $2\sqrt{3}$ rad/s).

As a concrete example of how we would match specific initial conditions, consider the response of the system to the initial conditions

$$x_1(0) = 1 \text{ mm}, \quad \dot{x}_1(0) = 0, \quad x_2(0) = 0, \quad \dot{x}_2(0) = 0 \tag{4.2.32}$$

Note that this initial deflection does not match either the first or the second eigenvector's shape. Thus we would predict the involvement of both eigenvectors in the subsequent motion. And we'd be right, because the solution for this case is

$$\phi_1 = 0, \quad \phi_2 = 0, \quad b_1 = .5 \text{ mm}, \quad b_2 = .5 \text{ mm} \tag{4.2.33}$$

The total response can be written as

$$X(t) = X_1 \cos(2t) + X_2 \cos(2\sqrt{3}t) \tag{4.2.34}$$

It's important to realize that even if the initial deflections of the two masses match only one particular eigenvector, both eigenvectors might be involved in the response because of the initial velocities. For instance, the initial conditions

$$x_1(0) = 1, \quad x_2(0) = 1, \quad \dot{x}_1(0) = 2\sqrt{3}, \quad \dot{x}_2(0) = \sqrt{3} \tag{4.2.35}$$

lead to the solution

$$X(t) = X_1 \cos(2t) + X_2 \cos(2\sqrt{3}t - \tfrac{\pi}{2}) \tag{4.2.36}$$

even though no trace of the second eigenvector showed up in the initial displacements. Thus both velocities and displacements must correspond to those supported by a particular eigenvector for that to be the sole contributor to the response.

Although we can solve 2×2 problems by hand, it's more convenient to let the computer do the work, especially as we move into more complicated problems. MATLAB is a very nice piece of software that allows problems such as ours to be solved quickly and easily. There are other software packages out there as well but MATLAB is very widely used across the country and I'll be using it in this book.

MATLAB has two basic routines that handle eigenvalue problems, both called "eig." If you type

$$\text{eig([A]) (cr)}$$

where (cr) indicates a carriage return, MATLAB will determine the solutions to the problem:

$$\dot{X} = [A]X \tag{4.2.37}$$

If you type

$$\text{eig([A], [B]) (cr)}$$

then MATLAB will solve

$$[B]\dot{X} = [A]X \tag{4.2.38}$$

Note that within MATLAB you can call your matrix m or M or $[M]$; it won't matter. Since typing M is quicker than typing $[M]$, it'll be easier to forget the brackets when in MATLAB. However, to remain consistent, when referring to a matrix in the discussion we'll use the bracket notation ($[M]$) but from now on, when giving directions for MATLAB use we'll use the simpler M form.

In MATLAB, you can solve for the eigenvectors and eigenvalues of a spring-mass system by typing

$$[v, d] = \texttt{eig}(K, M)$$

where K is the system's spring matrix and M is the mass matrix. MATLAB doesn't care (within limits) how big your system is. If M and K are $n \times n$ matrices, then v will be a matrix whose columns will be the system's eigenvectors, and d will be a diagonal matrix having the ω_i^2 values on the diagonal. Since these matrices correspond to each other, the second column of (v), for example, will be the eigenvector associated with the $d(2, 2)$ entry. Note that MATLAB will not order the natural frequency entries in any particular fashion; i.e., they won't appear from smallest to largest.

Example 4.1

Now that you've seen what to expect in 2 DOF systems, you can go ahead and use your spring to run your own 2 DOF experiment. First, you'll need to cut the spring into two pieces. Next, you'll need two masses. The more symmetrics they are, the better, since you'll want the motions to stay purely translational; rocking due to imbalances will just make the experiments harder.

You'll be doing two things in this experiment—experimentally verifying the existence of two eigenvectors (modes) and analytically predicting their natural frequencies from a couple of SDOF tests. Even more amazingly, we won't even need to know the actual mass or spring constants. The only additional item you'll need is a watch with a stopwatch function.

To start, attach one of the masses to one spring and the other to the other spring. To minimize confusion, we'll call the mass and spring of the first set m_1 and k_1, and the second set will be denoted by m_2 and k_2. Next, hold the free end of the spring against a fixed surface (like a door frame), pull down on the mass, and let go. The mass will then bob up and down. If you start your stopwatch at a high point of the motion and time the motion for five or ten cycles, you can calculate a pretty accurate estimate of the oscillation period by dividing the total time by the number of cycles that occurred. If you call this period t_1, then the $\frac{k_1}{m_1}$ is equal to $\left(\frac{2\pi}{t_1}\right)^2$. You should then repeat this procedure with m_2 and k_2, leading to an estimate of $\frac{k_2}{m_2}$. Finally, hook m_1 and k_2 together and determine $\frac{k_2}{m_1}$. What you'll now do is attach m_1 to the free end of k_2, as shown in Figure 4.2. If you hold the free end of k_1, you'll have a 2 DOF system (for vertical motions).

Getting this system to exhibit its first and second eigenmodes isn't difficult. Simply start moving your hand up and down as you did in the SDOF experiments and try to get a resonant response. If you start at a low frequency and slowly increase it, you'll reach a frequency for which very small motions of your hand produce large, in-phase motions of the masses (they both move in the same direction at all times). If you then move your hand faster, you'll reach a point at which the masses move in opposition to each other (one moves up while the other moves down). At this point your hand is oscillating at the system's second natural frequency. Because the damping is so low, the individual mode shapes should be quite apparent. Once the oscillations are established, you can hold your hand against a rigid support and watch the modal vibrations. Figure 4.3 shows what you should be seeing.

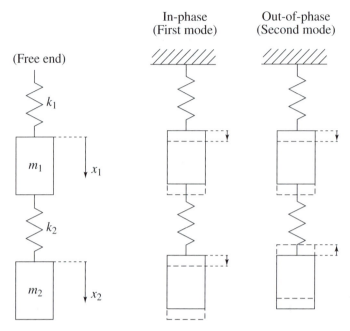

Figure 4.2
Experimental 2 DOF
system

Figure 4.3   Modal behavior of a physical
2 DOF system

All right, now that we've actually seen the eigenmodes in action, it's time to analytically predict what their frequencies should be (and verify the predictions, of course). The equations of motion for our system are given by

$$\begin{bmatrix} m_1 & 0 \\ 0 & m_2 \end{bmatrix} \begin{Bmatrix} \ddot{x}_1 \\ \ddot{x}_2 \end{Bmatrix} + \begin{bmatrix} k_1 + k_2 & -k_2 \\ -k_2 & k_2 \end{bmatrix} \begin{Bmatrix} x_1 \\ x_2 \end{Bmatrix} = \begin{Bmatrix} 0 \\ 0 \end{Bmatrix}$$

Now how can we progress any further without knowing the particular values of the masses and stiffnesses? By dividing through by the masses! If we divide the first equation by m_1 and the second by m_2, we'll have terms like $\frac{k_1}{m_1}$ and $\frac{k_2}{m_2}$. These are simply the squares of the natural frequencies associated with the simpler single-DOF problems we've already analyzed.

Going ahead and dividing by the masses gives us

$$\begin{bmatrix} 1 & 0 \\ 0 & 1 \end{bmatrix} \begin{Bmatrix} \ddot{x}_1 \\ \ddot{x}_2 \end{Bmatrix} + \begin{bmatrix} \frac{k_1}{m_1} + \frac{k_2}{m_1} & -\frac{k_2}{m_1} \\ -\frac{k_2}{m_2} & \frac{k_2}{m_2} \end{bmatrix} \begin{Bmatrix} x_1 \\ x_2 \end{Bmatrix} = \begin{Bmatrix} 0 \\ 0 \end{Bmatrix} \tag{4.2.39}$$

At this point, we have to part company, since I've got no idea what natural frequencies you have. But just for the record, I'll go through what I get when I run this test. For the masses, I used heavy doughnut-shaped pieces of steel (which came out of the weights of a grandfather clock). The periods of oscillation for the two SDOF oscillators were

$$t_1 = .864 \text{ second}$$

and

$$t_2 = .861 \text{ second}$$

which correspond to frequencies of

$$\omega_1 = 7.27 \text{ rad/s}$$

and

$$\omega_2 = 7.30 \text{ rad/s}$$

A ten-cycle average was used when obtaining these values.

If these results are substituted into (4.2.39), the equations become

$$\begin{bmatrix} 1 & 0 \\ 0 & 1 \end{bmatrix} \begin{Bmatrix} \ddot{x}_1 \\ \ddot{x}_2 \end{Bmatrix} + \begin{bmatrix} 106.2 & -53.3 \\ -53.3 & 53.3 \end{bmatrix} \begin{Bmatrix} x_1 \\ x_2 \end{Bmatrix} = \begin{Bmatrix} 0 \\ 0 \end{Bmatrix}$$

The characteristic equation is therefore

$$\omega^4 - 159.5\omega^2 + 2820 = 0$$

which has solutions

$$\omega_1 = 4.50 \text{ rad/s}$$

and

$$\omega_2 = 11.80 \text{ rad/s}$$

By exciting the experimental system at resonance, immediately placing the free end of the spring against the top of a door frame, timing 10 oscillations, and then averaging, the following time periods were obtained:

$$t_1 = 1.40 \text{ seconds}$$

and

$$t_2 = .538 \text{ second}$$

These correspond to experimentally derived natural frequencies of

$$\omega_{1_{\text{exp}}} = 4.49 \text{ rad/s}$$

and

$$\omega_{2_{\text{exp}}} = 11.68 \text{ rad/s}$$

Comparing the two sets of results, we see that the predicted first natural frequency was within .2% of the experimental value while the second was off by only 1%.

These excellent results weren't a fluke; hopefully you obtained equally accurate answers during your own test. What this experiment shows is that the modeling we've been talking about is really pretty useful and reflects reality quite well. Of course, you'll rarely be dealing with masses attached to springs in your professional life. But the basic conclusion still holds. The techniques we've learned (and the ones still to come) can do an amazing job of predicting real-world behavior as long as the assumptions that underlie them aren't violated.

Example 4.2

Problem   Determine the natural frequencies and eigenvectors for the system shown in Figure 4.1a: $m_1 = 2$ kg, $m_2 = 4$ kg, $k_1 = 40$ N/m, $k_2 = 100$ N/m, and $k_3 = 200$ N/m.

Solution   The equations of motion for the system are found from performing a force balance:

$$m_1\ddot{x}_1 = -k_1x_1 + k_2(x_2 - x_1)$$
$$m_2\ddot{x}_2 = -k_2(x_2 - x_1) - k_3x_2$$

which can be rewritten as

$$m_1\ddot{x}_1 + (k_1 + k_2)x_1 - k_2x_2 = 0$$
$$m_2\ddot{x}_2 + (k_2 + k_3)x_2 - k_2x_1 = 0$$

Putting this into matrix form and using the given parameter values yields

$$\begin{bmatrix} 2 & 0 \\ 0 & 4 \end{bmatrix} \begin{Bmatrix} \ddot{x}_1 \\ \ddot{x}_2 \end{Bmatrix} + \begin{bmatrix} 140 & -100 \\ -100 & 300 \end{bmatrix} \begin{Bmatrix} x_1 \\ x_2 \end{Bmatrix} = \begin{Bmatrix} 0 \\ 0 \end{Bmatrix}$$

This time I'll use MATLAB. Typing [v,d]=eig(K,M) where

$$M = \begin{bmatrix} 2 & 0 \\ 0 & 4 \end{bmatrix} \text{ and } K = \begin{bmatrix} 140 & -100 \\ -100 & 300 \end{bmatrix}$$

yields

$$\omega_1^2 = 37.056 \implies \omega_1 = 6.087 \text{ rad/s and } X_1 = \begin{Bmatrix} .8350 \\ .5502 \end{Bmatrix}$$

$$\omega_2^2 = 107.9 \implies \omega_1 = 10.39 \text{ rad/s and } X_1 = \begin{Bmatrix} -.7966 \\ .6045 \end{Bmatrix}$$

Now that we've investigated the eigenvectors a bit, which are a free response characteristic, we'll move on to consider some forced responses and point out how the natural frequencies and eigenvectors affect the solutions.

4.3 FORCED RESPONSE

If we include the applied forces shown in Figure 4.4, our equations of motion become

$$\begin{bmatrix} m_1 & 0 \\ 0 & m_2 \end{bmatrix} \begin{Bmatrix} \ddot{x}_1 \\ \ddot{x}_2 \end{Bmatrix} + \begin{bmatrix} k_1 + k_2 & -k_2 \\ -k_2 & k_2 + k_3 \end{bmatrix} \begin{Bmatrix} x_1 \\ x_2 \end{Bmatrix} = \begin{Bmatrix} f_1 \\ f_2 \end{Bmatrix} \tag{4.3.1}$$

For simplicity, assume that the forcing is in the form

$$f_1(t) = \bar{f}_1 \cos(\omega t)$$

$$f_2(t) = \bar{f}_2 \cos(\omega t) \tag{4.3.2}$$

that is, the magnitudes of the two forces can vary but the frequency is the same for each. As in the SDOF problem, when we have a sinusoidal forcing, we should expect a sinusoidal response. And since the problem has no damping, the output should be in phase or 180 degrees out of phase with the input. Of course, we could use complex exponentials to represent the input and output (and we'll do so for the damped case). But for now we'll stick with an explicit cosine forcing, since the math is simpler without any damping and the cosine solution has a more straightforward physical interpretation.

If we let the response be

$$x_1(t) = \bar{x}_1 \cos(\omega t)$$
$$x_2(t) = \bar{x}_2 \cos(\omega t) \tag{4.3.3}$$

and substitute this (along with the forcing) into (4.3.1) then we'll obtain

$$-\omega^2 \begin{bmatrix} m_1 & 0 \\ 0 & m_2 \end{bmatrix} \begin{Bmatrix} \bar{x}_1 \\ \bar{x}_2 \end{Bmatrix} + \begin{bmatrix} k_1 + k_2 & -k_2 \\ -k_2 & k_2 + k_3 \end{bmatrix} \begin{Bmatrix} \bar{x}_1 \\ \bar{x}_2 \end{Bmatrix} = \begin{Bmatrix} \bar{f}_1 \\ \bar{f}_2 \end{Bmatrix} \tag{4.3.4}$$

or

$$\left[[K] - \omega^2 [M] \right] \bar{X} = \bar{F} \tag{4.3.5}$$

where $F \equiv \{\bar{f}_1 \ \bar{f}_2\}^T$

Unlike (4.2.11), we now have a nonhomogeneous problem; i.e., we're not dealing with $[A]\bar{X} = O$ but with $[A]\bar{X} = \bar{F}$. And unlike the homogeneous case, $[A]$ must be invertible for a solution to exist. If $[A]^{-1}$ exists, then we'll premultiply by it to find $\bar{X} = [A]^{-1}\bar{F}$. For this 2 DOF example, we can work out the solution explicitly, but for more complex problems we'll need to rely on numerical solutions.

You'll recall from your linear algebra course or from Appendix D that the inverse of a 2×2 matrix is given by

$$\begin{bmatrix} a & b \\ c & d \end{bmatrix}^{-1} = \begin{bmatrix} d & -b \\ -c & a \end{bmatrix} \frac{1}{(ad - bc)} \tag{4.3.6}$$

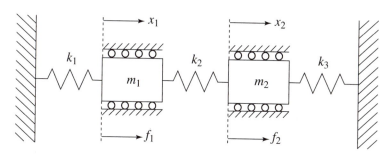

Figure 4.4   Forced 2 DOF spring-mass system

Thus our solution $\bar{X}$ is given by

$$\bar{X} = \frac{1}{\Delta(\omega)} \begin{bmatrix} k_2 + k_3 - m_2\omega^2 & k_2 \\ k_2 & k_1 + k_2 - m_1\omega^2 \end{bmatrix} \begin{Bmatrix} \bar{f}_1 \\ \bar{f}_2 \end{Bmatrix} \tag{4.3.7}$$

where $\Delta(\omega) \equiv (k_1 + k_2 - m_1\omega^2)(k_2 + k_3 - m_2\omega^2) - k_2^2$. The individual components of $\bar{X}$ are given by

$$\bar{x}_1 = \frac{(k_2 + k_3 - m_2\omega^2)\bar{f}_1 + k_2\bar{f}_2}{\Delta(\omega)} \tag{4.3.8}$$

and

$$\bar{x}_2 = \frac{k_2\bar{f}_1 + (k_1 + k_2 - m_1\omega^2)\bar{f}_2}{\Delta(\omega)} \tag{4.3.9}$$

Notice that the denominator for both terms, $\Delta(\omega)$, is the same, and if it is set to zero we have exactly the same equation that we used to solve for the system natural frequencies (4.2.17). Is this a bizarre coincidence? Of course not! This will invariably be the case when we find the forced response of a linear system such as this one. The denominator is always the characteristic polynomial for our system. If you'll recall what was so special about this polynomial, namely, that its roots were the natural frequencies of the system, you'll realize that they play an important role for the forced problem also. Since the characteristic polynomial is equal to zero at the system's natural frequencies, (4.3.8) and (4.3.9) tell us that the denominators go to zero at these frequencies. Thus, if the numerator isn't equal to zero (and it usually isn't), the response will go to infinity. What does this mean? It means that we've now got two resonance conditions, one for each natural frequency. For the SDOF system we had only one resonant frequency. Now, not surprisingly, we've got two. Figure 4.5 shows the response for $\bar{x}_1$ and $\bar{x}_2$ for a representative case (in case you're really interested, the parameters $m_1 = 1$, $m_2 = 1$, $k_1 = 1$, $k_2 = 2$, $k_3 = 2$, $\bar{f}_1 = 2$, and $\bar{f}_2 = -1$ were used). As you can see, the responses for both $\bar{x}_1$ and $\bar{x}_2$ are finite at zero frequency (as in the SDOF problem), go to infinity at the resonant frequencies of the system, and die down to zero as the forcing frequency becomes very large (also like the SDOF problem). Figure 4.5 also shows that the phase response is either 0 or π radians (i.e., a positive or negative response, respectively). Just as for the SDOF problem, the phase shifts π radians after each resonant frequency is passed.

The situation is much the same for higher order systems. For instance, Figure 4.6 shows the response of the first mass in the four-mass system shown in Figure 4.7. With four masses, we now have four natural frequencies, $\omega_1-\omega_4$. You can see from Figure 4.6 that the response goes to infinity at each of these frequencies and also that the phase shifts by π radians as we pass these frequencies (i.e., the response switches sign).

Calculating $\bar{x}_1$ and $\bar{x}_2$ with MATLAB is pretty straightforward. If you represented the quantities of interest as $K \equiv [K]$, $M \equiv [M]$, $om \equiv \omega$ and $F \equiv \bar{F}$ then you'd find the desired solution vector

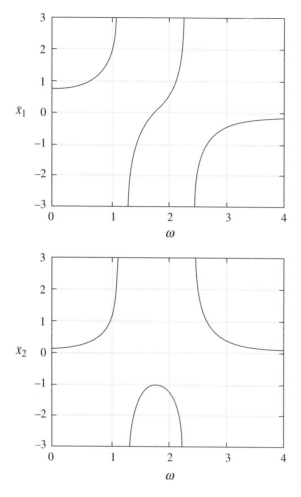

Figure 4.5 $\bar{x}_1$ and $\bar{x}_2$ versus ω

from entering

$$X = \mathtt{inv}(K - \mathtt{om} \wedge 2 * M) * F$$

Example 4.3

Problem   To see how we might use the forced response of an MDOF system, consider Figure 4.8.
We can think of this system as approximating an autonomous vehicle, to be used in unmanned explo-
rations of Mars. The two springs represent the front and rear suspensions, and the bar represents
the body. We'll assume that the disturbance represents the effects of an unbalanced force in the
engine. Our design goal will be to ensure that the vibrations are minimized at the location of the
main instrumentation. As a simplification, we'll assume that the inertia of the beam doesn't alter as
the instrumentation is moved. Although we'll work out the governing equations for arbitrary values of
the various parameters, for our numerical results we'll use $\rho = 50$ kg, $l = 1$ m, $\bar{I} = 16.\bar{6}$ kg·m^2

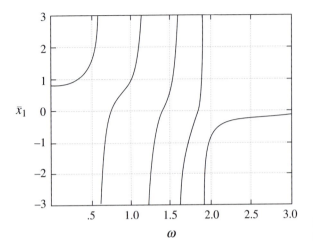

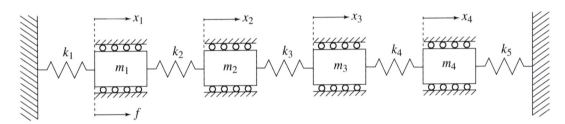

Figure 4.6 Response of the first mass of a four-mass system

Figure 4.7 Forced, 4 DOF spring-mass system

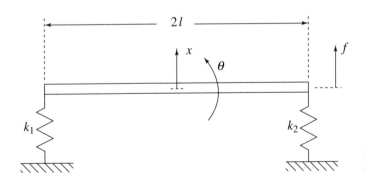

Figure 4.8 2 DOF model of an autonomous vehicle

(mass moment of inertia about the mass center), $k_1 = 30{,}000$ N/m, $k_2 = 20{,}000$ N/m, and $f = 100\cos(\omega t)$ N.

Solution To solve this problem, we first need to obtain the system's equations of motion. A convenient set of coordinates is the vertical translation x of the vehicle's center of mass and the rotation θ about that center. Summing moments and forces about the center of mass will

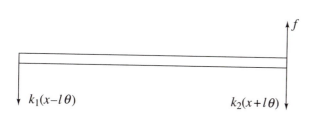

Figure 4.9   Free body diagram for an autonomous vehicle

give us (Figure 4.9)

$$m\ddot{x} = -k_1(x - l\theta) - k_2(x + l\theta) + f$$

and

$$\bar{I}\ddot{\theta} = -k_2(x + l\theta)l + k_1(x - l\theta)l + fl$$

Rearranging this yields:

$$m\ddot{x} + k_1(x - l\theta) + k_2(x + l\theta) = f$$
$$\bar{I}\ddot{\theta} + k_2(x + l\theta)l - k_1(x - l\theta)l = fl$$

or

$$m\ddot{x} + (k_1 + k_2)x + (k_2 - k_1)l\theta = f \tag{4.3.10}$$
$$\bar{I}\ddot{\theta} + (k_1 + k_2)l^2\theta + (k_2 - k_1)lx = fl \tag{4.3.11}$$

Although with more complicated systems we'd have to find our solutions numerically, we can get a closed-form solution when we've only got two equations to deal with. Assuming solutions of $x(t) = \bar{x}\cos(\omega t)$, $\theta(t) = \bar{\theta}\cos(\omega t)$ to go along with the forcing $f = \bar{f}\cos(\omega t)$, (4.3.10) and (4.3.11) will become

$$(k_1 + k_2 - \omega^2 m)\bar{x} - +(k_2 - k_1)l\bar{\theta} = \bar{f} \tag{4.3.12}$$

and

$$(k_2 - k_1)l\bar{x} + (k_1 l^2 + k_2 l^2 - \omega^2 \bar{I})\bar{\theta} = \bar{f}l \tag{4.3.13}$$

Solving (4.3.13) for $\bar{x}$ in terms of $\bar{\theta}$ gives

$$\bar{x} = \frac{\bar{f}l - (k_1 l^2 + k_2 l^2 - \bar{I}\omega^2)\bar{\theta}}{(k_2 - k_1)l} \tag{4.3.14}$$

Substituting (4.3.14) into (4.3.12) and solving for $\bar{\theta}$ yields

$$\bar{\theta} = \frac{\bar{f}l(2k_1 - m\omega^2)}{\Delta(\omega)} \tag{4.3.15}$$

where

$$\Delta(\omega) = (k_1 + k_2 - m\omega^2)(k_1 l^2 + k_2 l^2 - \bar{I}\omega^2) - l^2(k_2 - k_1)^2$$

Using (4.3.15) in (4.3.14) gives us

$$\bar{x} = \frac{\bar{f}(2k_1l^2 - \bar{I}\omega^2)}{\Delta(\omega)} \tag{4.3.16}$$

Now we're in a position to do some design work. We'll assume that the instruments are αl meters to the right of the bar's center. Thus the displacement is given by $y = x + \alpha l\theta$ for small angles of oscillation, where α runs from -1 to 1. We can use this information, together with (4.3.15) and (4.3.16), to express the displacement of the instruments as

$$y = \bar{f}\left(\frac{2k_1l^2(1+\alpha) - (\bar{I} + \alpha ml^2)\omega^2}{\Delta(\omega)}\right) \tag{4.3.17}$$

Our minimization problem is now pretty straightforward. Once we're given the frequency of motor vibration, we can use (4.3.17) to determine the placement at which y will be zero. And $y = 0$ in (4.3.17) means that the numerator must equal zero. Thus we'll have

$$2k_1l^2(1+\alpha) - (\bar{I} + \alpha ml^2)\omega^2 = 0$$

or, solving for α in terms of ω,

$$\alpha = \frac{\bar{I}\omega^2 - 2k_1l^2}{2k_1l^2 - \omega^2 ml^2}$$

If we're told that the forcing frequency is 55 rad/s, then the optimal placement of the instrument package will be $\alpha = .105\,l$, or .105 m to the right of center.

Example 4.4

Problem   In this example we'll consider another kind of problem. The system will be identical to that of Example 4.3. The difference in this case is that instead of a force applied, we'll now have a sensor mounted. The purpose of the sensor will be to tell us what's happening in the overall system. For specificity, we'll assume that the sensor is an accelerometer that reads vertical accelerations. The question before us is where *not* to place the sensor.

Solution   The way to approach this problem is to ask what the worst situation might be. We know that the system will support two modes of vibration. If we want complete knowledge of what's going on, the sensor must be able to respond to both these modes. Therefore the worst placement of the sensor would be those places on the bar for which no information is transmitted for one or both modes.

Rewriting (4.3.10) and (4.3.11) in matrix form, we have

$$\begin{bmatrix} m & 0 \\ 0 & \bar{I} \end{bmatrix} \begin{Bmatrix} \ddot{x} \\ \ddot{\theta} \end{Bmatrix} + \begin{bmatrix} k_1 + k_2 & (k_2 - k_1)l \\ (k_2 - k_1)l & (k_1 + k_2)l^2 \end{bmatrix} \begin{Bmatrix} x \\ \theta \end{Bmatrix} = \begin{Bmatrix} 0 \\ 0 \end{Bmatrix}$$

Thus our eigenvector equations are

$$\begin{bmatrix} k_1 + k_2 - m\omega^2 & (k_2 - k_1)l \\ (k_2 - k_1)l & (k_1 + k_2)l^2 - \bar{I}\omega^2 \end{bmatrix} \begin{Bmatrix} \bar{x} \\ \bar{\theta} \end{Bmatrix} = \begin{Bmatrix} 0 \\ 0 \end{Bmatrix}$$

Solving for the eigenvectors associated with the two natural frequencies (using the parameter values of Example 4.3) yields

$$\omega_1 = 30.7 \text{ rad/s}, \quad X_1 = \begin{Bmatrix} .9600 \\ .2799 \end{Bmatrix}, \quad \omega_2 = 55.3 \text{ rad/s}, \quad X_2 = \begin{Bmatrix} -.0967 \\ .9953 \end{Bmatrix}$$

Now we can ask where on the beam the deflection is equal to zero for each of these modes. The general displacement along the beam is the same as it was for Example 4.3, $x + \alpha l\theta$. For the first eigenvector this means that the displacement is proportional to $.9600 + .2799\alpha$, where α measures displacement to the right of the midpoint in meters. The displacement is going to be equal to zero when

$$\alpha = -\frac{.9600}{.2799} = -3.430$$

Is this a ridiculous answer? After all, it says that the zero point occurs off the bar, way to the left. Well, the answer is that it's not ridiculous at all; what it's telling us is that there aren't any points on the bar that don't move when the bar is oscillating in the system's first mode. Our accelerometer can be placed anywhere and still pick up the motion from the first mode. The situation, however, is different for the second mode. For this mode, the displacement is proportional to $-.0967 + .9953\alpha$. This is zero when $\alpha = \frac{.0967}{.9953} = .0972$. Thus there is a node (point of zero motion) .0972 m to the right of center. If we place our accelerometer there, then there is no chance for it to pick up the second mode's motion.

Our conclusion is that we should put the accelerometer anywhere but .0972 m to the right of center. Of course, physical considerations tell us that the displacement of the second mode is pretty small in that point's surrounding neighborhood as well. Thus we'd want to position the accelerometer a reasonable distance away.

The effect we've just seen is a very important one in vibrational analyses. For complicated systems, we'll have a large number of locations for which the response of a particular mode will be very small. Knowing where these vibrational nodes occur is crucial to place our sensors in effective locations. In the same way, if you want to control a particular mode then you wouldn't want to apply the force to a node either, since this would keep the force from influencing that mode.

4.4   VIBRATION ABSORBERS WITHOUT DAMPING

We've already seen how altering the mass, spring, and damping properties of an SDOF system will alter its response. This section will discuss an interesting twist in which we control the response of a sinusoidally forced mass by adding more spring and mass elements to the system. We won't need to add damping, and we won't be dissipating more energy to control the response. Instead, we'll distribute the vibrational energy, moving it from the original mass to the new one.

Devices that absorb energy in this way are found in a variety of applications. One common use is to reduce the vibrational response of high tension power lines. Figure 4.10a shows a photo of a typical powerline with attached vibrational absorbers, known as Stockbridge dampers. Figure 4.10b shows a close-up of the device, and Figure 4.10c shows a simplified diagram of the total powerline/vibration absorber system. Keep in mind that, as already mentioned, there is no energy dissipation taking place,

(*a*)

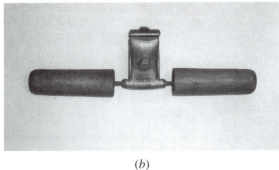

(*b*)

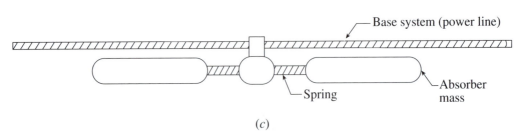

(*c*)

Figure 4.10 (*a*) Stockbridge dampers an powerlines, (*b*) Close-up of a stockbridge damper and (*c*) Elements comprising a stockbridge damper

in spite of the name. All that is happening is that energy which *was* vibrating the cable, will now vibrate the Stockbridge damper.

A simplified picture of a vibration absorber application is shown in Figures 4.11. Figure 4.11*a* indicates the original system, in which a spring-mass combination (m_1 and k_1) is acted upon by a

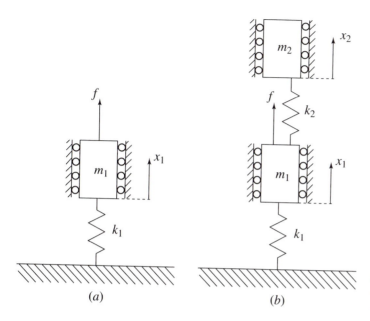

Figure 4.11 (*a*) Primary S-M system; (*b*) undamped vibration absorber, primary and secondary S-M systems

sinusoidal force. The vibration absorber consists of the mass m_2 and the spring k_2, with the total system illustrated in Figure 4.11*b*. We can picture the original system as representing any one of a variety of situations. For the purposes of illustration, let's presume that it represents an engine on its mount, and the sinusoidal force is the force due to a rotating imbalance in the engine. The equation of motion for the original system is given by

$$m_1\ddot{x} + k_1 x = \bar{f}\sin(\omega t) \tag{4.4.1}$$

and the solution is

$$x(t) = \frac{\bar{f}}{k_1 - m_1\omega^2}\sin(\omega t) \tag{4.4.2}$$

To be even more specific, let's assign values to m_1, k_1, $\bar{f}$, and ω. We'll let $m_1 = 100$ kg, $k_1 = 900$ kg/s^2, $\bar{f} = 1$ N, and $\omega = 3$ rad/s. What's the result of these choices? Well, since we've chosen the driving frequency to be equal to the system's natural frequency, we're in a resonance condition and the response becomes unbounded. Figure 4.12 shows the displacement response as a function of frequency. As you can see, the response goes to infinity at $\omega = 3$ rad/s.

This state of affairs is clearly *not* good. It is, however, a reasonable situation to consider. Even if the engine doesn't usually operate at the natural frequency of the engine/base combination, it may happen in some applications that this frequency must be passed through as the engine comes up to speed. Possibly this is also the frequency at which the engine operates most efficiently. Of course, one solution to the problem might be to redesign the suspension. Unfortunately, the suspension design might well be constrained by other design considerations. Thus, what we'll do instead is investigate how we can mitigate the vibration problem by adding a subsystem (m_2 and k_2) to the engine. This

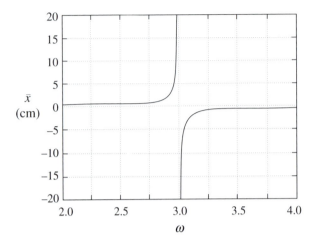

Figure 4.12 Response of primary mass before the addition of vibration absorber

subsystem will be attached to m_1 in the manner shown in Figure 4.11b. What we now have is a 2 DOF problem rather than the original SDOF one. The m_1/k_1 system is referred to as the primary system, and the m_2/k_2 system is known as the secondary system.

Since this problem is simply a special case of the more general problem worked out in the Section 4.3, we can substitute our particular values for m_1, k_1, etc. and obtain the solutions

$$\bar{x}_1 = \frac{(k_2 - m_2\omega^2)\bar{f}_1}{\Delta_1(\omega)} \tag{4.4.3}$$

and

$$\bar{x}_2 = \frac{k_2\bar{f}_1}{\Delta_1(\omega)} \tag{4.4.4}$$

with $\Delta_1(\omega) \equiv (k_1 + k_2 - m_1\omega^2)(k_2 - m_2\omega^2) - k_2^2$.

We can glean a wealth of observations from this particular case. First, we see from (4.4.3) that the response, $\bar{x}_1$, can be *zero* if the forcing frequency, ω, is equal to $\sqrt{\frac{k_2}{m_2}}$. Since we have complete freedom over how we choose m_2 and k_2, we can easily choose them such that $\frac{k_2}{m_2} = \frac{k_1}{m_1}$. Think of what this implies. It means that if our system used to be in a resonant condition because the driving frequency was equal to the SDOF's natural frequency $\left(\frac{k_1}{m_1}\right)$, the newly created 2 DOF system will actually have a zero amplitude response at that frequency. Thus, instead of reducing the amplitude from infinity to something finite, as the addition of damping would have done, we've actually reduced it to zero. Although it isn't obvious right now, adding a degree of freedom as we've done will always create one natural frequency above the original natural frequency and one below it. The frequency at which the zero amplitude response occurs will always lie between the two and thus can always be placed where the original natural frequency occurred.

Before moving beyond this point, we should think physically about what's going on here. If the amplitude $\bar{x}_1$ is zero, that means that m_1 is completely stationary. So we can view our system as not being the forced 2 DOF problem that it really is, but as the SDOF problem of a mass, connected by a spring to a stationary mass. This is shown in Figure 4.13. In this view, the forcing has disappeared (since it is applied to the stationary mass) and all we're left with is a free vibration problem.

We can also see why the mass m_1 remains stationary. Although it isn't *actually* fixed in place, we're forcing it at $\omega = \sqrt{\frac{k_1}{m_1}}$. As we've mentioned, this would be the natural frequency of oscillation of m_2 and k_2 if they had been attached to an immovable body. Thus the m_2/k_2 combination oscillates at just the right frequency to cancel out the external force, and this is exactly what happens. The phase of the absorber force is exactly opposed to that of the external forcing. By oscillating at the correct amplitude, the magnitude of the forcing is also matched. Thus m_1 experiences two forces acting on it that are exactly opposed to each other. The end result—no motion. These forces are illustrated in Figure 4.14.

Let's calculate this out just to be sure. The external force applied to m_1 is $\bar{f}\cos(\omega_n t)$, where $\omega_n^2 = \frac{k_1}{m_1} = \frac{k_2}{m_2}$. From (4.4.4) we have $\bar{x}_2 = \frac{k_2\bar{f}_1}{\Delta_1(\omega)}$. Since m_1 is stationary, the force acting on m_1 due to k_2 is equal to k_2x_2. Using $\omega = \omega_n$ to evaluate $\Delta_1(\omega)$ gives us $-\bar{f}\cos(\omega_n t)$ as the force being exerted on m_1 by k_2. Thus the forces due to the external forcing and the moving mass m_2 are exactly equal and opposed, resulting in no overall unbalanced force acting on m_1.

We can gain some additional insight by looking at two particular absorbers. Figure 4.15 shows the response of both the primary and absorber masses for the case $k_2 = 90$ N/m and $m_2 = 10$ kg. What this means is that we've changed the mass of the overall system by 10%. As you can see, $\bar{x}_1$ goes

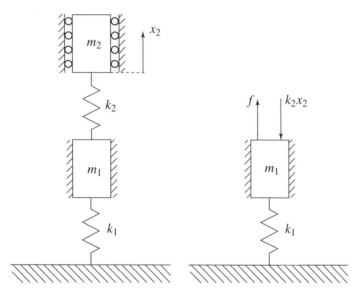

Figure 4.13 Physical equivalent to the effect of an absorber spring-mass

Figure 4.14 Opposed forces acting on primary mass

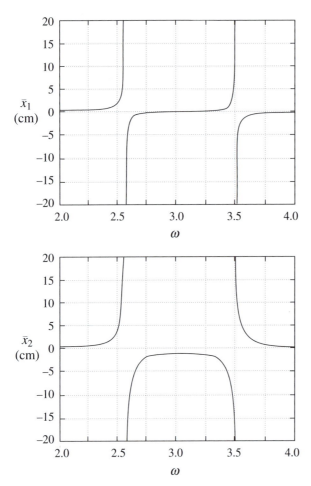

Figure 4.15   Response of vibration absorber with absorber mass equal to one-tenth of primary mass

through zero at $\omega = 3$ rad/s, as predicted. Furthermore, there is a reasonably broad band of frequencies about 3 rad/s for which the response level is quite low. Thus, even in the face of fluctuations in the exciting frequency, we'll still have only a small response. Figure 4.15 also shows the response of the absorber mass itself ($\bar{x}_2$). Again, as the equations predict, there is no frequency for which the amplitude is zero.

Although these results look pretty good, you might be wondering if we could do even better. After all, the only requirement for a vibration absorber to work is that $\sqrt{\frac{k_2}{m_2}}$ equal 3 rad/s. Although the 10% weight penalty we just used isn't terribly bad, wouldn't it be better to use even less mass? After all, designers are often very concerned about weight, and if the actual mass doesn't matter for a vibration absorber, then why not make it as small as possible.

Figure 4.16 shows what happens when we reduce both k_2 and m_2 by a factor of 10. In this case, the secondary mass is a hundred times smaller than the primary mass, meaning that we've hardly

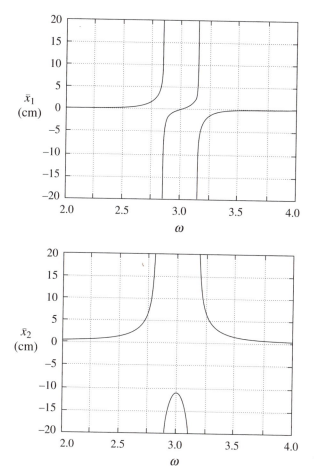

Figure 4.16 Response of vibration absorber with absorber mass equal to one one-hundredth of primary mass

changed the weight of the system at all. But as Figure 4.16 shows, we have to pay a price for the weight savings. First off, the two resonant frequencies are closer together than they were for the larger absorber. Instead of a relatively wide range of frequencies for which the absorber is effective, we now have a very tight range. Only very close to 3 rad/s do we have small vibrations for our primary mass. If the excitation frequency were to shift a bit, we could again end up with a vibration problem, since we'd then encounter resonance conditions around $\omega = 2.7$ rad/s and $\omega = 3.2$ rad/s.

The other downside is seen in the plot of $\bar{x}_2$ vs ω. By decreasing the mass and stiffness, we've increased the amplitude of $\bar{x}_2$ by a factor of 10 at $\omega = 3$ rad/s. This is, in fact, a direct consequence of our having reduced the spring stiffness: k_2 is the means through which the vibrational force is generated that counters the applied forcing. Since the force generated is equal to the stiffness times the amplitude of the motion, if you lower the stiffness by a factor of 10, then to compensate, you've got to increase the amplitude by the same factor.

These results show us that the familiar problem of trading off advantages and problems exists for vibration absorbers. To create a wide band of absorption, you need a heavy mass and a stiff spring. Along with the wide absorption band, you also have a small secondary mass motion. Therefore the packaging concerns aren't very significant but the weight concerns are. If you use a small mass and spring then you'll have a narrow absorption band, meaning that the absorber isn't terribly "robust" to changes in the parameters of the system. Furthermore, packaging will now be a problem because of the large deflections that the secondary mass will experience. As a designer, you'll have to balance these competing trends to come up with an acceptable overall design.

Example 4.5

Problem   Consider again the two-mass system of Figure 4.4. No force acts on m_1 and a force $f_2 = \bar{f} \cos(\omega t)$ acts on m_2. $k_2 = 10,000$ N/m, $k_3 = 0$, $m_1 = 20$ kg, and $m_2 = 10$ kg. What must k_2 be so that m_2 will be stationary when $\omega = 60$ rad/s?

Solution   For m_2 to be stationary, the applied force must be counteracting the free vibrations of the constrained system, as now illustrated.

The natural frequency of the constrained system is given by

$$\omega_n = \sqrt{\frac{k_1 + k_2}{m_1}}$$

Since ω must equal ω_n in a vibration absorber problem, we have

$$\omega = 60 = \sqrt{\frac{k_1 + k_2}{m_1}} = \sqrt{\frac{10,000 + k_1}{20}}$$

$$3600 = \frac{10,000 + k_1}{20}$$

$$k_1 = 62,000 \text{ N/m}$$

4.5   REAL BEHAVIOR OF A VIBRATION ABSORBER

We've just seen that vibrations can be avoided in a forced mass by the addition of a vibration absorber. However you might be a bit doubtful. After all, the second mass could counter the externally applied force only because it was oscillating exactly out of phase with this force. The obvious question is then, What if the force is *not* exactly out of phase with the second mass? In this case, we'll have an unbalanced force on the first mass, and it should move. How does the real world get around this problem?

Figure 4.17 shows the actual response (found from numerically integrating the equation of motion) of both the first and second masses of a vibration absorber problem for a system shown earlier (Figure 4.11b). Note that the primary mass does not remain stationary. However, it's important to realize that the vibrations are *bounded*. If the force and the second mass weren't correctly phase-aligned, we'd have a resonance condition and the oscillations would grow. Since we certainly don't

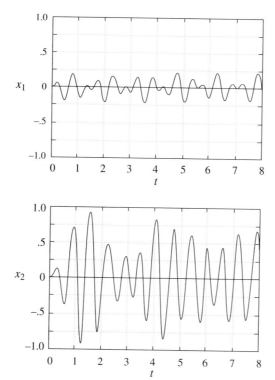

Figure 4.17 Actual response of undamped vibration absorber

have growing oscillations, the force and the second mass *are* correctly phase-aligned. The vibrations occur because with no damping in the system, the homogeneous solution (that due to initial conditions) never dies out; it "rings" forever. This is the vibration we're seeing—that due to initial conditions. To avoid this solution, we can carefully choose our initial conditions to obtain the vibration absorption we want. If the initial conditions are chosen so that the spring force is exactly opposing the applied force, then we'll see purely sinusoidal oscillations for x_2 and no motion for x_1.

Since we can't precisely set up the initial conditions we might need for a real-world implementation, must we assume that vibration absorbers don't actually work, that we'll always have transient vibrations due to initial conditions? Luckily, we don't have to assume this at all, and the reason was just alluded to. We'll run into this problem only in the absence of damping. Since purely undamped systems *never* exist in the real world, we should really add some damping to the system and see how it behaves. Figure 4.18 shows the same system as before with a small amount of linear damping added to the first mass. A typical response of this system is shown in Figure 4.19. Quite a different result! Now we see that after a transient phase, the system responds just as predicted, with a stationary first mass and a second mass that oscillates continually to cancel the external forcing, just what we'd have seen if the conditions had been perfectly set up.

There are only two problems left that we need to address. First, since we've changed the system by adding damping, how do we know that the system is behaving according to prediction? We predicted

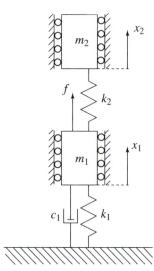

Figure 4.18   Vibration absorber with one damper

only that the *undamped* system, not a damped one, would act in the way shown. Second, just why is it that damping allowed the system to give us the response we predicted? The first question is most easily answered. Remember our physical discussion of the absorber problem: if the first mass is kept stationary, then we're left with an eigenvalue problem for the second mass. That's how we determined that the correct spring/mass ratio for the absorber should be one for which the natural frequency was equal to the forcing frequency of the external force. Luckily, since the damper is attached between the first mass and ground, it doesn't move if the first mass is stationary. So we've got the *same* eigenvalue problem as before. The damper is operative only if the first mass is moving; once it stops, the damper drops out of the picture. As you can see from looking at Figure 4.19, the first mass is certainly moving at the start of the numerical simulation. Thus the damper is operative and must in fact be responsible for bringing about the vibration cancellation that is going on later in the simulation. We're therefore stuck now with determining why this happens. The nice point about this second question is that it will shed a great deal of light on the action of damping in vibratory systems and will serve as a good motivation for the damping analyses to come in later sections.

The best way to approach this second question is to focus our attention on the *desired solution* (stationary first mass) and ask how perturbations $\Delta x_i(t)$ away from this solution will behave. Thus we'll assume that our solution looks like

$$x_1(t) = 0 + \Delta x_1(t) \tag{4.5.1}$$

$$x_2(t) = a\cos(\omega_2 t) + \Delta x_2(t) \tag{4.5.2}$$

for a forcing

$$f(t) = -ak_2\cos(\omega_2 t) \tag{4.5.3}$$

where $\omega_2 = \sqrt{\frac{k_2}{m_2}}$.

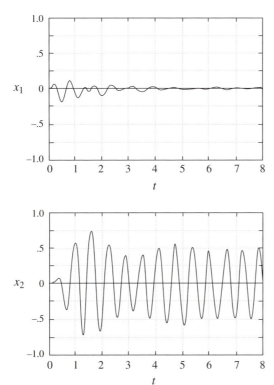

Figure 4.19 Actual response of damped vibration absorber

The approach we're using is a very important one. Since we're basing our assumed solution on a known, *steady state* solution, the added terms (Δx_1 and Δx_2) are the motions that occur *in addition* to the steady state solution, i.e., the transient behavior. Put another way, we want to capture the behavior of the system from $t = 0$ to some future time t_1. We already know that a steady state solution exists for which the first mass is stationary and the second acts to cancel the applied external force. We also know that the actual solution will have a transient part and a steady state part. Since we already know what the steady state part is, we can explicitly include it in our assumed solution, leaving the unknown Δx_i's to represent the transient solutions. This approach to finding the transient behavior of a system is quite common in nonlinear analyses and will be of equal use here.

Our governing equations of motion are

$$m_1\ddot{x}_1 + c_1\dot{x}_1 + (k_1 + k_2)x_1 - k_2 x_2 = f(t) \tag{4.5.4}$$

and

$$m_2\ddot{x}_2 + k_2 x_2 - k_2 x_1 = 0 \tag{4.5.5}$$

Substituting our assumed solutions into (4.5.4) and (4.5.5), we find

$$m_1\Delta\ddot{x}_1 + c_1\Delta\dot{x}_1 + (k_1 + k_2)\Delta x_1 - k_2(a\cos(\omega_2 t) + \Delta x_2) = -ak_2\cos(\omega_2 t) \tag{4.5.6}$$

and

$$m_2(-a\omega_2^2 \cos(\omega_2 t) + \Delta\ddot{x}_2) + k_2(a \cos(\omega_2 t) + \Delta x_2) - k_2\Delta x_1 = 0 \tag{4.5.7}$$

You can see that all the steady state solutions, along with the forcing, cancel out of the equations. (For instance, in (4.5.6) the $-k_2 a \cos(\omega_2 t)$ term on the left cancels out the $-ak_2 \cos(\omega_2 t)$ term on the right.) Thus, upon removing these known terms, we're left with equations that simply involve the Δx_i's:

$$m_1\Delta\ddot{x}_1 + c_1\Delta\dot{x}_1 + (k_1 + k_2)\Delta x_1 - k_2\Delta x_2 = 0 \tag{4.5.8}$$

and

$$m_2\Delta\ddot{x}_2 + k_2\Delta x_2 - k_2\Delta x_1 = 0 \tag{4.5.9}$$

Now, instead of the forced problem we had to deal with before, we've got a free vibration problem. If you examine these equations, you'll see that they're exactly the same equations you would have derived if you'd been concerned with free vibrations of Figure 4.11*b* in the absence of any external forcing. Thus we've made the very important observation that the transient motion in a linear, forced system is governed by the *free vibrations* of that system when no forcing is present.

We've already seen in our work with SDOF systems that damping alters the pure sinusoids that occur for undamped problems into damped sinusoids. The same thing occurs with MDOF systems. Rather than begin a detailed look at the general effect of damping in a system, we're going to postpone this topic until we've looked at normal forms in the next section . Thus we'll content ourselves with looking at the numerical response of (4.5.8) and (4.5.9). Figure 4.20 shows the response of these equations to an arbitrarily chosen set of initial conditions. As you can see, the responses simply oscillate and decay. Eventually, both Δx_1 and Δx_2 approach zero. What does this mean for our vibration problem? It means that as t becomes large, the transient solutions $\Delta x_1(t)$ and $\Delta x_2(t)$ die away, leaving the steady state solution in which the first mass is stationary, i.e., the desired solution for a vibration absorber. It's important to realize that there were no approximations involved in this analysis. Thus it is *always* true that the transient solution will die down as governed by (4.5.8) and (4.5.9), leaving the desired steady state solution.

Let's recap a bit to realize what we've gotten into. First, we've seen that damping, which always exists in the real world, can be quite helpful, not just because it allows disturbances to a system to die down, but also because it can be the mechanism through which passive vibrational control can be effected. In the case just examined, the vibrational response of the system with damping was the same as that predicted for the system without damping. We should now be motivated to ask if this is always true (it's not) and to try and understand the different ways in which damping affects the response of MDOF systems. This is (almost) the next item we'll be discussing. However, before trying to jump into this topic, it will be helpful to look more closely at two more items. First we'll generalize our vibration absorber observations to the multidimensional case. Then we'll look at the different ways of expressing general solutions, specifically, to see if we might be able to view MDOF systems in a

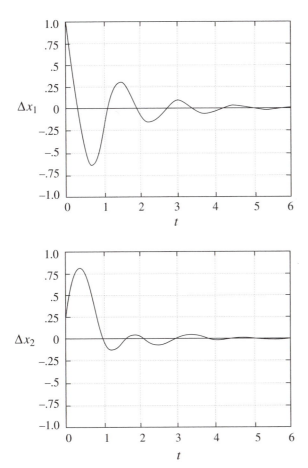

Figure 4.20 Transient solution behavior of a damped vibration absorber

simpler way, i.e., as sets of completely independent, single-degree-of-freedom systems. After we've tackled these issues, we'll return to the question of damping.

4.6 ZEROS IN A FORCED RESPONSE

In Section 4.4 we determined what frequencies of excitation will cause the response at the driven mass to equal zero for a 2 DOF system. The problem was stated differently, of course. We said that our original SDOF system had severe vibrations, and we determined how to reduce the vibrations to zero by the addition of a second mass-spring assembly. The end result, however, is the same as if we'd started with a 2 DOF system and asked what frequency of forcing, applied to the first mass, would result in that mass being stationary. You'll recall that in the process of solving the problem we noticed that, as far as the system response was concerned, a stationary mass was the same as a rigidly fixed mass. Thus the correct frequency of excitation turned out to be the natural frequency of the remaining mass and springs.

The same thing can be seen in MDOF systems. In an MDOF system, we have n masses. Our question will become, What frequency of force excitation, applied to a particular mass, will cause a particular mass in the system to have no motion? Consider the serial chain of mass and springs shown in Figure 4.21a. A sinusoidal force is applied to m_1, and we want to determine the frequencies of excitation at which m_1 will be stationary. Then we'll ask whether m_2 can also be stationary. After we've examined the different masses, we'll move the force and again see how the individual masses might be stationary, even though a sinusoidally varying force is being applied to the system.

Our first task is to see for what frequencies m_1 remains unmoving. Just as in the vibration absorber, we can realize that if m_1 is still, it's as if k_2 is attached to a rigid wall at the left in Figure 4.21b. We're left with a three-mass system: $m_2 - m_4$ with $k_2 - k_5$. Since this is a 3 DOF system, it will have three natural frequencies and three eigenvectors. This is the answer we're seeking. m_1 can be stationary if the applied force exactly opposes the forces that k_2 will transmit (which are due to the free vibrations of the three free masses).

Note that when we had two masses, we were able to get the response of one of the masses to be zero at only one frequency. Furthermore, we saw that the response of the second mass was never zero if the force was applied to the first. In this case we've seen that there are four masses in the system, and we can cause the forced mass to have zero response at three frequencies. For both cases, the forced mass can be zero at $n - 1$ frequencies, where there are n masses in the system.

What if we apply the force to the second mass, as shown in Figure 4.22a rather than to the first?, Are there $n - 1$ frequencies at which the second mass has zero response? If the second mass is stationary, then we've got an SDOF system on the left (made up of k_1, k_2, and m_1) and a 2 DOF system on the right (made up of k_3, k_4, k_5, m_3, and m_4) (Figure 4.22b). Thus we'll have one natural frequency from the left and two from the right. Again we've got $n - 1$ frequencies for which we get a zero amplitude response. This finding is a general one. When we have n masses, the number of frequencies at which the driven mass is stationary is $n - 1$.

What about if we move off from the driven mass? For instance, let's consider the situation of Figure 4.21a, where the force is applied to m_1. This time, let's suppose that m_2 is stationary. Since m_2 isn't moving (by assumption), we have a 2 DOF system to the right (m_3 and m_4) that can oscillate at one of its two natural frequencies as in Figure 4.21c. This vibration will cause a force on m_2 due to k_3. The only way to keep m_2 from moving in response to this force is to apply an equal and opposite force from the left. This force can easily be supplied by appropriately forcing m_1. Thus we have 2, or $n - 2$, zeros for m_2 when m_1 is forced. We saw the same thing for the 2 DOF vibration absorber. When the first mass was forced, the second mass had no zeros responses. For that case n was equal to 2 and therefore $n - 2 = 0$, telling us that there are no frequencies at which the second mass would be stationary.

If you look at m_3 instead of m_2 you'll see that the number of frequencies that yield a zero responses is equal to $n - 3$, i.e., there will be just one (Figure 4.21d). Finally, when you force the first mass and look at the mass furthest away (m_4 in this case) there will be no frequencies at which the response is zero.

All these deductions can be seen in Figure 4.23. This figure shows the actual response versus forcing frequency for each mass of the system of Figure 4.21a. For these plots all the m_i and k_i were

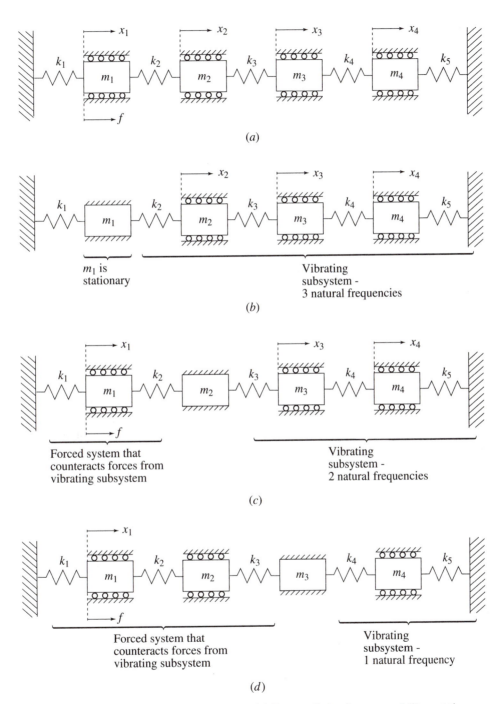

Figure 4.21 Serial chain of spring-mass elements: (*a*) Force applied to first mass and (*b*) m_1 stationary, force acting on m_1 (*c*) m_2 stationary, force acting on m_1 and (*d*) m_3 stationary, force acting on m_1

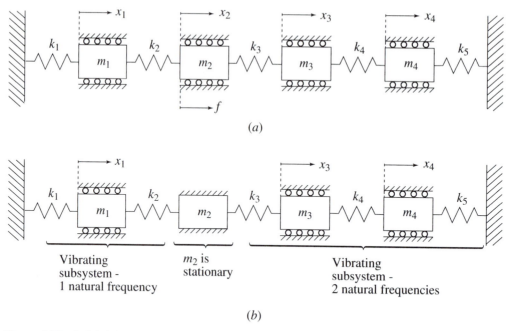

Figure 4.22   Serial chain of spring-mass elements: (*a*) force applied to second mass, (*b*) m_2 stationary, force acting on m_2

given unit values and the magnitude of f was also set equal to 1.0. Figure 4.23*a* shows the response of m_1 for a force applied to m_1. You can see that the response is equal to zero at three distinct frequencies. Moving from m_1 to m_2, Figure 4.23*b* loses one of the zeros. Instead of going through zero at around $\omega = 1.4$, the response approaches zero, but then turns around and moves away from zero and out to negative infinity. Figure 4.23*c* displays only a single zero. Note that this zero occurs at around $\omega = 1.4$. Thus we see that the zeros don't drop off in a regular fashion as we move away from the forced mass: m_2 (in Figure 4.23*b*) didn't have a zero at the same place that m_3 (Figure 4.23*c*) *does* have one. But m_3 has lost the zeros where m_2 had them. Lastly, Figure 4.23*d* displays no zeros.

These observations give us some insight into the design of our systems. We might well want a zero response somewhere in the system. Perhaps we need to mount a mirror and don't want any vibrations that would disturb it. Or it may be that a sensitive piece of equipment will be mounted in a particular place and we want to minimize any vibrations. The foregoing analysis will tell us how many frequencies there are that will permit a vibration absorption to take place.

If we call the forced mass our *actuated* mass and the mass whose output we're observing our *sensed* mass then what we've got is a sensor/actuator pair. By having the sensor and actuator be the same mass, we've created what is called a *collocated sensor/actuator pair*. The ability to be collocated is highly desirable in any control system, and the reason is pretty clear. What we normally want to do in a control system is keep some part of the system under control, usually implying that the part under control is kept at zero, even in the face of disturbances. Figure 4.23*d* shows the worst case of

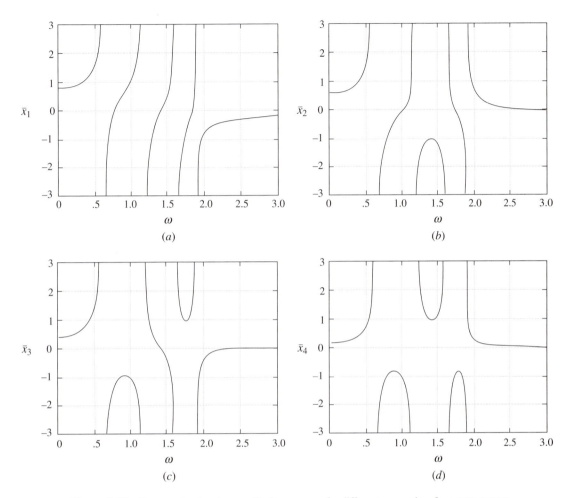

Figure 4.23   Four graphs showing amplitude responses for different masses in a four-mass system

noncollocation. In this situation the control force and sensing are located as far from each other as possible; they're separated by the dynamics of the entire system. This makes control very difficult, as reflected by the absence of any frequencies at which the response is zero.

The physical analogue of this situation is a golfer or a fly fisherman. In each case the person is holding the device (be it a flexible golf club or a really flexible fishing rod), and the part to be controlled is at the other end of the device (the club head or the end of the rod). We could model either device as a string of masses and springs and, if we assume that the person is watching the tip with a view to controlling the system, we'd have a completely noncollocated setup. (Interesting side note: You'll no doubt be saying—if you're a golfer or a fisherman—that people don't actually observe the tip of the implement when golfing or fishing. You're right. The reason is that the dynamics and control problems are too complex. They're beyond our abilities to control in any sort of feedback sense. That's why they're done in what's called an *open-loop* fashion. We simply learn by rote what the correct motions are and then repeat them mechanically.)

When running a modal test we'll also have a similar situation in that we have an exciter and sensor that we use to gather the modal data. We'll have either a collocated or a noncollocated setup depending upon whether sensor and exciter act on the same mass. Precise collocation usually is difficult to achieve, although a new type of sensor, a *self-sensing* actuator, has recently been developed and manages to achieve true collocation.

Example 4.6

Problem   Determine how many zeros to expect in the system shown in Figure 4.22a when the sensed mass is m_3. For this example $m_1 = .75$, $m_2 = .5$, $m_3 = .9$, $m_4 = .55$, $k_1 = 1.1$, $k_2 = 2$, $k_3 = 1$, $k_4 = 1$, and $k_5 = 3.3$. Masses are given in kilograms and stiffnesses in newtons per meter.

Solution   First we'll look to the right of m_3. Since m_3 is stationary, by implication the k_4, k_5, m_4 single-degree-of-freedom system could be oscillating at its natural frequency. This will impress a force on m_3 by means of k_4. If k_3 is impressing an equal and opposite force (because of the force f) then m_3 can remain stationary. That's one frequency. Are there others? Let's look to the left of m_3 to find out. For there to be another frequency, we're going to have m_2 be stationary as well as m_3. If m_2 wasn't stationary, then its motion would make m_3 move, violating our stationarity assumption. There is only one frequency at which m_2 could move without moving m_3, and that is the natural frequency of the k_4-k_5-m_4 system. However we've already taken that answer into account. Thus we conclude that if there is to be a second zero for m_3, m_2 must be stationary. Is this possible? Very definitely. The force on m_2 can be exactly counteracting free vibrations of the k_1-k_2-m_1 system. Physically, the k_1-k_2-m_1 system will be oscillating and the external force will exactly cancel the forces due to this oscillation that k_2 is impressing on m_2. Thus we see that m_2 and m_3 are both stationary. So, in fact, is m_4, since there's no way for it to get excited. This immediately tells us that we've found out something else, namely, that this zero for m_3 must also be a zero of m_2 and m_4.

Figure 4.24 shows the forced response for this system. As you can see, we've got just two frequencies that correspond to zero motion (around $\omega = 2$ rad/s and $\omega = 2.75$ rad/s.) Both zero frequencies that were predicted have shown up.

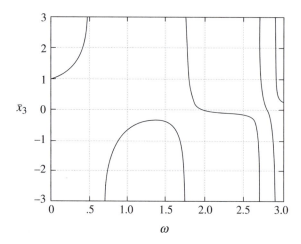

Figure 4.24   Graph showing zero amplitude for third mass, force applied to second mass

An interesting fact that can be seen in the preceding examples is that the number of zeros in a mass-spring chain is always equal to $n - (s+1)$, where n is the number of masses and s is the number of masses away the sensed mass is from the driven mass. In our first example the driven mass was m_1 and the sensed mass varied from m_1 to m_4 (Figure 4.23). When m_1 was the sensed mass, there was no separation between sensed and forced masses, therefore $s = 0$. The number of zeros was three, which matched the result predicted from evaluating $n - 1$. When the sensed mass was m_2, we had $s = 1$ and $n - (1+s) = 2$, matching the numerical results. In Example 4.6, the separation between the driving and forced mass was again equal to one, and $n - (s+1) = 2$, agreeing with the number of zeros shown in Figure 4.24. This formula also tells us to expect no zeros when the driven and sensed masses are at opposite ends of the chain. In this case $s = n - 1$ and $n - (s+1)$ is always equal to zero.

As an example of how this kind of analysis is being used [5], buildings can be analyzed with regard to active control during earthquakes by modeling the different floors as lumped masses, connected by spring/damper elements, with large mechanical actuators used to counter the effect of ground motion and render the floors stationary. The problem setup is essentially identical to the one discussed in this section.

4.7 PUTTING PROBLEMS INTO NORMAL FORM

So far, the way we've viewed 2 DOF systems has been reasonably similar to the way we've dealt with SDOFs. We've seen that 2 DOF systems support eigenvector responses, but we haven't particularly exploited this fact. Now we'll do so.

One obvious characteristic of the general 2 DOF systems we've seen so far is that the system equations are coupled. It's impossible to solve for the response of x_1 without at the same time solving for x_2 (where x_1 and x_2 are the two coordinates of our system). We even have names for the particular types of coupling we might run into. For instance, the following equation is said to be *mass coupled*, or *inertially coupled*:

$$\begin{bmatrix} 2 & -1 \\ -1 & 4 \end{bmatrix} \begin{Bmatrix} \ddot{x}_1 \\ \ddot{x}_2 \end{Bmatrix} + \begin{bmatrix} 5 & 0 \\ 0 & 8 \end{bmatrix} \begin{Bmatrix} x_1 \\ x_2 \end{Bmatrix} = \begin{Bmatrix} \bar{f}_1 \\ \bar{f}_2 \end{Bmatrix} \tag{4.7.1}$$

because the two equations are coupled through the $[M]$ matrix. In a like manner, (4.7.2) is *stiffness coupled* or *spring coupled* because the coupling comes through the $[K]$ matrix.

$$\begin{bmatrix} 1 & 0 \\ 0 & 2 \end{bmatrix} \begin{Bmatrix} \ddot{x}_1 \\ \ddot{x}_2 \end{Bmatrix} + \begin{bmatrix} 7 & -2 \\ -2 & 8 \end{bmatrix} \begin{Bmatrix} x_1 \\ x_2 \end{Bmatrix} = \begin{Bmatrix} \bar{f}_1 \\ \bar{f}_2 \end{Bmatrix} \tag{4.7.2}$$

A fully uncoupled set of equations would look like

$$\begin{bmatrix} 1 & 0 \\ 0 & 2 \end{bmatrix} \begin{Bmatrix} \ddot{x}_1 \\ \ddot{x}_2 \end{Bmatrix} + \begin{bmatrix} 7 & 0 \\ 0 & 8 \end{bmatrix} \begin{Bmatrix} x_1 \\ x_2 \end{Bmatrix} = \begin{Bmatrix} \bar{f}_1 \\ \bar{f}_2 \end{Bmatrix} \tag{4.7.3}$$

For this case, we see that the equation governing the response of x_1 has nothing to do with x_2's behavior. Thus we can solve either equation independently of the other; they're uncoupled.

It's clear that uncoupled equations are nice to have, since it's always easier to solve two independent equations than it is to solve a simultaneous set of equations. The question is, Can we somehow transform our particular equations into a form in which they're decoupled? Obviously, the answer is yes or I wouldn't be mentioning it. When the equations are fully decoupled we say the system is in *normal form*. To take a specific example, let's look at the problem illustrated in Figure 4.25. We'll consider the case for which $m_1 = m_2 = 1$ and $k_1 = k_2 = k_3 = 1$. The governing equations are:

$$\begin{bmatrix} 1 & 0 \\ 0 & 1 \end{bmatrix} \begin{Bmatrix} \ddot{x}_1 \\ \ddot{x}_2 \end{Bmatrix} + \begin{bmatrix} 2 & -1 \\ -1 & 2 \end{bmatrix} \begin{Bmatrix} x_1 \\ x_2 \end{Bmatrix} = \begin{Bmatrix} \bar{f}_1 \\ \bar{f}_2 \end{Bmatrix} \cos(\omega t) \tag{4.7.4}$$

This set of equations is stiffness coupled. To decouple them, we're going to define a new set of coordinates. Instead of x_1 and x_2, we'll express the equations of motion in terms of η_1 and η_2, where the η_i's are defined as

$$\eta_1 = \tfrac{1}{2}(\sqrt{2}x_1 + \sqrt{2}x_2) \tag{4.7.5}$$

and

$$\eta_2 = \tfrac{1}{2}(\sqrt{2}x_1 - \sqrt{2}x_2) \tag{4.7.6}$$

This can be expressed in matrix form as $H = [V]X$, where

$$H \equiv \begin{Bmatrix} \eta_1 \\ \eta_2 \end{Bmatrix}, \quad X \equiv \begin{Bmatrix} x_1 \\ x_2 \end{Bmatrix}, \quad \text{and} \quad [V] \equiv \frac{1}{\sqrt{2}} \begin{bmatrix} 1 & 1 \\ 1 & -1 \end{bmatrix} \tag{4.7.7}$$

At the moment, it shouldn't be clear why in the world we'd want to do this, but we're going to anyway. Later, we'll see why this particular transformation was chosen. Before moving on from this point, we should think about what this new coordinate set is all about. η_1 is simply $\sqrt{2}$ times the mean of x_1 and x_2, and η_2 is $\sqrt{2}$ times one-half of the difference between x_1 and x_2. Thus, if $x_1 = 1$ and $x_2 = 1$, then $\eta_1 = \sqrt{2}$ and $\eta_2 = 0$. Recall that the two eigenvectors for this system are

$$X_1 = \begin{Bmatrix} 1 \\ 1 \end{Bmatrix}, \quad X_2 = \begin{Bmatrix} 1 \\ -1 \end{Bmatrix} \tag{4.7.8}$$

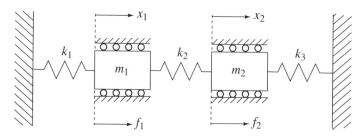

$m_1 = m_2 = 1 \text{ kg}$
$k_1 = k_2 = k_3 = 1 \text{ N/m}$

Figure 4.25   Forced, 2 DOF system

Thus, when only the first eigenvector is involved, $x_1 = x_2 = 1$ and therefore $\eta_1 = \sqrt{2}$ and $\eta_2 = 0$. Similarly, when only the second eigenvector is present, $\eta_1 = 0$ and $\eta_2 = \sqrt{2}$, since $x_1 = 1$ and $x_2 = -1$. Therefore the η coordinates are doing a great job of telling us which eigenvector is present in the response. When η_1 is nonzero and η_2 equals zero, we know that only the first eigenvector is involved. The same holds when $\eta_1 = 0$ and η_2 is nonzero, except that it's the second eigenvector that's now involved. And if both eigenvectors are present, the η's tell us how much of each is present.

This is *not* a trivial point. Previously, all we knew were the positions of m_1 and m_2. We could use this information to determine what was happening with regard to the eigenvectors of the system, but it required some calculation. Now we have an immediate reading with respect to the eigenvector's participation through the η coordinates. Of course in this case we don't know what the individual x_1 and x_2 responses are and we'd have to calculate them from our knowledge of the η's. We can get this information by multiplying by the inverse of the transformation matrix V, so that instead of $H = [V]X$ we have

$$X = [U]H \tag{4.7.9}$$

where

$$[U] = [V]^{-1} = \begin{bmatrix} \frac{1}{\sqrt{2}} & \frac{1}{\sqrt{2}} \\ \frac{1}{\sqrt{2}} & -\frac{1}{\sqrt{2}} \end{bmatrix} \tag{4.7.10}$$

If we go ahead and use (4.7.9) in (4.7.4), we'll get

$$[M][U]\ddot{H} + [K][U]H = \bar{F}\cos(\omega t) \tag{4.7.11}$$

Now there's one more thing we need to do. Rather than leave our equations in the form of (4.7.10) and (4.7.11), we're going to premultiply by $[U]^T$ (the transpose of $[U]$) to obtain

$$[U]^T[M][U]\ddot{H} + [U]^T[K][U]H = [U]^T\bar{F}\cos(\omega t) \tag{4.7.12}$$

We can express this more compactly as

$$[M']\ddot{H} + [K']H = \bar{F}'\cos(\omega t) \tag{4.7.13}$$

where

$$[M'] \equiv [U]^T[M][U], \quad [K'] \equiv [U]^T[K][U], \quad \text{and} \quad \bar{F}' \equiv [U]^T\bar{F} \tag{4.7.14}$$

If you carry through with the indicated operations you'll find that (4.7.13) has the following matrix entries:

$$[M'] = \begin{bmatrix} 1 & 0 \\ 0 & 1 \end{bmatrix}, \quad [K'] = \begin{bmatrix} 1 & 0 \\ 0 & 3 \end{bmatrix}, \quad \text{and} \quad \bar{F}' = \begin{bmatrix} \frac{1}{2}(\bar{f}_1' + \bar{f}_2') \\ \frac{1}{2}(\bar{f}_1' - \bar{f}_2') \end{bmatrix} \tag{4.7.15}$$

Note the good news—we've decoupled the equations. The equation for η_1 is completely independent of the equation for η_2 and vice versa. Our equations are in formal form and the η_i's are our normal coordinates. Furthermore, we can notice some interesting facts. First the mass matrix $[M']$ is the identity matrix. This isn't a big deal, since it was already equal to the identity matrix when we started. However, what's interesting is that we can work it so that *any* initial mass matrix is transformed into the identity matrix. Obviously this property will save storage space on a computer if it's true for any size system (and it is). Second, we can see that the entries in the $[K']$ matrix are exactly equal to the squares of the natural frequencies. It seems that this coordinate transformation is quite powerful. It makes the mass matrix equal to the identity matrix, produces a diagonal stiffness matrix having the square of each natural frequency on the main diagonal and, to top it off, gives us a direct readout of the modal character of the response through the η coordinates.

Having seen all there useful features, you may well be interested in why and how they come about, and how you can harness this transformation power for yourself. Amazingly, the procedure is pretty simple. Just look at the transformation matrix $[U]$ for a second and ask yourself what's familiar about the columns of this matrix. Yes, that's right! They're simply equal to the eigenvectors of the original system. That's almost all there is to it. We find the eigenvectors, put them side by side to define $[U]$, premultiply by $[U]^T$, and get our decoupled form. There are a couple of additional fine points, which we'll talk about as we rigorously show why this decoupling takes place. If you're completely uninterested in seeing this, you can skip to Section 4.9 now. (But you'll miss some good stuff!)

Example 4.7

Problem   Use the transformation matrix

$$[U] = \begin{bmatrix} .3606 & .7326 \\ .4541 & -.6785 \end{bmatrix}$$

to put the system

$$\begin{bmatrix} 2 & 1 \\ 1 & 2 \end{bmatrix} \begin{Bmatrix} \ddot{x}_1 \\ \ddot{x}_2 \end{Bmatrix} + \begin{bmatrix} 100 & -20 \\ -20 & 80 \end{bmatrix} \begin{Bmatrix} x_1 \\ x_2 \end{Bmatrix} = \begin{Bmatrix} 1 \\ 2 \end{Bmatrix} \cos(10t)$$

into normal form.

Solution

$$[U]^T [M][U] = \begin{bmatrix} 1 & 0 \\ 0 & 1 \end{bmatrix}$$

$$[U]^T [K][U = \begin{bmatrix} 22.95 & 0 \\ 0 & 110.4 \end{bmatrix}$$

$$[U]^T F = \begin{bmatrix} 1.27 \\ -.624 \end{bmatrix}$$

Thus we have

$$\begin{bmatrix} 1 & 0 \\ 0 & 1 \end{bmatrix} \begin{Bmatrix} \ddot{\eta}_1 \\ \ddot{\eta}_2 \end{Bmatrix} + \begin{bmatrix} 22.95 & 0 \\ 0 & 110.4 \end{bmatrix} \begin{Bmatrix} \eta_1 \\ \eta_2 \end{Bmatrix} = \begin{Bmatrix} 1.27 \\ -.624 \end{Bmatrix} \cos(10t)$$

4.8 ORTHOGONALITY OF SYSTEM EIGENVECTORS

You're probably all familiar with orthogonality as the term is applied to spatial vectors. It means that the vectors are oriented at right angles to each other. Mathematically, when two vectors are orthogonal, their dot product is zero. Additionally, one way of calculating the dot product of two column vectors A and B is by evaluating $A^T B$. These notions are very close to those that we'll develop in this section. The only twist is that instead of saying certain vectors are orthogonal just to each other, we'll see that the orthogonality conditions include the $[M]$ and $[K]$ matrices. Thus orthogonality is very tightly tied to the particular system we're examining.

We'll sneak up on these results in a very straightforward way, namely, by making a simple examination of our equations and then drawing some conclusions from them. For clarity, we'll derive our results with a 2 DOF example. Keep in mind though that the approach is applicable to a problem with any number of degrees of freedom, as we'll indicate after finishing with the 2 DOF example.

To begin, we'll look at our free vibration problem

$$[M]\ddot{X} + [K]X = 0 \tag{4.8.1}$$

We know now that the solutions to this problem define our system eigenvectors (X_1 and X_2) and natural frequencies (ω_1 and ω_2). Thus, writing (4.8.1) in terms of these known solutions gives us

$$-\omega_1^2[M]X_1 + [K]X_1 = 0 \tag{4.8.2}$$

and

$$-\omega_2^2[M]X_2 + [K]X_2 = 0 \tag{4.8.3}$$

Rearranging these equations gives us

$$\omega_1^2[M]X_1 = [K]X_1 \tag{4.8.4}$$

and

$$\omega_2^2[M]X_2 = [K]X_2 \tag{4.8.5}$$

What we've got here are two sets of vector equalities (the vector on the left equals the vector on the right). Clearly, we'll still have an equality if we premultiply (4.8.4) by X_2^T and (4.8.5) by X_1^T. It's not obvious yet *why* we'd want to do this, but there's certainly nothing that says we can't. So now we have

$$\omega_1^2 X_2^T[M]X_1 = X_2^T[K]X_1 \tag{4.8.6}$$

and

$$\omega_2^2 X_1^T[M]X_2 = X_1^T[K]X_2 \tag{4.8.7}$$

Now let's take the transpose of (4.8.7). (Yes, I know. This seems pointless. But it's not, so just hold on for another moment or two.) Transposing (4.8.7) gives us

$$\omega_2^2 X_2^T [M]^T X_1 = X_2^T [K]^T X_1 \tag{4.8.8}$$

Since the $[M]$ and $[K]$ matrices of spring-mass vibratory systems are symmetric, this is the same as

$$\omega_2^2 X_2^T [M] X_1 = X_2^T [K] X_1 \tag{4.8.9}$$

Take a second to compare (4.8.6) and (4.8.9). They are *identical* except that one involves ω_1 while the other contains ω_2. We're now in a position to be clever. Let's subtract (4.8.9) from (4.8.6). This will give us

$$(\omega_2^2 - \omega_1^2) X_2^T [M] X_1 = 0 \tag{4.8.10}$$

This is interesting! It tells us something fundamental about our eigenvectors. If $\omega_2 \neq \omega_1$, then $X_2^T [M] X_1$ must equal zero. (Otherwise (4.8.10) wouldn't be satisfied.) This is the start of our orthogonality condition. For vibrations problems, the eigenvectors aren't directly orthogonal to each other (i.e., it's not true that $X_1^T X_2 = 0$); rather, they're orthogonal with respect to the mass matrix $[M]$. But wait, there's even more. From (4.8.9) we see that if $X_2^T [M] X_1 = 0$, then $X_2^T [K] X_1$ must also equal zero. Thus the eigenvectors are orthogonal with respect to both the stiffness *and* the mass matrices. In other words, if $\omega_1 \neq \omega_2$ then $X_1^T [M] X_2 = X_1^T [K] X_2 = 0$.

The next thing we need to ask is whether the eigenvectors are also orthogonal to themselves. That is, does $X_1^T [M] X_1 = 0$? One result of linear algebra is that $X^T [A] X$ is never zero if $[A]$ is positive definite unless X is itself zero. Since $[M]$ is always a positive definite matrix, this means that $X_1^T [M] X_1 \neq 0$. The precise value of $X_1^T [M] X_1$ depends on both $[M]$ and X_1. $[M]$ is fixed but, as we've discussed, X_1 can be determined only up to a point. All multiples of our chosen X_1 are also valid eigenvectors. This actually can be turned to our advantage. Rather than always have this uncertainty about X_1, we can demand that it be chosen so that $X_1^T [M] X_1 = 1.0$. This is a type of normalization (known as *mass normalization*) and, if we choose each eigenvector such that its transpose multiplied by $[M]$ multiplied by the eigenvector equals 1.0, we've got an *orthonormal* set of vectors. The "ortho" refers to the vectors' orthogonality, and the "normal" refers to the fact that they're normalized. We'll use a tilde to indicate that the eigenvectors are normalized ($\tilde{X}_i$).

Getting the orthonormalized eigenvectors isn't too hard. Say that you've got an eigenvector X_i, which isn't mass normalized, and that

$$X_i^T [M] X_i = b_i \neq 1.0$$

The mass-normalized eigenvector will just be a scalar multiple of X_i

$$\tilde{X}_i = \alpha_i X_i$$

We know that the mass-normalized eigenvector satisfies

$$\tilde{X}_i^T [M] \tilde{X}_i = 1$$

and so

$$\alpha_i^2 X_i^T [M] X_i = 1$$

$$\alpha_i^2 b_i^2 = 1$$

$$\alpha_i = \frac{1}{\sqrt{b_i}}$$

Thus

$$\tilde{X}_i = \frac{1}{\sqrt{b_i}} X_i = \frac{1}{\sqrt{X_i^T [M] X_i}} X_i$$

And there we've got our mass normalized vector. Once we do this to all the vectors we'll have our orthonormal set of eigenvectors.

Getting back to our derivation, we next we need to ask what $\tilde{X}_1^T [K] \tilde{X}_1$ is equal to. Premultiplying (4.8.4) by $\tilde{X}_1^T$ yields

$$\omega_1^2 \tilde{X}_1^T [M] \tilde{X}_1 = \tilde{X}_1^T [K] \tilde{X}_1 \qquad (4.8.11)$$

Since $\tilde{X}_1^T [M] \tilde{X}_1$ is equal to 1.0, we see that $\tilde{X}_1^T [K] \tilde{X}_1$ is equal to ω_1^2. In exactly the same manner, $\tilde{X}_2^T [K] \tilde{X}_2$ is equal to ω_2^2.

Right about now is a good time to recall that one of our goals was to determine how to get a system into a decoupled form. Well, we've done it! All we need to do is make sure our transformation matrix $[U]$ is made up of our normalized system eigenvectors, i.e.,

$$[U] = [\tilde{X}_1 \vdots \tilde{X}_2] \qquad (4.8.12)$$

Using $X = [U] H$ in (4.8.1) and premultiplying by $[U]^T$ gives us

$$[U]^T [M][U] \ddot{H} + [U]^T [K][U] H = 0 \qquad (4.8.13)$$

which, in expanded form, is equal to

$$\begin{bmatrix} \tilde{X}_1^T [M] \tilde{X}_1 & \tilde{X}_1^T [M] \tilde{X}_2 \\ \tilde{X}_2^T [M] \tilde{X}_1 & \tilde{X}_2^T [M] \tilde{X}_2 \end{bmatrix} \begin{Bmatrix} \ddot{\eta}_1 \\ \ddot{\eta}_2 \end{Bmatrix} + \begin{bmatrix} \tilde{X}_1^T [K] \tilde{X}_1 & \tilde{X}_1^T [K] \tilde{X}_2 \\ \tilde{X}_2^T [K] \tilde{X}_1 & \tilde{X}_2^T [K] \tilde{X}_2 \end{bmatrix} \begin{Bmatrix} \eta_1 \\ \eta_2 \end{Bmatrix} = \begin{Bmatrix} 0 \\ 0 \end{Bmatrix} \qquad (4.8.14)$$

or, in terms of what we now know, these products are equal to

$$\begin{bmatrix} 1 & 0 \\ 0 & 1 \end{bmatrix} \begin{Bmatrix} \ddot{\eta}_1 \\ \ddot{\eta}_2 \end{Bmatrix} + \begin{bmatrix} \omega_1^2 & 0 \\ 0 & \omega_2^2 \end{bmatrix} \begin{Bmatrix} \eta_1 \\ \eta_2 \end{Bmatrix} = \begin{Bmatrix} 0 \\ 0 \end{Bmatrix} \qquad (4.8.15)$$

Although we derived these results using a 2 DOF model, it should be clear that the results are general and can be applied to any size system. Thus, for an n DOF system, our orthogonality conditions would read

$$X_i^T [M] X_j = X_i^T [K] X_j = 0, \quad i \neq j \qquad (4.8.16)$$

In this case, the matrices $[M]$ and $[K]$ are $n \times n$, X_i is the ith eigenvector, and X_j is the jth eigenvector. If the eigenvectors are mass normalized then

$$\tilde{X}_i^T [M] \tilde{X}_i = 1.0 \qquad (4.8.17)$$

and

$$\tilde{X}_i^T [K] \tilde{X}_i = \omega_i^2 \qquad (4.8.18)$$

We've now learned that the eigenvectors of our system are not in general orthogonal to each other, but rather are orthogonal with respect to the $[M]$ and $[K]$ matrices. To distinguish between this situation and the more common concept of orthogonal vectors, we'll make some definitions. If we've got a set of vectors that are orthogonal with respect to the $[M]$ matrix, we'll call them *mass orthogonal*. A set of vectors that are orthogonal to the $[K]$ matrix are *stiffness orthogonal*. If the set is simultaneously orthogonal to both the $[K]$ and $[M]$ matrices, then they are *system orthogonal*. Finally, if they are orthogonal to themselves, i.e.,

$$X_i^T X_j = 0 \qquad (4.8.19)$$

then they are *self-orthogonal*. Note that if $[M]$ is equal to a multiple of the identity matrix, then the eigenvectors are self-orthogonal, since with $[M] = \alpha[I]$ we have

$$X_i^T [M] X_j = \alpha X_i^T [I] X_j = \alpha X_i^T X_j = 0 \qquad (4.8.20)$$

This is a good time to make an observation. The eigenvectors of our system are orthogonal to the $[M]$ and $[K]$ matrices. As you might suspect, this implies that they are also independent. In fact, taken together, the eigenvectors form a very special set. If our system is n-dimensional (n physical coordinates needed to describe the system's configuration), then we can obtain *any* possible configuration by suitably combining the system eigenvectors. A simple example would be to look again at the 2 DOF system shown in Figure 4.1. If all the spring constants are identical and the masses are equal, then the eigenvectors will be $X_1 = \{1 \ 1\}^T$ and $X_2 = \{1 \ -1\}^T$. Any position of the two masses can be represented by these vectors. For instance, let's say the first mass has zero displacement and the second is moved to the right by 4 units. Then

$$\begin{Bmatrix} x_1 \\ x_2 \end{Bmatrix} = \begin{Bmatrix} 0 \\ 4 \end{Bmatrix} = 2 \begin{Bmatrix} 1 \\ 1 \end{Bmatrix} - 2 \begin{Bmatrix} 1 \\ -1 \end{Bmatrix} \qquad (4.8.21)$$

The same holds for any size system. A system's configuration can be represented by an appropriate combination of the eigenvectors. In this case, we say that the system eigenvectors *span* the configuration space of the system. It's worth noting that you don't need to use eigenvectors to span a system's configuration space. In fact, all you need is a set of independent vectors. Since the eigenvectors are independent, they work, but so will any number of other vector sets. In fact, this will be one of the main points when we introduce approximation techniques in Chapter 6.

We can show some more interesting results by using a little linear algebra. The first thing to realize is that our mass matrix is symmetric and positive definite. One linear algebra result tells us that such a matrix can always be written as the product of two identical matrices:

$$[M] = [M]^{\frac{1}{2}}[M]^{\frac{1}{2}} \tag{4.8.22}$$

If we use this fact in our equations of motion, we'll have

$$[M]^{\frac{1}{2}}[M]^{\frac{1}{2}}\ddot{X} + [K]X = 0 \tag{4.8.23}$$

If we define a new set of coordinates by

$$[M]^{\frac{1}{2}}X = H \tag{4.8.24}$$

then we'll have

$$[M]^{\frac{1}{2}}\ddot{H} + [K][M]^{-\frac{1}{2}}H = 0 \tag{4.8.25}$$

Premultiplying by $[M]^{-\frac{1}{2}}$ leaves us with

$$\ddot{H} + [M]^{-\frac{1}{2}}[K][M]^{-\frac{1}{2}}H = 0 \tag{4.8.26}$$

or

$$\ddot{H} + [A]H = 0 \tag{4.8.27}$$

where

$$[A] = [M]^{-\frac{1}{2}}[K][M]^{-\frac{1}{2}}$$

This shows us that we can always transform our equations into a form for which the mass matrix is equal to the identity matrix and the stiffness matrix is symmetric. We know the transformed stiffness matrix must be symmetric because $[M]^{-\frac{1}{2}}$ and $[K]$ are symmetric, and products of symmetric matrices of the foregoing form are themselves symmetric.

Now let's see what we can do here. We know that the solutions of this problem will support n natural frequencies and eigenvectors. If we put the eigenvectors together into a transformation matrix $[U]$, as suggested earlier, then we can write

$$-[U][\Omega^2] + [A][U] = 0 \tag{4.8.28}$$

where $[\Omega^2]$ is an $n \times n$ diagonal matrix having the squares of the system natural frequencies on the diagonal. $[U]$ is called the *modal matrix* for the system. Pulling the left-hand term onto the other side of the equality gives us

$$[U][\Omega^2] = [A][U] \tag{4.8.29}$$

and premultiplying by $[U]^{-1}$ gives us

$$[\Omega^2] = [U]^{-1}[A][U] \tag{4.8.30}$$

What we'll do now is take the transpose of (4.8.29) and show that there is a very special relationship between $[U]^T$ and $[U]^{-1}$. Taking the transpose of (4.8.29) gives us

$$[\Omega^2]^T[U]^T = [U]^T[A]^T \tag{4.8.31}$$

But remember that both $[\Omega^2]$ and $[A]$ are symmetric. Thus (4.8.31) is equivalent to

$$[\Omega^2][U]^T = [U]^T[A] \tag{4.8.32}$$

Postmultiplying (4.8.32) by $([U]^T)^{-1}$ gives us

$$[\Omega^2] = [U]^T[A]([U]^T)^{-1} \tag{4.8.33}$$

Now compare (4.8.33) and (4.8.30). What we have are essentially two expressions of a similarity transformation, between $[A]$ and $[\Omega^2]$. Since transformations like these are unique, we see by comparing terms that $[U]^T$ must be equal to a scalar constant times $[U]^{-1}$. If we use

$$[U]^T = \alpha[U]^{-1} \tag{4.8.34}$$

in (4.8.33) we'll obtain

$$[\Omega^2] = \alpha[U]^{-1}[A][U]\frac{1}{\alpha} \tag{4.8.35}$$

or

$$[\Omega^2] = [U]^{-1}[A][U] \tag{4.8.36}$$

Note that this is exactly the same as (4.8.30). So we've found an important fact, namely, that the transpose of the modal matrix is, within a scalar multiple, equal to the inverse of the modal matrix (when the associated mass matrix is an identity matrix). Knowing this certainly allows us to save time, since calculating a transpose is substantially easier than calculating an inverse.

Example 4.8

Problem   Find the modal matrix for the system

$$\begin{bmatrix} 3 & 0 & 0 \\ 0 & 3 & 0 \\ 0 & 0 & 3 \end{bmatrix}\begin{Bmatrix} \ddot{x}_1 \\ \ddot{x}_2 \\ \ddot{x}_3 \end{Bmatrix} + \begin{bmatrix} 2000 & -500 & 0 \\ -500 & 3000 & -200 \\ 0 & -200 & 2000 \end{bmatrix}\begin{Bmatrix} x_1 \\ x_2 \\ x_3 \end{Bmatrix} = \begin{Bmatrix} 0 \\ 0 \\ 0 \end{Bmatrix}$$

Show that $[U]^T = \alpha[U]^{-1}$.

Solution   Using $[U, d] = \text{eig}(K, M)$ in MATLAB produces the result

$$U = \begin{bmatrix} .3712 & .8511 & -.3714 \\ -.9166 & .3997 & 0 \\ .1485 & .3404 & .9285 \end{bmatrix}$$

Finding U^T (by typing U') yields

$$U^T = \begin{bmatrix} .3712 & -.9166 & .1485 \\ .8511 & .3997 & .3404 \\ -.3714 & 0 & .9285 \end{bmatrix}$$

while U^{-1} (from typing $\texttt{inv(U)}$) produces

$$U^{-1} = \begin{bmatrix} .3712 & -.9166 & .1485 \\ .8511 & .3997 & .3404 \\ -.3714 & 0 & .9285 \end{bmatrix}$$

Since for this case $U^T = U^{-1}$, we have $\alpha = 1$.

4.9 MORE ON NORMAL FORMS

Now that we've seen how to put our system into normal form, we can go ahead and tidy up all the remaining loose ends. To start, let's consider what to do when our problem has an external forcing. In this case, the equations of motion are given by

$$[M]\ddot{X} + [K]X = F \tag{4.9.1}$$

If we go ahead and use $X = [U]H$ to decouple this set of equations, we'll obtain

$$[M']\ddot{H} + [K']H = F' \tag{4.9.2}$$

where

$$[M'] \equiv [U]^T[M][U], \quad [K'] \equiv [U]^T[K][U], \quad \text{and} \quad [\bar{F}'] \equiv [U]^T F \tag{4.9.3}$$

This was derived a few pages back ((4.7.13) and (4.7.14)) and is simply rewritten here for convenience. If the modal matrix $[U]$ has been mass normalized then $[M']$ is equal to the identity matrix and $[K']$ is equal to $[\Omega^2]$. Thus our equations are given by

$$\ddot{H} + [\Omega^2]H = F' \tag{4.9.4}$$

This is a completely decoupled problem. In this form, our problem is exactly equivalent to n individual, second-order equations. Taking the ith one at random, we have

$$\ddot{\eta}_i + \omega_i^2 \eta_i = f_i' \tag{4.9.5}$$

where f_i' is simply the ith component of the modified forcing vector. We know from earlier work that η_i represents the degree of participation of the ith eigenvector in the response. Thus we can clearly see how the external force affects the various modes of the system.

Example 4.9

Problem Consider the 2 DOF example we recently looked at (in which $m_1 = m_2 = 1$ and $k_1 = k_2 = k_3 = 1$), the equations for which are given in (4.7.4). Put the system into normal form and analyze the resulting equations.

Solution We've already seen that, if the modal matrix is given by

$$[U] = \begin{bmatrix} \frac{1}{\sqrt{2}} & \frac{1}{\sqrt{2}} \\ \frac{1}{\sqrt{2}} & -\frac{1}{\sqrt{2}} \end{bmatrix}$$

Then

$$[M'] = \begin{bmatrix} 1 & 0 \\ 0 & 1 \end{bmatrix}, \quad [K'] = \begin{bmatrix} 1 & 0 \\ 0 & 3 \end{bmatrix}, \quad \text{and} \quad [F'] = \begin{bmatrix} \frac{1}{\sqrt{2}}(f_1 + f_2) \\ \frac{1}{\sqrt{2}}(f_1 - f_2) \end{bmatrix}$$

The equations are decoupled, in just the form we expected. Let $f_1 = \bar{f}_1 \cos(\omega t)$ and $f_2 = \bar{f}_2 \cos(\omega t)$. The two relevant equations of our normal-form representation are

$$\ddot{\eta}_1 + \eta_1 = \frac{1}{\sqrt{2}}(\bar{f}_1 + \bar{f}_2)\cos(\omega t)$$

and

$$\ddot{\eta}_2 + 3\eta_2 = \frac{1}{\sqrt{2}}(\bar{f}_1 - \bar{f}_2)\cos(\omega t)$$

The solutions to these problems are simply

$$\eta_1 = \frac{\frac{1}{\sqrt{2}}(\bar{f}_1 + \bar{f}_2)}{1 - \omega^2}\cos(\omega t)$$

and

$$\eta_2 = \frac{\frac{1}{\sqrt{2}}(\bar{f}_1 - \bar{f}_2)}{3 - \omega^2}\cos(\omega t)$$

This seems to make sense. If we realize that the first mode of the system is simply both masses moving together in unison, then the way to best excite it is to have both forces acting in unison, one on each mass. Of course, in general this doesn't happen. So what counts is how much force is being applied on average to the two masses. That's what the forcing component represents—an average force. Similarly, the second mode consists of both masses moving in an opposite direction. Thus, to best excite this mode, the forces must be opposed to each other.

Let's look at a couple of cases to make sure this is clear. If both forces are acting in the same direction and with the same magnitude, then they're aimed directly at the first mode. In this case $(\bar{f}_1 = \bar{f}_2 = \bar{f})$, our equations become

$$\ddot{\eta}_1 + \eta_1 = \sqrt{2}\bar{f}\cos(\omega t)$$

and

$$\ddot{\eta}_2 + 3\eta_2 = 0$$

Thus only the first equation is being forced; the second is completely unaffected. This means that the first mode will respond and the second mode will not.

Similarly, if the forces are of equal magnitude and opposed $(-\bar{f}_2 = \bar{f}_1 = \bar{f})$, we'll have

$$\ddot{\eta}_1 + \eta_1 = 0$$

and

$$\ddot{\eta}_2 + 3\eta_2 = \sqrt{2}\,\bar{f}\cos(\omega t)$$

In this case, the second mode is excited while the first is unaffected.

There are two fundamental observations we can draw from all this. The first is that it is possible to excite individual modes of a system and leave others totally unaffected. This fact is often used during modal tests of large systems, such as airliners. For instance, let's say you want to perform a modal test on a large airplane. Something this big is difficult to excite effectively. One way to get a reasonably sized modal response is to place shakers along the wings and body and then vibrate the plane in such a way as to couple into just a single vibrational mode. This will give you a large, well-defined response, just what you need for a good experimental test.

The second observation is that in general, *all* modes are excited when a system is forced. As we saw in Example 4.9, only for the special cases of $\bar{f}_1 = \bar{f}_2$ and $\bar{f}_1 = -\bar{f}_2$ did we avoid exciting one mode while getting a response from another. For any other combination of inputs, both modes will be involved in the total response. This fact is also of great use in modal testing. A *very* common way of running modal tests is to hit the structure with an instrumented hammer at a particular point. Since all modes are (usually) excited by such an input, a full range of spectral data can be acquired from this single test.

4.10   LINEAR DAMPING

We've already seen that damping isn't a trivial matter, even for the SDOF case. By approximating the actual damping by a linear damping characteristic, we kept the problem solvable and were able to draw some conclusions about how damped systems behave. And, as Section 2.6 showed, damping can often be quite helpful. So let's see how we can add damping to our repertoire.

Unfortunately, the addition of one or more additional degrees of freedom causes some real complications, even if we assume linear damping. To illustrate why this is so, look at Figure 4.26, a 2 DOF spring-mass-damper system. If the dampers are removed, we're left with our familiar spring-mass example. We know that for no damping, the system has the two eigenvectors $X_1 = \frac{1}{\sqrt{2}}\{1 \ \ 1\}^T$ and $X_2 = \frac{1}{\sqrt{2}}\{1 \ \ -1\}^T$. If the system is displaced into either of these configurations, it will oscillate forever in that mode.

It would be nice if, when damping is added, the modal character of the system were preserved. Thus, we'd expect a system that was displaced into the shape of an eigenvector, to find that it *stays* in that configuration, but with the oscillations dying down at a rate that was proportional to the damping

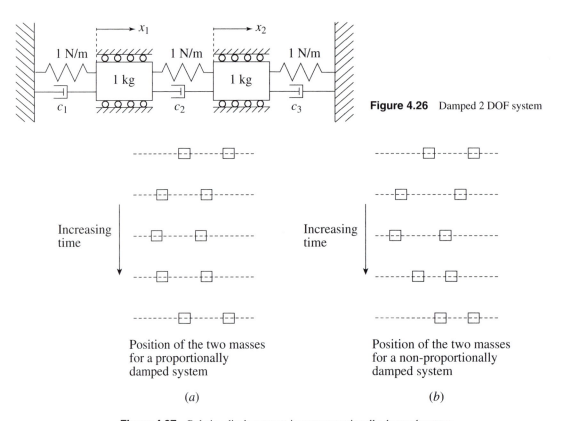

Figure 4.26 Damped 2 DOF system

(a)

(b)

Figure 4.27 Relative displacements in nonproportionally damped system

instead of continuing forever. Unfortunately, the actual behavior is most often not like this. What more often happens can be illustrated by looking at the behavior of the system shown in Figure 4.26 with the specific damping values $c_1 = 0$, $c_2 = .2$ N·s/m, and $c_3 = .1$ N·s/m. Figure 4.27*b* shows what the resultant vibrations look like if the damped system is first displaced into the first (undamped) eigenvector's shape. As you can see, the relative displacements of the two masses are definitely *not* remaining the same: they start out equal, but later are very different.

Interestingly, although the behavior we've just seen didn't give us the modal response we were looking for, a small change in the damping matrix can remedy the situation. Instead of the damping values we used earlier, we can use $c_1 = c_2 = c_3 = .1$ N·s/m. If the two masses are equally displaced (first undamped eigenvector) and then released, we'll see what we had originally hoped for (Figure 4.27*a*). The relative displacements of the masses will always be equal as the oscillation dies down.

What this means is that some damping matrices preserve the undamped modal character of the solution and others don't. The obvious step is then to ask when we can expect the solutions to behave one way or the other.

To make this determination, let's consider the n DOF problem shown in Figure 4.28. The equations of motion for this particular system are

$$[M]\ddot{X} + [C]\dot{X} + [K]X = F \tag{4.10.1}$$

where

$$[M] \equiv \begin{bmatrix} m_1 & 0 & \cdots & & 0 \\ 0 & m_2 & \cdots & & \vdots \\ \vdots & \cdots & m_{n-1} & & 0 \\ 0 & \cdots & & 0 & m_n \end{bmatrix} \tag{4.10.2}$$

$$[C] \equiv \begin{bmatrix} c_1 + c_2 & -c_2 & 0 & 0 & \cdots & & 0 \\ -c_2 & c_2 + c_3 & -c_3 & 0 & & & \vdots \\ 0 & -c_3 & & & & & \vdots \\ \vdots & & & & -c_{n-1} & & 0 \\ & & & -c_{n-1} & c_{n-1} + c_n & -c_n \\ 0 & \cdots & & 0 & -c_n & c_n + c_{n+1} \end{bmatrix} \tag{4.10.3}$$

$$[K] \equiv \begin{bmatrix} k_1 + k_2 & -k_2 & 0 & 0 & \cdots & & 0 \\ -k_2 & k_2 + k_3 & -k_3 & 0 & & & \vdots \\ 0 & -k_3 & & & & & \vdots \\ \vdots & & & & -k_{n-1} & & 0 \\ & & & -k_{n-1} & k_{n-1} + k_n & -k_n \\ 0 & \cdots & & 0 & -k_n & k_n + k_{n+1} \end{bmatrix} \tag{4.10.4}$$

and

$$F \equiv \begin{Bmatrix} f_1 \\ f_2 \\ \vdots \\ f_n \end{Bmatrix} \tag{4.10.5}$$

Although the system shown in Figure 4.28 has the particular $[M]$, $[C]$, and $[K]$ entries listed earlier, (valid for a serial chain of masses), the analysis we'll be undertaking will be valid for any other configuration of masses we might wish to examine. To simplify our analysis, we'll restrict ourselves to a 2 DOF example. As usual when examining the fundamental responses of a system, we'll consider the unforced case. To illustrate the effect of damping, we'll keep the $[M]$ and $[K]$ matrices fixed and vary $[C]$.

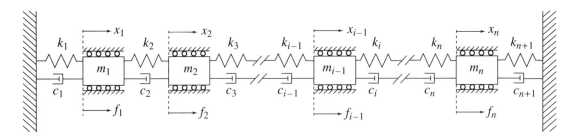

Figure 4.28   Forced n-DOF, spring-mass-damper system

The first issue that we must address is the question of how we'll be solving our problem. The equations of motion will be in the form

$$[M]\ddot{X} + [C]\dot{X} + [K]X = 0 \tag{4.10.6}$$

We could certainly attempt to solve this by assuming $X = \bar{X}e^{\lambda t}$ to obtain:

$$\left[\lambda^2[M] + \lambda[C] + [K]\right]\bar{X} = \{0\} \tag{4.10.7}$$

Unfortunately, solving this is not particularly easy. For instance, assume that we've got a 2 DOF system. With no damping, we'd have to solve a fourth order characteristic equation, *but* λ would appear only in powers of 2 (λ^2, λ^4, etc.). Thus we'd really have a quadratic in λ^2, not a fourth-order problem in λ. But now, with the addition of damping, when we take the determinant of $\lambda^2[M] + \lambda[C] + [K]$, we'll get an honest-to-goodness fourth-order equation. In a like manner, any n DOF problem will give us a $2n$th-order polynomial. And to make matters even more interesting, this polynomial will have complex solutions. This comes about because of the damping matrix's contribution ($\lambda[C]$). Thus we'd need to assume a complex solution ($\lambda = \lambda_r + i\lambda_i$) from the outset. We'd have to substitute this into our characteristic polynomial, set it equal to zero, separate all the real parts of the equation, and set them equal to zero while simultaneously setting the imaginary parts equal to zero as well. We'd then need to create a root-finder program that would solve these equations for us.

Clearly, this isn't overly easy, and it would be far preferable to use a prepackaged code. Unfortunately, there is no code that currently accepts problems in the form we're using (separate $[M]$, $[C]$, and $[K]$ matrices). Most often, a canned numerical code will want to see an eigenvalue problem in the form

$$\dot{X} = [A]X \tag{4.10.8}$$

or perhaps

$$[A]\dot{X} = [B]X \tag{4.10.9}$$

Certainly this is the way MATLAB expects to see the problem. The basic difference is that our equations are in second-order form (two differentiations with respect to time) as opposed to first order (one differentiation with respect to time). Thus, to use MATLAB or similar software, we need to get our equations into first-order form.

Those of you who have already taken a controls course will find this very familiar. The standard controls formulation is one that looks like

$$\dot{X} = [A]X + [B]U \tag{4.10.10}$$

This can be contrasted with our familiar forced representation:

$$[M]\ddot{X} + [K]X = F \tag{4.10.11}$$

The common theme behind our solution procedure for both these problems is to use linear algebra. Both our representation and the controls representation are cast as problems involving combinations of matrices and vectors, something that defines a linear algebra problem. The one advantage the controls community has over the vibrations folks is that they've *always* used the (4.10.10) formulation. We had the advantage as long as we were dealing with undamped problems, but now we're going to have to use the same form as the controls people if we want to get anywhere quickly. In fact, the MATLAB program was originally seized upon by the controls researchers as an excellent tool with which to solve their problems, and many applications have been written specifically for controls applications. Thus, all the development work has been done for us, and all we need to do is put our equations in a form MATLAB recognizes. When equations are in a form that looks like

$$[A]\dot{X} = [B]X \tag{4.10.12}$$

we say they are in *first-order form.*

One nice formulation that takes advantage of symmetry is to put our vibration problem into the form

$$[B]\dot{Y} = [A]Y \tag{4.10.13}$$

where

$$[B] \equiv \begin{bmatrix} -[K] & [0] \\ [0] & [M] \end{bmatrix}, \quad [A] \equiv \begin{bmatrix} [0] & -[K] \\ -[K] & -[C] \end{bmatrix}, \quad \text{and} \quad Y^T \equiv \{x_1 \ x_2 \ \dot{x}_1 \ \dot{x}_2\} \tag{4.10.14}$$

Another convention that's often used for systems that don't have too many degrees of freedom is the *state space form.* As an example of this representation, assume you're given the system equations

$$\ddot{x}_1 + 2\dot{x}_1 + 10x_1 - 5x_2 = 0$$

and

$$\ddot{x}_2 + 4\dot{x}_2 + 30x_2 - 5x_1 = 0$$

To put these equations in state space form you'd let $y_1 = x_1$, $y_2 = \dot{x}_1$, $y_3 = x_2$, and $y_4 = \dot{x}_2$. Two of the equations come from the definition of the y_i

$$\dot{y}_1 = y_2 \quad \text{and} \quad \dot{y}_3 = y_4$$

The other two come from the system's equations of motion. The total representation is given by

$$\begin{Bmatrix} \dot{y}_1 \\ \dot{y}_2 \\ \dot{y}_3 \\ \dot{y}_4 \end{Bmatrix} = \begin{Bmatrix} 0 & 1 & 0 & 0 \\ -10 & -2 & 5 & 0 \\ 0 & 0 & 0 & 1 \\ 5 & 0 & -30 & -4 \end{Bmatrix} \begin{Bmatrix} y_1 \\ y_2 \\ y_3 \\ y_4 \end{Bmatrix}$$

You're certainly free to choose a different representation if you wish. The main thing is to represent an n DOF vibration problem as $2n$ first-order differential equations. The next section will look more precisely at what sort of solution behavior we can expect.

Example 4.10

Problem Reexpress the following equations of motion in symmetric form and in standard state space form. Analyze them both with MATLAB and show that the final results are the same.

$$\begin{bmatrix} 2 & 0 \\ 0 & 1 \end{bmatrix} \begin{Bmatrix} \ddot{x}_1 \\ \ddot{x}_2 \end{Bmatrix} + \begin{bmatrix} 1000 & -400 \\ -400 & 2000 \end{bmatrix} \begin{Bmatrix} x_1 \\ x_2 \end{Bmatrix} = \begin{Bmatrix} 0 \\ 0 \end{Bmatrix}$$

Solution First we'll look at a state space form. Our state vector is given by

$$Y = \begin{Bmatrix} y_1 \\ y_2 \\ y_3 \\ y_4 \end{Bmatrix}$$

where $y_1 = x_1$, $y_2 = \dot{x}_1$, $y_3 = x_2$, and $y_4 = \dot{x}_2$.
Our equations of motion are

$$\dot{y}_1 = \dot{x}_1 = y_2$$

$$\dot{y}_2 = \ddot{x}_1 = \tfrac{1}{2}(-1000x_1 + 400x_2) = -500y_1 + 200y_3$$

$$\dot{y}_3 = \dot{x}_2 = y_4$$

$$\dot{y}_4 = \ddot{x}_2 = 400x_1 - 2000x_2 = 400y_1 - 2000y_3 \qquad (4.10.15)$$

or

$$\dot{Y} = [A]Y$$

where

$$[A] = \begin{bmatrix} 0 & 1 & 0 & 0 \\ -500 & 0 & 200 & 0 \\ 0 & 0 & 0 & 1 \\ 400 & 0 & -2000 & 0 \end{bmatrix}$$

Typing $[v, d] = \texttt{eig(A)}$ (where $\texttt{A} = [A]$) yields

$$v = \begin{bmatrix} -.0028 & -.0028 & .0457 & .0457 \\ -.1278i & .1278i & .9673i & -.9673i \\ .0219 & .0219 & .0118 & .0118 \\ .9916i & -.9916i & .2494i & -.2494i \end{bmatrix}$$

$$d = \begin{bmatrix} 45.2942i & 0 & 0 & 0 \\ 00 & -45.2942i & 0 & 0 \\ 0 & 0 & 21.1764i & 0 \\ 0 & 0 & 0 & -21.1764i \end{bmatrix}$$

We've got natural frequencies of 21.1764 and 45.2942 rad/s.
Next we'll use the symmetric formulation. Let

$$Z = \begin{Bmatrix} z_1 \\ z_2 \\ z_3 \\ z_4 \end{Bmatrix}$$

where $z_1 = x_1$, $z_2 = x_2$, $z_3 = \dot{x}_1$, and $z_4 = \dot{x}_2$. Then to

$$[B] = \begin{bmatrix} -1000 & 400 & 0 & 0 \\ 400 & -2000 & 0 & 0 \\ 0 & 0 & 2 & 0 \\ 0 & 0 & 0 & 1 \end{bmatrix}$$

$$[A] = \begin{bmatrix} 0 & 0 & -1000 & 400 \\ 0 & 0 & 400 & -2000 \\ -1000 & 400 & 0 & 0 \\ 400 & -2000 & 0 & 0 \end{bmatrix}$$

we apply $[w, e] = \texttt{eig(A, B)}$, which yields

$$v = \begin{bmatrix} .0028 & .0457i & .0457 & -.0028i \\ -.0219 & .0118i & .0118 & .0219i \\ -.1278i & .9673 & .9673i & .1278 \\ .9916i & .2494 & .2494i & -.9916 \end{bmatrix}$$

$$d = \begin{bmatrix} -45.2942i & 0 & 0 & 0 \\ 00 & -21.1764i & 0 & 0 \\ 0 & 0 & 21.1764i & 0 \\ 0 & 0 & 0 & 45.2942i \end{bmatrix}$$

You can see that the natural frequencies are the same (21.1764 and 45.2942 rad/s) and that the eigenvectors have the same entries (but rearranged owing to the difference in the order of the x_1, x_2, $\dot{x}_1$, $\dot{x}_2$ within them). We'll discuss the details of how the eigenvectors and natural frequencies are displayed in MATLAB in the following pages.

4.11 COMPARISON OF DAMPED EIGENSOLUTIONS

One of the more confusing aspects of eigenanalysis lies in the interpretation of the complex eigenvectors and eigenvalues that result from looking at damped systems. To shed some light on this problem, we present a carefully chosen succession of problems to illustrate what you can expect from typical analyses. All the calculations will be done using MATLAB, so feel free to double-check any of the results.

Example 4.11

Problem   To start, let's look at a problem that's already familiar—an undamped oscillator:

$$\begin{bmatrix} 1 & 0 \\ 0 & 2 \end{bmatrix}\begin{Bmatrix} \ddot{x}_1 \\ \ddot{x}_2 \end{Bmatrix} + \begin{bmatrix} 2 & -1 \\ -1 & 2 \end{bmatrix}\begin{Bmatrix} x_1 \\ x_2 \end{Bmatrix} = \begin{Bmatrix} 0 \\ 0 \end{Bmatrix}$$

Our usual analysis of these equations would be to assume motion of the form $Xe^{i\omega t}$, solving the characteristic equation (quadratic in ω^2) and then determining the eigenvectors. Doing so would lead to

$$\omega_1^2 = .634, \quad X_1 = \begin{Bmatrix} .731 \\ 1.0 \end{Bmatrix}$$

and

$$\omega_2^2 = 2.366, \quad X_2 = \begin{Bmatrix} 1.0 \\ -.366 \end{Bmatrix}$$

What will we get if we put our system into symmetric first-order form?

Solution   From (4.10.13) we'll have

$$[B] = \begin{bmatrix} -2 & 1 & 0 & 0 \\ 1 & -2 & 0 & 0 \\ 0 & 0 & 1 & 0 \\ 0 & 0 & 0 & 2 \end{bmatrix}$$

and

$$[A] = \begin{bmatrix} 0 & 0 & -2 & 1 \\ 0 & 0 & 1 & -2 \\ -2 & 1 & 0 & 0 \\ 1 & -2 & 0 & 0 \end{bmatrix}$$

If we then type

$$[v, d] = \texttt{eig(A, B)} \quad (\text{cr})$$

MATLAB will construct a modal matrix (v) and a matrix (d) containing the eigenvalues of the problem on the diagonal (and zeros as the off-diagonal entries). Note that the d(1, 1) will be the eigenvalue that corresponds to the first column of v, d(2, 2) will correspond to the second column, etc. Keep in mind that these column entries are eigenvector solutions.

Going ahead and typing

$$[v, d] = \mathtt{eig}(a, b) \quad (cr)$$

will get us:

$$v = \begin{bmatrix} .5118 & -.4621i & .5118 & -.4621i \\ -.1873 & -.6312i & -.1873 & -.6312i \\ -.7873i & -.3679 & .7873i & .3679 \\ .2882i & -.5026 & -.2882i & .5026 \end{bmatrix}$$

and

$$d = \begin{bmatrix} -1.5382i & 0 & 0 & 0 \\ 0 & -.7962i & 0 & 0 \\ 0 & 0 & 1.5382i & 0 \\ 0 & 0 & 0 & .7962i \end{bmatrix}$$

Note that we now have four eigenvalues and four eigenvectors. We can immediately notice a few things about the results. To start off, the entries in the d matrix are not in any particular order. First we've got $-1.5382i$, followed by $-.7962i$, $1.5382i$, and finally $.7962i$. We can recognize the magnitudes of these as ω_1 or ω_2. These are imaginary because the general solution assumed by MATLAB is of the form $Xe^{\lambda t}$. Thus the first entry indicates a time behavior of $e^{-i1.5382t}$, i.e., sinusoidal motion.

Why didn't MATLAB order these from smallest to highest, with the positive value first? The reason is that the programmers didn't think it was important to order them in this way. We'll see this kind of thing crop up continually: we might want an answer in a particular way, but the software isn't set up to give us that solution form. In MATLAB's case, the ordering of the eigenvalues is essentially random. Remember this! Canned routines are convenient, but it's very important to understand what the subroutines are doing, how they're doing it, and in what form they present the results. Failure to take the time to acquire such understanding will often cause pain and frustration.

Now let's examine the eigenvectors that correspond to the eigenvalues (found in the matrix v). We can see that the first eigenvector matches our predictions. We found the eigenmode of ω_2^2 to be $\{1. \quad -.366\}^T$. The first two entries of the first column of v are $\{.5118 \quad -.1873\}^T$. If we multiply both terms by 1.9539, we obtain $\{1. \quad -.366\}^T$, matching our original determination. Remember that eigenvectors are known only within a scalar multiple anyway; thus having to multiply MATLAB's results by a scalar is not a problem. We need look at only the first two entries because these correspond to x_1 and x_2, the physical displacements of our system. The last two entries, $-.7873i$ and $.2882i$, correspond to $\dot{x}_1$ and $\dot{x}_2$. We know that, given $X = \bar{X}e^{\lambda t}$, the velocity will simply be $\dot{X} = \lambda \bar{X}e^{\lambda t}$. Thus the velocity entries should differ from the displacement entries by λ. For our first eigenvector, the corresponding λ (the $d(1, 1)$ entry) is $-1.5382i$. Thus we'd expect $-1.5382i\{.5118 \quad -.1873\}^T$ to equal the third and fourth entries of the first column of v, i.e., $\{-.7873i \quad .2882i\}^T$. And, luckily for us, it does. Therefore we see that the latter half of each eigenvector is really redundant. Once we know the first half, along with the associated eigenvalue, we can compute the second half.

Example 4.12

Problem

Now we'll consider what happens if we add damping to our problem for which the damping matrix is a linear combination of $[M]$ and $[K]$. For this case, we'll let $[C] = .1[M] + .1[K]$:

$$\begin{bmatrix} 1 & 0 \\ 0 & 2 \end{bmatrix} \begin{Bmatrix} \ddot{x}_1 \\ \ddot{x}_2 \end{Bmatrix} + \begin{bmatrix} .3 & -.1 \\ -.1 & .4 \end{bmatrix} \begin{Bmatrix} \dot{x}_1 \\ \dot{x}_2 \end{Bmatrix} + \begin{bmatrix} 2 & -1 \\ -1 & 2 \end{bmatrix} \begin{Bmatrix} x_1 \\ x_2 \end{Bmatrix} = \begin{Bmatrix} 0 \\ 0 \end{Bmatrix}$$

What will an eigenanalysis yield for this problem?

Solution   Putting the equations into symmetric first-order form gives us

$$[B] = \begin{bmatrix} -2 & 1 & 0 & 0 \\ 1 & -2 & 0 & 0 \\ 0 & 0 & 1 & 0 \\ 0 & 0 & 0 & 2 \end{bmatrix}$$

and

$$[A] = \begin{bmatrix} 0 & 0 & -2 & 1 \\ 0 & 0 & 1 & -2 \\ -2 & 1 & -.3 & .1 \\ 1 & -2 & .1 & -.4 \end{bmatrix}$$

with eigensolutions

$$v = \begin{bmatrix} .5118 + .0041i & -.2126 - .4103i & .4681 + .2070i & .0248 - .4614i \\ -.1873 - .0015i & -.2904 - .5605i & -.1714 - .0758i & .0339 - .6303i \\ -.0799 - .7833i & -.3076 + .2019i & -.3952 + .6809i & .3634 + .0574i \\ .0292 + .2867i & -.4202 + .2758i & .1447 - .2492i & .4965 + .0783i \end{bmatrix}$$

and

$$d = \begin{bmatrix} -.1683 - 1.5290i & 0 & 0 & 0 \\ 0 & -.0817 - .7920i & 0 & 0 \\ 0 & 0 & -.1683 + 1.5290i & 0 \\ 0 & 0 & 0 & -.0817 + .7920i \end{bmatrix}$$

So, is this good or bad? Well, actually it's good, because the damped solution is behaving in the way we want. Of course, it's not crystal clear that this is so (because of the way MATLAB displays its results), so let's clarify it a bit.

First off, note that our eigenvalues have changed in an appropriate way. They were imaginary before. This was good because it indicated undamped oscillatory motion. Now they're complex. The imaginary component changed only slightly (because the damping wasn't large), and we've gotten a new real component. Specifically, the undamped solution $-1.5382i$ has now changed to $-.1683 - 1.529i$. The real component is negative for all four eigenvalues, indicating that all the solutions are going to decay in time. As expected, the roots are complex conjugates of each other. Therefore we can add them together to obtain sine and cosine solutions with an exponentially decaying envelope.

Now, on to the eigenvectors. Note that they are all different. For clarity, we'll take the complex vector and rotate it in the complex plane so that the first entry of each column is purely real. To do this, we need to first express the vectors in magnitude/phase notation. Next, we'll check what the phase happens to be for the first column entry and then subtract this phase from all entries in the column, including the first one. This will have the effect of rotating the entire vector. To avoid major boredom, we'll look at just one set of eigenvectors, those associated with the $-.1683 \pm 1.529i$ roots.

The easiest way to rotate the vector is to let MATLAB do the work. What we need to do is determine the magnitude and phase of the vector. Luckily, MATLAB has two functions called `abs` and `angle` that do just this. If our input vector is X then typing `abs(X)` returns the absolute value of each component in the vector and typing `angle(X)` returns the value of the phase associated with each entry. Thus to rotate the vector so that the first entry is real, we need simply use `angle`, record the phase angle (call it ϕ), and then multiply the vector by $\exp(-i * \phi)$.

Using `angle` on the first column of v gives us

$$\left\{ \begin{array}{r} .0080 \\ -3.1336 \\ -1.6724 \\ 1.4692 \end{array} \right\}$$

Thus we must remove a phase angle of .008 radian from each entry of our vector. Doing this (by multiplying by $\exp(-.008i)$ changes the first column of v to

$$\left\{ \begin{array}{c} .5118 \\ -.1873 \\ .7874e^{-1.6804i} \\ .2882e^{1.4612i} \end{array} \right\}$$

Note that the velocity terms (the third and fourth entries) have been expressed in magnitude/phase form to illustrate another point. The two entries should be exactly out of phase with each other, since the displacements are of opposite sign. This is exactly what we see: -1.6804 radians is π radians different from 1.4612 radians. Thus the two velocity entries are also exactly out of phase with each other.

Notice further that the first two entries are exactly the same as those for the undamped case. Not just close—but exact. This means that we've retained the same displacement shape for our eigenvector! Thus, if we displace the system into the configuration defined by this eigenvector, the masses will stay in this configuration as the overall motion dies down. You can verify that the third entry ($.7874e^{-1.6804i}$) is simply the first one (.5118), multiplied by the eigenvalue $-.1683 - 1.529i$. The same holds for the second and fourth entries. This shows that the third and fourth entries are still equal to the first and second entries multiplied by the associated eigenvalue.

Proceeding the same way with the third column of v transforms it to

$$\left\{ \begin{array}{c} .5118 \\ -.1873 \\ .7874e^{1.6804i} \\ .2882e^{-1.4612i} \end{array} \right\}$$

The only difference between this vector and the first is that the imaginary parts of the third and fourth entries have switched signs. This is expected, since the associated eigenvalues ($-.1683 - 1.529i$

and $-.1683 + 1.529i$) are complex conjugates of each other. If you wish, you can go ahead and verify that this behavior is exhibited by the remaining two eigenvectors.

In Example 4.12, the undamped eigenmode was preserved even when damping was added, because the damping was of a very special form. Recall that the damping matrix was defined as $[C] = .1[M] + .1[K]$. This is called *proportional damping* for the very obvious reason that $[C]$ is made up of components that are proportional to both the $[M]$ matrix and the $[K]$ matrix. Thus we define a proportionally damped system as one in which the damping matrix can be expressed by

$$[C] = \alpha[M] + \beta[K] \tag{4.11.1}$$

where α and β are real, scalar constants.

We can understand what's happening by applying our knowledge of normal forms. To start, let's consider

$$[M]\ddot{X} + [C]\dot{X} + [K]X = 0 \tag{4.11.2}$$

If we go ahead and use normal coordinates ($X = [U]H$) and then premultiply by $[U]^T$, we'll get

$$[U]^T[M][U]\ddot{H} + [U]^T[C][U]\dot{H} + [U]^T[K][U]H = 0 \tag{4.11.3}$$

Assuming that the $[U]$ matrix has been mass normalized, this becomes

$$[I]\ddot{H} + [U]^T[C][U]\dot{H} + [\Omega^2]H = 0 \tag{4.11.4}$$

where $[\Omega]$ contains the squares of the system's natural frequencies along its diagonal.

Now we'll use the fact that $[C] = \alpha[M] + \beta[K]$. Thus we have

$$[I]\ddot{H} + [U]^T[\alpha[M] + \beta[K]][U]\dot{H} + [\Omega^2]H = 0 \tag{4.11.5}$$

or

$$[I]\ddot{H} + \left[\alpha[U]^T[M][U] + \beta[U]^T[K][U]\right]\dot{H} + [\Omega^2]H = 0 \tag{4.11.6}$$

Since we already know $[U]^T[M][U]$ and $[U]^T[K][U]$, we can rewrite this equation as

$$[I]\ddot{H} + \left[\alpha[I] + \beta[\Omega^2]\right]\dot{H} + [\Omega^2]H = 0 \tag{4.11.7}$$

Notice that all the matrices are diagonal. Thus we have a set of n completely decoupled equations:

$$\ddot{\eta}_i + (\alpha + \beta\omega_i^2)\dot{\eta}_i + \omega_i^2\eta_i = 0, \quad i = 1, 2, \ldots, n \tag{4.11.8}$$

So there you have it. Using the *undamped* modal matrix produced a set of decoupled equations, each of which governs the motion of one particular eigenvector. Since there's no coupling between the eigenvectors, we can start the system off in the configuration of a given eigenvector and it will stay in that configuration. The rate of decay will be governed by the particular α, β, and ω_i parameters.

At this point, you may be tempted to ask what would happen with our eigenvalue analysis if we had input the normal-form matrices rather than the physical mass, spring, and damping matrices. We'll do just that in Example 4.13.

Example 4.13

Problem   Determine the results of using normal-form matrices rather than physical ones for the system of Example 4.12.

Solution   If we carry through the operations shown in (4.11.3), we'll obtain

$$\begin{bmatrix} 1.651 & 0 \\ 0 & 1.118 \end{bmatrix} \begin{Bmatrix} \ddot{x}_1 \\ \ddot{x}_2 \end{Bmatrix} + \begin{bmatrix} .270 & 0 \\ 0 & .376 \end{bmatrix} \begin{Bmatrix} \dot{x}_1 \\ \dot{x}_2 \end{Bmatrix} + \begin{bmatrix} 1.047 & 0 \\ 0 & 2.646 \end{bmatrix} \begin{Bmatrix} x_1 \\ x_2 \end{Bmatrix} = \begin{Bmatrix} 0 \\ 0 \end{Bmatrix}$$

Putting this into A, B form, we'll get

$$[B] = \begin{bmatrix} -1.047 & 0 & 0 & 0 \\ 0 & -2.646 & 0 & 0 \\ 0 & 0 & 1.651 & 0 \\ 0 & 0 & 0 & 1.118 \end{bmatrix}$$

and

$$[A] = \begin{bmatrix} 0 & 0 & -1.047 & 0 \\ 0 & 0 & 0 & -2.646 \\ -1.047 & 0 & -.270 & 0 \\ 0 & -2.646 & 0 & -.376 \end{bmatrix}$$

Note that each 2×2 block in the A and B matrices is diagonal, which is expected, since all the individual normal coordinate matrices are diagonal. Just to demonstrate that you don't have to mass-normalize everything, the precise $[U]$ matrix produced by MATLAB was used in the transformations. Since this matrix wasn't mass-normalized, we won't have an identity mass matrix for the η coordinate representation. The overall ratios, however, are correct.

If we go ahead and ask MATLAB to analyze calculate the eigensolution, we'll obtain

$$v = \begin{bmatrix} .782 & .782i & 0 & 0 \\ 0 & 0 & .5451 & -.545i \\ -.0639 - .620i & -.620 - .0639i & 0 & 0 \\ 0 & 0 & -.0917 - .833i & .8334 + .0917i \end{bmatrix}$$

and

$$d = \begin{bmatrix} -.0817 - .792i & 0 & 0 & 0 \\ 0 & -.0817 + .792i & 0 & 0 \\ 0 & 0 & -.1683 - 1.529i & 0 \\ 0 & 0 & 0 & -.1683 + 1.529i \end{bmatrix}$$

This is right on the money. Note that the displacement entries display complete decoupling, just as they should. Either the first or the second entry is nonzero for each eigenvector, while the other is zero. This tells us that our normal coordinates correspond to a single eigenvector, just as we expected. You can also see that MATLAB isn't going to give us a big break, even with such a well-ordered problem. The first and second columns of v, which would normally be written in the form

$$\begin{Bmatrix} a \\ 0 \\ \lambda_1 a \\ 0 \end{Bmatrix}$$

and

$$\begin{Bmatrix} 0 \\ a \\ 0 \\ \lambda_1^* a \end{Bmatrix}$$

don't quite fit the bill, as the second column of the MATLAB output has been multiplied by i. Again, this is because the software is written to accomodate the concerns of programmers, not to address the wishes of users doing vibrational analyses.

Unfortunately, it's probably pretty clear that the $[C]$ matrices are *not* proportionally damped more often than they are. So it's time to ask what happens to our solutions in the case of nonproportional damping. We'll be looking at three examples. First, we'll use our physical coordinates to analyze the case of nonproportional damping. Next, we'll transform into normal form and see what the solutions look like. Finally, we'll force the equations into a decoupled form to see the effect on the overall solution.

Example 4.14

Problem To start, we'll do an eigenanalysis on the physical system of the preceding examples, using the damping matrix

$$[C] = \begin{bmatrix} .6 & -.2 \\ -.2 & .3 \end{bmatrix}$$

The question is to determine what this does to our resulting solution.

Solution Our A and B matrices are

$$[B] = \begin{bmatrix} -2 & 1 & 0 & 0 \\ 1 & -2 & 0 & 0 \\ 0 & 0 & 1 & 0 \\ 0 & 0 & 0 & 2 \end{bmatrix}$$

and

$$[A] = \begin{bmatrix} 0 & 0 & -2 & 1 \\ 0 & 0 & 1 & -2 \\ -2 & 1 & -.6 & .2 \\ 1 & -2 & .2 & -.3 \end{bmatrix}$$

The eigenvectors and eigenvalues resulting from using these in `eig` are

$$v = \begin{bmatrix} .512 + .0584i & -.359 - .285i & .297 + .421i & -.0731 - .452i \\ -.172 - .0579i & -.538 - .335i & -.135 - .122i & -.0292 - .633i \\ -.0782 - .786i & -.203 + .304i & -.724 + .316i & .364 - .029i \\ -.0336 + .276i & -.232 + .450i & .225 - .164i & .506 + .0177i \end{bmatrix}$$

and

$$d = \begin{bmatrix} -.310 - 1.502i & 0 & 0 & 0 \\ 0 & -.0647 - .796i & 0 & 0 \\ 0 & 0 & -.310 + 1.502i & 0 \\ 0 & 0 & 0 & -.0647 + .796i \end{bmatrix}$$

Just as in Example 4.12, we can put the eigenvectors into magnitude/phase form and shift each of the vector entries by the appropriate phase to cause the first entry to be a real quantity. If our system supported real eigenvectors (as the proportionally damped case did), then both the first and second elements of the eigenvectors would have to be real.

Carrying out this procedure on the first column of v gives us

$$X_1 = \begin{Bmatrix} .515 \\ -.177 - .0396i \\ -.156 - .774i \\ -.0046 + .278i \end{Bmatrix}$$

Bad news. The second entry says that when x_1 is .515, x_2 is going to be $-.177 - .0396i$. Since this isn't a real quantity, it's pretty clear that the second mass is going to have a tough time doing this physically. In fact, it's impossible, since we're never going to see an imaginary displacement. The reason for the difficulty is that we don't have distinct, real eigenvectors that will decay smoothly to zero. The convenient view of real eigenvectors that decay at a particular rate is just not applicable in most damped systems. To solve for the overall vibrations, we'll have to use all the eigenvectors and combine them together with appropriate multipliers (undoubtedly complex ones). The total solution will involve only real displacements, but the individual eigenvectors no longer display purely real characteristics. The overall behavior is certainly real, but there is no discernible eigenvector anymore.

Example 4.15

Problem   Now we'll try and put the system into normal form. What we should expect to see is that the damping matrix doesn't diagonalize.

Solution   For this system, the A and B matrices are given by:

$$[B] = \begin{bmatrix} -.786 & 0 & 0 & 0 \\ 0 & -.641 & 0 & 0 \\ 0 & 0 & .332 & 0 \\ 0 & 0 & 0 & 1.01 \end{bmatrix}$$

and

$$[A] = \begin{bmatrix} 0 & 0 & -.786 & 0 \\ 0 & 0 & 0 & -.641 \\ -.786 & 0 & -.206 & -.0591 \\ 0 & -.641 & -.0591 & -.131 \end{bmatrix}$$

Keep in mind that to get $[A]$ and $[B]$ we took the undamped system eigenvectors, formed the modal matrix $[U]$, used $[U]$ to transform $[M]$, $[C]$, and $[K]$, and then assembled them into the larger $[A]$ and $[B]$ matrices to have a first-order system representation. We can see from the lower two by two block of $[A]$ that the undamped eigenvectors didn't decouple our $[M]$, $[C]$, and $[K]$. If $[U]$ had diagonalized our matrices, then the off-diagonal entries -.0591 would not have shown up.

If we use `eig` to find the eigenvalues and eigenvectors, we'll obtain

$$v = \begin{bmatrix} .526 + .143i & .147 + .525i & .0618 + .0149i & -0144 - .0619i \\ .0136 - .0248i & -.0247 + .0138i & .0712 - .776i & .776 - .0646i \\ .0514 - .835i & -.835 + .0585i & .0079 - .0502i & .0502 - .0075i \\ -.0415 - .0127i & -.0131 - .0414i & -.622 - .0065i & .0013 + .622i \end{bmatrix}$$

and

$$d = \begin{bmatrix} -.310 - 1.502i & 0 & 0 & 0 \\ 0 & -.310 + 1.502i & 0 & 0 \\ 0 & 0 & -.0647 - .796i & 0 \\ 0 & 0 & 0 & -.0647 + .796i \end{bmatrix}$$

As you can see, the columns in v are definitely not showing decoupling. All entries have finite values; there are no zero entries that indicate nonparticipation of one coordinate while another is active. Thus the eigenvectors (as expected) are not the same as the uncoupled system's eigenvectors.

The final topic is an interesting one to think about. Clearly, it would be nice if our system were decoupled. The problem is easier to solve this way, and decoupling supports the attractive notion of physically identifiable eigenvectors that govern the motion. Given that this is the case, some vibrations engineers wondered what the consequences might be of *forcing* the system to decouple. Since damping is such a difficult concept already, and our notion of purely viscous damping is already somewhat arguable, why not just throw out the damping terms that are spoiling the equations?

Actually, there are some more sophisticated ways to reorder the damping, but for our purposes it will be sufficient to examine the simplest approach. And it certainly is simple. What you want to do is try to put your system in normal form, as in the preceding example. Then, you simply cross out all terms that don't support proportional damping, i.e., all the off-diagonal terms of the transformed $[C]$ matrix. Example 4.16 demonstrates what this will buy us.

Example 4.16

Problem Construct the normal form of the equations from Example 4.15 and discard all off-diagonal damping terms. Examine the resulting solution.

Solution   Putting the system into normal form gives us

$$[B] = \begin{bmatrix} -.786 & 0 & 0 & 0 \\ 0 & -.641 & 0 & 0 \\ 0 & 0 & .332 & 0 \\ 0 & 0 & 0 & 1.01 \end{bmatrix}$$

and

$$[A] = \begin{bmatrix} 0 & 0 & -.786 & 0 \\ 0 & 0 & 0 & -.641 \\ -.786 & 0 & -.206 & 0 \\ 0 & -.641 & 0 & -.131 \end{bmatrix}$$

Note that we simply erased the old $A(3, 4)$ and $A(4, 3)$ entries from the $[A]$ matrix of Example 4.15. Because of this, the 2×2 submatrices that correspond to the transformed $[M]$, $[C]$, and $[K]$ matrices are diagonal. If we go ahead and use eig on these, we'll obtain

$$v = \begin{bmatrix} .541 + .0657i & .0657 + .541i & 0 & 0 \\ 0 & 0 & .249 + .742i & -.745 - .249i \\ -.0689 - .836i & -.836 - .0689i & 0 & 0 \\ 0 & 0 & .572 - .246i & .246 - .572i \end{bmatrix}$$

and

$$d = \begin{bmatrix} -.310 - 1.507i & 0 & 0 & 0 \\ 0 & -.310 + 1.507i & 0 & 0 \\ 0 & 0 & -.0648 - .794i & 0 \\ 0 & 0 & 0 & -.0648 + .794i \end{bmatrix}$$

Just compare these results with the exact case. The damped natural frequencies have hardly changed at all. For instance, the damped solution $-.310 - 1.502i$ from Example 4.15 is now $-.310 - 1.507i$. This isn't a very large difference. Furthermore, the eigenvectors are completely real again and nicely decoupled. If you compare the two cases, you'll see that crossing out those pesky off-diagonal damping terms didn't damage our accuracy all that much and allowed us to have a beautifully decoupled system.

The reason for the good results of Example 4.16 is that the off-diagonal terms were smaller than those on the diagonal. Now it's possible to come up with specially tailored examples in which it isn't safe to neglect small off-diagonal damping entries and also examples for which relatively large damping entries don't cause great problems. But the general rule of thumb is that when the off-diagonal elements are small in comparison to the diagonal ones, you won't induce a lot of error by neglecting them. The inverse of this view is also used often by researchers who identify the mass and spring matrices, run some modal tests, and them fit the best approximate *proportional* damping that matches the test results.

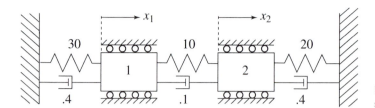

Figure 4.29   2 DOF
spring-mass-damper system

Example 4.17

Problem   Analyze the system shown in Figure 4.29. All the relevant parameter values are shown in the figure, where the units are kg, N·s/m, and N/m for the masses, dampers, and springs, respectively.

Solution   The equations of motion for this system are

$$\begin{bmatrix} 1 & 0 \\ 0 & 2 \end{bmatrix} \begin{Bmatrix} \ddot{x}_1 \\ \ddot{x}_2 \end{Bmatrix} + \begin{bmatrix} .5 & -.1 \\ -.1 & .5 \end{bmatrix} \begin{Bmatrix} \dot{x}_1 \\ \dot{x}_2 \end{Bmatrix} + \begin{bmatrix} 40 & -10 \\ -10 & 30 \end{bmatrix} \begin{Bmatrix} x_1 \\ x_2 \end{Bmatrix} = \begin{Bmatrix} 0 \\ 0 \end{Bmatrix}$$

This system has proportional damping ($[C] = .1[M] + .1[K]$), and so we should obtain purely real displacements for our eigenvectors. Furthermore, the eigenvectors should be the same as those of the undamped case. If we use MATLAB to find the eigenvectors and eigenvalues of the undamped case (using $[v, d] = \text{eig}(K, M)$) we will get

$$v = \begin{bmatrix} .9831 & .3489 \\ -.1830 & .9372 \end{bmatrix}$$

and

$$d = \begin{bmatrix} 41.8614 & 0 \\ 0 & 13.1386 \end{bmatrix}$$

Thus the first and second natural frequencies will be $\sqrt{13.1386} = 3.62$ rad/s and $\sqrt{41.8614} = 6.47$ rad/s, respectively. Just to make it easy to compare with our damped results, we'll rescale the eigenvectors (the columns of v) so that the first entry is equal to 1.0 for both cases. Dividing our first eigenvector by .3489 and our second by .9831 gives us

$$X_1 = \begin{Bmatrix} 1.0000 \\ 2.6861 \end{Bmatrix} \text{ and } X_2 = \begin{Bmatrix} 1.0000 \\ -.1861 \end{Bmatrix} \tag{4.11.9}$$

Now let's look at our damped problem. Putting our system into $[A]$, $[B]$ form yields

$$[B] = \begin{bmatrix} -40 & 10 & 0 & 0 \\ 10 & -30 & 0 & 0 \\ 0 & 0 & 1 & 0 \\ 0 & 0 & 0 & 2 \end{bmatrix}$$

and

$$[A] = \begin{bmatrix} 0 & 0 & -40 & 10 \\ 0 & 0 & 10 & -30 \\ -40 & 10 & -.5 & .1 \\ 10 & -30 & .1 & -.5 \end{bmatrix}$$

Typing $[v, d] = \text{eig}(A, B)$ gives

$$v = \begin{bmatrix} .1502 + .0012i & -.0620 - .0691i & .1443 + .0417i & .0397 - .0839i \\ -.0280 - .0002i & -.1664 - .1855i & -.0269 - .0078i & .1066 - .2253i \\ -.0312 - .9711i & -.2430 + .2325i & -.3070 + .9218i & .2993 + .1534i \\ .0058 + .1808i & -.6529 + .6245i & .0572 - .1716i & .8039 + .4122i \end{bmatrix}$$

and

$$d = \begin{bmatrix} -.2593 - 6.4648i & 0 & 0 & 0 \\ 0 & -.1157 - 3.6229i & 0 & 0 \\ 0 & 0 & -.2593 + 6.4648i & 0 \\ 0 & 0 & 0 & -.1157 + 3.6229i \end{bmatrix}$$

This looks pretty reasonable. The first two imaginary entries in d are -6.4648 and -3.6229, which are very close in magnitude to the undamped natural frequencies. The question is whether the associated eigenvectors have real displacement entries. We'll start by looking at the first column of v, which is the eigenvector associated with the eigenvalue $.2593 - 6.4648i$. Let's call this eigenvector $V1$. Currently, the displacement entries of $V1$ are complex. Since we know that MAT-LAB will happily multiply our eigenvectors by complex constants, we have to phase-shift the entire vector. Typing

$$\text{angle}(V1) \quad (cr)$$

will give us the output

$$\begin{Bmatrix} .0080 \\ -3.1336 \\ -1.6029 \\ 1.5387 \end{Bmatrix} \tag{4.11.10}$$

Thus we see that the first entry has a phase angle of .0080 radian. We can phase-shift the whole vector if we multiply it by

$$e^{-.008i}$$

This will rotate the first entry back onto the real axis. If all goes as expected, it should also move the second entry onto the real axis. To do this on MATLAB we need to type

$$V1 * \exp(-i * .008) \quad (cr)$$

which will return

$$\begin{Bmatrix} .1502 - 0i \\ -.0280 - 0i \\ -.0389 - .9798i \\ .0072 + .1807i \end{Bmatrix}$$

The first half of this vector corresponds to the x_1 and x_2 displacements. As hoped, the entries are both real. If we use the same normalization chosen for the unforced case (i.e., dividing both entries by .1502 so that the first entry equals 1.0), we obtain

$$\begin{Bmatrix} 1.0000 \\ -.1861 \end{Bmatrix}$$

which matches precisely with the second undamped eigenvector (4.11.9)

Just to be complete, we can go ahead and compute the other eigenvector using the second column of v (which we'll call $V2$). Typing `angle(V2)` will produce

$$\begin{Bmatrix} -2.302 \\ -2.302 \\ 2.3784 \\ 2.3784 \end{Bmatrix} \qquad (4.11.11)$$

Rotating the vector by 2.302 radians ($V2*exp(2.302*i)$) yields

$$\begin{Bmatrix} .0928 - 0i \\ .2492 + 0i \\ -.0107 - .3362i \\ -.0288 - .9030i \end{Bmatrix}$$

Finally, dividing the first two rows by .0928 gives us

$$\begin{Bmatrix} 1.0000 \\ 2.6861 \end{Bmatrix}$$

which matches up with the first undamped eigenvector.

Note that it was easy to see from (4.11.11) that the two entries both had the same phase for $V2$ (both displacement phases were –2.302 rad and both velocity phases were 2.3784 rad). Although it looked less obvious for $V1$ in (4.11.10) that the displacements and velocities were out of phase, if you look closely you'll see it. For instance, the phase for x_1 was .0080 rad and the phase for x_2 was –3.1336 rad. The difference between these is just π. Thus they're exactly out of phase with each other. The same argument holds for the velocities. Again, this is just the way MATLAB does things.

4.12 FORCED RESPONSE OF DAMPED SYSTEMS

We've looked at MDOF, forced, undamped systems and MDOF, unforced, damped systems, so all that remains is to examine MDOF, forced, damped systems. The equations of motion are given by

$$[M]\ddot{X} + [C]\dot{X} + [K]X = F \qquad (4.12.1)$$

where $F = \{f_1, f_2, \ldots\}^T$. The easiest way to solve this set of equations is to use complex exponential notation. Thus we'll substitute $F = \bar{F}e^{i\omega t}$ and $X = \bar{X}e^{i\omega t}$ into (4.12.1), which, after canceling out the common $e^{i\omega t}$ factor, gives us

$$\left[-\omega^2[M] + i\omega[C] + [K]\right]\bar{X} = \bar{F} \tag{4.12.2}$$

Solving for $\bar{X}$ yields

$$\bar{X} = \left[-\omega^2[M] + i\omega[C] + [K]\right]^{-1}\bar{F} \tag{4.12.3}$$

As (4.12.3) indicates, there will be a phase shift between the input $\bar{F}$ and the output $\bar{X}$. For instance, if $\bar{F}$ is purely real, then $\bar{X}$ can't be real since $\bar{F}$ is multiplied by the complex matrix $\left[-\omega^2[M] + i\omega[C] + [K]\right]^{-1}$. Just as in the SDOF cases, the response will lag the input. Thus, if we choose $\bar{F}$ to be real and are concerned with the real part of $\bar{F}e^{i\omega t}$, our forcing will be of the form $\bar{F}\cos(\omega t)$ and our output will be of the form $\bar{X}\cos(\omega t - \phi)$, where ϕ is greater than zero.

Example 4.18

Problem   Consider again Figure 4.28 for $n = 2$. We'll use the particular values $m_1 = 1$ kg, $m_2 = 2$ kg, $c_1 = .1$ N·s/m, $c_2 = .2$ N·s/m, $c_3 = .1$ N·s/m, $k_1 = 4$ N/m, $k_2 = 6$ N/m, and $k_3 = 16$ N/m. The forcing in newtons, will be given by

$$\begin{Bmatrix} f_1 \\ f_2 \end{Bmatrix} = \begin{Bmatrix} 1 \\ -1 \end{Bmatrix}\cos(3t)$$

Find the response of the system to this forcing.

Solution   The forcing is in the form of a cosine wave, i.e., the real part of e^{3it}. After for $\bar{X}$, using the complex exponential approach to solve, we'll need to take the real part to match the solution with the forcing.

For the given values, (4.12.3) becomes

$$\begin{Bmatrix} \bar{x}_1 \\ \bar{x}_2 \end{Bmatrix} = \begin{bmatrix} 1 + .9i & -6 - .6i \\ -6 - .6i & 4 + .9i \end{bmatrix}^{-1}\begin{Bmatrix} 1 \\ -1 \end{Bmatrix}$$

Taking the indicated inverse and multiplying out give us

$$\begin{Bmatrix} \bar{x}_1 \\ \bar{x}_2 \end{Bmatrix} = \begin{Bmatrix} .0604 - .01431i \\ -.1523 + .0219i \end{Bmatrix}$$

Keep in mind that this doesn't mean that the response is imaginary. Remember that we're looking for the real part of $\bar{X}e^{3it}$. Thus we need the real part of

$$\begin{Bmatrix} .0604 - .01431i \\ -.1523 + .0219i \end{Bmatrix}e^{3it}$$

or

$$\begin{Bmatrix} x_1(t) \\ x_2(t) \end{Bmatrix} = \begin{Bmatrix} .0604 \\ -.1523 \end{Bmatrix}\cos(3t) - \begin{Bmatrix} -.0143 \\ .0219 \end{Bmatrix}\sin(3t)$$

Thus the final response for $f_1 = \cos(3t)$, $f_2 = -\cos(3t)$ is given by

$$x_1 = .0604 \cos(3t) + .0143 \sin(3t)$$

and

$$x_2 = -.1523 \cos(3t) - .0219 \sin(3t)$$

Putting this into phase lag form, we have

$$x_1 = .0621 \cos(3t - 1.34)$$

and

$$x_2 = -.1539 \cos(3t - 1.43)$$

Note that the displacement values are given in meters.

Now that we've got an understanding of damping, it's time to return to the topic of vibration absorbers (Section 4.4). You'll recall that, for zero damping, we were able to cause the response of a forced main mass to be zero, at the expense of adding a secondary vibrational absorber to the system. For small secondary spring-mass systems, we had to endure large excursions of the absorber mass to attain vibration cancellation at the main mass. To reduce these excursions, you might think about adding a damper or two to the system. These would actually dissipate energy, rather than simply redirecting it to a vibration absorber. However, as we've seen, damping causes the responses to phase-shift. Thus it will probably be impossible to *exactly* cancel the driving force, since the response of the masses will have a phase shift with respect to the input force. So we'll end up with our familiar engineering trade-off: balancing reduced vibration levels of one part of the system at the expense of increasing the vibration levels of another part.

One additional goal of adding damping would be to widen the frequency range over which vibration reduction occurs. As we saw in Section 4.4, the range of vibration reduction can be quite narrow and depends directly on how large a secondary spring-mass system is added. Perhaps by adding damping we'll be able to widen this range, thus making the system more robust in the face of disturbance forces that might occur over a wide frequency range.

Figure 4.30 illustrates our basic damped absorber model. The disturbance force is acting on m_1 and k_2; m_2 and c_2 represent the secondary system. The equations of motion are given by

$$\begin{bmatrix} m_1 & 0 \\ 0 & m_2 \end{bmatrix} \begin{Bmatrix} \ddot{x}_1 \\ \ddot{x}_2 \end{Bmatrix} + \begin{bmatrix} c_1 + c_2 & -c_2 \\ -c_2 & c_2 \end{bmatrix} \begin{Bmatrix} \dot{x}_1 \\ \dot{x}_2 \end{Bmatrix} + \begin{bmatrix} k_1 + k_2 & -k_2 \\ -k_2 & k_2 \end{bmatrix} \begin{Bmatrix} x_1 \\ x_2 \end{Bmatrix} = \begin{Bmatrix} f_1 \\ 0 \end{Bmatrix} \qquad (4.12.4)$$

There's no problem in solving this; it's basically the same system we looked at in Example 4.18. We're not concerned with how to solve the problem here; rather, we want to know how the damping elements help (or hurt) our quest for vibration absorption. What we'll do here is show a few examples of how judiciously choosing the absorber parameters can yield good design solutions to our vibration problem. If we solve (4.12.4) by assuming a solution of the form $X = \bar{X}e^{i\omega t}$, we'll obtain

$$\bar{x}_1 = \frac{\bar{f}_1(k_2 - \omega^2 m_2 + i\omega c_2)}{\Delta(\omega)} \qquad (4.12.5)$$

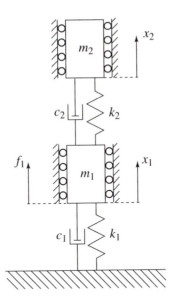

Figure 4.30 Damped vibration absorber

and

$$\bar{x}_2 = \frac{\bar{f}_1(k_2 + i\omega c_2)}{\Delta(\omega)} \tag{4.12.6}$$

where the system's characteristic determinant $\Delta(\omega)$ is given by

$$\Delta(\omega) = (m_1 + m_2)\omega^4 - i\omega^3\,(c_1 m_2 + c_2(m_1 + m_2)) - \omega^2\,(m_2 k_1 + (m_1 + m_2)k_2 + c_1 c_2)$$

$$+ i\omega(c_2 k_1 + c_1 k_2) + k_1 k_2 \tag{4.12.7}$$

To illustrate how damping affects our solutions, we'll look at a few graphical examples. Figure 4.31 shows the absolute value of the response for our base mass in the absence of any absorber and without damping. The magnitude of the force is 10 N, $m_1 = 10$ kg, and $k_1 = 1000$ N/m. In addition to looking at the amplitude of the response, we can examine the magnitude of the force that's transmitted to the base of the system. Since the only connection between m_1 and ground is the spring k_1, the magnitude of the force is given by $|k_1 x_1|$. This response is shown in Figure 4.32.

Now we'll add a vibration absorber to the system without any damping. The parameters will be $m_2 = 1$ kg and $k_2 = 100$ N/m. Plots of the amplitude response $\bar{x}_1$ and $\bar{x}_2$ are shown in Figure 4.33, while Figure 4.34 shows the magnitude of the transmitted force (again equal to $|k_1 \bar{x}_1|$). Here we see all the classic hallmarks of an undamped vibration absorber. At the original system's natural frequency (in this case 10 rad/s), we now have a zero response for the main mass. However, to accomplish this, the secondary mass has quite a sizable excursion (just under 10 cm). Thus we've traded off the infinite excursion we had originally for m_1 for no motion of m_1 and a relatively large motion of m_2. The transmitted force is zero at the same frequency that x_1 is equal to zero and grows in proportion to x_1 for all other frequencies (Figure 4.34).

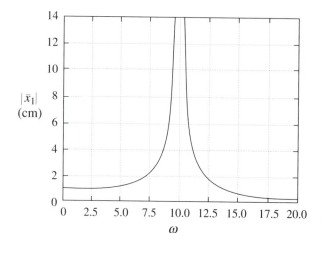

Figure 4.31 Amplitude response for a forced spring-mass system

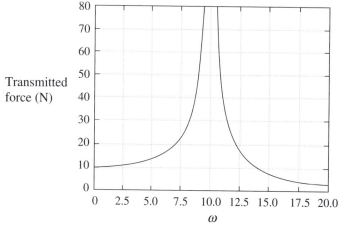

Figure 4.32 Transmitted force for a forced spring-mass system

The picture changes when we add damping. We'll let $c_1 = 2$ N·s/m and $c_2 = 2$ N·s/m. If we view the absorber and main mass systems are separate systems, these values correspond to ζ's of $\zeta_1 = .01$ and $\zeta_2 = .1$. Figure 4.35 shows how $|\bar{x}_1|$ and $|\bar{x}_2|$ look for this case. Certainly things have changed. Instead of going to zero, $|\bar{x}_1|$ is now finite near $\omega = 10$ rad/s (the original system's natural frequency). The resonances that used to exist below and above this frequency have been tamed, and reach amplitudes of only about 8 and 3 cm, respectively. This is certainly better than the infinite response we saw in Figure 4.33. The motions of the second mass have also been reduced, especially those around the undamped second resonant frequency (approximately 11.5 rad/s).

If we increase the damping in the secondary mass to 8 N·s/m, the response changes to that of Figure 4.36. The secondary mass's motion has been further reduced, but the maximum amplitude of the first mass hasn't really altered. One big qualitative change is that the two peaks we had before have now merged into a single peak. If we look at Figure 4.37, we can see how the transmitted force

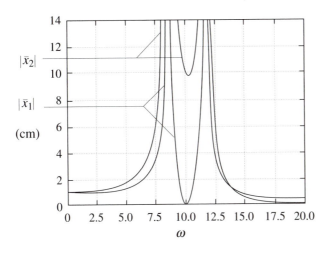

Figure 4.33   Undamped vibration absorber, amplitude response of both masses

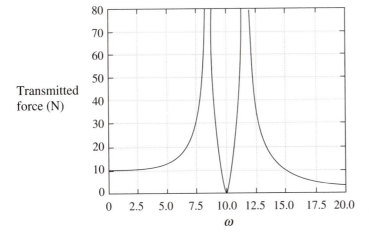

Figure 4.34   Undamped vibration absorber, transmitted force

was affected by our changes. $|\bar{f}_{T_1}|$ shows the transmitted force when c_2 was 2 N·s/m and $|\bar{f}_{T_2}|$ is the plot for the case of $c_2 = 8$ N·s/m. As you can see, the transmitted force has actually gone up around the original system's natural frequency for the higher value of damping.

What's the conclusion? Simply that optimizing a vibration absorber isn't a trivial job. By adding damping to the secondary absorber we can reduce m_2's excursions, which is good from a packaging viewpoint. We may have limited space to work with and excessive oscillations simply can't be permitted. However we just saw that reducing the level of vibration for this particular case also increased the transmitted forces. One of the goals of vibration absorption is to reduce these forces; indeed, for the undamped case the transmitted force is zero. If the vibrating object is an automobile engine, for example, then any transmitted vibration will excite the chassis, leading to higher vibration levels in the automobile cabin. This is clearly undesirable.

$c_1 = 2 \text{ N/m/s}$
$c_2 = 2 \text{ N/m/s}$

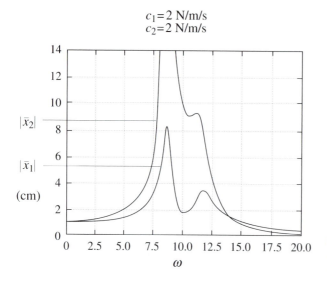

Figure 4.35   Damped vibration absorber, amplitude response of both masses

$c_1 = 2 \text{ N/m/s}$
$c_2 = 8 \text{ N/m/s}$

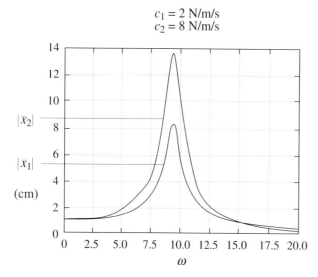

Figure 4.36   Damped vibration absorber, amplitude response of both masses

An additional goal in the design process might be to obtain a low amplitude response level over a wide frequency range. But in doing so we'd have to make sure we were not unduly increasing the transmitted force or the vibration amplitude of the second mass. Although we won't get into it, it's possible to try to optimize this problem, setting performance goals (like minimizing the vibration amplitude) subject to constraints (like keeping the secondary mass below 20% of the total mass). Those interested in pursuing this topic can refer to further discussion given elsewhere [6].

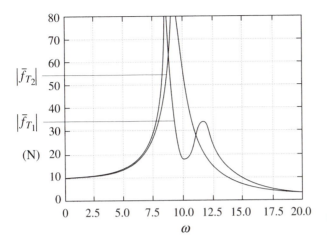

Figure 4.37   Damped vibration absorber, transmitted force

4.13 SYMMETRY OF MASS AND STIFFNESS MATRICES

In all our work so far, the mass and stiffness matrices have invariably been symmetric. This isn't an accident—it's a fundamental property of undamped vibrating structures. However it's also possible to get a bit confused because you can easily wind up with mass and stiffness matrices that appear nonsymmetric. This might make you doubt yourself, since you've come to expect $[K]$ and $[M]$ to be symmetric. Let's show how such apparent nonsymmetry can happen. Consider the system of Figure 4.38. We'll consider only vertical motions, thereby limiting the system to two degrees of freedom. Two different coordinate sets are illustrated in Figure 4.38. The x, θ set tracks the vertical motion of the beam's center of mass (x) and the beam's rotation (θ). The x_1, x_2 set gives us a direct reading of the springs' deflections. Either set is a reasonable one from which to find the system's equations of motion. The kinetic and potential energies in terms of x, θ are given by

$$KE = \frac{1}{2}m\dot{x}^2 + \frac{1}{2}\bar{I}\dot{\theta}^2 \tag{4.13.1}$$

and

$$PE = \frac{1}{2}k_2(x + l\theta)^2 + \frac{1}{2}k_1(x - l\theta)^2 \tag{4.13.2}$$

An application of Lagrange's equations gives us

$$\begin{bmatrix} m & 0 \\ 0 & \bar{I} \end{bmatrix} \begin{Bmatrix} \ddot{x} \\ \ddot{\theta} \end{Bmatrix} + \begin{bmatrix} k_1 + k_2 & l(k_2 - k_1) \\ l(k_2 - k_1) & l^2(k_1 + k_2) \end{bmatrix} \begin{Bmatrix} x \\ \theta \end{Bmatrix} = \begin{Bmatrix} 0 \\ 0 \end{Bmatrix} \tag{4.13.3}$$

As usual, the $[M]$ and $[K]$ matrices are symmetric. We'll be saying more about these equations in a bit, but for now let's carry on and find what they look like in terms of the x_1, x_2 coordinates. For this

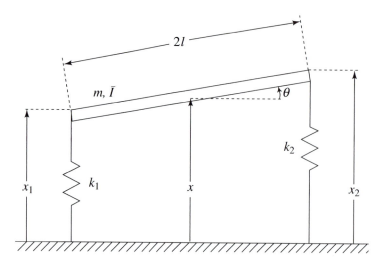

Figure 4.38　2 DOF system, with two illustrated coordinate systems

case, the energies are given by

$$KE = \frac{1}{2}m\left(\frac{\dot{x}_1 + \dot{x}_2}{2}\right)^2 + \frac{1}{2}\bar{I}\left(\frac{\dot{x}_2 - \dot{x}_1}{2l}\right)^2 \tag{4.13.4}$$

and

$$PE = \frac{1}{2}k_1x_1^2 + \frac{1}{2}k_2x_2^2 \tag{4.13.5}$$

Applying Lagrange's equations yields

$$\begin{bmatrix} \frac{m}{4} + \frac{\bar{I}}{4l^2} & \frac{m}{4} - \frac{\bar{I}}{4l^2} \\ \frac{m}{4} - \frac{\bar{I}}{4l^2} & \frac{m}{4} + \frac{\bar{I}}{4l^2} \end{bmatrix} \begin{Bmatrix} \ddot{x}_1 \\ \ddot{x}_2 \end{Bmatrix} + \begin{bmatrix} k_1 & 0 \\ 0 & k_2 \end{bmatrix} \begin{Bmatrix} x_1 \\ x_2 \end{Bmatrix} = \begin{Bmatrix} 0 \\ 0 \end{Bmatrix} \tag{4.13.6}$$

These matrices are also symmetric. So what's the point? Well, you'll note that we used a Lagrangian approach each time to find our equations. Let's see what happens if we apply Newton's laws instead of using the Lagrangian approach. We'll apply $f = ma$ and $\bar{I}\ddot{\theta} = M$, using the center of mass as our reference point. And just to be difficult, we'll use the x_1, x_2 coordinate set rather than the more immediate x, θ one. $f = ma$ gives us

$$m\left(\frac{\ddot{x}_1 + \ddot{x}_2}{2}\right) = -k_1x_1 - k_2x_2 \tag{4.13.7}$$

and $\bar{I}\ddot{\theta} = M$ yields

$$\bar{I}\left(\frac{\ddot{x}_2 - \ddot{x}_1}{2l}\right) = -k_2lx_2 + k_1lx_1 \tag{4.13.8}$$

Put this into matrix form by writing

$$
\begin{bmatrix} \frac{m}{2} & \frac{m}{2} \\ -\frac{\bar{I}}{2l} & \frac{\bar{I}}{2l} \end{bmatrix} \begin{Bmatrix} \ddot{x}_1 \\ \ddot{x}_2 \end{Bmatrix} + \begin{bmatrix} k_1 & k_2 \\ -k_1 l & k_2 l \end{bmatrix} \begin{Bmatrix} x_1 \\ x_2 \end{Bmatrix} = \begin{Bmatrix} 0 \\ 0 \end{Bmatrix}
\tag{4.13.9}
$$

Now we've got something. You'll note that both the $[M]$ and $[K]$ matrices are decidedly *not* symmetric. And yet we've already seen that the x_1, x_2 coordinates *do* yield symmetric matrices when we use Lagrange's equations. So what's happening? Are (4.13.6) and (4.13.9) really different equations, or is something more going on here? Logically, we know that they've got to be telling us the same thing because both Lagrange's equations and Newton's laws are perfectly reasonable ways to go about getting a system's equations of motion—the equations can't fundamentally change just because we use one approach or the other. So the change must not be fundamental. In fact, since they've got to be displaying the same physics, it's reasonable to assume that we can manipulate one set to yield the other. We can do this by first dividing the second row of (4.13.9) by l to produce

$$
\begin{bmatrix} \frac{m}{2} & \frac{m}{2} \\ -\frac{\bar{I}}{2l^2} & \frac{\bar{I}}{2l^2} \end{bmatrix} \begin{Bmatrix} \ddot{x}_1 \\ \ddot{x}_2 \end{Bmatrix} + \begin{bmatrix} k_1 & k_2 \\ -k_1 & k_2 \end{bmatrix} \begin{Bmatrix} x_1 \\ x_2 \end{Bmatrix} = \begin{Bmatrix} 0 \\ 0 \end{Bmatrix}
\tag{4.13.10}
$$

Next, we'll subtract the second row from the first to produce a new equation and then add the first and second rows together, giving us a second new equation (the new equations are equal to the difference and sum of the foregoing equations, respectively):

$$
\begin{bmatrix} \frac{m}{2} + \frac{\bar{I}}{2l^2} & \frac{m}{2} - \frac{\bar{I}}{2l^2} \\ \frac{m}{2} - \frac{\bar{I}}{2l^2} & \frac{m}{2} + \frac{\bar{I}}{2l^2} \end{bmatrix} \begin{Bmatrix} \ddot{x}_1 \\ \ddot{x}_2 \end{Bmatrix} + \begin{bmatrix} 2k_1 & 0 \\ 0 & 2k_2 \end{bmatrix} \begin{Bmatrix} x_1 \\ x_2 \end{Bmatrix} = \begin{Bmatrix} 0 \\ 0 \end{Bmatrix}
\tag{4.13.11}
$$

You can see that by dividing both rows of (4.13.11) by 2 we'll exactly replicate (4.13.6). Thus (4.13.6) and (4.13.9) *are* both giving us the same information; one is simply a linear combination of the other's equations. What we're seeing is simply an example of a coordinate transformation—going from one set of coordinates to another.

We still haven't addressed the problem of symmetry. Why is one formulation symmetric and the other not? The reason has to do with how the equations were found. Lagrange's equations work with the system's energies, quantities that are quadratic in nature (kinetic energy is proportional to velocity squared, spring potential is proportional to displacement squared, etc). By going through this route, the mass and stiffness matrix will always be symmetric in the way we've been led to expect. The Newtonian approach, however, is dealing with force and moment balances, and there's nothing that dictates a symmetric form. What *is* dictated is that you can make the equations symmetric by suitably rearranging them. And we can find out the appropriate rearrangement pretty simply.

As far as our current example is concerned, we would have obtained precisely the same equations shown in (4.13.9) if we'd started from (4.13.3) and simply used the substitutions

$$
x = \frac{x_1 + x_2}{2}
\tag{4.13.12}
$$

and

$$\theta = \frac{x_2 - x_1}{2l} \tag{4.13.13}$$

that is,

$$\begin{Bmatrix} x \\ \theta \end{Bmatrix} = \begin{bmatrix} \frac{1}{2} & \frac{1}{2} \\ -\frac{1}{2l} & \frac{1}{2l} \end{bmatrix} \begin{Bmatrix} x_1 \\ x_2 \end{Bmatrix} \tag{4.13.14}$$

We'll call the transformation matrix $[A]$ where

$$[A] = \begin{bmatrix} \frac{1}{2} & \frac{1}{2} \\ -\frac{1}{2l} & \frac{1}{2l} \end{bmatrix} \tag{4.13.15}$$

Thus we've got a matrix relationship between the two coordinates. Operations like this, however, don't preserve symmetry. Thus, if we start with a symmetric $[K]$ (which we did), switching coordinates will give us the new stiffness matrix $[K]'$, where

$$[K]' = [K][A]. \tag{4.13.16}$$

In general, $[K][A]$ will not be symmetric. Luckily, we can regain symmetry by premultiplying by $[A]^T$. That this is so can be seen in the following manner. Assume that a matrix $[K]$ is symmetric. We'll form a new matrix $[C]$ thus:

$$[C] = [A]^T [K][A] \tag{4.13.17}$$

Transposing both sides of (4.13.17) yields

$$[C]^T = [A]^T [K]^T [A] \tag{4.13.18}$$

Since we've already said that $[K]$ is symmetric, $[K]^T = [K]$, and so we can rewrite (4.13.18) as

$$[C]^T = [A]^T [K][A] \tag{4.13.19}$$

Equations (4.13.17) and (4.13.19) together imply that $[C] = [C]^T$. Thus we've retained symmetry by using the operation $[A]^T [K][A]$ instead of just sticking with $[K][A]$.

To see if this helps in the problem we're looking at, take the transpose of $[A]$:

$$[A]^T = \begin{bmatrix} \frac{1}{2} & -\frac{1}{2l} \\ \frac{1}{2} & \frac{1}{2l} \end{bmatrix} \tag{4.13.20}$$

What does premultiplying by this matrix do to our equations? Look at the first row. It takes one half of our first equation and subtracts $\frac{1}{2l}$ times our second equation. This is exactly what we did when rearranging the equations. The second row takes one half the first equation and then adds $\frac{1}{2l}$ times the second equation. Again, just what we did.

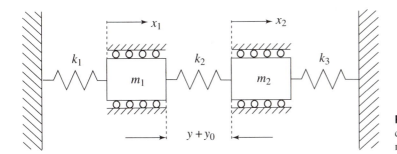

Figure 4.39 2 DOF system with coordinates showing absolute and relative motions

Thus we have our final observation. If we use Lagrange's equations to get our equations of motion, the spring and mass matrices will be symmetric. If we use Newton's laws, they might or might not be symmetric. If they aren't, it's because we've essentially performed half a transformation—the $[K][A]$ part. We would therefore need to premultiply by the appropriate $[A]^T$ to get to the symmetric form.

Example 4.19

Problem   As another example of how we can obtain nonsymmetric mass and stiffness matrices, consider the system shown in Figure 4.39. This is our standard two-mass, three-spring problem. We've seen repeatedly that if our coordinates tell us the absolute displacement of both m_1 and m_2, then our spring and mass matrices will be symmetric. However this time we're not going to use coordinates that tell us how much each mass has moved. Our first coordinate, x, does tell us the absolute motion of m_1. But the second coordinate, y, tells us the relative motion between m_1 and m_2. In Figure 4.39, y_0 represents the unstretched length of the physical spring between the two masses. Show that a Newtonian approach can be used to find the equations of motion in terms of these coordinates will yield nonsymmetric $[M]$ and $[K]$ matrices.

Solution   If we use these two coordinates to form the equations of motion, we'll obtain

$$m_1 \ddot{x} = -k_1 x + k_2 y$$

$$m_2 (\ddot{x} + \ddot{y}) = -k_2 y - k_3 (x + y)$$

The quantities $\ddot{x} + \ddot{y}$ and $x + y$ represent the absolute acceleration and displacement of m_2.

Putting these equations in matrix form gives us

$$\begin{bmatrix} m_1 & 0 \\ m_2 & m_2 \end{bmatrix} \begin{Bmatrix} \ddot{x} \\ \ddot{y} \end{Bmatrix} + \begin{bmatrix} k_1 & -k_2 \\ k_3 & k_2 + k_3 \end{bmatrix} \begin{Bmatrix} x \\ y \end{Bmatrix} = \begin{Bmatrix} 0 \\ 0 \end{Bmatrix}$$

As you can see, both $[M]$ and $[K]$ are nonsymmetric for these coordinates.

4.14 REPEATED FREQUENCIES AND ZERO FREQUENCIES

Although not the norm, we'll sometimes run into systems that have repeated natural frequencies. In any such system, there will be more than one distinct modal deflection that corresponds to a particular natural frequency. We won't go into great detail with regard to this subject, but we'll indicate what the basic phenomena are and discuss what such a situation means for our analyses.

We'll begin by returning to Figure 4.38, repeating for convenience the equations of motion for this system already found in Section 4.13;

$$\begin{bmatrix} m & 0 \\ 0 & \bar{I} \end{bmatrix} \begin{Bmatrix} \ddot{x} \\ \ddot{\theta} \end{Bmatrix} + \begin{bmatrix} k_1 + k_2 & l(k_2 - k_1) \\ l(k_2 - k_1) & l^2(k_1 + k_2) \end{bmatrix} \begin{Bmatrix} x \\ \theta \end{Bmatrix} = \begin{Bmatrix} 0 \\ 0 \end{Bmatrix} \qquad (4.14.1)$$

We'll now set both spring constants equal to k:

$$\begin{bmatrix} m & 0 \\ 0 & \bar{I} \end{bmatrix} \begin{Bmatrix} \ddot{x} \\ \ddot{\theta} \end{Bmatrix} + \begin{bmatrix} 2k & 0 \\ 0 & 2kl^2 \end{bmatrix} \begin{Bmatrix} x \\ \theta \end{Bmatrix} = \begin{Bmatrix} 0 \\ 0 \end{Bmatrix} \qquad (4.14.2)$$

What we see is that the equations are now decoupled. Since they are decoupled, x and θ must be the system's normal- coordinates. Oscillations in the x direction represent one mode and oscillations of θ represent the other mode. The natural frequency of the x mode is given by

$$\omega_1^2 = \frac{2k}{m} \qquad (4.14.3)$$

and the second natural frequency (associated with θ), is given by

$$\omega_2^2 = \frac{2kl^2}{\bar{I}} \qquad (4.14.4)$$

If we divide the first equation by m and the second by $\bar{I}$, our equations will be

$$\begin{bmatrix} 1 & 0 \\ 0 & 1 \end{bmatrix} \begin{Bmatrix} \ddot{x} \\ \ddot{\theta} \end{Bmatrix} + \begin{bmatrix} \omega_1^2 & 0 \\ 0 & \omega_2^2 \end{bmatrix} \begin{Bmatrix} x \\ \theta \end{Bmatrix} = \begin{Bmatrix} 0 \\ 0 \end{Bmatrix} \qquad (4.14.5)$$

Since we're already in normal form, the corresponding eigenvectors are $\{1 \quad 0\}^T$ (for ω_1) and $\{0 \quad 1\}^T$ (for ω_2).

Now here's the question. What happens if the parameters work out so that

$$\frac{2k}{m} = \frac{2kl^2}{\bar{I}} \qquad (4.14.6)$$

In this case we'd have a repeated natural frequency. The equations of motion would be given by

$$\begin{bmatrix} 1 & 0 \\ 0 & 1 \end{bmatrix} \begin{Bmatrix} \ddot{x} \\ \ddot{\theta} \end{Bmatrix} + \begin{bmatrix} \omega_1^2 & 0 \\ 0 & \omega_1^2 \end{bmatrix} \begin{Bmatrix} x \\ \theta \end{Bmatrix} = \begin{Bmatrix} 0 \\ 0 \end{Bmatrix} \qquad (4.14.7)$$

What happens if we try to analyze (4.14.7)? If we assume harmonic motion and set the determinant of $[K] - \omega^2[M]$ equal to zero, we'll obtain

$$(\omega_1^2 - \omega^2)(\omega_1^2 - \omega^2) = 0 \qquad (4.14.8)$$

This tells us that both natural frequencies are equal to ω_1. No problem there. But what about when we try to find the eigenvectors? If we assume an eigensolution of

$$\begin{Bmatrix} x \\ \theta \end{Bmatrix} = \begin{Bmatrix} \bar{x} \\ \bar{\theta} \end{Bmatrix} e^{i\omega_1 t} \tag{4.14.9}$$

and substitute this into (4.14.7), we'll get

$$\begin{bmatrix} 0 & 0 \\ 0 & 0 \end{bmatrix} \begin{Bmatrix} \bar{x} \\ \bar{\theta} \end{Bmatrix} = \begin{Bmatrix} 0 \\ 0 \end{Bmatrix} \tag{4.14.10}$$

This is a weird one. How can we figure out the eigenvectors if the entire matrix is composed of zeros? *Are* there any eigenvectors anymore? If we think logically about the problem, it seems that there would have to still be eigenvectors. Why? Well, our problem occurred only when ω_2 was made to equal ω_1. As long as ω_2 and ω_1 aren't equal, no matter how close they might be to each other, there are still distinct eigenvectors, $\{1 \quad 0\}^T$ and $\{0 \quad 1\}^T$; it doesn't make sense for them to suddenly disappear just because ω_2 has changed from being almost equal to ω_1 to being exactly equal to ω_1. Let's presume that these are both still eigenvectors of the problem. Do they satisfy (4.14.10)? Yes, they definitely do. And are they independent of each other? They certainly are. So they *are* eigenvectors for the system. The interesting point is that the system eigenvectors are no longer uniquely defined. For instance, what if we looked at $\{1 \quad 1\}^T$ and $\{1 \quad -1\}^T$?. These also satisfy (4.14.10) and are independent. So are $\{2 \quad 1\}^T$ and $\{1 \quad 2\}^T$. In fact, so are an infinite number of choices.

What's happened is that the problem has become a bit indeterminate. Since both the original eigenvectors have the same natural frequency, any initial displacement of the system will oscillate at ω_1.

Example 4.20

> **Problem**   Say that l is 1 m for the system shown in Figure 4.38 and the system has repeated natural frequencies, as just discussed. Assume initial conditions of $x_1 = 1$ cm and $x_2 = 0$. Find the response of the system.
>
> **Solution**   In this case we've got
>
> $$x_1 = x - l\theta = .01 \tag{4.14.11}$$
>
> and
>
> $$x_2 = x + l\theta = 0 \tag{4.14.12}$$
>
> The solution to this is $x = .5$ cm and $\theta = -.005$ rad. Since the translational and rotational frequencies are the same, the system stays in this configuration, as shown in Figure 4.40. Thus it could be thought that this initial deflection is itself an eigenvector. In the same way, any initial deflection will simply oscillate at ω_1, like an eigenvector.

Since our system is no longer telling us what the eigenvectors should be when we have repeated roots, we have to figure them out for ourselves. In a case such as the one we've just seen, it would be

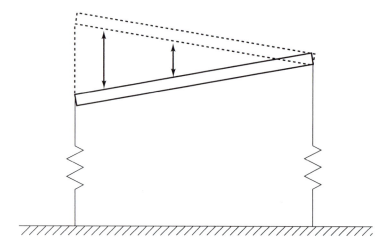

Figure 4.40 System with repeated natural frequencies

easiest to simply choose $\{1 \;\; 0\}^T$ and $\{0 \;\; 1\}^T$ as our eigenvectors. This way, there'd be no discontinuity in our eigenvectors for the cases of $\omega_1 = \omega_2$ and $\omega_1 \neq \omega_2$. In general, you'll want to have n independent eigenvectors for an n DOF problem. If you've got a repeated root, then you should construct two vectors that satisfy the problem and are independent of the other (well-determined) eigenvectors, as well as being independent of each other.

Up until this point, all the systems we've been analyzing have been attached to ground in some way. Although this is a pretty common occurrence in real life, it's also reasonably common to have systems that aren't connected to the ground. Airplanes and spacecraft come to mind immediately, as do boats and submarines. The complicating factor when dealing with systems of this sort is that, according to Newton, once they're set in motion they'll remain in motion until acted on by some force. This behavior gives rise to *zero frequencies,* i.e., natural frequencies that are equal to zero.

To see how these come about, we'll consider the system shown in Figure 4.41. The equations of motion for this system are given by

$$\begin{bmatrix} m & 0 \\ 0 & m \end{bmatrix} \begin{Bmatrix} \ddot{x}_1 \\ \ddot{x}_2 \end{Bmatrix} + \begin{bmatrix} k & -k \\ -k & k \end{bmatrix} \begin{Bmatrix} x_1 \\ x_2 \end{Bmatrix} = \begin{Bmatrix} 0 \\ 0 \end{Bmatrix} \qquad (4.14.13)$$

If you consider the $[K]$ matrix, you'll soon notice that it's different in a fundamental way from those we've seen so far; namely, its determinant is equal to zero. Thus $[K]$ isn't invertible. Note

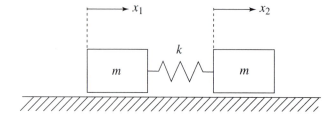

Figure 4.41 2 DOF system with a rigid body mode

that $[M]$ is always invertible, since the mass elements can't be equal to zero (unless you've set your problem up incorrectly). To see what this lack of invertibility does to the problem, we'll assume a solution of $\bar{X}e^{i\omega t}$ and substitute this into (4.14.13), leading to

$$\begin{bmatrix} k - m\omega^2 & -k \\ -k & k - m\omega^2 \end{bmatrix} \begin{Bmatrix} \bar{x}_1 \\ \bar{x}_2 \end{Bmatrix} = \begin{Bmatrix} 0 \\ 0 \end{Bmatrix} \tag{4.14.14}$$

Setting the determinant of the matrix to zero gives us

$$m^2\omega^4 - 2mk\omega^2 = 0 \tag{4.14.15}$$

or, factoring this into two parts and dividing through by m

$$\omega^2(m\omega^2 - 2k) = 0 \tag{4.14.16}$$

The two solutions are therefore

$$\omega_1^2 = 0 \tag{4.14.17}$$

and

$$\omega_2^2 = \frac{2k}{m} \tag{4.14.18}$$

The second solution corresponds to a "breathing" mode of the system. Both masses move out of phase with each other at a frequency of $\sqrt{\frac{2k}{m}}$. You can see this by solving (4.14.14) for $\omega^2 = \frac{2k}{m}$, which will give us the eigenvector solution $\{1 \quad -1\}^T$. The more relevant solution for us is the one corresponding to $\omega_1 = 0$. If we solve (4.14.14) for this case, we'll obtain the associated eigenvector $\{1 \quad 1\}$. This looks just like the first eigenvector of our familiar three-spring, two-mass problem when both masses had equal values, as did all the springs (4.7.8). However the difference between these two cases is that now the frequency of the solution is zero. This mode represents both masses moving off to the right or left, each being displaced the same amount as the other. Given an initial velocity, they'll keep moving until something stops them. Since there is no relative motion, this type of response is referred to as a *rigid body* mode, since this is the same response you'd see if the internal spring's value were infinite (making the system a rigid body).

You'll note that this particular example allowed rigid body translation in only one direction. Since a general rigid body has six degrees of freedom (three translations and three rotations), you should expect the equations of motion for such systems (satellites, aircraft, etc.) to have six zero frequency modes corresponding to these degrees of freedom. Although this complicates the analysis when one is interesting in controlling these systems, it doesn't really affect a general vibrational analysis. A modal analyzer will simply record these as zero frequency modes, and the transfer functions will show a peak as the driving frequency approaches zero.

4.15 INFLUENCE COEFFICIENTS

To finish up the chapter, we'll look at something called *influence coefficients*. The best way to let you know what these are is to examine them in a one dimensional setting, i.e., an SDOF spring-mass system (Figure 4.42). For such a system, we know that to displace mass m by an amount x requires a force f

$$f = kx \tag{4.15.1}$$

Similarly, if we apply a force f to the mass, it will deflect thus:

$$x = \frac{1}{k}f \tag{4.15.2}$$

If we define a as the inverse of k, our *stiffness influence coefficient*, then we have

$$x = af \tag{4.15.3}$$

where a is our *flexibility influence coefficient*. As you can see, the flexibility influence coefficient is simply the inverse of the stiffness influence coefficient. Thus, if we've got one, we can find the other. To obtain either one experimentally, we'd simply have to apply a force to the mass and observe its deflection or deflect it and have a force transducer handy that would tell us the force needed to cause that deflection. If you think about it, this is a nice way to determine the spring stiffness without needing to know the mass. We could place a known mass on our current mass, which would translate into a known force being exerted on the spring ($f = mg$). We could then measure the deflection and deduce that k is simply $\frac{f}{x}$.

Now that we understand what's what with the SDOF case, it's time to move onto the n DOF case. The same general characteristics will hold. This time, we're not going to be finding k but rather $[K]$, the system's stiffness matrix. We'll be calling this the *stiffness influence coefficient matrix*, and all

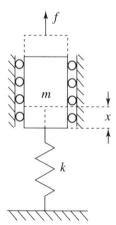

Figure 4.42 SDOF influence coefficients

the entries of $[K]$ will be the individual stiffness influence coefficients. We'll also be able to find $[A]$, which will be equal to $[K]^{-1}$. Not surprisingly, the elements of $[A]$ will be the flexibility influence coefficients and $[A]$ itself will be called the *flexibility influence coefficient matrix*. Just as in the SDOF case, knowing $[K]$ and how the n masses are displaced will allow us to determine the forces acting on the masses. Or, in a reversed sense, knowing what the applied forces are will allow us to calculate the displacements.

For clarity, let's look at a serial chain of masses (Figure 4.43). We'll be solving for the flexibility influence coefficients, $a_{i,j}$. Assume we apply a force to m_j. We know that all the masses will (in general) move into a new static configuration as a result of this force. We'll call $a_{i,j}$ the flexibility influence coefficient between mass i and mass j. It lets us determine how much the mass m_i will deflect if the force f_j is applied to m_j. The deflection due to f_j will simply be $a_{i,j}f_j$. Each time we put a new force on another mass, all the masses will again move to a new static configuration. But since the system is linear, we can just add up all the contributions to find the final deflection. Thus

$$x_i = a_{i,1}f_1 + a_{i,2}f_2 + a_{i,3}f_3 + \cdots + a_{i,n}f_n \tag{4.15.4}$$

This is simply saying that the deflection at m_i will be equal to the deflection caused by the first force acting on the first mass plus the deflection caused by the second force acting on the second mass, etc.

So far, so good. We can even realize that since we're summing over all the masses, this can be written as

$$x_i = \sum_{j=1}^{n} a_{i,j}f_j \tag{4.15.5}$$

This looks very much like summation notation for a matrix multiplied by a vector. Therefore all we need do is define

$$X \equiv \{x_1\ x_2\ \cdots\ x_n\}^T, \quad F \equiv \{f_1\ f_2\ \cdots\ f_n\}^T \tag{4.15.6}$$

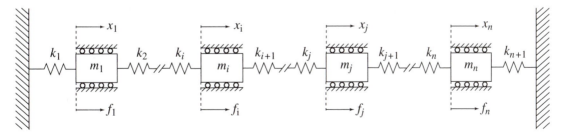

Figure 4.43 Serial mass-spring system

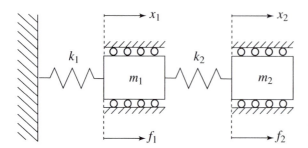

Figure 4.44 2 DOF flexibility influence coefficient example

and

$$[A] \equiv \begin{bmatrix} a_{1,1} & a_{1,2} & a_{1,3} & \cdots \\ a_{2,1} & a_{2,2} & a_{2,3} & \cdots \\ a_{3,1} & a_{3,2} & a_{3,3} & \cdots \\ \vdots & & & a_{n,n} \end{bmatrix} \tag{4.15.7}$$

and we can write (4.15.5) as

$$X = [A]F \tag{4.15.8}$$

In this form, our multi-degree-of-freedom result looks like the single-degree-of-freedom $x = af$ (with $a = \frac{1}{k}$). Let's do an example now just to see that it all works out in reality, and then we'll look at stiffness influence coefficients.

Example 4.21

Problem Determine the flexibility influence coefficients for the system shown in Figure 4.44.

Solution First, we'll apply a force f_1 to the first mass and let f_2 equal zero. Since the second mass is unrestrained, the change in displacement for m_1 is just

$$x_1^{f_1} = \frac{1}{k_1} f_1$$

The displacement is indicated by $x_1^{f_1}$ and not simply x_1, to indicate that it's the displacement of the ith mass due to the force f_1. We'll next need to find the displacement due to the force f_2 and add it to the first to find the total displacement. x_2 moves the same amount as x_1 because there's no additional spring affecting it, and so

$$x_2^{f_1} = \frac{1}{k_1} f_1$$

Now let's apply zero force to the first mass and f_2 to the second. We can see that the deflection of m_2 will depend upon the equivalent spring between it and ground, and the motion of m_1 will come from a static force balance. Beginning with m_2, the displacement is going to be equal to

$$x_2^{f_2} = \frac{1}{k_{eq}} f_2$$

where $k_{eq} = \frac{k_1 k_2}{k_1 + k_2}$ (recall our work involving parallel and series springs from Chapter 1). Thus the deflection is

$$x_2^{f_2} = \frac{k_1 + k_2}{k_1 k_2}$$

To find $x_1^{f_2}$, we'll need to do a force balance at m_1

$$k_1 x_1^{f_2} = (x_2^{f_2} - x_1^{f_2}) k_2$$

which, upon substitution of x_2 and solving for x_1, gives us

$$x_1^{f_2} = \frac{1}{k_1} f_2$$

Adding together the results of the two cases (f_1 finite, $f_2 = 0$, and $f_1 = 0$, f_2 finite) gives us

$$\left\{ \begin{array}{c} x_1 \\ x_2 \end{array} \right\} = \left[\begin{array}{cc} \frac{1}{k_1} & \frac{1}{k_1} \\ \frac{1}{k_1} & \frac{k_1 + k_2}{k_1 k_2} \end{array} \right] \left\{ \begin{array}{c} f_1 \\ f_2 \end{array} \right\}$$

where $x_1 = x_1^{f_1} + x_1^{f_2}$ and $x_2 = x_2^{f_1} + x_2^{f_2}$. Thus we see that

$$[A] = \left[\begin{array}{cc} \frac{1}{k_1} & \frac{1}{k_1} \\ \frac{1}{k_1} & \frac{k_1 + k_2}{k_1 k_2} \end{array} \right]$$

The first thing to comment about regarding the preceding example is that $a_{1,2}$ is equal to $a_{2,1}$. Is this expected? Well, you'll recall that the stiffness matrix $[K]$ is symmetric. And when we started this section the claim was made that $[A]$ is equal to the inverse of $[K]$. Thus, if this is true, shouldn't we expect symmetric flexibility influence coefficient matrices? And, assuming that $[K]$ is invertible, this will in fact be the case.

Since we've suggested that $[A]$ is equal to $[K]^{-1}$, it might be worth checking this out for the example we've just gone over. For this example, we can solve for the inverse exactly and find

$$\left[\begin{array}{cc} \frac{1}{k_1} & \frac{1}{k_1} \\ \frac{1}{k_1} & \frac{k_1 + k_2}{k_1 k_2} \end{array} \right]^{-1} = \left[\begin{array}{cc} k_1 + k_2 & -k_2 \\ -k_2 & k_2 \end{array} \right] \tag{4.15.9}$$

You can verify this result by multiplying the two matrices together and seeing that the result is indeed the identity matrix. As promised, the inverse we just found is exactly equal to the system's stiffness matrix. So now it's time to consider how we can find the stiffness influence coefficient matrix and see if it's the same as the stiffness matrix.

We'll again look at Figure 4.43. This time, we'll enforce given displacements of the masses and ask what forces are necessary to hold these positions. The stiffness influence coefficient $k_{i,j}$ will tell us what force is needed to hold the jth mass at a deflection of x_j according to

$$f_i = k_{i,j} x_j \tag{4.15.10}$$

Note that for this to hold, the deflections of all other masses must be held at zero. This is the analogue of what we did when determining the flexibility influence coefficients. For these, we applied a particular force to one mass and no force to the others. Now we're deflecting one mass and keeping all the others fixed. Just as in the preceding case, we can express the complete stiffness influence coefficient information by defining

$$X \equiv \{x_1 \ x_2 \ \cdots \ x_n\}, \quad F \equiv \{f_1 \ f_2 \ \cdots \ f_n\} \tag{4.15.11}$$

and

$$[K] \equiv \begin{bmatrix} k_{1,1} & k_{1,2} & k_{1,3} & \cdots \\ k_{2,1} & k_{2,2} & k_{2,3} & \\ k_{3,1} & k_{3,2} & k_{3,3} & \\ \vdots & & & k_{n,n} \end{bmatrix} \tag{4.15.12}$$

and writing

$$F = [K]X \tag{4.15.13}$$

Example 4.22

Problem   Again consider the system of Figure 4.44. Determine the stiffness influence coefficients.

Solution   Actually, it's easy enough to solve this problem immediately, by moving both m_1 and m_2 and determining the total force needed at m_1 and m_2 to hold those positions, but we'll do it the longer way just to be complete. First, we'll move m_1 by x_1 and hold m_2 fixed. In this case the springs are pushing back with a force equal to $(k_1 + k_2)x_1$, and so our applied force must exactly counter this. Thus $f_1 = (k_1 + k_2)x_1$. The k_2 spring is trying to push m_2 to the right with a force equal to $k_2 x_1$ and so the counteracting force must be $f_2 = -k_2 x_1$. Thus we have

$$\left\{ \begin{matrix} f_1^{x_1} \\ f_2^{x_1} \end{matrix} \right\} = \begin{bmatrix} k_1 + k_2 & 0 \\ -k_2 & 0 \end{bmatrix} \left\{ \begin{matrix} x_1 \\ 0 \end{matrix} \right\} \tag{4.15.14}$$

Next, we fix x_1 at zero and let the second mass move by x_2. Now the second spring is pulling m_1 to the right with a force equal to $k_2 x_2$, implying that $f_1 = -k_2 x_2$ and is also pulling m_2 to the left with the same force, leading to $f_2 = k_2 x_2$. This gives us

$$\left\{ \begin{matrix} f_1^{x_2} \\ f_2^{x_2} \end{matrix} \right\} = \begin{bmatrix} 0 & -k_2 \\ 0 & k_2 \end{bmatrix} \left\{ \begin{matrix} 0 \\ x_2 \end{matrix} \right\} \tag{4.15.15}$$

Adding together (4.15.14) and (4.15.15) (and letting $f_1 = f_1^{x_1} + f_1^{x_2}$, $f_2 = f_2^{x_1} + f_2^{x_2}$) gives us

$$\left\{ \begin{matrix} f_1 \\ f_2 \end{matrix} \right\} = \begin{bmatrix} k_1 + k_2 & -k_2 \\ -k_2 & k_2 \end{bmatrix} \left\{ \begin{matrix} x_1 \\ x_2 \end{matrix} \right\}$$

And, as we can see, the flexibility influence coefficient matrix is exactly equal to the system's stiffness matrix (not to mention equal to the inverse of $[A]$).

Now it's high time we answered a question. What's the point? Why do we care about influence coefficients? Well, think about this for a second. Let's say you've been given the assignment of performing an analysis on some system. You've got a good idea of what the dominant masses in the system are, but you're not so sure about the stiffnesses running around the system. How do you figure it out? Well, one way is to create a detailed model of the system, based on the stiffnesses of the material that comprises the structure, and then analyze this model. But you'd like to find an easy way to check your results. Or you might not want to go to the extensive trouble of building up a detailed model—you might prefer to get an idea of the stiffnesses from some simple experiments. Well, the influence coefficients help you do this. To find $[K]$ directly, you need to deflect the structure in a known manner and record the forces needed to do so. Then you can just read off $[K]$ as we did in Example 4.22. The difficulty with this approach is that it isn't really easy to deflect a continuous structure into a particular configuration and then figure out the forces needed—you would need large clamps that could move the structure in just the way you want, as well as some good force transducers.

The simpler way to get the same information is to go after the flexibility influence coefficient matrix and then invert it to find $[K]$. Obtaining $[A]$ experimentally isn't nearly as hard as directly finding $[K]$. All you need do is put a predetermined load at a point of the structure corresponding to where you're placing a lumped-mass approximation and then measure how the structure deflects. If the loading is vertical, just put a known mass on the structure and let gravity supply the force. By moving the force to all the lumped-mass positions, and reading off all the different deflections, you can build up your $[A]$ matrix. Inverting this then gives you $[K]$.

There's even another reason to be interested in the influence coefficients, although we won't be going into it in this book. As the number of masses in your system becomes larger, the system becomes more and more of a continuum, until finally we arrive at a fully continuous system. At this point, we'll no longer have influence coefficients but will rather have influence functions, functions that tell us how the system deflects due to any distribution of forces on the object. This is a very useful piece of information to have when trying to determine the response of a continuous system to a continuous loading.

4.16 HOMEWORK PROBLEMS

Section 4.2

4.1. Find the two natural frequencies and their associated eigenvectors for the system illustrated in Figure P4.1. $m_1 = 1 \times 10^{-3}$ kg, $m_2 = 10 \times 10^{-3}$ kg, $k_1 = 3 \times 10^3$ N/m, $k_2 = 3 \times 10^3$ N/m.

4.2. Find the three natural frequencies and their associated eigenvectors for the system illustrated in Figure P4.2. $m_1 = .5$ kg, $m_2 = .5$ kg, $m_3 = .02$ kg, $k_1 = 1 \times 10^3$ N/m, $k_2 = 1.5 \times 10^3$ N/m, $k_3 = .2 \times 10^3$ N/m.

4.3. You've been asked to add realism to a new cartoon. The action of interest is when Dopey Dog, having attached himself by means of an elastic bungie cord to a rocky outcropping, is dismayed

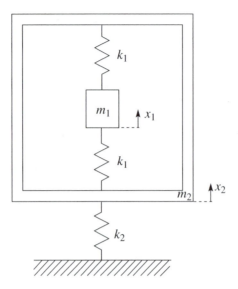

Figure P4.1

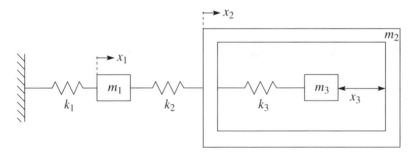

Figure P4.2

to see the outcropping come apart from the ledge and fall (Figure P4.3). Before the break Dopey was in static equilibrium below the ledge. What will the behavior of the Dopey/bungie cord/ledge system be during the fall? Treat the bungie cord as massless with a spring constant equal to 1000 N/m. Dopey's mass is 18 kg, the ledge's mass is 300 kg, and the unstretched spring is 1 m long.

4.4. Referring to Problem 4.3, how would the answers change if we ignore the dynamical effect of Dopey on the ledge and treat the ledge as being infinitely massive?

4.5. Again consider the system of Problem 4.3. If the connection between the ledge and Dopey is viewed as an inextensible, massless string, what will the dynamic response of the system be after the ledge breaks? Relate this behavior to the case of a spring connection between Dopey and the ledge (Problem 4.3).

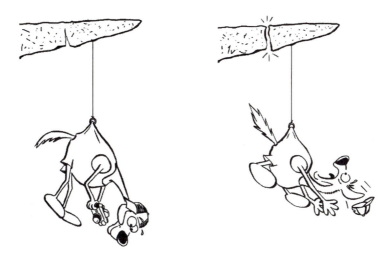

Figure P4.3

4.6. Find the natural frequencies and eigenvectors for the system shown in Figure P4.6. $m_1 = 15$ kg, $m_2 = 25$ kg, $k_1 = 120$ N/m, $k_2 = 200$ N/m, $k_3 = 50$ N/m.

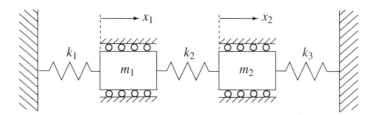

Figure P4.6

4.7. Solve for the response of the system illustrated in Figure P4.6 if the initial conditions are given by

$$\begin{Bmatrix} x_1(0) \\ x_2(0) \end{Bmatrix} = \begin{Bmatrix} 1 \\ 0 \end{Bmatrix}, \qquad \begin{Bmatrix} \dot{x}_1(0) \\ \dot{x}_2(0) \end{Bmatrix} = \begin{Bmatrix} 0 \\ 1 \end{Bmatrix}$$

$m_1 = 1$ kg, $m_2 = 2$ kg, $k = 3$ N/m, $k_2 = 2$ N/m, $k_3 = 4$ N/m.

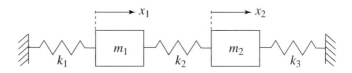

Figure P4.7

4.8. Find the eigenvectors and natural frequencies for the system illustrated in Figure P4.8. Comment on the physical behavior. $m_1 = 2$ kg, $m_2 = .02$ kg, $m_3 = 2$ kg, $k_1 = 1000$ N/m, $k_2 = 20$ N/m, $k_3 = 2000$ N/m, $k_4 = 20$ N/m, $k_5 = 1000$ N/m.

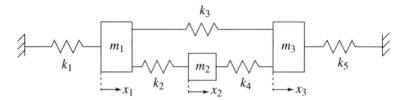

Figure P4.8

4.9. Figure P4.9 shows a double pendulum system, which also can be looked at as a model of a two-link robotic manipulator. Find the equations of motion about the system's stable equilibrium position ($\theta_1 = \theta_2 = 0$). Once you've found them, linearize the equations by assuming that all the angular deflections are small. Then calculate the system's natural frequencies and eigenvectors for the given parameter values. $m_1 = 2$ kg, $m_2 = 2$ kg, $l_1 = 1$ m, $l_2 = 1.5$ m, $g = 9.81$ m/s^2.

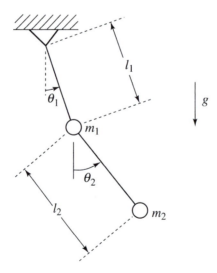

Figure P4.9

4.10. A nonuniform bar has a length of 2 m, a mass of 100 kg, and a moment of inertia about the mass center (located at the bar's geometric center) equal to 100 kg·m^2. The bar is supported at each end by a spring with spring constant equal to 1000 N/m. Analyze the natural frequencies and eigenvectors for this system and comment on why they are unusual. What do the system's modes look like? (See Figure P4.10.)

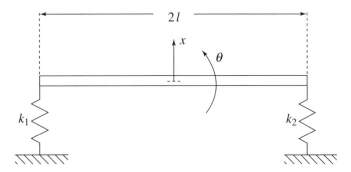

Figure P4.10

4.11. Derive the linearized equation of motion for the 2 DOF system shown in Figure P4.11. Find the natural frequencies and eigenvectors. $m_1 = 30$ kg, $m_2 = 2$ kg, $k_1 = 15$ N/m, $l = 2$ m.

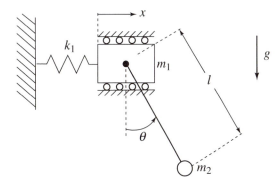

Figure P4.11

4.12. Find the equations of motion, linearize them, and find the natural frequencies and eigenvectors for the system illustrated in Figure P4.12. $m_1 = 2$ kg, $m_2 = 20$ kg, $m_3 = 1$ kg, $k_1 = 1000$ N/m, $l = 1$ m.

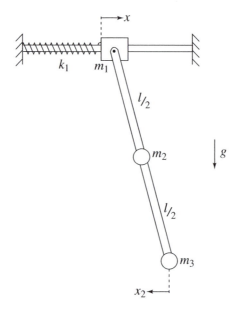

Figure P4.12

4.13. Consider the system illustrated in Figure P4.13. The entire mass is concentrated in three places; the rest of the rigid bar is massless. Find the system's equations of motion. Determine the eigenvectors and natural frequencies. $m_1 = 1$ kg, $m_2 = 1$ kg, $m_3 = 1$ kg, $k = 2$ N/m, $l = 1$ m.

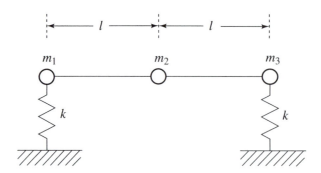

Figure P4.13

4.14. Determine the eigenvectors and natural frequencies for the system illustrated in Figure P4.14. The bar is uniform. Discuss the physical meaning of the responses. $m_1 = 100$ kg, $m_2 = 1500$ kg, $k_1 = 10,000$ N/m, $k_2 = 12,000$ N/m, $k_3 = 70,000$ N/m, $l = 4$ m.

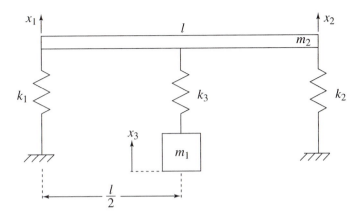

Figure P4.14

4.15. Analyze the system shown in Figure P4.13 for $m_1 = 2$ kg, $m_2 = 1$ kg, and $m_3 = 1$ kg, and comment on how (and why) the natural frequencies and eigenvectors differ from those of Problem 4.13. $k = 2$ N/m and $l = 2$ m.

4.16. For the following n DOF system

$$[M]\ddot{X} + [K]X = 0$$

Is it possible to have $[K]X_i = \{0\}$? (X_i is the ith eigenvector of the system.) If so, under what circumstances? What would this imply about $[K]$?

4.17. You're told that the three eigenvectors for a three-mass vibratory system, corresponding to three distinct natural frequencies, are

$$X_1 = \begin{Bmatrix} 4 \\ 2 \\ 2 \end{Bmatrix}, \quad X_2 = \begin{Bmatrix} 1 \\ 0 \\ -1 \end{Bmatrix}, \quad \text{and } X_3 = \begin{Bmatrix} 3 \\ 2 \\ 3 \end{Bmatrix}$$

Is this believable? Why or why not?

4.18. For the system shown in Figure P4.18, determine which modal deflections make physical sense and which don't. Explain how you came to your conclusions.

$$X_1 = \begin{Bmatrix} 0 \\ 1 \\ 1 \\ 0 \end{Bmatrix}, \quad X_2 = \begin{Bmatrix} 1 \\ 1 \\ 1 \\ 1 \end{Bmatrix}, \quad \text{and } X_3 = \begin{Bmatrix} 2 \\ 1 \\ -1 \\ -2 \end{Bmatrix}$$

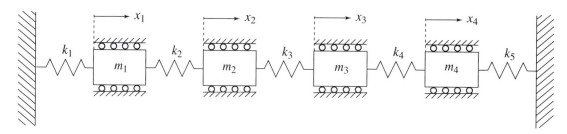

Figure P4.18

4.19. Given

$$[K]^{-1}[M] = \begin{bmatrix} 4 & 3 & 6 \\ 1 & 3 & 1 \\ 1 & 1 & 3 \end{bmatrix}$$

try to determine whether $X_1 = \{3 \;-4\; 1\}^T$ is an eigenvector of the spring-mass system satisfying

$$[M]\ddot{X} + [K]X = 0$$

Can this be done?

4.20. Vitally important data with regard to the stiffness and mass of a TOP SECRET device have been lost. The inventor blewup herself, and many of her notes, to smithereens in her latest experiment. All that was recovered from her journal is the fragment shown in Figure P4.20. From this information, can you deduce the rest of the stiffness matrix? Assume that $[K]$ is symmetric.

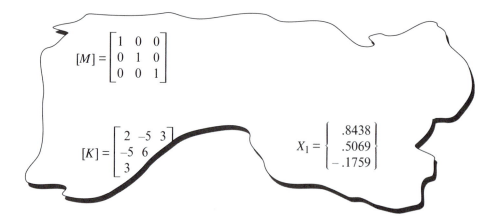

Figure P4.20

4.21. Solve the following eigenvalue problem and write down the general solution for arbitrary initial conditions.

$$[M]\ddot{X} + [K]X = 0$$

$$M = \begin{bmatrix} 1 & 0 \\ 0 & 1 \end{bmatrix}; K = \begin{bmatrix} 2 & -1 \\ -1 & 3 \end{bmatrix}; X = \begin{Bmatrix} x_1 \\ x_2 \end{Bmatrix}$$

4.22. The system illustrated in Figure P4.22a shows two disks that are supported by identical thin wire rods. The rods can be treated as massless spring elements. Shown in Figure P4.22b is the result of a bending test on one of the rods. If $I_1 = 15$ kg·m^2 and $I_2 = 20$ kg·m^2, determine the natural frequencies and mode shapes for rotational motions of the disks.

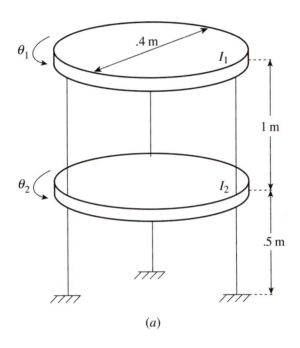

(a)

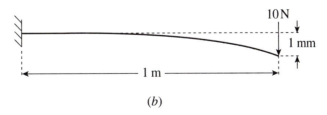

(b)

Figure P4.22

4.23. Are $X_1 = \begin{Bmatrix} 1 \\ 2 \\ 3 \end{Bmatrix}$, $X_1 = \begin{Bmatrix} 2 \\ -1 \\ 1 \end{Bmatrix}$, and $X_1 = \begin{Bmatrix} 5 \\ -0 \\ 5 \end{Bmatrix}$ a possible set of independent eigenvectors?

4.24. Problem 1.66 in Chapter 1 looked at the motion of a constrained particle in a tube. Now we're going to go beyond this model by adding a second degree of freedom. Figure P4.24 shows the system we'll be considering. If the springs are identical (as shown), one possible use of this model would be to approximate the first mode of vibration for a water tower of the sort shown on the left. If the spring constant of each spring is equal to $\frac{k}{2}$ and the mass is given by m, determine the response of the system to an initial displacement. Is the motion stable? What will the free vibrations look like?

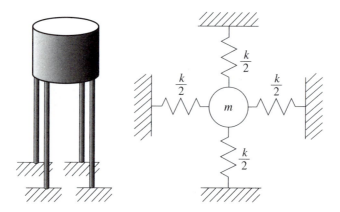

Figure P4.24

4.25. Consider again the system of Figure P4.24. This time, the spring stiffnesses are not identical, as shown in Figure P4.25. What will the response now be to an initial displacement? Is the motion stable? What do the motions look like?

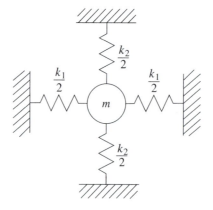

Figure P4.25

4.26. In this problem, we'll be moving a bit beyond Problem 4.24. We'll be considering the same system, but this time we'll let the entire system rotate at an angular rate equal to Ω, as shown in Figure P4.26. We'll require that $k_1 = k_2$. This system will demonstrate the fundamental response characteristics we'd see if we examined a rotating disk, such as a computer hard disk. It can also be used as a model for rotating vibration problems, such as those found in pumps. Once again, you should determine the response of the system to initial displacements and/or velocities. Comment on the motion's stability and compare your results to those of Problem 4.24.

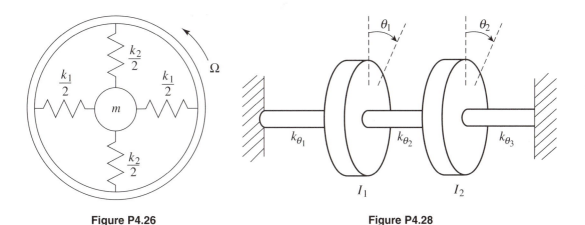

Figure P4.26   Figure P4.28

4.27. The system of Problem 4.26 is now further refined by dropping the requirement that $k_1 = k_2$. This is the most general problem of all those we've looked at so far. If k_1 goes to infinity, we've got Problem 1.66 from Chapter 1. If $k_1 = k_2$, then we've got Problem 4.26. This model is applicable to spinning disk problems, such as those involving computer hard drives. It is also a simplified model for a helicopter on its landing gear. The fact that k_1 isn't necessarily equal to k_2 reflects the different stiffnesses a helicopter would experience in pitch-vs-roll motions. It could also model turbomachinery, in which the bearing support stiffness isn't identical in all directions. Just as in Problem 4.26, determine the system's response to different initial conditions. Comment on the stability of the response and compare the results to those of Problems 1.66, 4.26 (if they've been assigned).

4.28. In the system illustrated in Problem 4.28, the three rods have rotational stiffness (k_{θ_1}, k_{θ_2}, and k_{θ_3}) but should be treated as massless (zero rotational inertia). Given the values for k_{θ_i} and I_i, find the system's natural frequencies and associated eigenvectors. $k_{\theta_1} = 8.0 \times 10^5$ N·m, $k_{\theta_2} = 12.0 \times 10^5$ N·m, $k_{\theta_3} = 5.0 \times 10^5$ N·m, $I_1 = 10$ kg·m^2, $I_2 = 30$ kg·m^2.

4.29. Consider again the system of Problem 4.28. Determine how the natural frequencies and eigenvectors change as I_2 is varied from 0 to 60 kg·m^2. Plot the variation of the natural frequencies as a function of I_2. What do the eigenvectors look like as I_2 approaches 0 and ∞?

4.30. Consider once more the system of Problem 4.28. Determine how the natural frequencies and eigenvectors change as k_{θ_2} is varied from zero to 10×10^5 N·m. Plot the variation of the natural frequencies as a function of k_{θ_2} What do the eigenvectors look like as k_{θ_2} approaches ∞ and 0?

4.31. Consider a final time the system of Problem 4.28. Determine the natural frequencies and eigenvectors if k_{θ_1} is set equal to zero.

4.32. Figure P4.32 shows a scrap of information found in an inventor's wastebasket. What can you deduce about the $[M]$ and $[K]$ matrices and the eigenvectors and eigenvalues of the problem the inventor was working on?

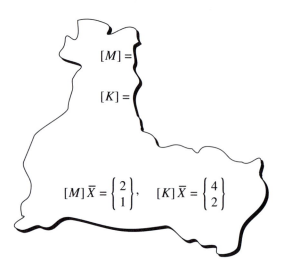

$$[M] = \qquad$$

$$[K] = \qquad$$

$$[M]\bar{X} = \begin{Bmatrix} 2 \\ 1 \end{Bmatrix}, \quad [K]\bar{X} = \begin{Bmatrix} 4 \\ 2 \end{Bmatrix}$$

Figure P4.32

4.33. Given that

$$[M] = \begin{bmatrix} \frac{2}{16} & \frac{1}{16} \\ \frac{1}{16} & \frac{5}{32} \end{bmatrix} \quad \text{and } [K] = \begin{bmatrix} \frac{26}{32} & \frac{3}{32} \\ \frac{3}{32} & ? \end{bmatrix}$$

with eigenvectors

$$X_1 = \begin{Bmatrix} 1 \\ 2 \end{Bmatrix} \quad \text{and } X_2 = \begin{Bmatrix} -3 \\ 2 \end{Bmatrix}$$

deduce the unknown element of $[K]$ as well as ω_1^2 and ω_2^2. Verify your answer by solving the resulting eigenproblem.

4.34. Consider the system shown in Figure P4.28. Let $k_{\theta_3} = 0$. Determine the equations of motion. The rotational inertias are I_1 and I_2 and the angular stiffnesses equal k_{θ_1} and k_{θ_2} as in Figure P4.34. What are the system natural frequencies? $k_{\theta_1} = 4000$ N·m, $k_{\theta_2} = 6000$ N·m, $I_1 = 1$ kg·m^2, $I_2 = 2$ kg·m^2.

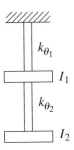

Figure P4.34

4.35. Determine the eigenvalues and eigenvectors of the following system: $a = b = .5$ m, $m_1 = m_2 = 1$ kg, $k = 1$ N/m, $g = 5$ m/s^2 (Obviously the system, shown in Figure P4.35, isn't on Earth.)

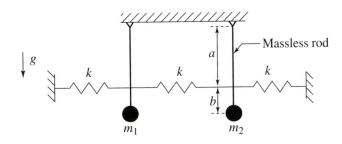

Figure P4.35

4.36. Find the natural frequencies and the associated eigenvectors for

$$\begin{bmatrix} 2 & -1 \\ -1 & 3 \end{bmatrix} \begin{Bmatrix} \ddot{x}_1 \\ \ddot{x}_2 \end{Bmatrix} + \begin{bmatrix} 8 & -1 \\ -1 & 8 \end{bmatrix} \begin{Bmatrix} x_1 \\ x_2 \end{Bmatrix} = \begin{Bmatrix} 0 \\ 0 \end{Bmatrix}$$

4.37. Figure P4.37 gives three snapshots of the vibrational response of a 2 DOF system undergoing free vibration. One possible explanation of the response is that we're seeing the response corresponding to a particular eigenvector, $X_1 = \begin{Bmatrix} 1 \\ 2 \end{Bmatrix}$, which has a natural frequency of π rad/s. What is another explanation?

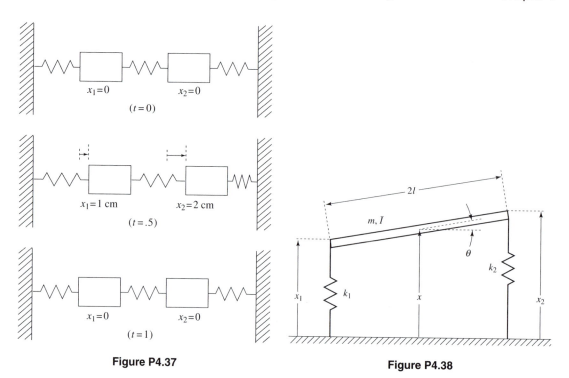

Figure P4.37   **Figure P4.38**

4.38. Consider the system shown in Figure P4.38. Calculate the kinetic and potential energies using the x and θ coordinates and then use Lagrange's equations to obtain equations of motion. Comment on the couplings in the energies and the equations of motion. Next, determine the kinetic and potential energies using x_1, x_2 coordinates, and use Lagrange's equations to find the equations of motion. Again, comment on the couplings in both the energies and the equations of motion. Now take the equations of motion you found for the x and θ coordinates and use the substitutions $x_1 + x_2 = 2x$ and $x_2 - x_1 = 2l\theta$ to transform them into equations involving x_1 and x_2.

     Are the resulting mass and stiffness matrices symmetric? Do the equations look like those you found using Lagrange's equations? Explain how you can transform from this set of equations into the symmetric form you found from Lagrange's equations.

4.39. Consider the system shown in Figure P.4.39. For one of the system eigenvectors, m_2 is completely stationary. Determine for which natural frequency this response occurs and explain physically why it happens. $m_1 = 4$ kg, $m_2 = 20$ kg, $m_3 = 2$ kg, $k_1 = 14$ N/m, $k_2 = 2$ N/m, $k_3 = 6$ N/m, $k_4 = 2$ N/m.

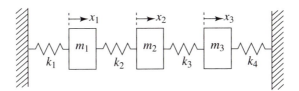

Figure P4.39

4.40. What are the eigenvectors and natural frequencies for the system illustrated in Figure P4.40? Note that the x_1 and x_2 coordinates reflect relative motion of the masses, not absolute motion. The unstretched length of each of the three springs is l. $k_3 = 1000$ N/m, $k_2 = 2000$ N/m $k_1 = 1000$ N/m, $m_3 = .1$ kg, $m_2 = .2$ kg, $m_1 = .1$ kg.

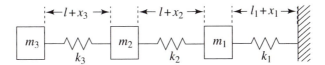

Figure P4.40

4.41. Solve Problem 4.40 using the more usual x_1, x_2, x_3 coordinates referenced to the individual masses. Show that the natural frequencies are the same and that the eigenvectors represent the same physical motions.

4.42. Find the linearized equations of motion for the system illustrated in Figure P4.42 (small θ, only linear terms retained). Calculate the natural frequencies and eigenvectors for the given parameter value. (m_1 is a uniform bar hinged m_2 and of length l). $m_1 = 2$ kg, $l = .5$ m, $m_2 = 3$ kg, $k_1 = 100$ N/m, $k_2 = 200$ N/m, $a = .3$ m.

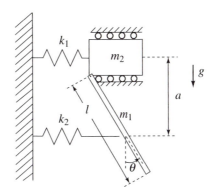

Figure P4.42

4.43. Determine whether $X_i = \begin{Bmatrix} 1 \\ 1 \\ 1 \\ 1 \end{Bmatrix}$ is an eigenvector for the system having the following $[M]$ and $[K]$

matrices:

$$[M] = \begin{bmatrix} 1 & 0 & 0 & 0 \\ 0 & 2 & 0 & 0 \\ 0 & 0 & 3 & 0 \\ 0 & 0 & 0 & 1 \end{bmatrix}; \quad [K] = \begin{bmatrix} 3 & -1 & -1 & -1 \\ -1 & 3 & -2 & 0 \\ -1 & -2 & 4 & -1 \\ -1 & 0 & -1 & 2 \end{bmatrix}$$

without solving the eigenvalue problem

$$| [K] - \omega^2 [M] | = 0$$

Section 4.3

4.44. In the next three problems we'll be examining how an accelerometer's finite mass and stiffness properties affects its measurements. We'll analyze the vibrating structure in Figure P4.44, which has mass $m_1 = .6$ kg and support stiffness $k_1 = 1000$ N/m. An external force acts on the mass and is equal to $\bar{f} \sin(\omega t)$. The accelerometer body has a mass $m_2 = .005$ kg, the moving element has mass $m_2 = .001$ kg, and the internal spring has stiffness $k_2 = 10,000$ N/m. Show that the magnitude of the acceleration response of the m_1 / k_1 structure to an applied forcing at 30 rad/s (no accelerometer attached) is equal to 1.9565 m/s^2.

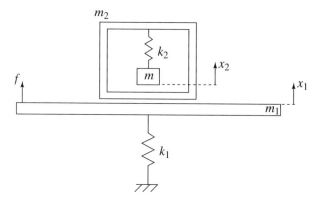

Figure P4.44

4.45. Consider again the system shown in Figure P4.44. What will the accelerometer show the acceleration to be? Assume that $\bar{f} = 1$ N. This will require a 2 DOF analysis of the combined accelerometer/base structure. What is the error induced by the accelerometer's dynamics?

4.46. Consider again the system of Figure P4.44. Most of the error produced by using the accelerometer to measure the plate's acceleration is induced by the accelerometer's mass.

(a) Include the accelerometer's mass along with m_1 and calculate the acceleration response due to $f(t)$.

(b) What's the error between solving the complete 2 DOF problem (Problem 4.45) and using a lumped-mass approximation for the accelerometer?

4.47. To analyze the effects of the motions of an automobile on its tires, we approximate the car by a bar of uniform mass (1500 kg) with a moment of inertia equal to 2000 kg · m^2. The bar is 4 m long. The tire/suspension assembly is approximated by a pair of springs, with spring constant e qual to 11,697 N/m^2 for each one. If the road profile is sinusoidal, with an amplitude of 20 cm and a wavelength of 10 m, as shown in Figure P4.47, at what speed will the car experience resonance? What mode is this resonance associated with?

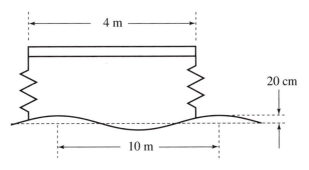

Figure P4.47

4.48. How will the answer for Problem 4.47 change if the wavelength of the road is shortened to 8 m?

4.49. For what road wavelength will the car in Problem 4.47 experience purely translational motion only (no rotation)?

4.50. The system shown in Figure P4.50 is more involved than that of Problem 4.47 because the front and rear springs of the vehicle model are different. Find the equations of motion, determine the natural frequencies, and comment on whether it is possible to find a vehicle speed for which rotational motions are not present.

The car body is treated as a uniform bar with linear density equal to 250 kg/m. A lumped mass of 150 kg is located at the back end of the bar (approximating a rear engine configuration). $k_1 = 15,000$ N/m, $k_2 = 10,000$ N/m, the road profile's amplitude is 2 cm, and the road profile's wavelength is 8 m.

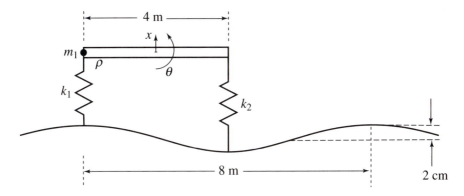

Figure P4.50

4.51. Can you deduce from Figure P4.51 the number of degrees of freedom of the tested system and the type of sensor used to acquire the data? The mass was direct-forced.

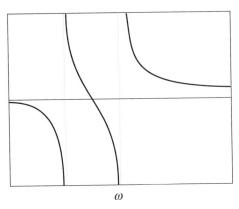

ω

Figure P4.51

4.52. Could the plot shown in Figure P4.52*a* be a physically correct transfer function for the displacement response for the system shown in Figure P4.52*b*? Why or why not?

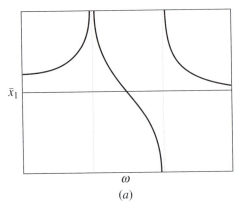

$\bar{x}_1$

ω

(a)

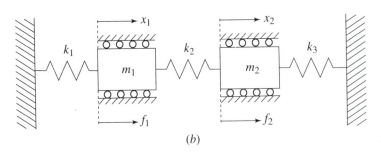

(b)

Figure P4.52

4.53. Use Figure P4.53 to explain in detail how you would solve for the steady state response for the system shown in Figure P4.52*b*. Write down the equations of motion and indicate the solution procedure you would follow. Could you use numerical integration to verify your solution? (Don't actually determine the Fourier coefficients.)

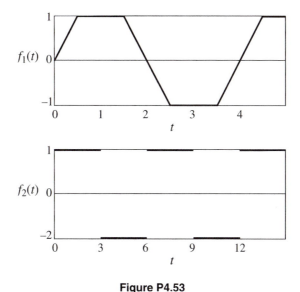

Figure P4.53

4.54. Some buildings can be analyzed as rigid bodies with flexible ground constraints when subjected to loadings due to earthquakes. Figure P4.54 illustrates such a model. Assume that the horizontal motion of the ground is given by $x = \bar{x}\sin(\omega t)$ and derive the equations governing the structure's response. Treat the structure as a uniform bar, and neglect gravity.

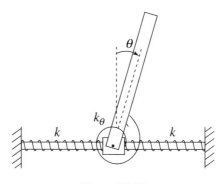

Figure P4.54

4.55. During the filming of a movie, a bus was launched through the air, supposedly clearing a gap in a highway overpass. To protect the stunt driver, bungie cords were employed to restrain his body and isolate him from the shock of the landing impact. This situation can be modeled as shown in Figure P4.55, where m_1 is the stunt driver, k_1 is the equivalent spring constant of all the restraining bungie cords, m_2 is the mass of the bus, and k_2 is the bus's suspension stiffness. If $m_1 = 75$ kg, $m_2 = 10,000$ kg, $k_1 = 73,500$ N/m, and $k_2 = 3.3 \times 10^6$ N/m, determine the natural frequencies and associated eigenvectors for the system. Next, determine the relative motion of m_1 with respect to m_2 if the bus is dropped from a height of 2 m. To simplify the analysis, ignore the effect of the static offset due to gravity.

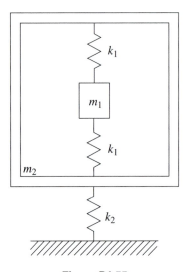

Figure P4.55

4.56. Find the amplitude of θ as a function of the forcing amplitude $\bar{f}$ and ω for the system illustrated in Figure P4.56, with $f = \bar{f} \cos(\omega t)$. Also, determine an appropriate m and k_2 combination so that when $\omega = 10$ rad/s, the amplitude of $\theta = 0$.

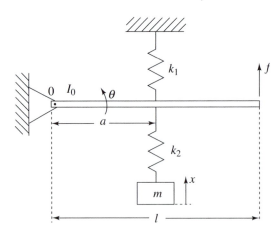

Figure P4.56

4.57. Consider again the system shown in Figure P4.52b, with $f_1 = 0$ and $f_2 = \bar{f}\cos(\omega t)$. Which of the transfer functions illustrated corresponds to $\frac{\bar{x}_1}{\bar{f}}$ and which to $\frac{\bar{x}_2}{\bar{f}}$. Why?

4.58. Consider the system shown in Figure P4.58 . Let $n = 3$, $k_1 = 10$ N/m, $k_2 = 20$ N/m, $k_3 = 30$ N/m, and $k_4 = 0$. All $c_i = 0$, $m_1 = 4$ kg, $m_2 = 2$ kg, $m_3 = 5$ kg. $f_1 = 0$, $f_2 = \cos(1.2t)$, and $f_3 = 3\sin(5.45t)$ (units in newtons). Solve for $x_1(t)$ and plot $x_1(t)$ vs t for 10 seconds.

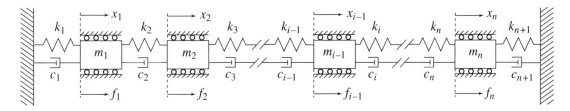

Figure P4.58

Section 4.4

4.59. Return to the system shown in Figure P4.52b. Let $k_1 = 40,000$ N/m, $k_2 = 20,000$ N/m, $f_1 = \bar{f}\sin(\omega t)$, and $f_2 = 0$. Can you find an appropriate value for k_3 so that m_1 is stationary when $\omega = 80$ rad/s? $m_1 = 50$ kg and $m_2 = 40$ kg.

4.60. At what frequency of forcing will the mass m_2 in Figure P4.60 be stationary? What will the forcing amplitude be equal to if m_1 is limited to an excursion of 3 mm? $m_1 = .4$ kg, $m_2 = .8$ kg, $k = 3000$ N/m.

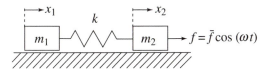

Figure P4.60

4.61. Consider the system shown in Figure P4.61. Determine the smallest secondary mass that you could use if your design constraint is that the magnitude of x_2's motion must be no more than 10 cm. $m_1 = 1000$ kg, and $k_1 = 36,000$ N/m; the magnitude of the forcing is .7 N, and the forcing frequency is 6 rad/s.

4.62. Consider Problem 4.61. What we'll now assume is that the forcing frequency increases by only .5%. What effect will this have on the response of both x_1 and x_2?

4.63. In this problem we'll see what happens for a particular vibration absorber of the type shown in Figure P4.61 in two different situations. $m_1 = 10$ kg and $k_2 = 4000$ N/m. We'll consider a vibration absorber that's just as massive as the primary mass, giving us $m_2 = 10$ kg and $k_2 = 4000$ N/m. Determine the amplitude of the primary and secondary masses when $\omega = 20$ rad/s and the forcing magnitude is to 10 N. Then remove k_2 and simply join m_2 and m_1 together. What is the resulting magnitude of oscillation? For this case, was is there any benefit to using the spring k_2 if the packaging goal was to minimize the overall vibration amplitude of the system?

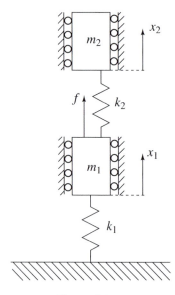

Figure P4.61

4.64. You're asked to analyze the washing machine illustrated in Figure P4.64. The mass m_1 represents the mass of the clothes washer, m_0 represents the largest bunch of clothes likely to be found in the washer, r represents the radius of the washer's drum, and k_1 represents the spring stiffness of the washer's supporting feet. The spin cycle operates at 14 rad/s. Design a spring-mass vibration absorber to be added to the washer so that the amplitude of vibration of the system is less than 3 mm over a range of 4 rad/s centered at the operating frequency. $r = .05$ m, $m_0 = 1$ kg, $m_1 = 9$ kg, $k_1 = 2250$ N/m.

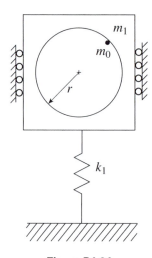

Figure P4.64

4.65. Consider a system similar to that of Figure P4.64. A vibration absorber m_2/k_2 is to be added to the housing (Figure P4.65) to quell vibrations from imbalanced loads and k_2 is given at 1500 N/m. $m_0 = 1$ kg, $m_1 = 35$ kg, $r = .25$ m, $\omega = 16$ rad/s. What must m_2 be, and how large an excursion will m_2 experience?

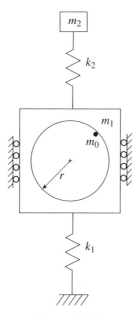

Figure P4.65

4.66. Figure P4.66 shows a transfer function that corresponds to the system of Figure P4.52*b*. What variables might this transfer function be measuring? For instance, it might be the transfer function between f_1 and x_2 or between f_2 and $\dot{x}_2$, etc. Justify your answers.

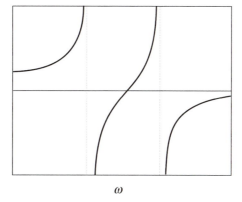

Figure P4.66

Section 4.6

4.67. Consider the system considered earlier in Figure P4.18 but with the force applied to the fourth mass instead of the first. For what ω's will x_2 be equal to zero?

4.68. For what frequencies will $x_3 = 0$ for the system illustrated in Figure P4.68?

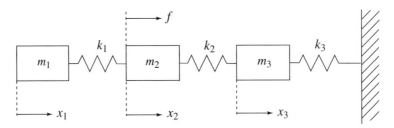

Figure P4.68

4.69. Is there a finite frequency ω for which both m_2 and m_4 are stationary for the system illustrated in Figure P4.69? $(\bar{f} \neq 0)$

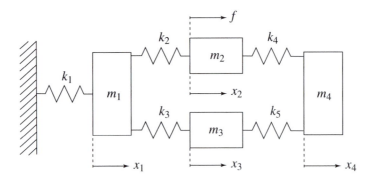

Figure P4.69

4.70. In an n DOF serial chain of lumped spring-mass elements, give the maximum number of frequencies at which a mass can be forced for which the response of that or some other mass of the system is zero. Give the minimum number. Explain the differences between the two cases.

4.71. What restrictions are there on the values of the m_1 and k_i if we wish m_1 to remain stationary when a force $f = \bar{f} \cos(30t)$ is applied to m_4?

4.72. We're given a forced, spring-supported bar. A forcing frequency is needed that will cause x (the vertical displacement of the bar's center of mass) to be zero at the same time that θ (the bar's rotation) is nonzero; i.e., we'll see pure rotation only about the system's center of mass. (Don't consider any initial-condition-induced vibrations; look only at the forced solution.) You will discover that such a solution, although seemingly easy to find, isn't attainable. Why not? $m = 4$ kg, $f = \bar{f} \cos(\omega t)$, $\bar{I} = 1$ kg·m^2, $k_1 = 500$ N/m, $k_2 = 700$ N/m, $l = .5$ m.

4.73. In Problem 4.72 we tried unsuccessfully to find a forcing frequency for which $x = 0$ and $\theta \neq 0$. Modify the system so that the force doesn't act on the left end of the bar but rather is applied .2 m to the left of the midpoint. Can a frequency that be found for this problem that satisfies $x = 0$ and $\theta \neq 0$? (*Hint:* It can.) How does this problem differ from Problem 4.72? (That is, why can we achieve $x = 0$, $\theta \neq 0$ in this case but not in earlier one?) $m = 4$ kg, $f = \bar{f} \cos(\omega t)$, $\bar{I} = 1$ kg·m^2, $k_1 = 500$ N/m, $k_2 = 700$ N/m.

Sections 4.7 and 4.8

4.74. When will the eigenvectors of an n DOF system be geometrically orthogonal to one another?

4.75. Consider the system shown in Figure P4.52b. Let $m_1 = 1$ kg, $m_2 = 5$ kg, $k_1 = 100$ N/m, $k_2 = 200$ N/m, and $k_3 = 50$ N/m; $f_1(t) = 1.8318 \cos(10t)$ and $f_2(t) = -1.3145 \cos(10t)$ (forces given in newtons). Put the system into normal form and show that under the given forcing, the system responds in the second mode (the eigenvector associated with the highest frequency).

4.76. Mass-normalize the given vectors with respect to the given mass matrix

$$[M] = \begin{bmatrix} 2 & 0 & 0 \\ 0 & 5 & 0 \\ 0 & 0 & 1 \end{bmatrix}, \quad X_1 = \begin{Bmatrix} 1 \\ 1 \\ 2 \end{Bmatrix}, \quad X_2 = \begin{Bmatrix} 1 \\ -1 \\ 2 \end{Bmatrix}, \quad X_3 = \begin{Bmatrix} 2 \\ -1 \\ 2 \end{Bmatrix}$$

4.77. Use the nondiagonal mass matrix $[M]$ to put the two given vectors into mass-normalized form.

$$[M] = \begin{bmatrix} 1 & -1 & 0 & 2 \\ -1 & 2 & -1 & 0 \\ 0 & -1 & 4 & -2 \\ 2 & 0 & -2 & 7 \end{bmatrix}, \quad X_1 = \begin{Bmatrix} 1 \\ 2 \\ 1 \\ 2 \end{Bmatrix}, \quad X_2 = \begin{Bmatrix} 2 \\ 1 \\ -1 \\ 3 \end{Bmatrix}$$

4.78. Demonstrate that the system modes are orthogonal with respect to the $[M]$ and $[K]$ matrices for the following system:

$$\begin{bmatrix} 10 & 0 & 0 \\ 0 & 2 & 0 \\ 0 & 0 & 4 \end{bmatrix} \begin{Bmatrix} \ddot{x}_1 \\ \ddot{x}_2 \\ \ddot{x}_3 \end{Bmatrix} + \begin{bmatrix} 500 & -50 & 0 \\ -50 & 400 & -20 \\ 0 & -20 & 100 \end{bmatrix} \begin{Bmatrix} x_1 \\ x_2 \\ x_3 \end{Bmatrix} = \begin{Bmatrix} 0 \\ 0 \\ 0 \end{Bmatrix}$$

4.79. Given $[M]$ and the following two eigenvectors, can you solve for X_3?

$$[M] = \begin{Bmatrix} 2 & 0 & 0 \\ 0 & 2 & 0 \\ 0 & 0 & 2 \end{Bmatrix}, \quad X_1 = \begin{Bmatrix} \frac{1}{2} \\ \frac{1}{\sqrt{2}} \\ \frac{1}{2} \end{Bmatrix}, \quad X_2 = \begin{Bmatrix} \frac{1}{\sqrt{2}} \\ 0 \\ -\frac{1}{\sqrt{2}} \end{Bmatrix}$$

4.80. Given $[M]$ and the eigenvector X_1, is it possible to determine the other two?

$$[M] = \begin{Bmatrix} 4 & 0 & 0 \\ 0 & 3 & 0 \\ 0 & 0 & 1 \end{Bmatrix}, \quad X_1 = \begin{Bmatrix} 1 \\ 1 \\ 1 \end{Bmatrix}$$

Section 4.9

4.81. Decouple the following equations of motion.

$$\begin{bmatrix} 2 & 0 & 0 \\ 0 & 2 & 0 \\ 0 & 0 & 4 \end{bmatrix} \begin{Bmatrix} \ddot{x}_1 \\ \ddot{x}_2 \\ \ddot{x}_3 \end{Bmatrix} + \begin{bmatrix} 100 & -15 & 0 \\ -15 & 200 & -80 \\ 0 & -80 & 100 \end{bmatrix} \begin{Bmatrix} x_1 \\ x_2 \\ x_3 \end{Bmatrix} = \begin{Bmatrix} 1 \\ 0 \\ -1 \end{Bmatrix}$$

4.82. Put the following system into decoupled form.

$$\begin{bmatrix} 10 & 0 & 8 \\ 0 & 12 & 0 \\ 8 & 0 & 34 \end{bmatrix} \begin{Bmatrix} \ddot{x}_1 \\ \ddot{x}_2 \\ \ddot{x}_3 \end{Bmatrix} + \begin{bmatrix} 10 & 0 & 0 \\ 0 & 200 & 0 \\ 0 & 0 & 16 \end{bmatrix} \begin{Bmatrix} x_1 \\ x_2 \\ x_3 \end{Bmatrix} = \begin{Bmatrix} 1 \\ 0 \\ 0 \end{Bmatrix}$$

4.83. Put the following system into normal form (Use a mass-normalized modal matrix.)

$$\begin{bmatrix} 10 & 0 & 18 \\ 0 & 12 & 0 \\ 18 & 0 & 34 \end{bmatrix} \begin{Bmatrix} \ddot{x}_1 \\ \ddot{x}_2 \\ \ddot{x}_3 \end{Bmatrix} + \begin{bmatrix} 1000 & -400 & 0 \\ -400 & 2000 & -400 \\ 0 & -400 & 1250 \end{bmatrix} \begin{Bmatrix} x_1 \\ x_2 \\ x_3 \end{Bmatrix} = \begin{Bmatrix} -1 \\ 3 \\ 5 \end{Bmatrix}$$

4.84. Find the eigenmodes and natural frequencies for the following system. Comment on the results.

$$\begin{bmatrix} 2 & 0 & 0 \\ 0 & 3 & 0 \\ 0 & 0 & 1 \end{bmatrix} \begin{Bmatrix} \ddot{x}_1 \\ \ddot{x}_2 \\ \ddot{x}_3 \end{Bmatrix} + \begin{bmatrix} 1 & -1 & 0 \\ -1 & 2 & -1 \\ 0 & -1 & 1 \end{bmatrix} \begin{Bmatrix} x_1 \\ x_2 \\ x_3 \end{Bmatrix} = \begin{Bmatrix} 0 \\ 0 \\ 0 \end{Bmatrix}$$

4.85. Find the eigenmodes and natural frequencies for the following system. Form the modal matrix and use it to put the system into normal-form.

$$\begin{bmatrix} 30 & -20 & 16 \\ -20 & 59 & -34 \\ 16 & -34 & 70 \end{bmatrix} \begin{Bmatrix} \ddot{x}_1 \\ \ddot{x}_2 \\ \ddot{x}_3 \end{Bmatrix} + \begin{bmatrix} 50 & -20 & 30 \\ -20 & 80 & -65 \\ 30 & -65 & 200 \end{bmatrix} \begin{Bmatrix} x_1 \\ x_2 \\ x_3 \end{Bmatrix} = \begin{Bmatrix} 0 \\ 0 \\ 0 \end{Bmatrix}$$

4.86. Put the following equations of motion into a mass-normalized normal form.

$$4\ddot{x}_1 - 2\ddot{x}_2 + 4x_1 - x_2 = \sin(3t)$$

$$8\ddot{x}_2 - 2\ddot{x}_1 + 5x_2 - x_1 = \cos(t) + 4\cos(3t)$$

4.87. Given the following system, determine the magnitude of the first mode's response. Make sure to mass-normalize the system modes.

$$\begin{bmatrix} 1 & 0 \\ 0 & 1 \end{bmatrix} \begin{Bmatrix} \ddot{x}_1 \\ \ddot{x}_2 \end{Bmatrix} + \begin{bmatrix} 2 & -1 \\ -1 & 2 \end{bmatrix} \begin{Bmatrix} x_1 \\ x_2 \end{Bmatrix} = \begin{Bmatrix} 2 \\ 3 \end{Bmatrix} \cos(2t)$$

4.88. What is the physical interpretation of normal coordinates?

Section 4.10 and 4.11

4.89. A 3 DOF system has the following mass and stiffness matrices.

$$[M] = \begin{bmatrix} 4 & 0 & 0 \\ 0 & 2 & 0 \\ 0 & 0 & 1 \end{bmatrix}, \quad [K] = \begin{bmatrix} 6 & -2 & 0 \\ -2 & 8 & -1 \\ 0 & -1 & 3 \end{bmatrix}$$

Using the knowledge that the ω_i^2 for the system are 1.2929, 2.7071, and 4.5, determine the complex eigenvalues that result from adding linear damping to the system for which the damping matrix is given by $[C] = .1[M] + .05[K]$. Determine the correct result *without* actually solving the eigenvalue problem for the damped system.

4.90. The first three entries of X_1, the first eigenvector for an unknown 3 DOF system, correspond to the displacement of the three masses, and the last three correspond to their velocities. The undamped natural frequency for this mode is equal to 1. Given that the system damping is of the proportional type, determine what the damping factor ζ is for this eigenvector.

$$X_1 = \begin{Bmatrix} 1+i \\ 2+2i \\ 1+i \\ -1.4+.2i \\ -2.8+.4i \\ -1.4+.2i \end{Bmatrix}$$

4.91. Take Problem 4.69 ($f = 0$) and add proportional damping of the form

$$[C] = .05[M] + .1[K]$$

Find the eigenvectors and natural frequencies. Mass-normalize the eigenvectors and phase-shift them so that the first entry is real. Are the other displacement entries real also? Use the mass-normalized eigenvectors to construct the normal-form equations $m_i = 1$ kg, $k_1 = 2$ N/m, $k_2 = k_5 = 1$ N/m, $k_3 = k_4 = 3$ N/m.

Section 4.12

4.92. Determine the forced response of the following system.

$$\begin{bmatrix} 3 & 0 \\ 0 & 2 \end{bmatrix} \begin{Bmatrix} \ddot{x}_1 \\ \ddot{x}_2 \end{Bmatrix} + \begin{bmatrix} .6 & -.1 \\ -.1 & .2 \end{bmatrix} \begin{Bmatrix} \dot{x}_1 \\ \dot{x}_2 \end{Bmatrix} + \begin{bmatrix} 20 & -4 \\ -4 & 40 \end{bmatrix} \begin{Bmatrix} x_1 \\ x_2 \end{Bmatrix} = \begin{Bmatrix} 20 \\ 0 \end{Bmatrix} \sin(5t)$$

4.93. Determine the forced response of the following system.

$$
\begin{bmatrix} 2 & 0 & 0 \\ 0 & 4 & 0 \\ 0 & 0 & 10 \end{bmatrix} \begin{Bmatrix} \ddot{x}_1 \\ \ddot{x}_2 \\ \ddot{x}_3 \end{Bmatrix} + \begin{bmatrix} .6 & -.1 & 0 \\ -.1 & .8 & -.2 \\ 0 & -.2 & 1 \end{bmatrix} \begin{Bmatrix} \dot{x}_1 \\ \dot{x}_2 \\ \dot{x}_3 \end{Bmatrix} + \begin{bmatrix} 70 & -40 & 0 \\ -40 & 90 & -10 \\ 0 & -10 & 140 \end{bmatrix} \begin{Bmatrix} x_1 \\ x_2 \\ x_3 \end{Bmatrix}
$$

$$
= \begin{Bmatrix} 30 \\ 30 \\ 20 \end{Bmatrix} \sin(8t)
$$

4.94. Determine the forced response of the following system.

$$
\begin{bmatrix} 20 & 0 & 0 \\ 0 & 40 & 0 \\ 0 & 0 & 2 \end{bmatrix} \begin{Bmatrix} \ddot{x}_1 \\ \ddot{x}_2 \\ \ddot{x}_3 \end{Bmatrix} + \begin{bmatrix} 2. & -.9 & 0 \\ -.9 & 1.8 & -.2 \\ 0 & -.2 & 1.6 \end{bmatrix} \begin{Bmatrix} \dot{x}_1 \\ \dot{x}_2 \\ \dot{x}_3 \end{Bmatrix} + \begin{bmatrix} 450 & -20 & 0 \\ -20 & 80 & -20 \\ 0 & -20 & 1125 \end{bmatrix} \begin{Bmatrix} x_1 \\ x_2 \\ x_3 \end{Bmatrix}
$$

$$
= \begin{Bmatrix} 2000 \\ -1000 \\ 2000 \end{Bmatrix} \cos(20t)
$$

4.95. Form the equations of motion for the system of Figure P4.2 using a Newtonian as well as a Lagrangian approach, and show that, although the equations differ in form, they possess the same physical solutions.

Section 4.14

4.96. Take Problem 4.69 ($f = 0$), set $k_1 = 0$, and use MATLAB to calculate the system eigenvectors and natural frequencies. Do you obtain the expected rigid body mode? Do the elastic frequencies go up or down from the case of $k_1 = 1000$ N/m? Do the frequency changes make sense?

4.97. Consider the system shown in Figure P4.97. Find its natural frequencies and eigenvectors. The left-hand element is a uniform cylinder, and the spring k is attached to the cylinder through a frictionless pivot. The cylinder rolls without slipping. $m_1 = 1000$ kg, $m_2 = 1500$ kg, $k = 500$ N/m, $r = .9$ m.

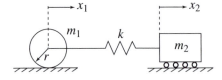

Figure P4.97

4.98. We've seen that the $[K]$ and $[M]$ matrices can be nonsymmetric if the equations are derived from a Newtonian framework for the appropriate set of coordinates. Realizing this, is the following $[K]$ matrix possible for the system illustrated in Figure P4.98. Vertical translation and rotation of the rigid bar, and vertical translation of m_1, are allowed.

$$K = \begin{bmatrix} 2 & -1 & 3 \\ 4 & 7 & -2 \\ 2 & -1 & 3 \end{bmatrix}$$

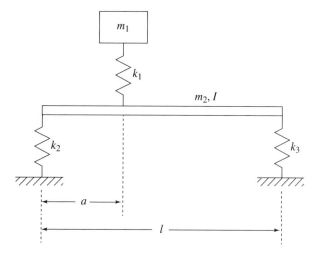

Figure P4.98

4.99. Is $X_i = \{1\ 2\ 2\ 1\}^T$, $\omega_i = 0$, a possible eigenvector/natural frequency pair for the system shown earlier in Figure P4.18? If $k_1 = k_5 = 0$? $m_1 = 10$ kg, $m_2 = 5$ kg, $m_3 = 7$ kg, $m_4 = 9$ kg, $k_2 = 100$ N/m, $k_3 = 200$ N/m, $k_4 = 400$ N/m.

4.100. Consider again the system of Figure P4.6. Let $k_1 = k_3 = 0$, $k_2 = k$, and $m_1 = m_2 = m$; $k = 2$ N/m and $m = 1$ kg. Find the equations of motion and determine the natural frequencies and eigenvectors. Can you put this system into normal form?

4.101. Determine the natural frequencies and eigenvectors associated with the system illustrated in Figure P4.101. $m = 1$ kg, $k = 2$ N/m, $l = 1$ m, $\bar{I} = 1$ kg·m^2. Vertical displacements and rotations are allowed.

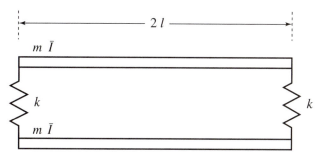

Figure P4.101

Section 4.15

4.102. Consider the system shown in Figure P4.102. Determine the displacement of m_1 and m_2 for $f_1 = 1$ N, $f_2 = 3$ N, and $f_3 = -1$ N; $k_1 = 1$ N/m, $k_2 = 2$ N/m, $k_3 = 4$ N/m, $k_4 = 1$ N/m, and $k_5 = 3$ N/m.

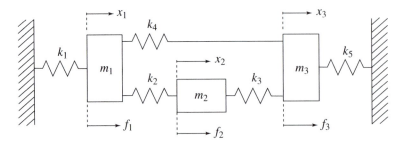

Figure P4.102

4.103. Determine the displacement of the three masses for the system shown in Figure P4.102 if $f_1 = .1$ N and all other forces are zero.

4.104. If we want to model the dynamics of a spider on its web, we'll need to find the flexibility influence coefficient of the web. As a simplified model of the web, consider the system shown in Figure P4.104. Show that the orientation of the "web" doesn't matter and that when acted on by a force $m_{\text{spider}}g$, the center of the web will deflect by $\frac{m_{\text{spider}}g}{2k}$.

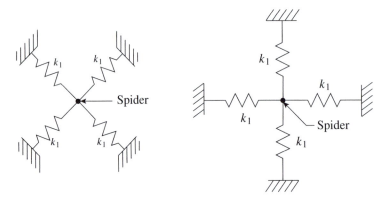

Figure P4.104 2 orientations of the web

4.105. What are the equations of motion for the system illustrated in Figure P4.105? Use the results of Problem 4.104.

4.106. What forces must be applied to x_1, x_2, and x_3 to induce displacements of $x_1 = .4$ cm, $x_2 = -.2$ cm, and $x_3 = 0$ for the system illustrated in Figure P4.106? $k_1 = 12,000$ N/m, $k_2 = 9000$ N/m, $k_3 = 1000$ N/m, $k_4 = 5000$ N/m.

4.107. In Figure P4.107, find the stiffness influence coefficient matrix and then use it to determine the steady state deflections of m_1, m_2 and m_3 in response to loads $f_1 = 10$ N, $f_2 = 2$ N, and $f_3 = 1$ N. $k_1 = 100$

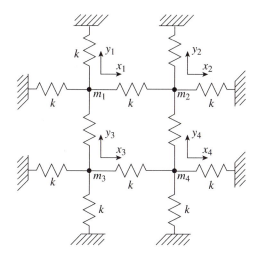

Figure P4.105

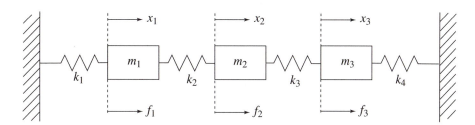

Figure P4.106

N/m, $k_2 = 200$ N/m, and $k_3 = 50$ N/m. Verify your answer through a force balance. (The nonstandard direction on f_2 is just to keep you awake.)

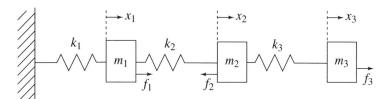

Figure P4.107

4.108. What forces must be applied to x_1, x_2, and x_3 to induce displacements of $x_1 = 1$ cm, $x_2 = 2$ cm, and $x_3 = 1$ cm for the system illustrated in Figure P4.108? $k_1 = 2000$ N/m, $k_2 = 10,000$ N/m, $k_3 = 1500$ N/m, $k_4 = 400$ N/m.

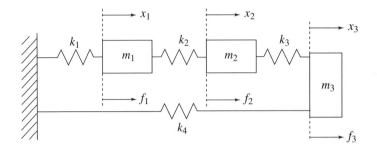

Figure P4.108

4.109. Consider the system of Figure P4.106. Let $k_4 = 0$. Determine what the relationship is between the forces applied at m_1 and m_2 if the goal is to keep m_3 from moving with the constraint that $f_3 = 0$.

4.110. Find the flexibility and stiffness influence coefficients for the system shown in Figure P4.110. Consider torsional motions only. Assume that the rod simply supplies torsional stiffness. The $\bar{I}_i$ are the moments of inertia of the disks about their center of rotation, and the GJ_i are the values of the rod's torsional rigidity. (See Appendix B for lumped spring constants.) Verify that $[K] = [A]^{-1}$

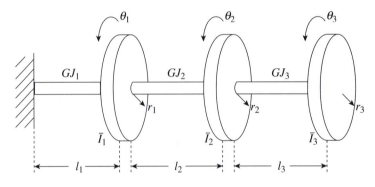

Figure P4.110

4.111. Find the stiffness and flexibility matrices for the beam shown in (a). Treat the beam as massless. Next, write down the equations of motion for the system shown in (b), ignoring the beam's mass but using the equivalent lumped stiffnesses found in (a).

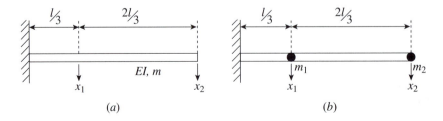

Figure P4.111

5

Distributed Systems

5.1 INTRODUCTION

Up until now the systems we've been examining have been composed of easily identified mass and spring elements. The masses and springs always appeared discretely, never in a distributed or continuous manner. But in the real world, such idealized structures never exist. The fact of the matter is that real-world structures are *always* continuous. Nobody can construct a true point mass; all real bodies have finite dimensions and so are characterized by their density or mass per unit volume. Similarly, ideal springs exist only in our minds. In the simple experiments we conducted in Chapters 1, 2, and 4, we treated the physical spring as a pure spring. But it is clear that any physical spring always has mass, a mass that we neglected in our analysis.

So, if we can never construct a system made up of just masses and springs, why did we study them? In the first place, although we can never have systems whose mass and spring elements occur in a precisely discrete manner, we can often approximate real systems as being discrete with only a negligible loss in accuracy. In addition, the mathematics associated with discrete systems is a bit easier to deal with. And finally, we'll be introducing methods that explicitly include continuous responses, yet put the overall system into a discrete framework.

The point is that we may find ourselves wanting to analyze realistic systems having continuous distributions of mass, elasticity, and damping. The problem is that except in some special cases, we cannot solve continuous problems exactly. Thus we'll find ourselves forced to consider approximate solutions. And we'll see that the approximate solutions fall into two distinct classes: those that discretize the original system into a number of discrete mass and stiffness elements (surprise—back to lumped elements!) and those that approximate the system's response by a finite number of mode shapes. We've already seen that MDOF systems possess a finite number of modes, i.e., free vibration configurations of the mass displacements that occur at the system's natural frequencies. What we'll see in this chapter is that continuous systems also have modes, but in this case they have an infinite number of them. It so happens that the modal approach (that of approximating the system's response through a finite number of mode shapes) is very widely used and forms the basis of modal analysis, the topic of Chapter 8.

Although there are several types of system that can be solved exactly, we'll look at only two of them in this chapter and leave some others to Appendix A (which interested readers can peruse at their leisure). The underlying mathematical analysis is exactly the same for all these problems, hence once you've solved one you can pretty much solve the rest as well. The two examples presented in this chapter are longitudinal vibrations in a bar and transverse vibrations of a beam. The bar example will also be used later (in Chapter 6) to illustrate how one can approximate a continuous system by a lumped one. What's interesting is that in their mathematical form, the bar equations are *exactly* the same as those that govern the vibrations of a tensioned string, which are identical to those that govern the torsional motions of a circular rod, which happen to be precisely the same as those governing the behavior of acoustic waves in a one-dimensional duct. Thus, once you've solved one of these problems, you've solved them all—the only things that change are the meaning of the dependent variable (transverse deflection of a string, rotation of the circular rod, etc.) and the meaning of the particular physical quantities (tension in the string, torsional stiffness of the rod, etc.).

We'll also be looking at transverse motions of a beam because this problem is very, very common in practical applications—almost all structures contain beams or beamlike elements. Since you'll be practicing engineering in the real world, it'll be helpful to have some knowledge of beam behavior. It'd be nice if the beam equations were also exactly the same as those of the string, rod, bar, etc., but unfortunately they're not. Luckily, the differences aren't too bad, as we'll soon see.

Before we get involved with some actual continuous systems, you should think about what the differences are between a vector and a function, since these are the relevant quantities for discrete and continuous problems, respectively. Let's consider position as the variable that we're measuring. In this case, a vector is simply a string of numbers. The position of the numbers within the vector tell us where on the structure we are, and the numbers themselves tell us how much motion is taking place at a particular position.

Take, for example, the deflection of a string. A simple way to represent the string's position would be to use three pieces of data: the end deflections and the deflection in the middle. A result for a string that's tied down on both its ends would be

$$
\left\{
\begin{array}{c}
0 \\
1 \\
0
\end{array}
\right\}
\tag{5.1.1}
$$

This vector tells us that the leftmost end's deflection (the first entry) is zero, the middle deflection is 1 (the second entry) and the right end's deflection (the third entry) is zero. This is depicted in Figure 5.1, where the middle deflection is taken to be 1 cm and the string's length is taken to be 1 m.

We can now ask ourselves what the difference would be if, instead of having only 3 positions along the string, we had 101. If these were again uniformly distributed, then the second entry in (5.1.1) would become the 51st entry in our new position vector. It would still be the middle entry, but we'd now have many additional entries. A plot of the new vector is shown in Figure 5.2. You'll note that we've got a lot more data about what's going on at other points, but our knowledge of what's happening at $x = .5$ m is the same as it was. If we had an interest in the behavior at $x = .25$ m and were only using the 3-entry vector, we'd have had to estimate the answer by using the first and second entries and averaging them, giving us an estimate of .5 cm. With the greater precision that the 101-entry vector gives us, we can simply read the result off as the value associated with the 25th entry of the vector: .707 cm.

You'll notice that the plot of the vector shown in Figure 5.2 looks an awful lot like half of a sine wave. If we used a 1001-entry vector instead of an 101-entry one, the plot would look like that shown in Figure 5.3. This is now virtually indistinguishable from half a sine. What's happening here? The answer is that our vector is "trying" to become a function. As we add more and more entries, it "looks" closer and closer to a continuous function. That's all a function really is! It's a very long vector. In fact, it's a vector in which the number of entries has gone to infinity. Pretty neat, eh? This means that,

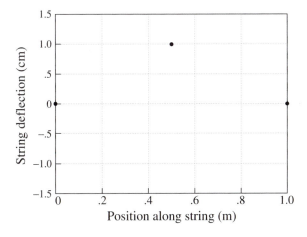

Figure 5.1 Pictorial representation of a 3-element vector

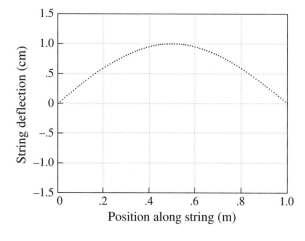

Figure 5.2 Pictorial representation of a 101-element vector

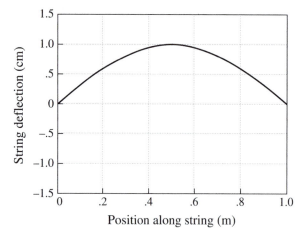

Figure 5.3 Pictorial representation of a 1001-element vector

as far as our computers are concerned, there is *no* difference between vectors and functions, since the computers can only deal with finite strings of information. Mathematically there's a difference (since mathematics has no problems with infinity), but practically we'll find that large vectors and functions are handled in very similar ways.

You should keep this little demonstration in mind as you encounter continuous systems in this and following chapters. All the continuous operations can really be viewed as discrete operations in which all the operations go to their limits. For example, summation of vector entries will mutate into integrations as we go from discrete to continuous descriptions. The distance between two entries of a position vector will go to a differential distance dx as the vector becomes bigger and bigger and finally becomes a function. If you have a problem following what's being done on the continuous problem, try thinking about it in a discrete form and see if that makes it clearer.

5.2 FREE VIBRATION OF A BAR (ROD, STRING, ETC.)

As the heading says, this section looks at bar vibrations, among other applications. Appendix A shows the equations of motion for several systems, all of which are identical in form although different in the specific parameters used. We'll perform a full derivation of the equations of motion only for a bar, realizing that the approach for all the other applications is essentially the same.

To begin studying bar vibrations, we must derive the equations of motion for our system. The bar is of length l and we identify locations along the bar with the coordinate x (Figure 5.4). Since the bar is deformable, individual elements will move from their initial positions if a force is applied. The amount of movement will be different at different points along the bar, and so we should use $\xi(x, t)$ to indicate these deformations. However, this would quickly become tiring because we'd have to write a lot of $\xi(x, t)$'s in the coming pages. Thus we'll simply write ξ. Just keep in mind that this refers to a function that is both position and time dependent. Also, be clear that the bar's motion is horizontal in this case; it's motion *along* the bar (stretching and compressing of the bar).

Please make sure the distinction between x and ξ is clear in your mind. x indicates a position *on the bar*. If you were asked to put a pencil mark on the bar 5 cm from the left, then the pencil mark is at $x = 5$ cm. $\xi(x, t)$ indicates *how much that pencil mark moves* if the bar is vibrating. The x and t dependences indicate that the deformation depends on the location along the bar as well as on time.

In your deformable bodies course you learned that the relation between applied an force, f, and the resultant deformation of the bar is given by

$$f = EA(x)\xi_x \tag{5.2.1}$$

where E is Young's modulus and $A(x)$ is the bar's cross-sectional area. This result means that the spatial rate of change of deformation is proportional to the applied force.

If we separate a small element from the bar, as shown in Figure 5.5, and assume that the motion occurs purely in the longitudinal direction, then we can find the equation of motion by balancing the inertial and applied forces:

$$m(x)\ddot{\xi}dx = [EA(x) + (EA(x))_x dx][\xi_x + \xi_{xx}dx] - EA(x)\xi_x \tag{5.2.2}$$

Note that $m(x)$ denotes the mass per unit length. Multiplying this equation out and canceling the $EA(x)\xi_x$ terms yields

$$m(x)\ddot{\xi}dx = (EA(x)_x)\,\xi_x dx + EA(x)\xi_{xx}dx + (EA(x))_x\,\xi_{xx}(dx)^2 \tag{5.2.3}$$

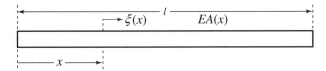

Figure 5.4   Schematic of a bar

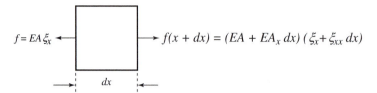

Figure 5.5   Differential element of a bar

We've got a common factor of dx, and canceling it out leaves us with

$$m(x)\ddot{\xi} = (EA(x)_x)\,\xi_x + EA(x)\xi_{xx} + (EA(x))_x\,\xi_{xx}dx \qquad (5.2.4)$$

Keep in mind that the analysis we are doing is an infinitesimal one; i.e., the element we've taken out of the bar is of only infinitesimal length. If we take the limit of dx (the length of the element) going to zero, then the $(EA(x))_x\,\xi_{xx}dx$ term will go to zero. Thus we can eliminate this term, and we are left with

$$m(x)\ddot{\xi} = EA(x)_x\xi_x + EA(x)\xi_{xx} \qquad (5.2.5)$$

which can be written more compactly as

$$m(x)\ddot{\xi} = \frac{\partial}{\partial x}\left(EA(x)\xi_x\right) \qquad (5.2.6)$$

To keep the analysis simple, we will assume that $m(x)$ and $EA(x)$ are constant. Thus, even though in general they might vary along the beam, in our case they do not. The basic reason for using this simplifying assumption is that letting these terms vary along the beam moves this from the "simple" category and into the "just about impossible" category. Luckily, Chapter 6 shows how to deal with these more difficult cases. Note that the second term of (5.2.5) disappears under this assumption, since the spatial variation has been set to zero.

Dropping the x dependence from our notation and using the fact that EA and m are constant gives us

$$m\ddot{\xi} = EA\xi_{xx} \qquad (5.2.7)$$

Dividing by m serves to group all the material parameters together and leaves us with

$$\ddot{\xi} = \frac{EA}{m}\xi_{xx} \qquad (5.2.8)$$

At this point we need to invoke a new mathematical technique to get any further. The technique we'll use is called the *separation of variables* method. You may have seen this approach in a math class and it may have seemed a bit obscure. Hopefully, this example will clear up any confusion.

Simply put, the method requires us to assume that $\xi(x, t)$ can be written in the form

$$\xi(x, t) = \mathbf{x}(x)\mathbf{t}(t) \qquad (5.2.9)$$

where $\mathbf{x}(x)$ and $\mathbf{t}(t)$ are as yet undetermined functions of position and time. Although this is often a very reasonable assumption, it should be stressed that it is *only an assumption* at this point. We'll have to carry on and see if we eventually get a reasonable answer. If we do, then we'll know the assumption was justified. As a quick demonstration that this assumption doesn't always have to be satisfied, just try and put $xt + x^2$ into this form.

Substituting (5.2.9) into (5.2.8) yields

$$\frac{EA}{m}\mathbf{x}_{xx}\mathbf{t} = \mathbf{x}\ddot{\mathbf{t}} \tag{5.2.10}$$

If we divide by $\mathbf{xt}$, we'll obtain

$$\frac{EA}{m}\frac{\mathbf{x}_{xx}}{\mathbf{x}} = \frac{\ddot{\mathbf{t}}}{\mathbf{t}} \tag{5.2.11}$$

From this result we see that the term on the left (which only depends on position) has to be equal to the term on the right (which only depends on time). This is the point that is invariably the most confusing. Ask yourself how two quantities can be equal when each depends upon a different dependent variable. Clearly, this is a tough one. For instance, the term on the left might conceivably be equal to $\sin(2x)$ and the term on the right equal to t^3. But then they *can't* be equal to each other for arbitrarily varying values of x or t. In fact, functions that depend on completely different variables can equal each other only if they *don't vary at all*—if they're equal to a constant. Thus we need to require that

$$\frac{EA}{m}\frac{\mathbf{x}_{xx}}{\mathbf{x}} = c \tag{5.2.12}$$

and

$$\frac{\ddot{\mathbf{t}}}{\mathbf{t}} = c \tag{5.2.13}$$

as well.

Taking (5.2.13) first, we multiply by $\mathbf{t}$ and pull both terms to the same side of the equation

$$\ddot{\mathbf{t}} - c\mathbf{t} = 0 \tag{5.2.14}$$

At this point it makes sense to replace c with $-\omega^2$. Why? Well, think about it for a moment. The system we're looking at has spring and mass properties. We've already seen what springs and masses do when perturbed from equilibrium—they oscillate. There's no logical reason to expect the bar to behave any differently. In fact, one way to think of a bar is simply as a collection of many small masses connected to many small springs. If we use the particular form $-\omega^2$ for our constant, then (5.2.14) will be in the form of an oscillator equation, exactly analogous to the oscillators we first saw in Chapter 1.

Replacing c with $-\omega^2$ gives us

$$\ddot{\mathbf{t}} + \omega^2\mathbf{t} = 0 \tag{5.2.15}$$

which, as we know, has solutions

$$\mathbf{t}(t) = a_1 \cos(\omega t) + a_2 \sin(\omega t) \tag{5.2.16}$$

What this tells us is that the bar undergoes sinusoidal motion at a frequency ω. As far as we know right now, any ω is as good as any other; there's nothing to indicate that the value of ω isn't completely arbitrary. However, as we'll see next, the actual choices for ω are quite restricted and are closely tied to the physical configuration of the system.

Since there's nothing more to be learned from (5.2.13), we'll move on to (5.2.12). Once we've substituted $-\omega^2$ for c, we see that multiplying (5.2.12) by $\frac{m}{EA}\mathbf{x}$ and pulling all terms to the left-hand side leaves us with

$$\mathbf{x}_{xx} + \frac{m\omega^2}{EA}\mathbf{x} = 0 \tag{5.2.17}$$

Note that perhaps surprisingly, this is in the form of an oscillator equation. So, not only does the bar oscillate in time, it now appears that it oscillates in space as well. Very interesting. Just to keep the notation simple, let's define a new variable, β,

$$\beta \equiv \omega\sqrt{\frac{m}{EA}} \tag{5.2.18}$$

so that (5.2.17) becomes

$$\ddot{\mathbf{x}}_{xx} + \beta^2\mathbf{x} = 0 \tag{5.2.19}$$

with solutions

$$\mathbf{x}(x) = b_1\cos(\beta x) + b_2\sin(\beta x) \tag{5.2.20}$$

At this point we seem to be stuck. We have to determine what β is equal to (which will tell us ω from (5.2.18)), but we don't seem to have enough information yet to do this. This is because we can't find the precise form of the solution until we've fully pinned down the physics of the problem. And we haven't done that yet. Perhaps you've already noticed that no mention has been made of the boundary conditions at the bar ends. Is the rod fixed to a wall? Is it free to move around? Until we decide on what's happening at the bar ends, we'll be unable to progress further. And this makes sense, since it seems obvious that the bar's response should be different for different boundary conditions. So let's pick fixed-fixed conditions, the situation we'd encounter if the rod was built into rigid walls on both ends (Figure 5.6). In this case the ends can't move longitudinally and so

$$\xi(0, t) = 0 \tag{5.2.21}$$

and

$$\xi(l, t) = 0 \tag{5.2.22}$$

Since $\xi(x, t) = \mathbf{x}(x)\mathbf{t}(t)$, we see that these boundary conditions imply that

$$\mathbf{x}(0) = \mathbf{x}(l) = 0 \tag{5.2.23}$$

Evaluating $\mathbf{x}(x)$ at $x = 0$ leaves us with $\mathbf{x}(0) = b_1$, since the sine term goes to zero and the cosine to 1. Since our boundary condition states that $\mathbf{x}(0) = 0$, we see that b_1 must be zero. Therefore

Figure 5.6 Bar with restrained ends

we're left with

$$\mathbf{x}(x) = b_2 \sin(\beta x) \tag{5.2.24}$$

Evaluating this expression at $x = l$ and using the remaining boundary condition yields

$$\mathbf{x}(l) = b_2 \sin(\beta l) = 0 \tag{5.2.25}$$

There are two ways (5.2.25) can be satisfied. The first is to set b_2 to zero. The only problem with this is that this would make $\mathbf{x}$ equal to zero, and so $\xi(x, t)$ would equal zero. This is clearly the trivial solution—no motion at all. Not very exciting. The other way in which (5.2.25) can be satisfied is for $\sin(\beta l)$ to equal zero. If we look at a plot of $\sin(\beta l)$ versus βl (Figure 5.7), we see that this condition occurs for an infinite number of nonzero β values. If we identify these values as $\beta_1, \beta_2, \ldots$, we have

$$\beta_n = \frac{n\pi}{l}, \quad n = 1, 2, \ldots, \infty \tag{5.2.26}$$

Thus our spatial solution is

$$\mathbf{x}(x) = b_2 \sin(\beta_n x) \tag{5.2.27}$$

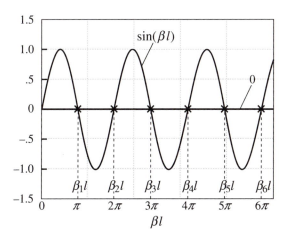

Figure 5.7 $\beta_n l$ values for a fixed-fixed bar

If we recall that $\beta = \omega\sqrt{\frac{m}{EA}}$, we see that we can now find the possible frequencies of oscillation for our system, i.e.,

$$\omega_n = \beta_n\sqrt{\frac{EA}{m}} \qquad (5.2.28)$$

We have an infinite set of β_n values and an associated infinite set of ω_n's. We'll call $\sin(\beta_n x)$ the modes of our system and n the mode number.

Let's examine some of these solutions more closely. The first value of β (represented as β_1) is $\frac{\pi}{l}$. The mode shape associated with this β value is

$$\gamma(x) = \sin\left(\frac{\pi x}{l}\right) \qquad (5.2.29)$$

Note that $\gamma(x)$ will be used from now on to indicate an eigenfunction. This distinguishes it from the more general $\mathbf{x}(x)$, which is used to indicate the general spatial part of a solution before the particular boundary conditions have been applied. Also, when it's being used in a complicated equation, the (x) will be dropped to keep the expression clear.

This mode is shown in Figure 5.8. Note that it indicates zero deflection at the bar ends and a maximal deflection in the middle of the bar. The overall time-varying behavior of this mode is given by

$$\xi(x, t) = a_1 \sin\left(\frac{\pi x}{l}\right) \cos(\omega_1 t + \phi_1) \qquad (5.2.30)$$

where $\omega_1 = \sqrt{\frac{EA}{m}}\frac{\pi}{l}$. The constants a_1 and ϕ_1 are necessary to match the particular initial conditions of the problem.

Figure 5.9 shows what the bar deformations would be like over one oscillation cycle. The period of the oscillation is T ($T = \frac{2\pi}{\omega_1}$). For simplicity, a_1 is taken to be 1.0 and ϕ_1 is set to zero. It is clear from this plot that the deflected shape is always in the form of a half sine wave—only the magnitude changes. As time progresses from zero, the deflected positions reduce in magnitude, going to zero at

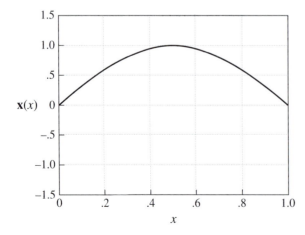

Figure 5.8 First mode of a fixed-fixed bar

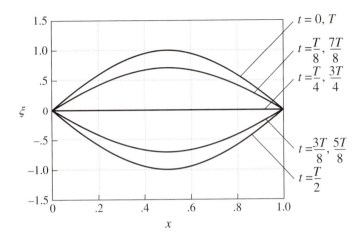

Figure 5.9 Time behavior of fixed-fixed bar's first mode

$t = \frac{\pi}{2\omega_1}$, reaching a maximum negative deflection at $t = \frac{\pi}{\omega_1}$, and returning to the original deflections at $t = \frac{2\pi}{\omega_1}$.

This is an exact analogue of the discrete eigenvector behavior first seen in Chapter 4. If the bar is deflected into the first mode and then released, it will continue to oscillate with the same continuous modal deflection, and at a frequency equal to ω_1, just as discrete systems can oscillate in particular modal vectors. The difference between this continuous system and those of Chapter 4 is that in this case we have an infinite number of mode shapes. And this makes sense. A continuous system, such as the bar we're looking at, can be thought of as a collection of an infinite number of infinitesimal mass and spring elements. Since the number of mode shapes depends upon the number of masses in a discrete system, we should expect an infinite number of modes when we've got an infinite number of masses. In fact, part of the motivation behind the approximate methods (discussed in Chapter 6) is to try to approximate the infinite number of mass elements of a continuous structure by a finite number of masses.

The remaining mode shapes for our bar are simply sines with more and more included segments (Figure 5.10). As the mode shape becomes more "wiggly," the frequency of oscillation also rises. This is a characteristic of vibrating systems—more "wiggles" in the mode shape implies a higher frequency of oscillation.

Now that we've gotten all our mode shapes pinned down, we can determine the system's response to a set of initial conditions. The general solution is given by a summation of all the possible modal responses:

$$\xi(x, t) = \sum_{n=1}^{\infty} a_n \sin\left(\frac{n\pi x}{l}\right) \cos(\omega_n t + \phi_n) \tag{5.2.31}$$

The particular values of the a_n and ϕ_n will depend on the particular initial conditions. In general, we'll have a some initial displacement, $f(x)$, and velocity, $g(x)$, for all points of the bar. Thus,

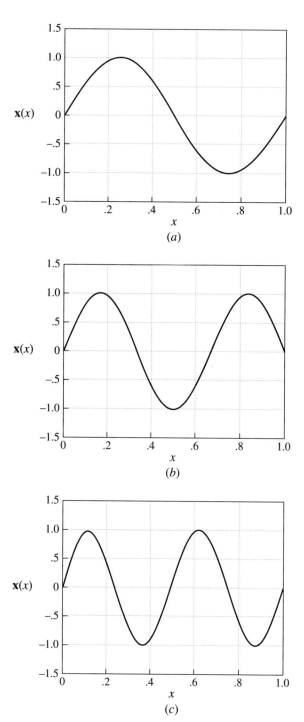

Figure 5.10 Second, third, and fourth modes of a fixed-fixed bar

evaluating (5.2.31) at $t = 0$ will give us the two equations

$$f(x) = \sum_{n=1}^{\infty} a_n \sin\left(\frac{n\pi x}{l}\right) \cos(\phi_n) \tag{5.2.32}$$

and

$$g(x) = -\sum_{n=1}^{\infty} a_n \omega_n \sin\left(\frac{n\pi x}{l}\right) \sin(\phi_n) \tag{5.2.33}$$

We'll further explore the meaning of these summations by looking at a particular set of initial conditions. The most immediately understandable initial conditions are those in which the system under consideration has been displaced into some initial configuration and then released. In this case, all the ϕ_n's will equal zero. This can be seen from an examination of (5.2.33). If all the velocity of all points along the bar are zero at $t = 0$, then having all ϕ_n's equal zero makes $g(x)$ zero, since all the individual terms in the summation will be zero. We're then left with

$$f(x) = \sum_{n=1}^{\infty} a_n \sin\left(\frac{n\pi x}{l}\right) \tag{5.2.34}$$

Note that the $\cos(\phi_n)$ is 1.0 since all the ϕ_n's are zero.

If we look at (5.2.34), we see that it is *exactly* in the form of a Fourier series! The $\sin(n\pi x)$ terms are precisely the harmonics we would use if we were trying to construct a Fourier series to approximate a physical displacement that had to start at zero (at $x = 0$) and end at zero (at $x = l$). The a_n are the unknown Fourier coefficients. And, since we now know how to deal with Fourier series problems, we are fully capable of determining the a_n for this problem.

Although this example was a bit easier to study than the most general problem, there are no real difficulties with the fully general problem (for which both $f(x)$ and $g(x)$ are nonzero), other than an increased workload.

Let's be clear now about what we've got here. A quick glance at Appendix A (come on, a quick glance won't kill you) shows us that as advertised at the start of the chapter, we now know the free responses for a circular rod, a tensioned string, and a one-dimensional acoustic duct. So let's say you were interested in building a guitar. You can now analytically predict the necessary length, materials, and tension needed for the strings to get the particular pitch you're after. You can also evaluate the differences in pitch that will exist between different string materials. Thus you have a guide to whether you'd want to use nylon or steel for your strings. Or let's say you've decided to go into the pipe organ business. You can now figure out how long your pipes should be for them to give the right notes. Or maybe you want to determine the value of EA experimentally. If so, you can set up a bar experiment like the one described here and use your transducers to observe the response of the system to an initial excitation, record the frequencies of the response, and by correlating these with your theoretical results (which involve EA), deduce what EA must be for your test specimen.

Bottom line—this stuff isn't trivial by any means, and you're in a good position to do quite a bit of engineering just with the results we've gotten so far. However, our life won't be complete until we've looked at a beam. So, on to the next section we go (after an example or two).

Example 5.1

Problem Find the eigenfunctions and natural frequencies of the spring restrained, tensioned string shown in Figure 5.11. $\rho = .002$ kg/m, $T = 21$ N/m, $l = 1$ m, $k = 25$ N/m.

Solution From Appendix A we find that the equation of motion for a uniform, tensioned string is given by

$$T y_{xx} = \rho \ddot{y}$$

Comparing this to (5.2.8) shows that the correspondence between the rod equation and the string equation is

$$\xi \leftrightarrow y$$
$$m \leftrightarrow \rho$$
$$EA \leftrightarrow T$$

Thus we can immediately write the general solution as

$$\mathbf{x}(x) = b_1 \cos(\beta x) + b_2 \sin(\beta x)$$

where $\beta \equiv \omega \sqrt{\frac{\rho}{T}}$.

The boundary condition at $x = l$ is $y(l, t) = 0$. At $x = 0$ we need to have a force balance between the deflected spring and the tensioned string. If we sketch the attachment point A in a deflected position, these forces can be drawn as shown in Figure 5.12.

For small angles of θ the approximations

$$\sin(\theta) \approx \theta \approx \frac{dy}{dx}$$

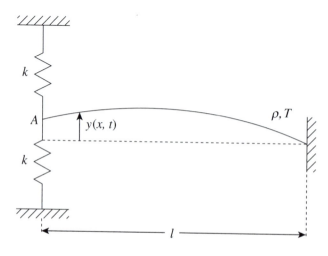

Figure 5.11

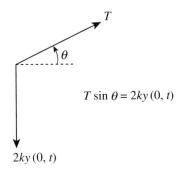

$$T \sin \theta = 2ky(0, t)$$

$2ky(0, t)$

Figure 5.12

hold. Using this gives us the boundary condition

$$T\frac{dy(0, t)}{dx} = 2ky(0, t)$$

as our left-end boundary condition. Using

$$\mathbf{x}(x) = b_1 \cos(\beta x) + b_2 \sin(\beta x)$$

and

$$\mathbf{x}_x(x) = \beta \left(-b_1 \sin(\beta x) + b_2 \cos(\beta x)\right)$$

and evaluating at the two end points gives us the new boundary conditions

$$\beta T b_2 = 2k b_1 \tag{5.2.35}$$

$$b_1 \cos(\beta l) + b_2 \sin(\beta l) = 0 \tag{5.2.36}$$

Equation (5.2.35) implies that $b_2 = \frac{2k}{\beta T} b_1$ which, when substituted into (5.2.36), leads to

$$\tan(\beta l) = -\frac{T\beta l}{2kl}$$

Graphically this looks like the plot in Figure 5.13.

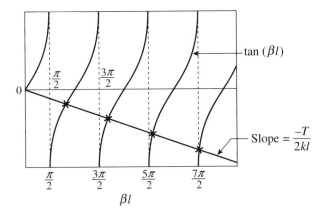

Figure 5.13

As you can see, reducing $\frac{T}{2kl}$ toward zero (infinitely stiff springs) moves the $\beta_i l$ roots toward multiples of π. This corresponds to a fixed-fixed solution, analogous to the bar example worked in the beginning of the chapter.

For the given values of T, k, ρ, and l, we have to solve

$$\tan(\beta) = -.42\beta$$

the first two solutions of which are $\beta_1 = 2.3605$ and $\beta_2 = 5.1458$.

These correspond to natural frequencies of

$$\omega_1 = \beta_1 \sqrt{\frac{T}{\rho}} = 2.3605 \sqrt{\frac{.21}{.002}} = 241.9\,\text{rad/s}$$

$$\omega_2 = \beta_2 \sqrt{\frac{T}{\rho}} = 5.1458 \sqrt{\frac{.21}{.002}} = 527.3\,\text{rad/s}$$

Since

$$\mathbf{x}(x) = b_1 \cos(\beta x) + \left(\frac{2k}{\beta T}\right) b_1 \sin(\beta x)$$

$$= b_1 \left(\cos(\beta x) + \frac{2k}{\beta T} \sin(\beta x)\right) \tag{5.2.37}$$

we'll finally obtain (ignoring the constant b_1)

$$\mathbf{x}_1(x) = \cos(2.3605x) + 1.0087 \sin(2.3605x)$$

$$\mathbf{x}_2(x) = \cos(5.1458x) + .4627 \sin(5.1458x)$$

as the first two eigenfunctions.

Example 5.2

Problem   Determine the β equation for a uniform string that's fixed on one end and has both a spring and a mass attached to the other (only vertical motions allowed), as in Figure 5.14

Solution   The left boundary condition is given by $y(0, t) = 0$. The right boundary condition is found from a force balance, as in Figure 5.15.

Summing forces in the vertical direction gives

$$m\ddot{y}(l, t) = -ky(l, t) - T \sin(\theta)$$

Since $\theta = \tan^{-1}\left(\frac{dy}{dx}\right)$ and $\frac{dy}{dx}$ is small, we have

$$\ddot{y}(l, t) = -ky(l, t) - T \frac{dy(l, t)}{dx}$$

From Example 5.1 we know that

$$\mathbf{x}(x) = b_1 \cos(\beta x) + b_2 \sin(\beta x)$$

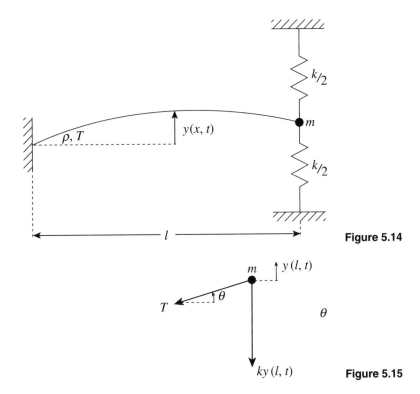

Figure 5.14

Figure 5.15

where $\beta \equiv \omega\sqrt{\frac{\rho}{T}}$. Using $\mathbf{x}(0) = 0$ yields $b_1 = 0$. Applying the right boundary condition gives

$$-m\omega^2\mathbf{x}\Big|_{x=l} = -k\mathbf{x}\Big|_{x=l} - T\frac{dy}{dx}\Big|_{x=l}$$

$$-m\omega^2\sin(\beta l) = -k\sin(\beta l) - T\beta\cos(\beta l)$$

Using $\omega^2 = \beta^2\frac{T}{\rho}$ and simplifying, we have

$$\tan(\beta l) = \frac{-T\beta}{k - \frac{T}{\rho}\beta^2}$$

5.3 FREE VIBRATION OF A BEAM

In Section 5.2, we saw how to analyze the free vibration of a uniform bar. If the bar materials are known exactly and are completely uniform, then our results are theoretically exact also. Now we'll look at a more complicated question, namely, how the transverse vibrations of a beam behave.

Almost everyone reading this book has already had a course in deformable bodies (that's the good news), and almost everyone has forgotten just what was covered in the course (that's the bad news). If you're one of the fortunate few who can recall the governing equations of a beam, congratulations.

You can skip a few pages. But for the rest of you, follow along for a bit as we rediscover the wonderful world of beams.

A system we know how to handle that supports transverse vibrations (as the beam does) is a tensioned string. From Appendix A we see that the governing equations are just like those of a bar, something we're familiar with. Basically, the difference between a beam and a string is that a string is limp and floppy unless it's under tension, whereas a beam is stiff even without any tensile load. It's this additional characteristic, an internal bending stiffness, that makes the beam equations more complex. We'll derive the complete equation of motion, including applied forces here, but we won't deal with the effect of forcing immediately.

Figure 5.16 shows what our beam looks like and also shows an infinitesimal element taken out of the beam. We shall be deriving the simplest beam model, namely the Bernoulli-Euler beam model, in this section. There are two basic simplifications involved in this model, and the first has to do with the shape of the deformed element. You'll note that the beam element is drawn as a parallelogram. Actually, the drawing would be more accurate if we drew this piece as a curved element. In this way we'd be including the shear deformation of the beam. Thus, we've reached our first approximation. We're going to assume that the deformations due to bending are much larger than those due to shear, and so we'll be neglecting shear deformation effects. A more complete model, called a Timoshenko beam model, includes this additional effect. As long as the beam doesn't "wiggle" too much, this approximation is a good one. That is, it will start to break down as we go to higher and higher modes of vibration.

The next step in determining the equations of motion for the beam is to simply apply to the beam element what you've learned in your dynamics class. We're allowing only vertical motions of the beam, so we'll derive the equation of motion for the vertical direction only. Also, the element could conceivably rotate—which would require us to use one of the several formulas that deal with moments

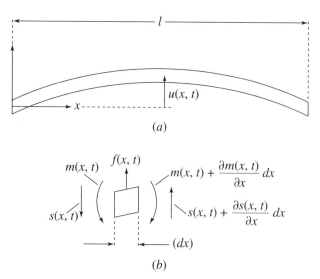

Figure 5.16 Differential analyses of a beam

and rotation. At this point we'll invoke our second assumption. We'll assume that the rotational inertia of our beam element is negligibly small compared with the translational inertia. As long as our beam is much longer than it is high, this will be a reasonable approximation. However, as the beam becomes "stubbier," we'll start to see error creep in because of our assumption. Once again, Timoshenko beam theory will include this effect. As a rule of thumb, as long as the length is about 10 times greater than the height, neglecting rotational inertia will be a reasonable assumption.

Applying a force balance to the differential element dx of Figure 5.16 will give us

$$f(x, t)dx + \left(s(x) + \frac{\partial s(x)}{\partial x} dx \right) - s(x) = \rho(x)\ddot{u}\, dx \tag{5.3.1}$$

where $s(x)$ represents the shear force within the beam and $\rho(x)$ is the mass per unit length along the beam. Canceling out $s(x)$ and dividing through by the common factor of dx leaves us with

$$f(x, t) + \frac{\partial s(x)}{\partial x} = \rho(x)\ddot{u} \tag{5.3.2}$$

Next, let's look at the rotational part of the problem. Summing moments (in the z direction) about the center of mass of our differential element will give us

$$m(x) + \frac{\partial m(x)}{\partial x} dx - m(x) + \left(s(x) + \frac{\partial s(x)}{\partial x} dx \right) \frac{dx}{2} + s(x) \frac{dx}{2} = 0 \tag{5.3.3}$$

where $m(x)$ is the moment acting on the element. There's no inertial term on the right because we're ignoring rotational inertia by assumption. Canceling out the m's, dividing by dx, and taking the limit of dx going to zero then gives us

$$\frac{\partial m(x)}{\partial x} = -s(x) \tag{5.3.4}$$

Using (5.3.4) in (5.3.2) yields

$$\rho(x)\ddot{u} = f(x, t) - \frac{\partial^2 m(x)}{\partial x^2} \tag{5.3.5}$$

Now we need only one more piece of information, namely, the relationship between m and u. Flipping back through whatever text you used in your deformable bodies course, you'll find that this relationship is

$$m(x) = EI(x)\frac{\partial^2 u(x)}{\partial x^2} \tag{5.3.6}$$

where $EI(x)$ is the bending stiffness of the beam. Combining (5.3.6) and (5.3.5) gives us our final beam equation:

$$\rho(x)\ddot{u} = -(EI(x)u_{xx})_{xx} + f(x, t) \tag{5.3.7}$$

The difference between this equation and the one governing the bar's behavior is that instead of two partial differentiations with respect to x, the beam equation has four. If you ignore this detail, you'll see that it's simply an equation that equates an inertial force (in this case $\rho\ddot{u}$) with the forces due to the internal workings of the beam (those involving EI) and any externally applied loads $f(x,t)$.

Now that we've got the equation of motion, we can set $f(x,t)$ to zero and solve for the free vibration of the beam. Just as we did for the bar problem, we're going to use separation of variables, i.e., we'll assume that the part of the motion depending on time can be separated out from that part depending on position. If we let $u(x,t) = \mathbf{x}(x)\mathbf{t}(t)$ and divide by $\rho\mathbf{x}\mathbf{t}$ (5.3.7) becomes

$$\frac{\ddot{\mathbf{t}}}{\mathbf{t}} = -\frac{(EI(x)\mathbf{x}_{xx})_{xx}}{\rho(x)\mathbf{x}} \tag{5.3.8}$$

(Note that all the spatially dependent components are now grouped together.)

Just as we did in the bar example, we'll assume that the physical properties of the beam are constant over the beam. Thus (5.3.8) simplifies to

$$\frac{\ddot{\mathbf{t}}}{\mathbf{t}} = -\frac{EI\mathbf{x}_{xxxx}}{\rho\mathbf{x}} \tag{5.3.9}$$

Once again, a time-dependent quantity can equal a spatially varying quantity on if they're both equal to the same constant. Since we now expect the time-dependent behavior to be in the form of an oscillator, we can immediately set the constant equal to $-\omega^2$, giving us the two equations

$$\frac{\ddot{\mathbf{t}}}{\mathbf{t}} = -\omega^2 \tag{5.3.10}$$

and

$$-\frac{EI\mathbf{x}_{xxxx}}{\rho\mathbf{x}} = -\omega^2 \tag{5.3.11}$$

Equation (5.3.10) can be rearranged to give us the same oscillator equation we found with the bar example,

$$\ddot{\mathbf{t}} + \omega^2\mathbf{t} = 0 \tag{5.3.12}$$

The new discovery is the equation we obtain from (5.3.11),

$$\frac{EI}{\rho}\mathbf{x}_{xxxx} - \omega^2\mathbf{x} = 0 \tag{5.3.13}$$

If we rearrange this a bit, we can get it into a form that resembles an oscillator equation

$$\mathbf{x}_{xxxx} - \frac{\omega^2\rho}{EI}\mathbf{x} = 0 \tag{5.3.14}$$

except that we have four differentiations with respect to x instead of two and we've got a minus sign instead of a plus on the second term.

If we're to get any further, we're going to have to figure out what the solutions to (5.3.14) might be. Before we do this, let's clean up the notation just a bit. Rather than writing $\frac{\omega^2\rho}{EI}$ all the time, we'll replace it with β^4. You'll recall we did something similar in the bar example. Thus we're left with

$$\mathbf{x}_{xxxx} - \beta^4 \mathbf{x} = 0 \tag{5.3.15}$$

The solutions have to be functions that, when differentiated four times, give us back the original function (multiplied by some constant). This is because we have to subtract $\beta^4 \mathbf{x}$ from it, which is simply the original function times β^4. Luckily, there are four well-known functions that operate this way: sin, cos, sinh, and cosh! So even though the equation looks more complicated that a simple oscillator, it has the exact same solutions (sin and cos) and also gives us two new solutions (sinh and cosh). More specifically, our general solution is

$$\mathbf{x}(x) = b_1 \cos(\beta x) + b_2 \sin(\beta x) + b_3 \cosh(\beta x) + b_4 \sinh(\beta x) \tag{5.3.16}$$

You'll note that this solution, when differentiated four times, is exactly cancelled out by β^4 times itself, thus satisfying (5.3.15). For those who don't recall what sinh and cosh are, they're simply given by

$$\sinh(x) = \frac{e^x - e^{-x}}{2} \tag{5.3.17}$$

and

$$\cosh(x) = \frac{e^x + e^{-x}}{2} \tag{5.3.18}$$

Thus both functions get very large for large values of x. Only for small x is there much difference between them, with $\sinh(x)$ going to 0 as x goes to zero and $\cosh(x)$ going to 1.

What we have here is simply a richer choice of possible mode shapes than we had with the bar problem. The sin and cos terms give us the "wiggle" in our solutions, while the sinh and cosh terms give us solutions that get large quickly as x increases. By putting these components together, we can find the solutions for beam vibrations having any possible set of boundary conditions. And now, having mentioned boundary conditions, we should probably move on and examine this subject.

Because the beam has bending stiffness, its boundary conditions are more complicated than those found in the bar example. In that case, only one boundary condition was needed at each end of the bar to fully describe the physical setup of the problem. But now we're going to need *two* boundary conditions at each end. Figure 5.17 shows some of the possible boundary conditions commonly associated with beam vibration problems. Figure 5.17a represents a clamped (or cantilevered) boundary condition. For this one we visualize the beam as being immovable, locked into a wall. Thus the appropriate boundary conditions are $u(0, t) = \frac{\partial u(0,t)}{\partial x} = 0$; i.e., the displacement and slope at the clamped end are zero. The next possibility (Figure 5.17b) is a pinned boundary condition. In this case we're constraining the displacement to be zero ($u(0, t) = 0$), but we allow finite slopes. What *aren't* allowed are moments. This is one of the results that should have shown up in your deformable bodies course or perhaps in a dynamics course. If you didn't see

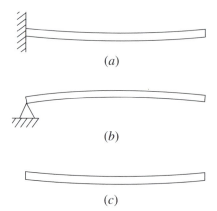

(a)

(b)

(c)

Figure 5.17 (a) Clamped boundary condition; (b) pinned boundary condition; (c) free boundary condition

it there, you know it now—pinned joints imply zero moments. So $\frac{\partial^2 u(0,t)}{\partial x^2} = 0$. The final example (Figure 5.17c) is that of the free boundary condition. In this case we're not constraining the displacement or slope—both are in general nonzero. But the moment and shear at this point must equal zero, i.e.,

$$\frac{\partial^2 u(0,t)}{\partial x^2} = 0 \quad \text{and} \quad \frac{\partial^3 u(0,t)}{\partial x^3} = 0$$

Taking a moment to look at both the string problem and the beam problem reveals something important. For the string problem, the dependent variable is differentiated twice with respect to x (y_{xx}). Also, the problem has two boundary conditions, with one applied at each end. Two differentiations and two boundary conditions. In the beam problem we see that the greatest number of differentiations is four (u_{xxxx}) and the problem has four boundary conditions, with two at each end. Four differentiations and four boundary conditions. There seems to be a pattern here, namely, that the number of boundary conditions equals the highest number of differentiations of the dependent variable. Since half the boundary conditions are applied to one end and half to the other, you'll often see people characterize a continuous problem as being of order $2p$, where $2p$ is the highest number of differentiations and p is therefore the number of boundary conditions you'll need to apply at each end. (Keep in mind that we're talking about one-dimensional problems here; the same general concept holds for plates and other higher dimensional problems.)

We should also take this time to give names to the different kinds of boundary condition we've just encountered. When a boundary condition involves the geometric configuration of the structure, like the displacement or the slope at an end point, we call it a *geometric boundary condition*. The other type of boundary condition involves the shear force or the moment at an end point, a so-called *natural boundary condition*. A pinned end of a beam therefore involves both types: the geometric boundary condition is the one that says $u = 0$, and the natural boundary condition is the one that says $u_{xx} = 0$. The distinction between boundary conditions will be important when we look at approximation techniques in Chapter 6.

Example 5.3

Problem   In this example we'd like to determine the response of a free-free beam, i.e., one for which both the left and right end are unrestrained. Thus this might be considered to be the simplest model for the behavior of an airplane or satellite.

Solution   We must keep in mind that this is an unrestrained support condition (like floating in space), and so we'll have rigid body motion as well as vibrations. The analysis to follow is concerned solely with the vibrations due to the internal stiffness of the beam. Any rigid body modes can be solved for separately if needed, a problem we'll consider after we've looked at the flexible modes of the system. For transverse vibrations of the sort we're looking at, they'll correspond to rigid translation in the u direction and rigid rotation.

All we need do now is apply the boundary conditions

$$\frac{\partial^2 u(0,t)}{\partial x^2} = \frac{\partial^2 u(l,t)}{\partial x^2} = \frac{\partial^3 u(0,t)}{\partial x^3} = \frac{\partial^3 u(l,t)}{\partial x^3} = 0$$

to (5.3.16). Differentiating (5.3.16) the appropriate number of times, applying the boundary conditions, and using the fact that $\cos(0) = 1$, $\sin(0) = 0$, $\cosh(0) = 1$, and $\sinh(0) = 0$ gives us

$$\beta^2(-b_1 + b_3) = 0$$

$$\beta^3(-b_2 + b_4) = 0$$

$$\beta^2 \left(-b_1 \cos(\beta l) - b_2 \sin(\beta l) + b_3 \cosh(\beta l) + b_4 \sinh(\beta l) \right) = 0$$

and

$$\beta^3 \left(b_1 \sin(\beta l) - b_2 \cos(\beta l) + b_3 \sinh(\beta l) + b_4 \cosh(\beta l) \right) = 0$$

The first two equations tell us that $b_1 = b_3$ and $b_2 = b_4$ (since letting $\beta = 0$ is clearly an uninteresting solution). Using this bit of knowledge, and factoring out the β terms, lets us rewrite the third and fourth equations as

$$b_1 \left(\cosh(\beta l) - \cos(\beta l) \right) + b_2 \left(\sinh(\beta l) - \sin(\beta l) \right) = 0$$

$$b_1 \left(\sin(\beta l) + \sinh(\beta l) \right) + b_2 \left(\cosh(\beta l) - \cos(\beta l) \right) = 0 \qquad (5.3.19)$$

or

$$\begin{bmatrix} (\cosh(\beta l) - \cos(\beta l)) & (\sinh(\beta l) - \sin(\beta l)) \\ (\sin(\beta l) + \sinh(\beta l)) & (\cosh(\beta l) - \cos(\beta l)) \end{bmatrix} \begin{Bmatrix} b_1 \\ b_2 \end{Bmatrix} = \begin{Bmatrix} 0 \\ 0 \end{Bmatrix}$$

You'll note that, just when it seemed we'd left the exciting field of finite-dimensional eigenvalue problems, they're back, since we've got something in the form $[A]X = 0$. And from what we've already learned, we know that there'll be a solution only if $\|[A]\| = 0$. Taking this determinant gives us

$$(\cosh(\beta l) - \cos(\beta l)) (\cosh(\beta l) - \cos(\beta l)) - (\sin(\beta l) + \sinh(\beta l)) (\sinh(\beta l) - \sin(\beta l)) = 0$$

Expanding this out and using the trigonometric identities $\sin^2 + \cos^2 = 1$, along with $\cosh^2 - \sinh^2 = 1$, we'll finally obtain

$$\cos(\beta L) \cosh(\beta L) = 1$$

Thus, the particular values of β, those that will determine the mode shapes and associated oscillation frequencies from our beam, are found from this single equation. We'll denote these particular solutions as β_n. Figure 5.18 plots both $\cos(\beta l)\cosh(\beta l)$ and 1. These two curves intersect at $\cos(\beta l)\cosh(\beta l) = 1$, indicating the associated values of $\beta_n l$.

You can see from Figure 5.18 that the β values no longer increase in a regular manner, as they did for the bar problem (where $\beta_n = \frac{n\pi}{l}$). The first three nonzero values of $\beta_n l$ are 4.7300, 7.8532, and 10.9956. Although these don't differ from each other by a constant amount, the difference *approaches a constant* as we go to higher and higher modes. This is actually predictable if we think about what's happening with the cos and cosh terms. The cosh is getting big very quickly as the mode number increases, and thus cos has to be very, very small for their product to equal 1. Therefore the roots approach those β's at which $\cos(\beta l)$ is zero, namely, $\frac{(2n-1)\pi}{2}$. It's only when the cosh isn't really huge (as in the lower modes) that the roots move substantially away from the $\cos(\beta l) = 0$ positions.

Now we can find the exact mode shape associated with each of the $\beta_n l$ values.

Using (5.3.19) to solve for b_1 in terms of b_2 gives us

$$b_2 = b_1 \frac{\sin(\beta_n l) + \sinh(\beta_n l)}{\cos(\beta_n l) - \cosh(\beta_n l)} \tag{5.3.20}$$

Since $b_1 = b_3$ and $b_2 = b_4$, (5.3.16) becomes

$$\mathbf{x}(x) = b_1\left(\cos(\beta_n x) + \cosh(\beta_n x)\right) + b_2\left(\sin(\beta_n x) + \sinh(\beta_n x)\right)$$

Making use of (5.3.20) then gives us

$$\gamma(x) = b_1\left(\cos(\beta_n x) + \cosh(\beta_n x) + \frac{\sin(\beta_n l) + \sinh(\beta_n l)}{\cos(\beta_n l) - \cosh(\beta_n l)}\left(\sin(\beta_n x) + \sinh(\beta_n x)\right)\right) \tag{5.3.21}$$

Now (5.3.21) looks pretty complicated because it is complicated. But the important thing to remember is that it is fundamentally no different from the simple sine and cosine solutions we found for the bar problem. It's simply the mathematical representation of the actual mode shapes for our problem. It seems so complicated only because we're not used to dealing with combinations of sines and

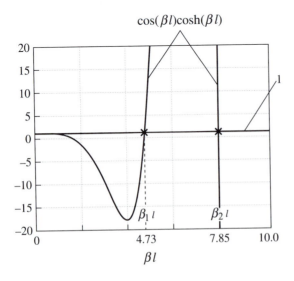

Figure 5.18 $\beta_n l$ values for a free-free beam

hyperbolic sines and such. And if you look at Figure 5.19, you'll see that the system's modes look like quite reasonable mode shapes, just the sort you might expect in a freely vibrating beam. (Keep in mind that the plots are massively exaggerated in scale; the actual vibration amplitudes would normally be too small to be visible to the naked eye.)

Even though the equations are complex, it doesn't really matter in a practical sense unless you're planning on evaluating them by hand. But if you ever plan to really manipulate these modes, you'll undoubtedly use a computer (which won't be at all upset because there's more going on here than just sines and cosines).

Actually, now that we're on the subject, it's a good idea to consider just what would happen if (5.3.21) were entered into a computer program. The reason for wanting to look at this is that there are pitfalls associated with beam solutions, pitfalls that have caused hours of frustration to many students who didn't think carefully enough about potential problems before running their programs.

The main problem lies in the behavior of cosh and sinh. Recall two things:

$$\cosh(x) = \frac{e^x + e^{-x}}{2} \quad \text{and} \quad \sinh(x) = \frac{e^x - e^{-x}}{2}$$

Both these functions grow very large, very quickly. Since the argument of our mode shape depends on βx, it's clear that for large β's, we'll have large cosh and sinh terms. Furthermore, as their arguments grow, the difference between a cosh and a sinh goes to zero. A glance at the exponential components of $\sinh(\beta x)$ and $\cosh(\beta x)$ shows that the difference between them is simply $e^{-\beta x}$, something that goes to zero very quickly as βx increases.

Look carefully at the mode shape given by (5.3.21). If we make the approximation that $\sin(\beta_n l) + \sinh(\beta_n l)$ is basically equal to $\sinh(\beta_n l)$ for large values of β_n and $\cos(\beta l) - \cosh(\beta l)$ is basically equal to $-\cosh(\beta_n l)$, then their ratio is just about equal to 1. So we're left with $\mathbf{x} = b_1[\cos(\beta x) + \cosh(\beta x) - (\sin(\beta x) + \sinh(\beta x))]$. Since $\sinh(\beta_n x)$ and $\cosh(\beta_n x)$ are basically equal for large β_n, $\mathbf{x}$ is approximately equal to $b_1[\cos(\beta_n x) - \sin(\beta_n x)]$. Although this isn't actually a good enough approximation, it is sufficient to illustrate some points. The first thing to realize is that a computer is going to have a much easier time computing $\cos(\beta_n x) - \sin(\beta_n x)$ than it will with the more exact expression. Thus we'd save computation time by using the approximate solution. Second, the computer will actually be unable to calculate the exact expression if too high a mode is being considered. For instance, $\cosh(30)$ is roughly 5.34×10^{12}. When you're dealing with numbers this large, even the best computers will start to have difficulties. Why? Well, consider that you're asking the computer to compute $\cos(\beta_n l) - \cosh(\beta_n l)$. If $\beta_n l$ is 30, then you're asking the computer to accurately add together -5.34×10^{12} plus something that's around 1. There's no way you're going to do that and still retain accuracy. And take the ratio itself (which we've been saying is around 1). Well, we have to be pretty darn sure what it's precisely equal to, since we're multiplying by $\sinh(\beta_n x)$, which is going to be huge for values of x away from 0.

Finally, we've got to be concerned that some of our arguments involve $\beta_n l$ while others involve $\beta_n x$. Since x goes from 0 to l, it's clear that $\cosh(\beta_n x)$ will vary between zero and very large values. Thus it's not true everywhere on the beam that $\cosh(\beta_n x)$ is approximately equal to $\sinh(\beta_n x)$ and both are much greater than $\cos(\beta_n x)$ and $\sin(\beta_n x)$; it's true only well away from $x = 0$.

What we have to do in a case like this is carefully construct an approximation to the actual mode shape that takes everything we've been talking about into account. Luckily for us, most computers will be fine for the first half-dozen modes or so. Thus it isn't necessary to get into this any further unless you're really interested in getting an excellent grasp of modal analysis or are doing detailed analyses. If you have an interest in reading further, Reference 2 goes into more detail on how to construct appropriate approximations.

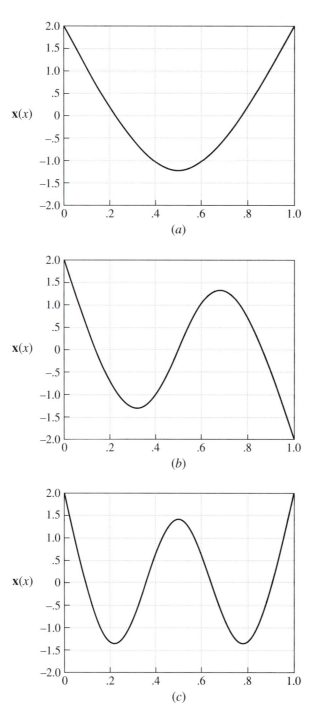

Figure 5.19 (*a*) First mode of a free-free beam; (*b*) second mode of a free-free beam; and (*c*) third mode of a free-free beam

Before leaving this section, it's time to make a mention of rigid body modes. All the elastic modes we've looked at so far were associated with nonzero values of βl. However, we can see from Figure 5.18 that $\beta l = 0$ is also a valid solution. To see what this implies in terms of modal responses, we need to go back to (5.3.15). If $\beta l = 0$ and l isn't zero, then the unforced equation governing the spatial deflections becomes

$$\mathbf{x}_{xxxx} = 0 \tag{5.3.22}$$

This is straightforward to integrate, and doing so gives us

$$\mathbf{x}(x) = c_1 x^3 + c_2 x^2 + c_3 x + c_4 \tag{5.3.23}$$

The boundary conditions $\mathbf{x}_{xx} = \mathbf{x}_{xxx} = 0$ at $x = 0$ mean that $c_1 = c_2 = 0$. This leaves us with $\mathbf{x} = c_3 x + c_4$. This also satisfies the boundary conditions at $x = l$. Since we've got two undetermined constants, we can define two mode shapes. A reasonable choice is to break up the possible motion into a rigid body translational mode and rigid body rotational mode. Thus, we can set the first mode equal to a constant d_1 and the second equal to $d_2(x - \frac{l}{2})$. In this way the first mode is symmetric and the second is antisymmetric with respect to the center of the beam. These modes are quite important when one is actually analyzing freely supported systems, such as airliners or spacecraft, since they represent motion that occurs at zero frequency. The zero frequency implies that, once the object is pushed into one of these modes, it will continue moving in that way. Thus, if a satellite is given an initial translational velocity, it will keep translating. Since this is often undesirable, we have to keep track of these motions (through our rigid body modes) so we can be sure to leave them unexcited and/or control them.

With the knowledge you now have, you can address a wide range of problems. For instance, imagine that you're working for a tennis racket manufacturer and you want to improve your racket line. Your first dynamics course may have covered how you can find the center of percussion, i.e., the point of a rigid bar at which impact forces will not impart reaction forces to a supporting pivot (which in this case would be your hand). But this analysis completely ignores the flexibility of tennis rackets. You'll want to consider what the modes are for a tennis racket. If you started your analysis by assuming that the racket is well modeled by a beam, you could use free–free boundary conditions (on the assumption that the inertial forces of the ball-racket interaction outweigh the forces of your hand). Then you could go through an analysis similar to the one we've just done. Finally, knowing the modes of the racket, you could look at them and determine where they're the greatest and where they're equal to zero. Presumably, you'd like the point at which you grasp the racket to be where a node (point of zero deflection) is located, since this probably would be the most comfortable for the player. Of course, if you wanted to be really accurate, you'd use something more involved than a uniform beam model, but you'd get an easy first estimate from using the simple beam approximation.

If we denote the mode shapes of our problem by $\gamma_n(x)$, then the general solution is given by

$$u(x, t) = \sum_{n=1}^{\infty} \gamma_n(x) \cos(\omega_n t + \phi_n) \tag{5.3.24}$$

Equation (5.3.24) is exactly the same form as (5.2.31), and thus we'd use exactly the same approach in matching any initial conditions. Actually, *all* solvable continuous vibration problems will have free responses of this form. The general response will be a summation of the system's eigenmodes, each oscillating at its particular natural frequency. Only the form of the modes will change. For a clamped-clamped beam, we'd still end up with (5.3.24), except for this case the γ's would represent the clamped-clamped modes of the beam, not the free-free ones. If we were looking at plate vibrations, we'd have to solve a plate equation, which is definitely more complex than the beam equation. But we'd still end up with an infinite number of eigenfunctions, each with its own vibration frequency. For this case we'd also have a $\gamma_{m,n}(x, y)$ because the mode shape would have both an x and y dependence, due to the structure being a two-dimensional plate rather than a one-dimensional beam.

5.4 CONTINUOUS SYSTEMS—FORCED VIBRATION

Although we've pretty well covered how to calculate the free vibrations of a continuous system, we haven't yet touched on what we'd do in the face of an external forcing. Just as in the finite-dimensional systems we've already looked at, the response of our systems to an applied forcing will very much depend upon the system's free response characteristics. Thus, having figured out how to get the system's natural modes and frequencies, it's time to turn to the forced problem.

You'll recall that in the finite-dimensional case, applying a force to one of the system's discrete masses excited (in general) all the system modes. A particular eigenmode wasn't excited only when it had a zero in the modal response that was associated with the mass being excited. Otherwise, all modes were excited, and the amplitude of each contributing mode depended upon the amplitude of the forcing and on how close the forcing frequency was to that mode's natural frequency. Well, the same thing occurs with continuous systems. The only thing that's a bit more involved is the case of exciting *only* one particular mode. For the finite-dimensional case we could apply a force to *every* mass and force the system at the correct frequency and amplitudes to excite just a particular mode. Doing this isn't inconceivable if the number of masses isn't too high. But for a continuous system, we'd need an *infinite* number of actuators, since the mode is a distributed one. Clearly this isn't possible. However, it's still possible to get pretty close to just exciting a single mode, if appropriate care is taken.

Let's consider the forced beam problem (all others being similar in approach). We'll rewrite (5.3.7) as follows:

$$\rho(x)\ddot{u} = -(EI(x)u_{xx})_{xx} + f(x, t) \tag{5.4.1}$$

We're generally concerned with sinusoidal forcing and, since we know how to break down any periodic signal into a summation of sinusoids and apply them individually (from Section 3.2), we'll just concern ourselves with a single sinusoidal force,

$$\rho(x)\ddot{u} = -(EI(x)u_{xx})_{xx} + \bar{f}(x)\cos(\omega t) \tag{5.4.2}$$

Now we're going to exploit some of what we know about eigenfunctions. We know from our finite-dimensional work that any possible vector can be represented by an appropriate summation of the system's eigenvectors. We called this spanning the configuration space. The same is true for continuous problems. Thus we shall represent the force distribution over the beam by an infinite summation of the systems's eigensolutions. Therefore, instead of $\bar{f}(x)$, we'll have the following form:

$$\rho(x)\ddot{u} = -(EI(x)u_{xx})_{xx} + \left(\sum_{n=1}^{\infty} \bar{f}_n \gamma_n(x)\right) \cos(\omega t) \tag{5.4.3}$$

In this expression, $\bar{f}_n$ represents the amount of mode $\gamma_n(x)$ that's present in $\bar{f}$. Thus by adding up all these terms, we'll obtain $\bar{f}$.

Just as we did in the finite-dimensional case, we can represent the response as a summation of all the eigenmodes of the system. We'll represent $u(x, t)$ by

$$u(x, t) = \sum_{n=1}^{\infty} a_n(t)\gamma_n \tag{5.4.4}$$

As advertised earlier, the (x) is being dropped to avoid cluttering the integrals that follow. Remember, though, that the γ's are x dependent.

The $a_n(t)$ in (5.4.4) take the role that the generalized coordinates, η, took in our finite-dimensional studies (Section 4.9). What we're saying here is that each eigenmode will contribute to the overall response, and the amount of that contribution will be controlled by each $a_n(t)$.

Substituting (5.4.4) into (5.4.3), we obtain

$$\rho(x)\sum_{n=1}^{\infty} \ddot{a}_n(t)\gamma_n = -\left(EI(x)\sum_{n=1}^{\infty} a_n(t)\gamma_{n_{xx}}\right)_{xx} + \left(\sum_{n=1}^{\infty} \bar{f}_n \gamma_n\right) \cos(\omega t) \tag{5.4.5}$$

Very messy indeed. We've got a single equation that involves *all* the eigenmodes of the system and it doesn't look all that solvable. We need to get a *set* of equations, rather than the summation of terms we're currently looking at. To do this, we'll use the trick we've seen already in Section 4.8, i.e., we'll exploit orthogonality. All the eigenmodes are orthogonal (just as the finite-dimensional ones were). We'll show this mathematically in Section 5.5, but for now you can just take my word for it. To apply orthogonality to this problem, we'll multiply (5.4.5) by one of the eigenmodes (γ_m) and integrate the equation from $x = 0$ to $x = l$:

$$\int_0^l \rho(x)\gamma_m \sum_{n=1}^{\infty} \ddot{a}_n(t)\gamma_n dx = -\int_0^l \gamma_m \left(EI(x)\sum_{n=1}^{\infty} a_n(t)\gamma_{n_{xx}}\right)_{xx} dx$$

$$+ \int_0^l \gamma_m \left(\sum_{n=1}^{\infty} \bar{f}_n \gamma_n\right) \cos(\omega t) dx \tag{5.4.6}$$

Whoa! This looks even worse than before. But is it really? If a system's eigenmodes are all orthogonal, then we should get some simplifications. Again, if you want the details you'll have to wait until the next section. However, if you're content to believe me then you can be amazed when I tell you that almost every term in the above integrals of (5.4.6) goes to zero. Orthogonality for a beam problem means that

$$\int_0^l \rho(x)\gamma_m\gamma_n dx = 0 \qquad (5.4.7)$$

and

$$\int_0^l \left(EI(x)\gamma_m\gamma_{n_{xx}}\right)_{xx} dx = 0 \qquad (5.4.8)$$

whenever m and n are different. Therefore, only one term out of each integral will remain—the one associated with γ_m. Thus we're left with

$$\ddot{a}_m(t) \int_0^l \rho(x)\gamma_m^2 dx = -a_m(t) \int_0^l \gamma_m \left(EI(x)\gamma_{m_{xx}}\right)_{xx} dx$$

$$+ \bar{f}_m \cos(\omega t) \int_0^l \gamma_m^2 dx, \quad m = 1, \dots, \infty \qquad (5.4.9)$$

For now we won't pursue some of the further simplifications that will be introduced in Section 5.5 and will content ourselves with a few observations. First, you'll note that we have an uncoupled set of equations to solve. Furthermore, it's now eminently solvable. The integral terms may look complicated, but they're just equal to scalar constants. Once you've specified what l, EI, ρ, and the $\gamma_m(x)$'s are, you just integrate and get some result. Thus what we really have is

$$m_m\ddot{a}_m(t) = -k_m a_m(t) + \bar{f}_m \cos(\omega t)c_m, \quad m = 1, \dots, \infty \qquad (5.4.10)$$

c_m, m_m, and k_m are the constants that result from numerically integrating the spatial integrals. Bringing the second term to the left gives us

$$m_m\ddot{a}_m(t) + k_m a_m(t) = c_m \bar{f}_m \cos(\omega t), \quad m = 1, \dots, \infty \qquad (5.4.11)$$

We've *finally* gotten down to something reasonable. All we're left with is the old familiar, forced oscillator problem, and the dependent variable is $a_m(t)$. We've got an equivalent mass (m_m) and an equivalent spring constant (k_m), along with a forcing magnitude ($c_m \bar{f}_m$) and a forcing frequency of ω. So just as in the finite-dimensional case, all we've got to do is determine the response magnitude of each eigenmode due to the applied forcing, multiply by the associated eigensolution, and then sum the solutions up, as shown in (5.4.4).

Example 5.4

Problem   Determine the response of a tensioned, uniform, fixed-fixed string to the distributed forcing

$$f = \bar{f} \cos(\omega t)$$

where ω is not equal to any of the string's natural frequencies.

Solution   From Appendix A we find the equation of motion of a uniform string to be

$$f(x, t) + T y_{xx} = \rho \ddot{y}$$

The system is mathematically the same as the fixed-fixed bar examined in Section 5.2, and thus we know that the eigenfunctions are of the form

$$\gamma_n(x) = \sin\left(\frac{n\pi x}{l}\right)$$

Following (5.4.4) we can express the string's displacement as

$$y(x, t) = \sum_{n=1}^{\infty} a_n(t) \gamma_n$$

and from (5.4.9) (using the string equation instead of the bar) we have

$$\rho \ddot{a}_m(t) \int_0^l \gamma_m^2 dx = T a_m(t) \int_0^l \gamma_{m_{xx}} \gamma_m dx$$

$$+ \bar{f}_m \cos(\omega t) \int_0^l \gamma_m^2 dx, \quad m = 1, \ldots, \infty \qquad (5.4.12)$$

A Fourier series decomposition of $\bar{f}$ in terms of the system eigenfunctions yields

$$\bar{f} = \sum_{n=1}^{\infty} \bar{f}_n \sin\left(\frac{n\pi x}{l}\right)$$

$$\bar{f}_n = \frac{4\bar{f}}{n\pi}, \quad n = 1, 3, 5, \ldots$$

Note that only the odd order harmonics show up in the decomposition.
Using this expansion in (5.4.12) and evaluating the spatial integrals yields

$$\rho \ddot{a}_n + \frac{T(n\pi)^2}{l^2} a_n = \frac{4\bar{f}}{n\pi} \cos(\omega t), \quad n = 1, 3, 5, \ldots$$

or

$$\ddot{a}_n + \omega_n^2 a_n = f_n \cos(\omega t), \quad n = 1, 3, 5, \ldots \qquad (5.4.13)$$

where $\omega_n^2 = \frac{T(n\pi)^2}{\rho l^2}$ and $f_m = \frac{4\bar{f}}{n\pi\rho}$. The solution to (5.4.13) is

$$a_n(t) = \frac{f_n}{\omega_n^2 - \omega^2} \cos(\omega t)$$

and therefore the complete response is given by

$$y(x,t) = \sum_{n=1,3}^{\infty} \frac{f_n}{\omega_n^2 - \omega^2} \sin\left(\frac{n\pi x}{l}\right)\cos(\omega t)$$

5.5 ORTHOGONALITY OF EIGENFUNCTIONS

As we saw in Section 5.4, the concept of orthogonality exists for continuous systems just as it does for finite-dimensional systems. Having seen that, it is now appropriate to delve a bit more deeply into the concept. To clearly see what's going on with continuous systems, let's review what we've already learned about orthogonality for discrete systems. Back in Section 4.8 we saw that the eigenvectors of a discrete vibration problem usually aren't orthogonal to each other (in a geometric sense) but rather are orthogonal with respect to each other *and* the stiffness or mass matrices, i.e.,

$$X_i^T[M]X_j = 0 \tag{5.5.1}$$

and

$$X_i^T[K]X_j = 0 \tag{5.5.2}$$

where $[M]$ and $[K]$ are the mass and stiffness matrix, respectively, and the X_i's are eigenvectors for the system. Only in the special case for which $[M]$ or $[K]$ is proportional to the identity matrix did we have

$$X_i^T X_j = 0 \tag{5.5.3}$$

Since the mass and stiffness distributions were closely tied up in the concept of orthogonality, it makes sense that they'll still be involved in the continuous case. The only difference is that this time they'll be continuous distributions, rather than lumped quantities.

One of the advantages we had in the finite-dimensional problems was that they all were expressed in the same way:

$$[M]\ddot{X} + [K]X = 0 \tag{5.5.4}$$

where $[M]$ and $[K]$ were $n \times n$ matrices. Therefore once we had demonstrated orthogonality for one $[M]$ and $[K]$ we were done; all other combinations were conceptually the same. Not so, however, for the continuous case. When we deal with continuous systems, we're dealing with partial differential equations, and there are many qualitatively different partial differential equations that apply to vibratory systems. Rather than go to a correct but abstract operator formulation, we'll look at a few basic continuous problems and indicate how any other particular problem can be handled.

We'll start by examining the bar equation:

$$m(x)\ddot{\xi} = (EA(x)\xi_x)_x \tag{5.5.5}$$

If we go through the separation of variables procedure, we can find the natural frequencies ω_i and eigenfunctions $\bar{\xi}_i$ and, substituting them into (5.5.5), obtain

$$-m(x)\omega_i^2\bar{\xi}_i = \frac{\partial}{\partial x}\left(EA(x)\frac{\partial\bar{\xi}_i}{\partial x}\right) \tag{5.5.6}$$

This equation also holds for $\bar{\xi}_j$, the jth eigenfunction

$$-m(x)\omega_j^2\bar{\xi}_j = \frac{\partial}{\partial x}\left(EA(x)\frac{\partial\bar{\xi}_j}{\partial x}\right) \tag{5.5.7}$$

We'll now multiply (5.5.6) by $\bar{\xi}_j$ and integrate along the beam while also multiplying (5.5.7) by $\bar{\xi}_i$ and integrating along the beam as well, giving us

$$-\int_0^l m(x)\omega_i^2\bar{\xi}_j\bar{\xi}_i \, dx = \int_0^l \bar{\xi}_j\frac{\partial}{\partial x}\left(EA(x)\frac{\partial\bar{\xi}_i}{\partial x}\right)dx \tag{5.5.8}$$

and

$$-\int_0^l m(x)\omega_j^2\bar{\xi}_i\bar{\xi}_j \, dx = \int_0^l \bar{\xi}_i\frac{\partial}{\partial x}\left(EA(x)\frac{\partial\bar{\xi}_j}{\partial x}\right)dx \tag{5.5.9}$$

Next, we want to subtract (5.5.9) from (5.5.8):

$$(\omega_j^2 - \omega_i^2)\int_0^l m(x)\bar{\xi}_i\bar{\xi}_j dx = \int_0^l \bar{\xi}_j\frac{\partial}{\partial x}\left(EA(x)\frac{\partial\bar{\xi}_i}{\partial x}\right)dx$$
$$- \int_0^l \bar{\xi}_i\frac{\partial}{\partial x}\left(EA(x)\frac{\partial\bar{\xi}_j}{\partial x}\right)dx \tag{5.5.10}$$

Just to keep a grip on where we're going, refer back to the finite-dimensional work. We did a similar thing back in Section 4.8 when we premultiplied the equations by X_j^T and X_i^T and then subtracted them. We ended up with a quadratic that involved the $[M]$ matrix, which was multiplied by the difference of two natural frequencies ($\omega_2^2 - \omega_1^2$ for the specific example we were examining). It then turned out that the right-hand side (in that case the difference between two quadratic expressions involving $[K]$) was equal to zero. Thus we found the orthogonality relation that $X_i^T[M]X_j = 0$ for $i \neq j$.

We're going to find just about the same result here. The left-hand side is a quadratic relation between ξ_i, ξ_j, and the mass distribution $m(x)$. It's multiplied by the difference between two natural frequencies. All that's left is to show that the right-hand side equals zero. To do this, we'll use a mathematical operation that we'll be using quite a bit in Chapter 6: integration by parts. When we integrate by parts, we use the fact that

$$\int_0^l u \, dv = uv\Big|_0^l - \int_0^l v \, du \tag{5.5.11}$$

Our right-hand sides are just begging to be integrated by parts. For instance, if we look at the first term after the equality sign in (5.5.10) we see that identifying $\bar{\xi}_j$ as u and $\frac{\partial}{\partial x}\left(EA(x)\frac{\partial\bar{\xi}_i}{\partial x}\right)dx$ as dv

allows us rewrite it as

$$\int_0^l \bar{\xi}_j \frac{\partial}{\partial x} \left(EA(x) \frac{\partial \bar{\xi}_i}{\partial x} \right) dx = \bar{\xi}_j \left(EA(x) \frac{\partial \bar{\xi}_i}{\partial x} \right) \bigg|_0^l - \int_0^l EA(x) \frac{\partial \bar{\xi}_i}{\partial x} \frac{\partial \bar{\xi}_j}{\partial x} dx \qquad (5.5.12)$$

In the same manner, the last term on the right-hand side of (5.5.10) can be written as

$$\int_0^l \bar{\xi}_i \frac{\partial}{\partial x} \left(EA(x) \frac{\partial \bar{\xi}_j}{\partial x} \right) dx = \bar{\xi}_i \left(EA(x) \frac{\partial \bar{\xi}_j}{\partial x} \right) \bigg|_0^l - \int_0^l EA(x) \frac{\partial \bar{\xi}_i}{\partial x} \frac{\partial \bar{\xi}_j}{\partial x} dx \qquad (5.5.13)$$

If we substitute both these results into (5.5.10), we'll obtain

$$(\omega_j^2 - \omega_i^2) \int_0^l m(x) \bar{\xi}_i \bar{\xi}_j \, dx = \bar{\xi}_j \left(EA(x) \frac{\partial \bar{\xi}_i}{\partial x} \right) \bigg|_0^l - \bar{\xi}_i \left(EA(x) \frac{\partial \bar{\xi}_j}{\partial x} \right) \bigg|_0^l \qquad (5.5.14)$$

The integrals terms are identical, thus they canceled each other out, leaving us with just the boundary condition terms. Well, what if the beam is fixed at both ends? Then $\bar{\xi}_i$ and $\bar{\xi}_j$ are both zero, thus making the right-hand side zero. And what if the beam is fixed at one end and free at the other? Then the displacement is zero at one end and the slope is zero at the other, and the boundary terms again go to zero. In fact, you can put in whatever combination of boundary terms you'd like. When you evaluate the right-hand side, you'll see that both terms either are zero or exactly cancel each other. Therefore we're left with what we were hoping for, namely,

$$(\omega_j^2 - \omega_i^2) \int_0^l m(x) \bar{\xi}_i \bar{\xi}_j \, dx = 0 \qquad (5.5.15)$$

As in the discrete case, we realize that if all the natural frequencies are different, then $\omega_i^2 - \omega_j^2 \neq 0$ and therefore

$$\int_0^l m(x) \bar{\xi}_i \bar{\xi}_j dx = 0 \qquad (5.5.16)$$

This is our continuous bar analogue to $X_i^T [M] X_j = 0$. This orthogonality condition doesn't involve the eigenfunctions alone but also includes the mass distribution, just as it did in the discrete case. And whereas the discrete case involved vector/matrix multiplications, here we have multiplications of the functions followed by integration over the length of the bar. Notice that if $m(x)$ is constant, then it factors out of the integral and we're left with

$$\int_0^l \bar{\xi}_i \bar{\xi}_j dx = 0 \qquad (5.5.17)$$

In this case the functions are actually orthogonal to each other, just as the discrete vectors X_i are for mass matrices, which are just constant multiples of the identity matrix.

Finally, using this orthogonality relation in (5.5.9) gives us our other orthogonality relation:

$$\int_0^l \bar{\xi}_i \frac{\partial}{\partial x} \left(EA(x) \frac{\partial \bar{\xi}_j}{\partial x} \right) dx = 0 \qquad (5.5.18)$$

5.6 HOMEWORK PROBLEMS

Note: Assume constant material properties for all the following problems.

Section 5.2

5.1. Determine the boundary conditions for the system illustrated in Figure P5.1. T is the tension in the string and ρ is the linear mass density.

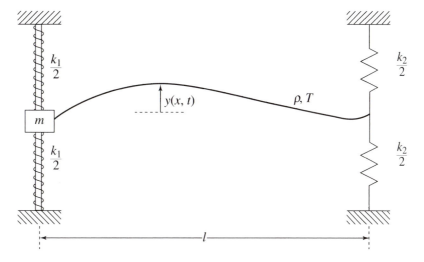

Figure P5.1

5.2. Find the natural frequencies and eigenfunctions for the, spring-restrained string shown in Figure P5.2. Let $k_1 = 90$ N/m, $k_2 = 150$ N/m, $T = 1500$ N, $l = 1$ m, and $\rho = .1$ kg/m. Show that the frequencies go to $\frac{n\pi}{l}\sqrt{\frac{T}{\rho}}$ for $k_1 = k_2 = 0$ and $k_1 = k_2 = \infty$ and to $\left(\frac{2n-1}{2}\right)\frac{\pi}{l}\sqrt{\frac{T}{\rho}}$ for $k_1 = 0$, $k_2 = \infty$ and $k_2 = \infty$, $k_1 = 0$.

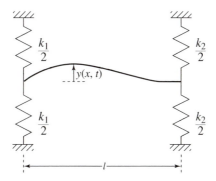

Figure P5.2

5.3. Assume that you wish to use a noncontacting sensor to observe the odd modes of oscillation of a string. The string has uniform mass and tension, is fixed at both ends, and is 1 m long. Would it be better to place the sensor near $x = .5$ m or near $x = \frac{\sqrt{2}}{2}$ m? Why?

5.4. Figures P5.4 shows two snapshots in time of a tensioned string undergoing free vibration. Explain how such a situation could arise if the system modes were proportionally damped. Explain how it could also occur under nonproportional damping.

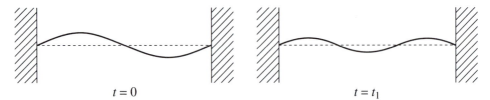

$t = 0$ $\qquad\qquad\qquad\qquad$ $t = t_1$

Figure P5.4

5.5. When discussing lumped spring responses in Chapter 1 it was mentioned that the exact solution for the first natural frequency for the system illustrated in Figure P5.5 was .9836 rad/s. Verify this fact and show that for high frequencies, the natural frequencies, tend toward $\pi \sqrt{\frac{EA}{m}}$. $m = 1$ kg/m, $EA = 10$ N, $m_1 = 10$ kg, $l = 1$ m.

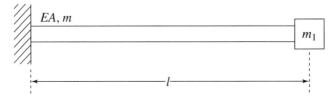

EA, m

m_1

l

Figure P5.5

5.6. By how much will the pitch of a note change if the density of the corresponding guitar string is halved and the tension is doubled?

5.7. The G string of a violin has a mass of .25 g and a length of 38 cm. What is the tension needed to get the string to the proper pitch? (Assume that the G below middle C occurs at 196 Hz.) How many pounds is this equivalent to?

5.8. If you increase the tension of a guitar string by 10%, by how much does the pitch alter?

5.9. It is often difficult to get a really good clamped boundary condition experimentally. Assume that insufficient clamping conditions have caused the effective beam length to go to $l + dl$, rather than the l you planned for in Figure P5.9. How much will your natural frequencies be altered by? Can you use this shift as a tool for determining whether your beam is clamped properly?

5.10. Solve for the eigenfunctions and eigenfrequencies of a uniformly tensioned string for which the boundary conditions are $w_x(0) = w_x(l) = 0$, i.e., zero slope at each end.

5.11. Is it possible for a tensioned, spring-restrained string, oscillating at one of its natural frequencies, to have the modal deflection illustrated in Figure P5.11?

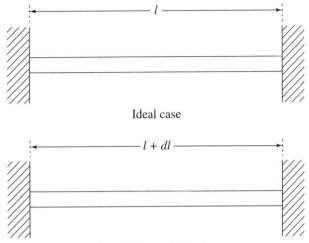

Ideal case

Actual clamped situation

Closeup of clamping condition

Figure P5.9

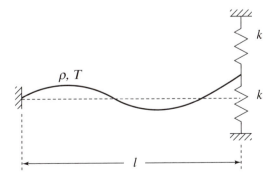

Figure P5.11

5.12. The natural frequencies of the system illustrated in Figure P5.12a are displayed versus a system parameter in Figure P5.12b, c. What are the possible parameters that might reasonably appear on the x axis of each plot? Why?

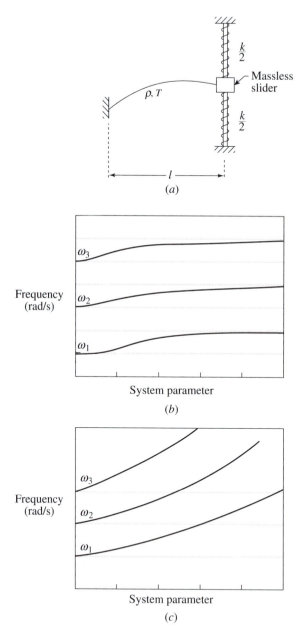

Figure P5.12

5.13. Determine the correct boundary conditions for the system shown in Figure P5.12a and indicate graphically where the β solutions lie.

5.14. Derive the β equation that will let you solve for the β values of the system shown in Figure P5.14. Consider longitudinal motions of the bar/lumped masses.

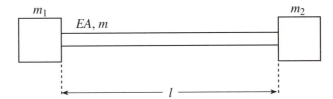

Figure P5.14

5.15. Determine the correct boundary conditions for the system shown in Figure P5.5 and indicate graphically where the β solutions lie.

5.16. Let's look more closely at Problem 5.13. Let $l = 1$ m, $T = 100$ N, and $\rho = .002$ kg/m. Show that the β solutions move from those of a restrained-free string $\left(\left(\frac{2n-1}{2}\right)\pi\right)$ to those of a restrained-restrained string ($n\pi$), as k is increased from 0 toward ∞.

5.17. We'll look more closely at Problem 5.15 in this exercise. Show that as m_1 varies from 0 toward ∞, the $\beta_i l$ values move from $\left(\frac{2n-1}{2}\right)\pi$ to $n\pi$, i.e., from the fixed-free to the fixed-fixed solutions. Let $m = .2$ kg/m, $l = 2$ m, and $EA = 1000$ N.

5.18. Show that the β values for a vertically unrestrained, tensioned string are the same as those for a fully restrained, tensioned string (Figure P5.18). Are the eigenfunctions also identical?

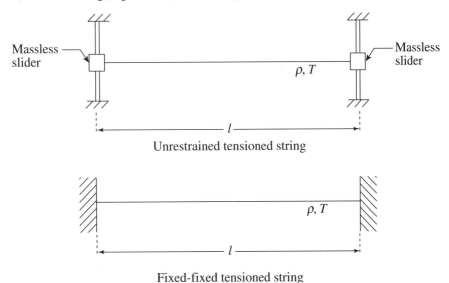

Unrestrained tensioned string

Fixed-fixed tensioned string

Figure P5.18

5.19. Show how to calculate the natural frequencies of a tensioned string (Figure P5.19) for which the linear mass density is constant at ρ_0 from $x = 0$ to $x = l_1$ and constant at $\rho = \rho_1$ from $x = l_1$ to $x = l$.

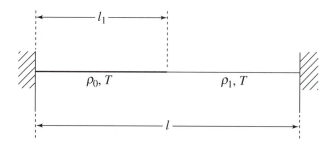

Figure P5.19

5.20. What is the response of a fixed-fixed tensioned string to the following initial conditions?

$$y(x) = \sin\left(\frac{2\pi x}{l}\right) \quad \text{at } t = 0$$

$$\dot{y}(x) = \sin\left(\frac{\pi x}{l}\right) \quad \text{at } t = 0$$

5.21. What is the free response of a fixed-free bar given the following initial conditions?

$$\xi(x, 0) = ax \quad \text{at } t = 0$$

$$\dot{\xi}(x, 0) = 0 \quad \text{at } t = 0$$

5.22. Consider Problem 5.19. Show that if $\rho_0 = \rho_1$, you obtain the expected uniform, tensioned string results.

5.23. Consider Problem 5.19. Show that if ρ_0 is much greater than ρ_1 you'll obtain natural frequencies corresponding to vibrations of a string having tension T, density ρ_1, and length $l - l_1$. Explain why such a result occurs. Show also that natural frequencies

$$\omega_n = \sqrt{\frac{T}{\rho_0}} \frac{(2n-1)\pi}{2l_1}$$

are obtained, and explain what modes they correspond to.

5.24. Determine the β equation for the system shown in Figure P5.24 (i.e., the equation that lets you find the β_n for the eigenfunctions).

5.25. Determine the response of a fixed-fixed bar to the following initial conditions:

$$\xi(x, 0) = 0$$

$$\dot{\xi}(x, 0) = b\sin\left(\frac{3\pi x}{l}\right)$$

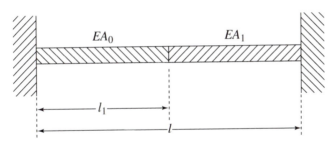

Figure P5.24

5.26. Calculate and plot the response at $x = .5\,l$ and $.75\,l$ for a pinned-pinned string having the following initial conditions:

$$y(x, 0) = 0$$

$$\dot{y}(x, 0) = \sin\left(\frac{\pi x}{l}\right) - \sin\left(\frac{2\pi x}{l}\right)$$

that is, solve for the particular values of the general response

$$y(x, t) = \sum_{n-1}^{N} a_n \sin(\beta_n x) \cos(\omega_n t - \phi_n)$$

Plot the responses for $N = 1$ and $N = 2$, i.e., a one- and two-term approximation. Comment on any differences between the two cases.

5.27. If the fundamental frequency of a fixed-fixed tensioned string should equal 440 Hz (A above middle C) and the string is 1 m long and has a mass of .009 kg, what must the tension be?

5.28. How do the natural frequencies differ for a 1 m long bar made of steel, aluminum, and copper? Assume fixed-free boundary conditions.

5.29. A bar is thrown into a rigid wall as shown below in Figure P5.29. It contacts the wall with a speed of v m/s at $t = 0$. Determine the response of the bar for $t > 0$. Assume that the right end of the bar remains attached to the wall. $l = 1$ m.

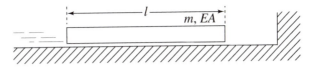

Figure P5.29

5.30. A tensioned string is struck in an impulsive manner at $t = 0$ and $x = .5l$. The initial conditions resulting from the impact are given by

$$y(x) = 0 \quad \text{at } t = 0$$

$$\dot{y}(x) = a\delta(x - .5l) \quad \text{at } t = 0$$

Determine the string's response for $t > 0$.

5.31. Determine the first four natural frequencies for the system illustrated in Figure P5.31. $\rho = 9.2 \times 10^3$ kg/m^3, $G = 8.5 \times 10^{10}$ N/m^2, $J = 1$ m^4, $l = 1.5$ m.

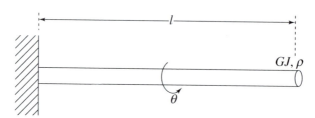

Figure P5.31

5.32. Determine the percentage change in the first three natural frequencies for torsional motion if a circular disk is attached to the end of the circular rod shown in Figure P5.32. $I_0 = .2$ kg$\cdot$ m^2, $l = 1$ m, $G = 7.3 \times 10^{11}$ N/m^2, $\rho = 7.2 \times 10^3$ kg/m^3, $J = 4 \times 10^{-7}$ m^4.

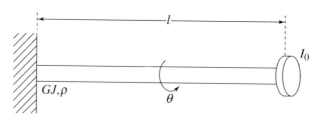

Figure P5.32

5.33. Figure P5.33 below shows part of a machine in which a torsional spring-restrained, circular shaft is moved by a sinusoidally varying torque applied at x_1. At what frequency of excitation will the system stop being well approximated as a rigid shaft in a spring restraint? $l = 1$ m, $G = 6.2 \times 10^{11}$ N/m^2, $\rho = 6.4 \times 10^3$ kg/m^3, $J = 2 \times 10^{-7}$ m^4, $k_\theta = 1.5 \times 10^5$ N$\cdot$m/rad.

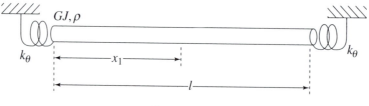

Figure P5.33

5.34. In Problem 5.33 we looked at a spring-restrained circular shaft. Numerically determine the lowest frequency oscillation mode, and show that it matches the prediction you'd obtain from a lumped analysis.

5.35. Consider a fixed-free circular rod made of steel. What is lower, the first torsional or the first longitudinal natural frequency? $G = 7.9 \times 10^{10} \, \text{N/m}^2$, $E = 2.1 \times 10^{11} \, \text{N/m}^2$, $\rho = 7.8 \times 10^3 \, \text{kg/m}^2$.

5.36. Bicycle designers worry about lightness and stiffness (among other concerns) when designing their frames. Determine which is torsionally stiffer, a smaller diameter tube or a larger one. Use a constant mass of metal each time. The first tube will have an inside diameter of 3 cm and a wall thickness of .5 mm. The second tube will have an inside diameter of 4 cm. We'll determine stiffness by calculating the first natural frequency of a fixed-free tube upon which we've attached a lumped inertia I_0. $G = 7.9 \times 10^{10} \, \text{N/m}^2$, $\rho = 7.8 \times 10^3 \, \text{kg/m}^3$, $l = 1 \, \text{m}$, $I_0 = 5 \times 10^{-3} \, \text{kg·m}^2$.

5.37. Assume that you're employed as a bicycle designer and your task is to oversee the move from steel to aluminum. Your task is to decrease the tubing's weight by 10% and at the same time increase its torsional stiffness. Consider a fixed-fixed tube, 1 m long. The wall thickness of the steel tube is .5 mm, and the inside radius is 12.5 mm.

   (a) Calculate the natural frequencies of both a steel tube and a tube of the same dimensions made of aluminum.

   (b) Calculate the torsional stiffness for each tube from part (a) (torsional stiffness $= GJ$).

   (c) Increase the inside radius of the aluminum tube by 30% and determine what the wall thickness must be to attain a 10% weight reduction over the steel tube.

   (d) How much stiffer is the tube designed in part (c) than the steel tube?

5.38. Is there any difference in the torsional natural frequencies of a solid circular rod and a hollow circular shaft, assuming that they have the same length and mass? Consider the case of fixed-free boundary conditions.

5.39. Consider again the system of Figure P5.33. Let the torsional spring on the left end be infinitely stiff. How large must k_θ be for the first torsional natural frequency of this system to be within 5% of the first torsional natural frequency for fixed-fixed conditions? $l = 1 \, \text{m}$, $G = 6 \times 10^{10} \, \text{N/m}^2$, $J = 1 \times 10^{-8} \, \text{m}^4$, $\rho = 7 \times 10^3 \, \text{kg/m}^3$.

5.40. Putting a lumped rotational inertia on the end of a fixed-free, hollow, circular shaft will produce natural frequencies lower than those of the original system. How large a rotational inertia can be used before the first frequency shifts by 10%? The inner radius of the shaft is 1 cm, the wall thickness is 1 mm, the length is 2 m, $G = 5.7 \times 10^{10} \, \text{N/m}^2$, and $\rho = 6.7 \times 10^3 \, \text{kg/m}^3$.

5.41. Consider again the system of Problem 5.12. How large a spring constant k should be used to bring the first natural frequency to 185 rad/s from its value when $k = 0$? $l = 2 \, \text{m}$, $T = 1000 \, \text{N}$, $\rho = .02 \, \text{kg/m}$.

5.42. The system in Figure P5.42 has natural frequencies that are essentially equal to the fixed-fixed ones, as well as a low frequency, "rigid body" mode. How small can m_1 get before the second natural frequency has increased to 10% above its original value? $l = 3\text{m}$, $\rho = .02 \, \text{kg/m}$, $T = 3000 \, \text{N}$, $m_1 = 1000 \, \text{kg}$.

5.43. The system illustrated in Figure P5.43 has natural frequencies of $\omega_1 = 1000$ rad/s, $\omega_2 = 3000$ rad/s, $\omega_3 = 5000$ rad/s, ... for $m_1 = 0$. As m_1 is increased, the frequencies drop. Can a value of m_1 be found such that $\omega_i = 2100$ rad/s and $\omega_{i+1} = 2700$ rad/s, for some i? Why or why not?

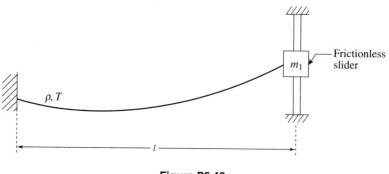

Figure P5.42

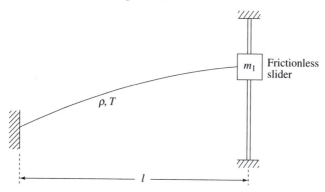

Figure P5.43

5.44. What are the first three natural frequencies for the system illustrated in Figure P5.44 (longitudinal motion)? $EA = 2 \times 10^7$ N/m, $\rho = 3 \times 10^3$ kg/m³, $l = 1$ m, $A = 1 \times 10^{-4}$ m², $k = 5 \times 10^5$ N/m.

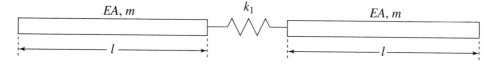

Figure P5.44

5.45. Treating the system illustrated in Figure P5.45 as two identical bars joined by a lumped spring, determine the first two β values for the system. $EA = 2 \times 10^7$ N/m, $\rho = 3 \times 10^3$ kg/m³, $l = 1$ m, $A = 1 \times 10^{-4}$ m², $k = 5 \times 10^5$ N/m.

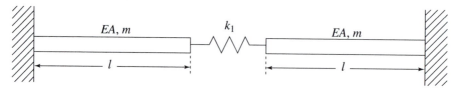

Figure P5.45

5.46. Show analytically that the odd harmonics (1st, 3rd, 5th, etc.) of the lowest natural frequency for a fixed-fixed string of the length l are exactly equal to the natural frequencies of a string of length $l/2$ that is both fixed at one end and vertically unrestrained at the other. Explain why this is the case.

5.47. Why can't the modal response illustrated in Figure P5.47a exist as a free vibration eigenfunction of a fixed-spring restrained tensioned string as in Figure P5.47b.

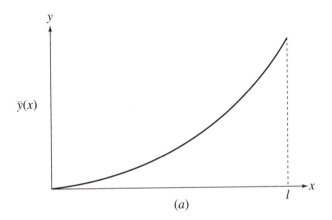

(a)

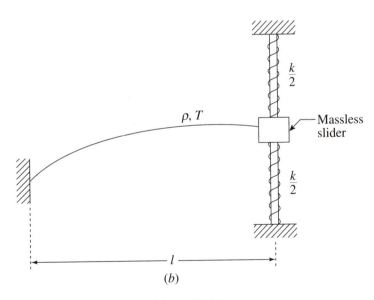

(b)

Figure P5.47

5.48. In Figure P5.48, can you predict any of the natural frequencies for system A, given the natural frequencies of system B?

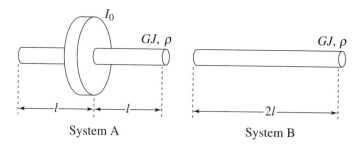

System A   System B

Figure P5.48

5.49. What is the largest change in the natural frequencies that you'd expect if to the original tensioned string of Figure P5.49a you added a spring restraint as in Figure P5.49b?

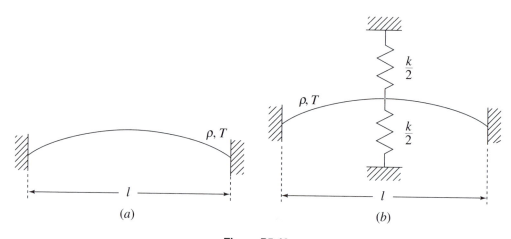

(a)   (b)

Figure P5.49

5.50. Solve for the eigenfunctions of a fixed-free bar. Use this information to find the strain (first spatial derivative with respect to x). Where is the strain a maximum for the eigenfunction? Where is it a minimum? How would you use this information if you were using strain gauges as sensors?

Section 5.3

5.51. Show that the β values (hence the natural frequencies) are the same for both a free-free beam and a clamped-clamped beam.

5.52. Determine the first four βl values for a beam like that in Figure P5.52: pinned on one end and both pinned and torsionally spring restrained at the other end.

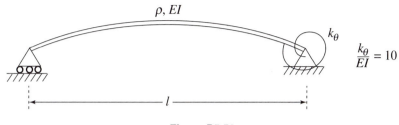

Figure P5.52

5.53. Figure P5.53 shows a pinned-pinned beam on a resilient foundation. Calculate the natural frequencies and eigenfunctions. The stiffness per unit length of the foundation, k, is 175 N/m/m, $l = 2$ m, $EI = 22.4$ N/m², $\rho = .624$ kg/m.

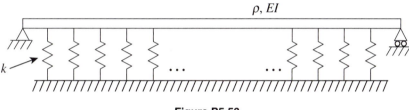

Figure P5.53

5.54. What is the equation of motion for free vibration of a beam with nonconstant stiffness and nonconstant mass distribution?

5.55. Often the boundary conditions for a beam are not completely fixed, and thus an accurate model should include discrete torsional spring restraints at the ends, as shown in Figure P5.55. Derive an expression that allows you to solve for the system's β values. There's no need to obtain a single equation; leaving the problem in matrix form is sufficient.

Figure P5.55

5.56. Derive the equation of motion for a beam under tension, T, as in Figure P5.56. Assume constant T, EI, and ρ. Then derive the equation needed to solve for the beam's natural frequencies.

Figure P5.56

5.57. The eigenfunctions for a fixed-fixed bar were given by $\sin(n\pi x)$ and the natural frequencies were evenly spaced: $\omega_n = \beta_n \sqrt{\frac{EA}{m}}$, where $\beta_n = n\pi$. Show that the eigenfunctions for a pinned-pinned beam are also given by $\sin(n\pi x)$. Are the natural frequencies also evenly spaced?

5.58. What sort of beam deflection would you have if the displacement, slope, moment, and shear were zero at $x = 0$, with no conditions specified at $x = l$?

5.59. Derive another (equivalent) form for the mode shape of the beam in Example 5.3.

Section 5.4

5.60. Calculate the response of a fixed-free bar if the applied forcing is of the form

$$f(x, t) = \overline{f} \sin\left(\frac{3\pi x}{2l}\right) \sin(\omega t)$$

for which ω is not equal to any of the bar's natural frequencies.

5.61. Determine the forced response of a tensioned string to an applied forcing of the form shown in the following equation. ω isn't equal to any of the string natural frequencies. The boundary conditions of the string are given by $y(0, t) = y(l, t) = 0$.

$$f(x, t) = \overline{f}\delta(x - x_0)\cos(\omega t)$$

5.62. What is the forced response of a fixed-free circular rod subjected to a time varying torque at $x = x_1$?

$$\tau = \overline{\tau}\delta(x - x_1)\cos(\omega t)$$

5.63. Determine the general response of a pinned-pinned beam when subjected to a distributed force of the form

$$f(x) = 3\sin(\pi x)\sin(5t) - 2\sin(2\pi x)\sin(6t).$$

Assume constant beam properties and let $l = 1$ m.

5.64. What is the response of a pinned-pinned uniform beam to a concentrated, time varying load $f(t)$ applied to the beam's midpoint? Assume that ω_f doesn't correspond to any of the natural frequencies of the system in Figure P5.64.

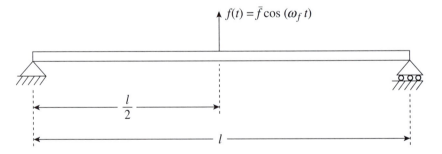

Figure P5.64

5.65. What will the response be of a free-free beam (length l, mass per unit length ρ, bending stiffness EI) to a transverse distributed force of $\bar{f}$ N/m? Note that the force is constant, both in time and with respect to x.

Section 5.5

5.66. Show that the eigenfunctions of a vertically unrestrained, tensioned string are orthogonal to each other.

5.67. Show that the eigenfunctions of a string that's fixed at one end and vertically unrestrained at the other are mutually orthogonal.

5.68. Show that the eigenfunctions of a uniform, unrestrained torsional rod are orthogonal. Ignore the zero-frequency, rigid body solution.

5.69. Numerically solve for two different eigenfunctions for the system shown in Figure P5.69, and demonstrate numerically that they satisfy orthogonality. $l = 1.0$ m, $EA = 2.0 \times 10^7$ N/m, $m = 0.8$ kg, $k = 5.0 \times 10^5$ N/m, $m_1 = .3$ kg

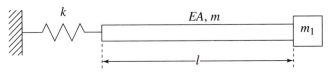

Figure P5.69

5.70. Numerically calculate the first two eigenfunctions of a fixed-fixed bar and verify that they satisfy orthogonality. $EA = 8.0 \times 10^6$ N, $m = 2.0$ kg/m, $l = 1.0$ m.

5.71. Demonstrate that the eigenfunctions of the fixed-free, uniform circular rod in Figure P5.71 satisfy orthogonality.

Figure P5.71

5.72. Consider the system of Figure P5.69. Set m_1 equal to zero. Numerically calculate the first two eigenmodes and demonstrate that they satisfy orthogonality. $EA = 5.0 \times 10^6$ N, $k = 7.787 \times 10^6$ N/m, $l = 1.0$ m, $m = 1.0$ kg/m.

5.73. Consider the system shown in Figure P5.5. Show that the first two numerically calculated modes satisfy orthogonality (i.e., generate the modes numerically (vector representation) and numerically evaluate $\int_0^l m\phi_i\phi_j dx + m_1\phi_i(1)\phi_j(1))$. $m = 1$ kg/m, $l = 1$ m, $EA = 1.3 \times 10^6$ N, $m_1 = 1.5$ kg.

5.74. Show that the first two modes of the system illustrated in Figure P5.74 satisfy orthogonality. (Note that the first mode is a rigid body mode.) Numerically calculate the β's and the eigenfunctions and numerically verify orthogonality. $G = 7 \times 10^9 \, \text{N/m}^2$, $J = 2 \times 10^{-3} \, \text{m}^4$, $\rho = 3 \times 10^3 \, \text{kg/m}^3$, $l = 1$ m, $I_1 = 12 \, \text{kg} \cdot \text{m}^2$, $I_2 = 12 \, \text{kg} \cdot \text{m}^2$.

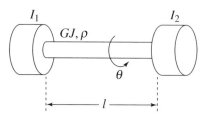

Figure P5.74

5.75. Consider again the system of Figure P5.74. Set I_2 equal to zero. Show that orthogonality holds by numerically generating the first two eigenfunctions for the system and then numerically evaluate

$$\int_0^l \rho J \phi_i(x) \phi_j(x) dx + I_1 \phi_i(0) \phi_j(0)$$

$G = 7 \times 10^9 \, \text{N/m}^2$, $J = 2 \times 10^{-3} \, \text{m}^4$, $\rho = 3 \times 10^3 \, \text{kg/m}^3$, $l = 1$ m, $I_1 = 12 \, \text{kg} \cdot \text{m}^2$.

6

Approximate Solution Methods

6.1 INTRODUCTION

Although we've learned a lot in the preceding pages, it should also be clear that many problems are still beyond our analytical abilities. Anything that is more complex than a group of lumped spring-mass systems or a simple continuous system, such as a string, is currently beyond our skills. The simple addition of a nonuniform mass to a bar or a spatially varying stiffness in a beam makes the problem too difficult for closed-form analysis. These are the problems that we now address. Our solutions will not be exact, but merely approximate. Luckily, since we'll be able to create these approximate solutions to whatever degree of accuracy we desire, they'll meet all our needs. With the techniques to be introduced in the following pages, you'll be able to analyze quite general systems efficiently and accurately.

6.2 LUMPED APPROXIMATIONS

This section is probably the shortest in the book. The main focus of this chapter is on so-called modal approximations, i.e., methods that presume the existence of distributed modes for a structure. In the opposite view, one would pay no attention to whether distributed modes might exist but instead would break the problem into little chunks. The finite element method is a well-known approach that takes this second point of view. In this approach the actual continuous body is broken into simpler units (elements) for which displacement/stress relations can be easily calculated. These elements are then joined together in a mesh to approximate the original system. The technique is very widely used and is usually taught as a separate course at the first-year graduate level. Many books are available that are devoted solely to the subject.

A different approach is to forget continuous elements entirely and try to approximate the continuous system by a collection of springs and masses (and perhaps dampers). In this way the system would look like an MDOF lumped system, something we already know how to solve. This approach will be the focus of this section.

Here's how you might proceed for a particular example. Let's assume you're examining a uniform bar. You might decide to approximate the distributed mass by n equal lumped masses located along the length of the bar. You could find the springs that connect the masses by determining the static deflection of the bar under a fixed load at the free end (the other end being restrained) and then determine what combination of discrete springs would yield this overall stiffness. As the number of masses and springs are increased, the quality of the approximation would improve, assuming you've been reasonable in assigning the mass and spring values.

The advantage of this approach is that it can be used to analyze quite general systems. The complexity of the continuous system isn't a big issue except insofar as it will generate a large number of lumped masses (which of course impacts the ease of solution).

Example 6.1

Problem   Use a lumped approach to determine the first two natural frequencies for a uniform tensioned string (fixed-fixed boundary conditions). Compare the results to the exact analytical predictions. Let the length l of the string be .9 m, the total mass m_t be .05 kg, and the tension T be 1000 N/m.

Solution   We'll first divide the mass of the string into n equal portions. Thus, instead of having a string of length .9 m we'll have n sections, each $\Delta = \frac{.9}{n}$ m long. All these sections will do is contribute a stiffness to the problem—they'll have no mass of their own. Between each pair of string sections we'll put a lumped mass. If we imagine each mass as a little round ball, we can view each string section as having half a ball on each end. If this is the case, then to maintain consistency we'd have to put a half-ball of mass at each of the string end points. Of course, these lumped masses won't move, since they're fixed in place owing to the boundary conditions. Nonetheless it's important to include them, since ignoring them would move onto the interior of the string than should be there more mass. This model is shown in Figure 6.1.

Counting the two half masses at the end points, we have a total of $n + 1$ complete "balls," each of mass $m = \frac{m_t}{n+1}$ kg. Since the string sections are massless, they'll always be line segments, and we can therefore determine the forces acting on each mass element in a straightforward way. If we concentrate our attention on the ith mass element, we can see that the equations of motion are given by

$$m\ddot{y}_i = \left(\frac{y_{i+1} - y_i}{\Delta}\right) T - \left(\frac{y_i - y_{i-1}}{\Delta}\right) T$$

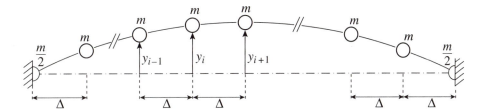

Figure 6.1 Lumped mass approximation of a tensioned string

We've used the approximation that the component of vertical force due to the tension is proportional to the angular deflection (small-angle approximation) and that the angle is approximated by the difference between the vertical deflections of two neighboring masses divided by the distance between them.

Rearranging gives us

$$m\ddot{y}_i - \frac{T}{\Delta}y_{i-1} + \frac{2T}{\Delta}y_i - \frac{T}{\Delta}y_{i+1} = 0$$

This can be expressed in matrix form as

$$\begin{bmatrix} m & 0 & 0 & \cdots \\ 0 & m & 0 & \cdots \\ 0 & 0 & m & \cdots \\ \vdots & \vdots & \vdots & \ddots \end{bmatrix} \begin{Bmatrix} \ddot{y}_1 \\ \ddot{y}_2 \\ \ddot{y}_3 \\ \vdots \end{Bmatrix} + \begin{bmatrix} \frac{2Tn}{l} & -\frac{Tn}{l} & 0 & \cdots \\ -\frac{Tn}{l} & \frac{2Tn}{l} & -\frac{Tn}{l} & \cdots \\ 0 & -\frac{Tn}{l} & \frac{2Tn}{l} & \cdots \\ \vdots & \vdots & \vdots & \ddots \end{bmatrix} \begin{Bmatrix} y_1 \\ y_2 \\ y_3 \\ \vdots \end{Bmatrix} = \begin{Bmatrix} 0 \\ 0 \\ 0 \\ \vdots \end{Bmatrix}$$

Dividing all the equations by m and reexpressing Δ in terms of l and n gives us

$$\begin{bmatrix} 1 & 0 & 0 & \cdots \\ 0 & 1 & 0 & \cdots \\ 0 & 0 & 1 & \cdots \\ \vdots & \vdots & \vdots & \ddots \end{bmatrix} \begin{Bmatrix} \ddot{y}_1 \\ \ddot{y}_2 \\ \ddot{y}_3 \\ \vdots \end{Bmatrix} + \frac{(n+1)^2 T}{lm_t} \begin{bmatrix} 2 & -1 & 0 & \cdots \\ -1 & 2 & -1 & \cdots \\ 0 & -1 & 2 & \cdots \\ \vdots & \vdots & \vdots & \ddots \end{bmatrix} \begin{Bmatrix} y_1 \\ y_2 \\ y_3 \\ \vdots \end{Bmatrix} = \begin{Bmatrix} 0 \\ 0 \\ 0 \\ \vdots \end{Bmatrix}$$

If we perform an eigenanalysis on this set of equations using $n = 10$, we'll obtain the following first and second natural frequency estimates: $\tilde{\omega}_1 = 466.7$ rad/s and $\tilde{\omega}_2 = 924.0$ rad/s. The actual natural frequencies of the continuous system are given by

$$\omega_n = n\pi \sqrt{\frac{T}{m_t l}}$$

which yields results of $\omega_1 = 468.3$ rad/s and $\omega_2 = 936.6$ rad/s. As you can easily observe, the lumped approach did a good job of approximating the continuous system's natural frequencies.

As you might surmise, this lumped approach can be used in many ways. There is nothing to stop you from using nonuniform lumped masses, for instance. Thus if you wanted each mass element to be spaced uniformly along a structure, you might vary the amount of mass at each location to accurately reflect the mass distribution in the original structure.

6.3 RAYLEIGH'S QUOTIENT

One of the most amazing people in the field of vibrations was Lord Rayleigh. He was born wealthy but decided to devote his life to science (a pretty rare occurrence). At age 35, he published the *Theory of Sound* [10] in which he laid out the fundamental principles of both acoustics and vibrations. Of course, much of this material was already known, but Lord Rayleigh added a great deal of his own and was the first to put it down on paper in an easily readable form. Once you've finished the book you're now reading, you might want to find a copy of the *Theory of Sound* and read Volume One, the volume concerned with vibrations. You'll find a good bit that is very similar to what you're learning now.

Although we'll run into Rayleigh again, later in the chapter, I mention him now because we're going to discuss what's called Rayleigh's quotient. The idea behind Rayleigh's quotient is relatively straightforward. Basically, we'd like to be able to find a system's natural frequencies without having to solve the associated eigenvalue problem. Now you might think this isn't possible; it would be like getting something for nothing. However, if we already know the eigensolutions to a problem that's *close* to the one we're currently interested in, we can exploit this knowledge to our advantage. We'll consider discrete systems first and then move on to continuous ones in Section 6.5.

If we're dealing with an undamped, lumped mass system, we know that the equations of motion are of the form

$$[M]\ddot{X} + [K]X = O \tag{6.3.1}$$

We can recognize that this is an eigenvalue problem, and thus we know the solutions are given by

$$-\omega_i^2[M]X_i + [K]X_i = O \tag{6.3.2}$$

where X_i is the ith eigenvector and ω_i is the ith natural frequency of oscillation. What we'll do next is premultiply both sides of this equation by $\frac{1}{2}X_i^T$:

$$-\frac{1}{2}\omega_i^2 X_i^T[M]X_i + \frac{1}{2}X_i^T[K]X_i = O \tag{6.3.3}$$

It's worth discussing why we might want to do this at this point. For ease, let's look at a 3 DOF system (Figure 6.2). A 2 DOF system would be even easier, but we'll be looking at some characteristics that require at least three degrees of freedom in a few pages, so we'll stick with 3 DOF.

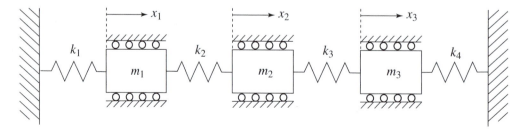

Figure 6.2   3 DOF spring mass system

We know that the potential and kinetic energies of the system can be written as

$$PE = \tfrac{1}{2}k_1 x_1^2 + \tfrac{1}{2}k_2(x_2 - x_1)^2 + \tfrac{1}{2}k_3(x_3 - x_2)^2 + \tfrac{1}{2}k_4 x_3^2 \tag{6.3.4}$$

and

$$KE = \tfrac{1}{2}m_1 \dot{x}_1^2 + \tfrac{1}{2}m_2 \dot{x}_2^2 + \tfrac{1}{2}m_3 \dot{x}_3^2 \tag{6.3.5}$$

These are all quadratic terms; i.e., everything is raised to the second power. Thus we can write them in matrix form as

$$PE = \tfrac{1}{2}\{x_1 \ x_2 \ x_3\}\begin{bmatrix} k_1 + k_2 & -k_2 & 0 \\ -k_2 & k_2 + k_3 & -k_3 \\ 0 & -k_3 & k_3 + k_4 \end{bmatrix}\begin{Bmatrix} x_1 \\ x_2 \\ x_3 \end{Bmatrix} \tag{6.3.6}$$

and

$$KE = \tfrac{1}{2}\{\dot{x}_1 \ \dot{x}_2 \ \dot{x}_3\}\begin{bmatrix} m_1 & 0 & 0 \\ 0 & m_2 & 0 \\ 0 & 0 & m_3 \end{bmatrix}\begin{Bmatrix} \dot{x}_1 \\ \dot{x}_2 \\ \dot{x}_3 \end{Bmatrix} \tag{6.3.7}$$

More compactly, this is just

$$PE = \tfrac{1}{2}X^T[K]X \tag{6.3.8}$$

and

$$KE = \tfrac{1}{2}\dot{X}^T[M]\dot{X} \tag{6.3.9}$$

where $[M]$ and $[K]$ are our familiar mass and stiffness matrices. If we go one step further and ask what the energies would be if the system was oscillating at one of its natural frequencies, say ω_i, then (6.3.8) and (6.3.9) would become

$$PE = \tfrac{1}{2}X_i^T[K]X_i \cos^2(\omega_i t) \tag{6.3.10}$$

$$KE = \tfrac{1}{2}\omega_i^2 X_i^T[M]X_i \sin^2(\omega_i t) \tag{6.3.11}$$

where X_i is the ith eigenvector and $X = X_i \cos(\omega_i t)$ describes the motion.

Both these energies are time-varying functions that oscillate between zero and a maximum value. As a glance at (6.3.10) and (6.3.11) reveals, the maximum value of the potential energy is $\tfrac{1}{2}X_i^T[K]X_i$, while the maximum kinetic energy is $\tfrac{1}{2}\omega_i^2 X_i^T[M]X_i$. Using an overbar to represent the maximum, we therefore have

$$\overline{PE} = \tfrac{1}{2}X_i^T[K]X_i \tag{6.3.12}$$

and

$$\overline{KE} = \tfrac{1}{2}\omega_i^2 X_i^T[M]X_i \tag{6.3.13}$$

We're going to need another term in a second, so we'll introduce it now. If we factor the ω^2 out of (6.3.13) we'll obtain what we'll call the *zero-frequency kinetic energy*, which we'll indicate by a zero superscript:

$$\overline{KE}^0 = \tfrac{1}{2} X_i^T [M] X_i \qquad (6.3.14)$$

Now compare (6.3.12) and (6.3.13) with (6.3.3). We can see that the first term of (6.3.3) is just the maximum kinetic energy of the system and the second is the maximum potential energy. So what we've done by premultiplying by X_i^T is to change what were force terms into energy terms. If we now divide (6.3.3) by $X_i^T[M]X_i$ we'll have

$$\omega_i^2 = \frac{\tfrac{1}{2} X_i^T [K] X_i}{\tfrac{1}{2} X_i^T [M] X_i} \qquad (6.3.15)$$

What this tells us is that the square of a system's natural frequency is *always* equal to the ratio of the system's maximum potential energy to the system's zero-frequency kinetic energy. Since we're concerned with the ratio of the two energies, the $\tfrac{1}{2}$ factors aren't a problem; we can just as well factor them out,

$$\omega_i^2 = \frac{X_i^T [K] X_i}{X_i^T [M] X_i} \qquad (6.3.16)$$

The more general form of (6.3.16), one that simply depends on an arbitrary input vector, is given by

$$R_Q(X) = \frac{X^T [K] X}{X^T [M] X} \qquad (6.3.17)$$

where R_Q stands for, not surprisingly, Rayleigh's quotient. Any given problem has an infinite number of values for the Rayleigh's quotient, since changing X will change $R_Q(X)$. Notice that simply scaling the input vector doesn't alter Rayleigh's quotient; i.e., $R_Q(X) = R_Q(nX)$, where n is any constant. This is in keeping with our notions of eigensolutions—and the fact that the eigenvectors can be determined only up to a multiplicative constant. It is important to remember that Rayleigh's quotient will produce a continuum of values, depending upon X. For the special cases of the input X being equal to an eigenvector, Rayleigh's quotient will equal the square of the natural frequency associated with that eigenvector.

Equation (6.3.17) is very significant, not just for our discussion of Rayleigh's quotient but for our later work in approximation methods. Basically, it's a way of getting the eigenvalues for our problem without solving an eigenvalue problem. If (and that's a *big* if right now) someone hands us the actual eigenvectors of our problem, then simply substituting them into (6.3.17) will give us the associated eigenvalue (or equivalently, the square of the natural frequency).

As it now stands, this may not seem like a really big deal. First, you have to wonder how often someone will conveniently hand you the system eigenvectors. And if you don't have the exact eigenvectors to work from, but only an approximation of them, who's to say that (6.3.17) won't just give an inaccurate estimate of the natural frequencies? Well, the good news is that there's a bit more to (6.3.17) than there initially appears to be. In the first place, design invariably involves an iterative process in which a system is "tweaked" in various ways to improve its performance. Generally, you know a good deal about the system before the "tweaking" starts. In a similar vein, improvements of an existing product often consist of small modifications to the basic unit, modifications that are, again, meant to improve performance in some way. Here also, the basic system is well understood before modifications begin. Thus there is a very good chance that the system eigenmodes and natural frequencies have already been catalogued and are ready for use in later analyses. Since the changes to the existing system are generally not too large, it seems reasonable that the eigenmodes of the system will not change drastically. Therefore the old eigenmodes make very reasonable starting estimates for the modified system's eigenmodes.

So let's say that someone gives you (or you estimate for yourself) an approximation to the ith eigenvector of the modified system. You then take this approximation and use it in (6.3.17). You obviously won't obtain ω_i^2, because the only way to get that would be to have used the *exact* ith eigenvector. But, what you *will* obtain is an extremely good estimate of ω_i^2. In fact, the error in the estimate of ω_i^2 will be far less than the error in your approximation of the ith eigenvector. That's the power of (6.3.17); it gives a good quality estimate for the natural frequency even if your input eigenvector estimate isn't very good.

Example 6.2

Problem   Let's look at the system shown in Figure 6.3 to illustrate how well Rayleigh's quotient can work. Figure 6.3a represents our original system. We then impose some changes in the system based on design considerations and wind up with Figure 6.3b. Use Rayleigh's quotient to determine how the natural frequencies have changed .

Solution   The original system has the following mass and stiffness matrices

$$[M] = \begin{bmatrix} 10 & 0 & 0 \\ 0 & 10 & 0 \\ 0 & 0 & 10 \end{bmatrix}, \quad [K] = \begin{bmatrix} 200 & -100 & 0 \\ -100 & 200 & -100 \\ 0 & -100 & 200 \end{bmatrix}$$

which support the eigenvectors

$$X_1 = \begin{Bmatrix} .5000 \\ .7071 \\ .5000 \end{Bmatrix}, \quad X_2 = \begin{Bmatrix} .7071 \\ .0000 \\ -.7071 \end{Bmatrix}, \quad \text{and} \quad X_3 = \begin{Bmatrix} .5000 \\ -.7071 \\ .5000 \end{Bmatrix}$$

with $\omega_1^2 = 5.858$ rad/s, $\omega_2^2 = 20.00$ rad/s, and $\omega_3^2 = 34.14$ rad/s.
The modified system (indicated by a hat) has mass and stiffness matrices

$$[\hat{M}] = \begin{bmatrix} 11 & 0 & 0 \\ 0 & 9 & 0 \\ 0 & 0 & 12 \end{bmatrix}, \quad [\hat{K}] = \begin{bmatrix} 175 & -90 & 0 \\ -90 & 185 & -95 \\ 0 & -95 & 185 \end{bmatrix}$$

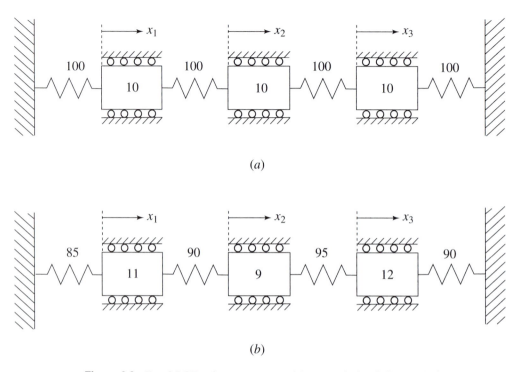

Figure 6.3 Two 3 DOF spring-mass system: (a) unperturbed and (b) perturbed

which support the eigenvectors

$$\hat{X}_1 = \begin{Bmatrix} .5130 \\ .6831 \\ .5198 \end{Bmatrix}, \quad \hat{X}_2 = \begin{Bmatrix} .7308 \\ .0216 \\ -.6823 \end{Bmatrix}, \quad \text{and} \quad \hat{X}_3 = \begin{Bmatrix} .4315 \\ -.8064 \\ .4045 \end{Bmatrix}$$

with $\hat{\omega}_1^2 = 5.014$ rad/s, $\hat{\omega}_2^2 = 15.67$ rad/s, and $\hat{\omega}_3^2 = 31.20$ rad/s.

Keep in mind that this problem was designed to be simple, and so there's no real reason to use Rayleigh's quotient; it isn't a big deal to solve the problem analytically. That's why both solutions have been solved exactly—so we can see how well Rayleigh's quotient works. Notice that the eigenvalues of the two problems (the square of the natural frequencies) have changed substantially, as have the eigenvectors.

What we'll do now is use the original eigenvectors as estimates of the new eigenvectors and utilize Rayleigh's quotient to determine the new natural frequencies. Thus, for the first natural frequency estimate (which we'll denote by a tilde), we have

$$\tilde{\omega}_1^2 = R_Q(X_1) = \frac{X_1^T [\hat{K}] X_1}{X_1^T [\hat{M}] X_1}$$

We're trying to estimate the new natural frequency by using a good guess for the eigenvector and our knowledge of what the new mass and stiffness matrices actually are. Using the foregoing values, what we'll find is that

$$\tilde{\omega}_1^2 = 5.043 \text{ rad/s}$$

If we go ahead and determine the estimates of the second and third natural frequencies, using X_2 and X_3, respectively, we'll obtain

$$\tilde{\omega}_2^2 = 15.65 \text{ rad/s} \quad \text{and} \quad \tilde{\omega}_3^2 = 30.57 \text{ rad/s}$$

Compare these results with the squares of the actual natural frequencies (5.014, 15.67, and 31.20 rad/s). The results are fantastically good! 5.043 rad/s is off only by half a percent from 5.014 rad/s, 15.65 rad/s is off by .1%, and 30.57 rad/s is off by 2%. Rayleigh's quotient gave us absolutely excellent estimates of the modified natural frequencies and did it using eigenvectors that were noticeably off from the actual eigensolutions.

Having seen that Rayleigh's quotient does this nifty job of producing answers that are be than the data we feed in, it's time to find out why this happens. The answer has a lot to do with what's happening in a geometric sense with the system's eigenstructure. This is a bit complicated to visualize, so we'll break the problem down and look at bits of it individually before putting it all together.

First let's see how Rayleigh's quotient varies as we use different amounts of a particular eigenvector. Figure 6.4 shows Rayleigh's quotient for our original system (6.2). In these plots we're using the actual eigenvectors of the system and simply changing how much is used. As you can see, the plots are flat; it makes no difference whether we use a lot or a little. This is completely in line with our notion of eigenvectors. All that matters are the relative displacements, not the actual magnitudes.

Now we'll go on to something more interesting. In Figure 6.5, we're again plotting Rayleigh's quotient. However this time our estimated vector consists of one eigenvector (X_1) plus a varying amount of another. More specifically, in Figure 6.5a we're letting our estimated vector equal $X_1 + a_2 X_2$ while keeping a_3 (the amount of the third eigenvector) at zero. When a_2 is zero, Figure 6.5a shows that R_Q gives us the correct value for ω_1^2, namely, 5.858. We expect the graph to exactly equal the square of the first natural frequency when a_2 equals zero, and it does. The real question is what happens if a_2 isn't zero. As we add more and more of a contamination vector (by letting a_2 grow as in Figure 6.5a); the R_Q estimate increases. Near the origin, the plot looks quadratic, and it levels off for high values of a_2. The same qualitative behavior is seen in Figure 6.6b where we're contaminating X_1 with X_3.

This is behavior is always present; the *lowest* value that Rayleigh's quotient can produce is the square of the first natural frequency. No input vector can produce a lower output. Therefore one extremely inefficient way to determine the modified system's first natural frequency would be to try every conceivable trial vector. The one giving you the lowest value for Rayleigh's quotient will be the first eigenvector. Of course, nobody would ever use such an approach, but at least we know we could if all else failed.

The next set of plots (Figure 6.6) shows how Rayleigh's quotient varies when we let the estimated vector equal $X_3 + a_1 X_1$ (Figure 6.6a) and then let it equal $X_3 + a_2 X_2$ (Figure 6.6b). We see that the plots have maximums when the estimated vector is simply X_3: i.e., the value decreases as we add

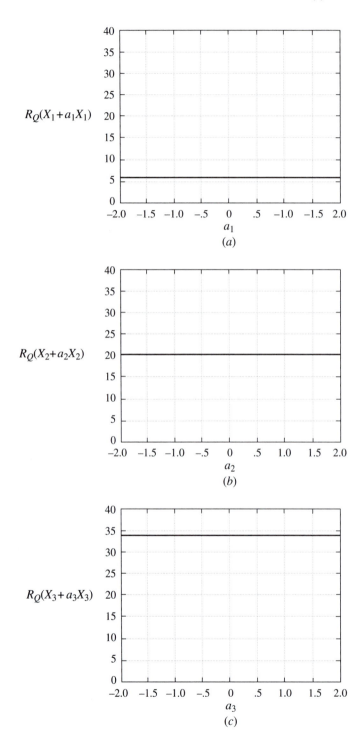

$R_Q(X_1 + a_1 X_1)$

a_1

(a)

$R_Q(X_2 + a_2 X_2)$

a_2

(b)

$R_Q(X_3 + a_3 X_3)$

a_3

(c)

Figure 6.4 Behavior of Rayleigh's quotient about each eigenvector when perturbed by that eigenvector

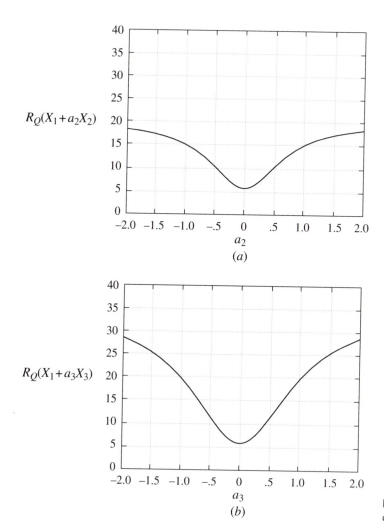

$R_Q(X_1 + a_2X_2)$

a_2

(a)

$R_Q(X_1 + a_3X_3)$

a_3

(b)

Figure 6.5 Behavior of Rayleigh's quotient about the first eigenvector

components of X_1 or X_2 to it. Thus we have another nugget of information, namely, that the *highest* possible value for Rayleigh's quotient is equal to the square of the largest natural frequency (in this case, the third natural frequency).

Now it's time for the weirder result. The first two cases were nicely behaved. Contaminating the first eigenvector (the one associated with the *lowest* natural frequency) with any other eigenvector simply made Rayleigh's quotient increase, while adding some other eigenvector to the *last* eigenvector (the one associated with the highest natural frequency) caused Rayleigh's quotient to decrease. These are the two simplest cases—contaminating the first or the last eigenvector. The behavior is more complex for any other eigenvector. Of course, in this example there is only one other eigenvector to consider, since it's only a three-mass system. But for more complex systems, all eigenvectors

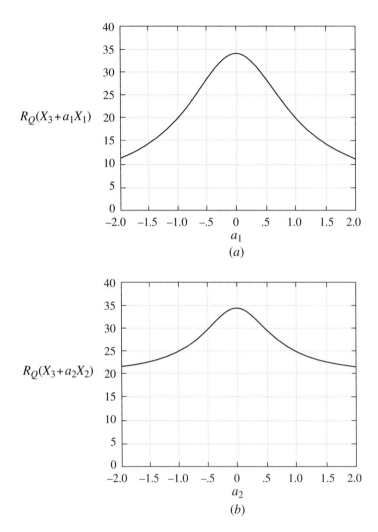

$R_Q(X_3 + a_1 X_1)$

(a)

$R_Q(X_3 + a_2 X_2)$

(b)

Figure 6.6   Behavior of Rayleigh's
quotient about the third eigenvector

other than the ones corresponding to the lowest or highest natural frequency will display the kind of
behavior we'll see next.

Figure 6.7 shows how Rayleigh's quotient changes as we contaminate the second eigenvector
with either the first eigenvector ($X_2 + a_1 X_1$: Figure 6.7a) or the third eigenvector ($X_2 + a_3 X_3$:
Figure 6.7b). We can see from these plots that Rayleigh's quotient *decreases* when we add some of
the first eigenvector (X_1) but *increases* if we add some of the third eigenvector (X_3). This actually
makes sense if you think about what Rayleigh's quotient does. As we've seen, Rayleigh's quotient
spits out a scalar that corresponds to the system's natural frequencies if the input is an eigenvector. We
know that the second natural frequency is less than the third and more than the first. By adding some
of X_3 to our estimated vector, we're reorienting the estimated vector toward the third eigenvector.

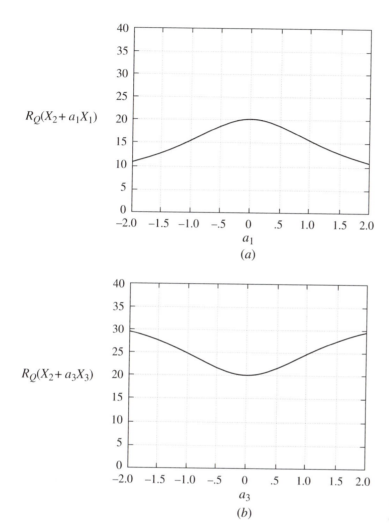

$R_Q(X_2 + a_1X_1)$

a_1

(a)

$R_Q(X_2 + a_3X_3)$

a_3

(b)

Figure 6.7 Behavior of Rayleigh's quotient about the second eigenvector

If we add a lot of X_3, then the second becomes trivial in comparison, thus making the estimated vector approximate X_3. And Rayleigh's quotient for the third eigenvector is *bigger* than for the second. So we should expect Rayleigh's quotient to increase as we add more and more of X_3. Otherwise, how would the result smoothly increase to the higher value?

The same logic holds when we add some of X_1. In this case Rayleigh's quotient will end up smaller as we get more and more of the first eigenvector involved, and so we'd expect it to decrease steadily toward this lower value.

Of course, it's more complicated if we add a little of *both* remaining eigenvectors to an initial eigenvector. When the initial eigenvector is the first or the last, we still get a minimum or maximum (Figure 6.8), since both lead in the same direction (toward a lower or higher value, depending on

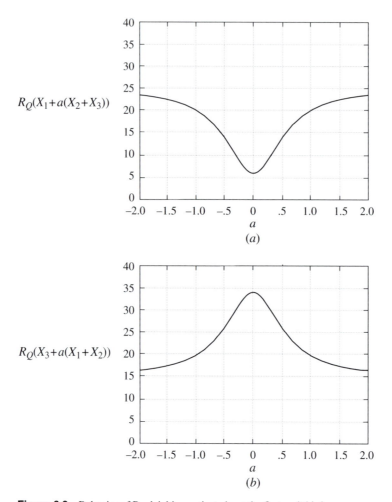

Figure 6.8 Behavior of Rayleigh's quotient about the first and third eigenvectors when perturbed by a combination of both remaining eigenvectors

whether the initial eigenvector is the highest or lowest one). But for the case of the second eigenvector we've got two competing trends. Adding some X_1 will cause Rayleigh's quotient to decrease, but adding some X_3 will cause it to increase. Therefore the overall result will depend upon how much of each is added. As an example, Figure 6.9 illustrates how Rayleigh's quotient will vary as equal amounts of X_1 and X_3 are added to X_2. Note that the result is *no* variation: the effect of X_1 is exactly canceled by that of X_3, resulting in no change to Rayleigh's quotient.

    This behavior is the same as we'd see in higher dimensional systems for all eigenvectors except those associated with the lowest and highest natural frequencies. Assuming that the natural frequencies have been ordered from lowest to highest, perturbing off X_i with some of X_j will cause the Rayleigh's quotient to increase if j is greater than i and decrease if j is less than i.

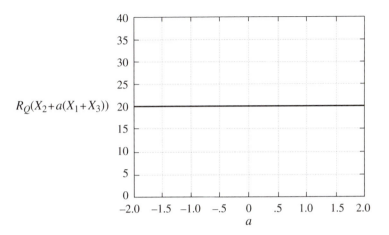

$R_Q(X_2 + a(X_1 + X_3))$

Figure 6.9 Behavior of Rayleigh's quotient about the second eigenvector when perturbed by a combination of both remaining eigenvectors

Time for the final observation. You may recall from your multivariable calculus class that when a function has a minimum, the derivative at that minimum with respect to the independent variable is zero. The same holds true at a maximum and also at inflection points. We can see this clearly in Figure 6.10. The slope of Rayleigh's quotient versus a_1, a_2, etc. is always zero at the minimum. For the case of X_2 (Figure 6.7) we still have a slope of zero with respect to the a_i's; the only thing different is that we have both a maximum and a minimum, depending upon which independent variable we're looking at. Even when we're looking at combinations of the a_i's (Figure 6.8), we still have a zero slope when the a_i's equal zero. Let's take a closer look at a particular case. Figure 6.11, an enlargement of Figure 6.5*b*, shows how Rayleigh's quotient varies when $a_2 = 0$ and $X_1 + a_3 X_3$ is the input. If we're near $a_3 = 0$, then, because the slope is zero, small changes in a_3 don't appreciably alter Rayleigh's quotient. The point labeled A on the graph shows the change in Rayleigh's quotient when $a_3 = .10$. As you can see, the value doesn't change much at all from what it was when $a_3 = 0$. This is a fundamental property of minima, maxima, or inflection points. Such points are called stationary points because the dependent function is essentially stationary about them and doesn't change (much) as the dependent variable is altered.

This is the result we've been looking for! If we have a halfway decent quess for a modified system's eigenvector, then our estimated vector has a lot of the actual eigenvector in it, along with some contamination from the other eigenvectors. What we've now found out is that this contamination doesn't really alter the result very much, as long as contamination isn't excessive. Therefore pretty good guesses about the eigenvectors will give us *very good* estimates of the natural frequencies.

Having numerically demonstrated all this, it's time to look at the mathematics to see *why* Rayleigh's quotient behaves as it does. We'll start by examining (6.3.17) more closely. We already know that the eigenvectors of a system span the system's configuration space. Thus, any vector that we come up with can be represented by an appropriate combination of the system's eigenvectors. We also know from our recent work that Rayleigh's quotient is stationary about the system's eigenvectors.

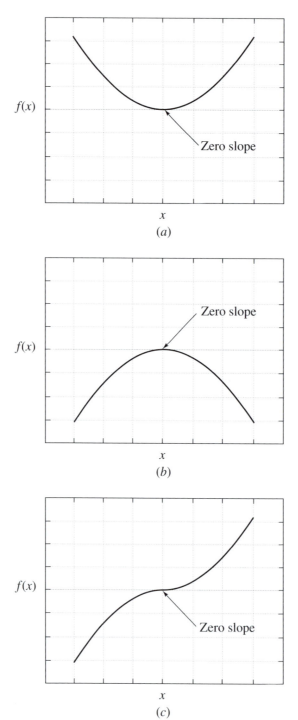

Figure 6.10 Zero-slope behavior when a function is evaluated at a minimum, a maximum, and an inflection point

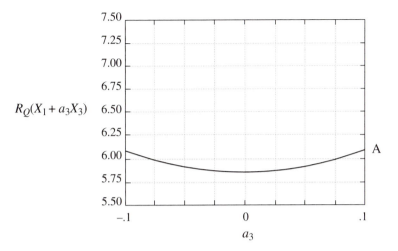

Figure 6.11   Behavior of Rayleigh's quotient about the first eigenvector (enlarged)

We'll assume that our estimated vector is given by

$$\hat{X} = a_1 X_1 + a_2 X_2 + \cdots + a_{j-1} X_{j-1} + X_j + a_{j+1} X_{j+1} + \cdots + a_n X_n \tag{6.3.18}$$

that is,

$$\hat{X} = \sum_{i=1}^{n} a_i X_i \tag{6.3.19}$$

where $a_j = 1$. We'll further assume that the remaining a_i's are small ($a_i \ll 1$). Thus $\hat{X}$ is mainly equal to the jth eigenvector X_j, with only small contamination from the other eigenvectors. This is the definition of a *good* estimate: it's close to, but not exactly equal to, the actual eigenvector. We don't have to pin down the a_i's exactly; it's enough to know that they're small.

Our next step is to see what Rayleigh's quotient looks like in terms of our estimate $\hat{X}$. Using (6.3.19) in (6.3.17) gives us

$$R_Q(\hat{X}) = \frac{\left(\sum_{i=1}^{n} a_i X_i^T \right) [K] \left(\sum_{i=1}^{n} a_i X_i \right)}{\left(\sum_{i=1}^{n} a_i X_i^T \right) [M] \left(\sum_{i=1}^{n} a_i X_i \right)} \tag{6.3.20}$$

For clarity, we can explicitly expand the summation out:

$$R_Q(\hat{X}) = \frac{(a_1 X_1^T + a_2 X_2^T + \cdots)[K](a_1 X_1 + a_2 X_2 + \cdots)}{(a_1 X_1^T + a_2 X_2^T + \cdots)[M](a_1 X_1 + a_2 X_2 + \cdots)} \tag{6.3.21}$$

We now need to recall that the eigenvectors of a system are orthogonal with respect to the $[K]$ and $[M]$ matrices. Therefore many of the multiplications in (6.3.21) will simply yield zero. Let's also decide that our eigenvectors have been normalized, so that

$$X_i^T[M]X_i = 1 \qquad (6.3.22)$$

This step isn't actually necessary, it just makes the mathematics a bit clearer. If the vectors have been normalized in this way, then we know from Chapter 4 that

$$X_i^T[K]X_i = \omega_i^2 \qquad (6.3.23)$$

If we now use the orthogonality properties on (6.3.21) we'll be left with

$$R_Q(\hat{X}) = \frac{a_1^2\omega_1^2 + a_2^2\omega_2^2 + \cdots + \omega_j^2 + \cdots a_n^2\omega_n^2}{a_1^2 + a_2^2 + \cdots + 1.0 + \cdots + a_n^2} \qquad (6.3.24)$$

The denominator can be rearranged slightly, as follows:

$$R_Q(\hat{X}) = \frac{a_1^2\omega_1^2 + a_2^2\omega_2^2 + \cdots + \omega_j^2 + \cdots a_n^2\omega_n^2}{1 + a_1^2 + a_2^2 + \cdots} \qquad (6.3.25)$$

Since all the a_i's are small, the denominator is in the form

$$1 + \varepsilon \qquad (6.3.26)$$

where ε is small. You may recall that the inverse of $1 + \varepsilon$ is given by

$$\frac{1}{1 + \varepsilon} = 1 - \varepsilon + \varepsilon^2 + \cdots \qquad (6.3.27)$$

If we use this approximation in (6.3.25) we can rewrite Rayleigh's quotient as

$$R_Q(\hat{X}) = \left(a_1^2\omega_1^2 + a_2^2\omega_2^2 + \cdots + \omega_j^2 + \cdots a_n^2\omega_n^2\right)\left(1 - (a_1^2 + a_2^2 + \cdots) + (a_1^2 + a_2^2 + \cdots)^2 - \cdots\right) \qquad (6.3.28)$$

All we'll need to be interested in are the terms up to second order (terms like a_1^2), and so when we expand out (6.3.28) we'll neglect all higher order terms. Therefore

$$R_Q(\hat{X}) \approx \omega_j^2 + a_1^2(\omega_1^2 - \omega_j^2) + a_2^2(\omega_2^2 - \omega_j^2) + \cdots + a_{j-1}^2(\omega_{j-1}^2 - \omega_j^2)$$
$$+ a_{j+1}^2(\omega_{j+1}^2 - \omega_j^2) + \cdots + a_n^2(\omega_n^2 - \omega_j^2) \qquad (6.3.29)$$

Let's take a second to think about what we've got. What we see is that our Rayleigh's quotient is equal to the square of the jth natural frequency plus error terms. Furthermore, all the error terms are *second-order terms*! Even though our input vector had first-order errors (terms like $a_1 X_1$), no equivalent errors are showing up in our Rayleigh's quotient approximation. If they had appeared we'd be seeing terms like $a_1 \omega_1^2$ and $a_2 \omega_j^2$. This is what gives us our great accuracy. The errors in the input depend on the a_i's, which are small. Well, if the a_i's are small, then the values of a_i^2 are absolutely tiny.

Equation (6.3.29) shows that our approximate value is equal to the actual eigenvalue we're seeking (the ω_j^2 term) plus some tiny errors that are proportional to the a_i^2's. Any decent guess leads to an excellent approximation, just as we've already seen.

Finally, can (6.3.29) also predict the precise kind of behavior we saw in Figures 6.5–6.9? Absolutely. Let's first consider what we get if we're looking for the first natural frequency. In this case, $j = 1$ and (6.3.29) becomes

$$R_Q(\hat{X}) = \omega_1^2 + a_2^2(\omega_2^2 - \omega_1^2) + a_3^2(\omega_3^2 - \omega_1^2) + \cdots \tag{6.3.30}$$

Since ω_1 is the lowest natural frequency, all the error terms are positive (the a_i^2 are always positive and the $\omega_i^2 - \omega_1^2$ terms are also positive, since $\omega_i > \omega_1$). Since all the error terms are positive, we see that R_Q is a minimum around the first eigenvector; all contaminating vectors can serve only to increase its value.

If we go to the other end of the spectrum and look at the highest natural frequency, then

$$R_Q(\hat{X}) = \omega_n^2 + a_1^2(\omega_1^2 - \omega_n^2) + a_2^2(\omega_2^2 - \omega_n^2) + \cdots \tag{6.3.31}$$

For this case R_Q is a maximum, since any contamination will reduce its value (the a_i^2 terms are still positive, and all the differences between natural frequencies are negative).

For any other natural frequency, we'll have some positive and some negative terms. This can be seen in (6.3.29). Whenever $i > j$, the $a_i^2(\omega_i^2 - \omega_j^2)$ terms will be positive (causing positive errors) and whenever $i < j$, the $a_i^2(\omega_i^2 - \omega_j^2)$ terms will be negative (causing negative errors).

All these characteristics can be seen in Figures 6.5–6.9. The only thing to keep in mind when looking at these figures and our mathematical results is that our entire analysis of Rayleigh's quotient has based on input vectors that are close to the actual eigenvectors. The results are therefore valid only for small a_i, as stated earlier. The plots in Figures 6.5–6.9 hold for both small and large values of a_i. Thus the analysis and figures will match for small values of a_i but diverge as the a_i get too large. For instance, Figure 6.5b shows how Rayleigh's quotient behaves for the first eigenvector, with some contamination from the third. Our mathematical analysis says that the plot should look like

$$R_Q(\hat{X}) = \omega_1^2 + a_3(\omega_3^2 - \omega_1^2) \tag{6.3.32}$$

which it does, for small values of a_3. However, as a_3 increases, the plot levels off, approaching ω_3^2, not a surprising result because X_3 becomes the dominant component of $\hat{X}$ as a_3 increases. Figure 6.12 superimposes both these solutions, showing how the two systems behave similarly when a_3 is small and diverge as the magnitude of a_3 grows.

One additional observation comes directly from looking at Rayleigh's quotient. Since Rayleigh's quotient is equal to $\frac{X^T[K]X}{X^T[M]X}$, we can see that increasing $[K]$, while holding $[M]$ fixed, will increase the resulting frequency estimates. Thus increasing $[K]$, which means increasing the stiffness in the system, results in higher natural frequencies. Similarly, increasing the mass of the system (leading to larger values of $[M]$) serves to decrease the natural frequencies $\left(\text{since Rayleigh's quotient is proportional to } \frac{1}{X^T[M]X}\right)$. This is in line with our intuition and with results we've already seen.

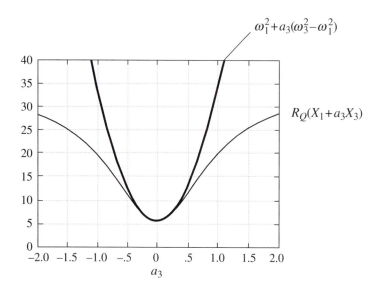

$$\omega_1^2 + a_3(\omega_3^2 - \omega_1^2)$$

$R_Q(X_1 + a_3 X_3)$

Figure 6.12 Comparison of Rayleigh's quotient and a second-order polynomial approximation

6.4 RAYLEIGH-RITZ METHOD: DISCRETE SYSTEMS

Rayleigh's quotient is a powerful tool for estimating system natural frequencies, but we can go quite a bit further to improve our results. The first person to realize this was Walter Ritz, and for this reason, the method is called the Rayleigh-Ritz method. The real power of the Rayleigh-Ritz method lies in how it allows us to analyze complicated continuous problems. To illustrate how the method works, however, we're first going to confine our attention to finite-dimensional systems. This will make it much easier to grasp what the mathematics is doing for us. Only after we've mastered a finite-dimensional application of the method will we move on to the more physically meaningful continuous problems.

Okay, back to the method. Ritz looked at the results we've just finished deriving and asked himself if there was any way to extract even more information from them. His logic went something like this:

> Rayleigh's quotient gives us a good way of estimating a system's fundamental frequency. If I use a pretty good estimate for the first eigenvector, I'll get an excellent approximation. But I'd like to know more than just the fundamental frequency; I'd like several natural frequencies, and I'd like to know their associated eigenvectors. Is there a way to use more than one estimated vector, and mix these estimated vectors around in an appropriate way, so that I end up with estimates for several natural frequencies and several eigenvectors? And can I work it so that, as I use more and more estimated vectors, my results get better and better?

Well, what Ritz decided, after a bit of rumination, was that the fundamental reality of Rayleigh's quotient was its *stationarity* when perturbed around any of the system's actual eigenvectors.

(We studied this fact in detail in Section 6.3.) So Ritz decided to generalize the concept of an estimated vector. Rather than use just a single vector as an estimated vector, Ritz used several. Thus, he first decided on a *set* of trial vectors, $\hat{X}_1$, $\hat{X}_2$, ... Keep in mind that (unless you're real lucky in choosing your trial vectors), these trial vectors are *not* the eigenvectors. Each $\hat{X}_i$ is simply your best estimate of the system's eigenvectors. They might be good estimates or they might be lousy. What we'll see is that, either way, you'll get good results from the Rayleigh-Ritz method as long as you use enough trial vectors. However, the better your initial estimates, the quicker and the more easily you'll get a good final result. Also, it's important to make sure that the trial vectors are independent of each other. In the work to follow, we'll be adding the vectors together to get new vectors, something that wouldn't work too well if we had dependent vectors floating around.

Once he'd decided on his trial vectors, Ritz decided to work *as if* these were really the system eigenvectors (which, again, they aren't). But *IF* they actually were the eigenvectors, then Rayleigh's quotient would be stationary with respect to perturbations around them. This is the key to Ritz's breakthrough. He chose to treat his trial vectors as if they were the eigenvectors, and thus he proposed to *force* Rayleigh's quotient to be stationary with respect to perturbations about them. Again, this is a crucial point. Rayleigh's quotient is *not* generally stationary with respect to perturbations about some arbitrary set of trial vectors; rather, it's stationary with respect to the *actual* eigenvectors. Ritz's conceptual leap was to realize that if he forced Rayleigh's quotient to be stationary with respect to his trial vectors, it would result in combinations of trial vectors that were as close as possible to the actual eigenvectors.

Since the trial vectors are independent, it is certainly possible to add them together to get new vectors. And it's also true that if we knew what the actual eigenvectors looked like, we could add them up in such a way as to best approximate the eigenvectors. The Rayleigh-Ritz method will figure out for us how to do this, without actually ever knowing what the real eigenvectors are. Pretty good trick, no?

Here's how it's done. We know that after we've appropriately added together our trial vectors to approximate an eigenvector, the resultant vector will look like this:

$$\text{Est} = \sum_{i=1}^{n} b_i \hat{X}_i \tag{6.4.1}$$

I'm really sorry about all the subscripts and superscripts, the hats and tildes, and whatnot. They're necessary, but I know they're a pain to keep track of. That's why I won't introduce yet another one to stand for "the best estimate of the eigenvector that's constructed from the trial vectors." I'll simply call it Est, for "estimated eigenvector." I hope this makes it a little easier to follow. The $\hat{X}_i$ are our trial vectors and the b_i are the coefficients that cause the overall summation to best approximate the eigenvector.

The only tricky part, therefore, is to determine exactly what these b_i are equal to. If we know that, then we're done, since we'd just use (6.4.1) to find Est. As mentioned, it's the process of forcing Rayleigh's quotient to be stationary that will give us these b_i's.

If we write out Rayleigh's quotient in terms of our estimated eigenvector, we'll have

$$R_Q(\text{Est}) = \frac{\text{Est}^T[K]\text{Est}}{\text{Est}^T[M]\text{Est}} \tag{6.4.2}$$

However, if we look closely at what Est is, we see that the real unknowns are the b_i's. So, since the only real unknowns in Rayleigh's quotient are the b_i's, we can indicate this by writing (6.4.2) as

$$R_Q(\text{Est}) = R_Q(b_i) \tag{6.4.3}$$

Equation (6.4.3) states that the unknown quantities are the b_i's, nothing more. Having said this, we'll refrain from writing the b_i dependency in explicitly, since this would make the following derivations more cluttered than they already are. Just keep in mind that when you see R_Q, this means the Rayleigh's quotient that depends on all the unknown b_i's, as well as the particular $\hat{X}_i$ we've chosen (which are already known).

It's now time to make Rayleigh's quotient stationary. Recall that the definition of stationarity is that the first partial derivatives with respect to each of the independent variables must be zero. This is exactly what we saw in Figures 6.5–6.9. We looked at Rayleigh's quotient around each of the eigenvectors and saw that as we varied our trial vector in the direction of another eigenvector, holding the others fixed, the value of Rayleigh's quotient didn't change. This is precisely what is meant by the first partial derivatives being equal to zero. We therefore require

$$\left.\frac{\partial R_Q}{\partial b_i}\right|_{\text{all } b_i=0} = 0, \quad i = 1, 2, \ldots, n \tag{6.4.4}$$

for every b_i.

Since both the numerator and denominator of Rayleigh's quotient depend on the b_i's, we'll have to use the chain rule to perform this differentiation. If we write Rayleigh's quotient as

$$R_Q = \frac{\text{Num}}{\text{Den}} \tag{6.4.5}$$

where Num $= \text{Est}^T[K]\text{Est}$ and Den $= \text{Est}^T[M]\text{Est}$, then

$$\frac{\partial R_Q}{\partial b_i} = \frac{\partial\left[\text{Num}\,(\text{Den})^{-1}\right]}{\partial b_i} = \frac{-\text{Num}\frac{\partial \text{Den}}{\partial b_i} + \text{Den}\frac{\partial \text{Num}}{\partial b_i}}{\text{Den}^2} = 0 \tag{6.4.6}$$

The value of Den is always positive, since it involves the positive masses of the system and quadratic (positive) combinations of the trial vectors. Consequently, we can satisfy (6.4.6) only if

$$-\text{Num}\frac{\partial \text{Den}}{\partial b_i} + \text{Den}\frac{\partial \text{Num}}{\partial b_i} = 0 \tag{6.4.7}$$

or if, after dividing by Den,

$$\frac{\partial \text{Num}}{\partial b_i} - R_Q\frac{\partial \text{Den}}{\partial b_i} = 0 \tag{6.4.8}$$

Now we have to figure out what partial differentiation of the numerator and denominator gives us. If we focus on the first term, we've got

$$\frac{\partial \text{Num}}{\partial b_i} = \frac{\text{Est}^T[K]\text{Est}}{\partial b_i}$$

$$= \frac{\partial}{\partial b_i}\left[(b_1\hat{X}_1^T + b_2\hat{X}_2^T + \cdots)[K](b_1\hat{X}_1 + b_2\hat{X}_2 + \cdots)\right]$$

$$= \frac{\partial}{\partial b_i}[b_1^2\hat{X}_1^T[K]\hat{X}_1 + b_2^2\hat{X}_2^T[K]\hat{X}_2 + \cdots$$

$$+ b_1b_2\hat{X}_1^T[K]\hat{X}_2 + b_2b_1\hat{X}_2^T[K]\hat{X}_1$$

$$+ b_1b_3\hat{X}_1^T[K]\hat{X}_3 + b_3b_1\hat{X}_3^T[K]\hat{X}_1$$

$$+ b_2b_3\hat{X}_2^T[K]\hat{X}_3 + b_3b_2\hat{X}_3^T[K]\hat{X}_2 + \cdots] \qquad (6.4.9)$$

What we see in (6.4.9) is a series of terms involving only one particular eigenvector (the $\hat{X}_i^T[K]\hat{X}_i$ terms) and another series involving all the cross-terms (terms like $\hat{X}_i^T[K]\hat{X}_j$). Since we're differentiating with respect to only one of the b_i at a time, most of the terms will drop out, leaving us with

$$\frac{\partial \text{Num}}{\partial b_i} = 2b_i\hat{X}_i^T[K]\hat{X}_i$$

$$+ b_1\hat{X}_1^T[K]\hat{X}_i + b_1\hat{X}_i^T[K]\hat{X}_1$$

$$+ b_2\hat{X}_2^T[K]\hat{X}_i + b_2\hat{X}_i^T[K]\hat{X}_2$$

$$+ b_3\hat{X}_3^T[K]\hat{X}_i + b_3\hat{X}_i^T[K]\hat{X}_3$$

$$+ \cdots \qquad (6.4.10)$$

This can be made a bit more compact by remembering that $[K] = [K]^T$ and $[M] = [M]^T$. Since the terms $\hat{X}_i^T[K]\hat{X}_j$ are just scalar quantities, they are equal to their own transpose. Thus $\left(\hat{X}_i^T[K]\hat{X}_j\right)^T = \hat{X}_j^T[K]^T\hat{X}_i = \hat{X}_j^T[K]\hat{X}_i$. So we see that each of the members of the cross-product pairs in (6.4.10) are equal to the same value, leading to

$$\frac{\partial \text{Num}}{\partial b_i} = 2b_i\hat{X}_i^T[K]\hat{X}_i$$

$$+ 2b_1\hat{X}_1^T[K]\hat{X}_i$$

$$+ 2b_2\hat{X}_2^T[K]\hat{X}_i$$

$$+ 2b_3\hat{X}_3^T[K]\hat{X}_i$$

$$+ \cdots \qquad (6.4.11)$$

You'll note that there's a pretty clear pattern shaping up in these results. To see this, let's first make a couple of new definitions. Rather than writing out all these $\hat{X}_i^T[K]\hat{X}_j$ terms, we'll say that

$$\mathcal{K}_{ij} \equiv \hat{X}_i^T[K]\hat{X}_j \tag{6.4.12}$$

and

$$\mathcal{M}_{ij} \equiv \hat{X}_i^T[M]\hat{X}_j \tag{6.4.13}$$

By combining (6.4.1) and (6.4.5) and making use of (6.4.12) and (6.4.13), we can write out the denominator and numerator of Rayleigh's quotient in the following form

$$\text{Num} = \sum_{i=1}^{n}\sum_{j=1}^{n} b_i b_j \mathcal{K}_{ij} \tag{6.4.14}$$

and

$$\text{Den} = \sum_{i=1}^{n}\sum_{j=1}^{n} b_i b_j \mathcal{M}_{ij} \tag{6.4.15}$$

Written like this, it is clear that Rayleigh's quotient depends only upon the b_i, all the $\mathcal{K}_{ij}$'s and $\mathcal{M}_{ij}$'s being known constants.

Using this new $\mathcal{K}_{ij}$ and $\mathcal{M}_{ij}$ notation, we can rewrite (6.4.11) as

$$\frac{\partial \text{Num}}{\partial b_i} = 2(b_1 \mathcal{K}_{i1} + b_2 \mathcal{K}_{i2} + b_3 \mathcal{K}_{i3} + \cdots) \tag{6.4.16}$$

which can be written as

$$\frac{\partial \text{Num}}{\partial b_i} = 2\begin{bmatrix} \mathcal{K}_{i1} & \mathcal{K}_{i2} & \mathcal{K}_{i3} & \cdots \end{bmatrix} \begin{Bmatrix} b_1 \\ b_2 \\ b_3 \\ \vdots \end{Bmatrix} \tag{6.4.17}$$

We can now follow exactly the same procedure for $\frac{\partial \text{Den}}{\partial b_i}$. Since the denominator and numerator are identical in form, the only difference being the use of $[M]$ instead of $[K]$, we can immediately realize that

$$\frac{\partial \text{Den}}{\partial b_i} = 2\begin{bmatrix} \mathcal{M}_{i1} & \mathcal{M}_{i2} & \mathcal{M}_{i3} & \cdots \end{bmatrix} \begin{Bmatrix} b_1 \\ b_2 \\ b_3 \\ \vdots \end{Bmatrix} \tag{6.4.18}$$

So, we can use (6.4.16) and (6.4.18) to write the ith equation (6.4.8) as

$$
\begin{bmatrix} \mathcal{K}_{i1} & \mathcal{K}_{i2} & \mathcal{K}_{i3} & \cdots \end{bmatrix} \begin{Bmatrix} b_1 \\ b_2 \\ b_3 \\ \vdots \end{Bmatrix} - R_Q \begin{bmatrix} \mathcal{M}_{i1} & \mathcal{M}_{i2} & \mathcal{M}_{i3} & \cdots \end{bmatrix} \begin{Bmatrix} b_1 \\ b_2 \\ b_3 \\ \vdots \end{Bmatrix} = 0 \qquad (6.4.19)
$$

This is obviously true for any i, and so we can write out all n equations by filling in all the other rows of $[\mathcal{M}]$ and $[\mathcal{K}]$:

$$
\begin{bmatrix} \mathcal{K}_{11} & \mathcal{K}_{12} & \vdots \\ \mathcal{K}_{12} & \mathcal{K}_{22} & \vdots \\ \cdots & \cdots & \mathcal{K}_{nn} \end{bmatrix} \begin{Bmatrix} b_1 \\ b_2 \\ \vdots \\ b_n \end{Bmatrix} - R_Q \begin{bmatrix} \mathcal{M}_{11} & \mathcal{M}_{12} & \vdots \\ \mathcal{M}_{12} & \mathcal{M}_{22} & \vdots \\ \cdots & \cdots & \mathcal{M}_{nn} \end{bmatrix} \begin{Bmatrix} b_1 \\ b_2 \\ \vdots \\ b_n \end{Bmatrix} = \begin{Bmatrix} 0 \\ 0 \\ \vdots \\ 0 \end{Bmatrix} \qquad (6.4.20)
$$

or, more compactly,

$$
[\mathcal{K}]B - R_Q[\mathcal{M}]B = 0 \qquad (6.4.21)
$$

What does this look like? An eigenvalue problem! (What else?) If we realize that R_Q is an unknown scalar, and the elements of B are also unknown, then we see we've got $n + 1$ unknowns and n equations—just like all of our previous eigenvalue problems. If we further realize that R_Q is our approximation to the square of the system's natural frequencies, then (6.4.21) looks exactly like our old friend $[M]\ddot{X} + [K]X = 0$, since the eigenvalue form of this problem is given by $[K]X_i - \omega_i^2[M]X_i = 0$.

Now, let's think a bit about what (6.4.21) is telling us. If we use only one trial vector, then (6.4.21) becomes a scalar equation. In fact, it becomes exactly equivalent to the Rayleigh's quotient approach. If we add another trial vector, then the matrices in (6.4.21) are 2×2. Three trial vectors makes the problem three-dimensional. And so on. What we've got is an eigenvalue problem that has dimensions equal to the number of trial vectors we're using. Let's assume that we want to use three trial vectors. Then, when we solve the associated eigenvalue problem, we'll find three values for the system's natural frequencies (corresponding to the three roots of the characteristic equation for the problem). For each of these solutions, we'll have an associated eigenvector. Note that this is not a *system* eigenvector. Rather, it tells us what the b_i's should be to best approximate a system eigenvector by means of our trial functions. The following examples will illustrate this point.

Example 6.3

Problem   We'll start by looking at a simple example, one that's very easy to analyze analytically and will therefore allow us to compare the Rayleigh-Ritz approach against known results. The system we'll consider is shown in Figure 6.13 and is simply that well-used example, a serial chain of masses and springs. For this example we'll let all the springs constants be equal to 1 N/m and all the masses be equal to 1 kg.

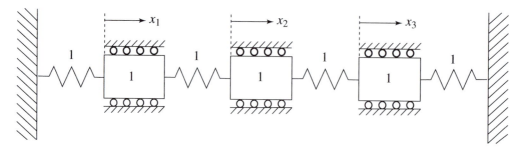

Figure 6.13   3 DOF spring-mass system, spatially symmetric masses, and springs

Solution   We know that the eigenvalue problem is given by

$$\left[[K] - \omega^2[M]\right] X = \{0\}$$

and has solutions $\omega_1^2 = .5858$ (rad/s)^2, $\omega_2^2 = 2.000$ (rad/s)^2, and $\omega_3^2 = 3.4142$ (rad/s)^2, with associated eigenvectors

$$X_1 = \begin{Bmatrix} 1 \\ \sqrt{2} \\ 1 \end{Bmatrix}, \quad X_2 = \begin{Bmatrix} 1 \\ 0 \\ -1 \end{Bmatrix}, \quad X_3 = \begin{Bmatrix} 1 \\ -\sqrt{2} \\ 1 \end{Bmatrix} \qquad (6.4.22)$$

Notice that these eigenvectors are somewhat special owing to the system's configuration. Since the system has identical masses and identical springs, it will produce eigenvectors that are symmetric with respect to the midpoint of the vector or antisymmetric ones. Thus X_2, an antisymmetric eigenvector, is fundamentally different from X_1 and X_3, which are symmetric. We'll see that this will affect our Rayleigh-Ritz results.

To begin, we'll use

$$\hat{X}_1 = \begin{Bmatrix} 1 \\ 2 \\ 1 \end{Bmatrix}$$

as our initial trial vector. If we calculate $\mathcal{M}_{11}$ and $\mathcal{K}_{11}$ as suggested in (6.4.12) and (6.4.13) we'll find that

$$\mathcal{M}_{11} = 6$$

and

$$\mathcal{K}_{11} = 4.$$

The Rayleigh-Ritz equation is then

$$6 - 4R_Q = 0$$

or $R_Q = .667$ (rad/s)^2. Since our estimated trial vector differed quite a bit from the actual eigenvector, we obtained only a reasonably accurate eigenvalue estimate for the actual value of .5858 (rad/s)^2. We'll now move on to an analysis featuring two trial vectors.

Example 6.4

Problem   Use two trial vectors to analyze the problem of Example 6.3.

Solution   This time, we'll say that

$$\hat{X}_1 = \begin{Bmatrix} 1 \\ 2 \\ 1 \end{Bmatrix} \quad \text{and} \quad \hat{X}_2 = \begin{Bmatrix} 1 \\ 0 \\ -1 \end{Bmatrix}$$

This case is special in two ways. For one thing, we've added an antisymmetric vector to our trial set, which formerly consisted of only a symmetric vector. Since the original system is symmetric, it can be broken into two simpler parts, the symmetric parts being independent of the antisymmetric ones, and so it would seem that adding this vector won't do us any good. Since the actual first eigenvector is symmetric, adding an antisymmetric correction by means of the second trial vector shouldn't improve the estimate.

In addition, this second vector happens to be equal to the second eigenvector. Therefore there is no reason for the Rayleigh-Ritz method to add anything to it; it's already as close as it can get to the correct vector. In the same way, adding it to the first vector shouldn't improve the estimate for the first natural frequency because the eigenvectors are independent of each other. And because all the eigenvectors are fundamentally independent of one another, if you're trying to approximate the first eigenvector, adding other eigenvectors to your approximation can't bring you any closer.

Having said all that, let's see what we actually get. Our $[\mathcal{K}]$ and $[\mathcal{M}]$ matrices are

$$[\mathcal{M}] = \begin{bmatrix} 6 & 0 \\ 0 & 2 \end{bmatrix} \quad \text{and} \quad [\mathcal{K}] = \begin{bmatrix} 4 & 0 \\ 0 & 4 \end{bmatrix}$$

and so our eigenvalue problem becomes

$$\begin{bmatrix} 4 & 0 \\ 0 & 4 \end{bmatrix} \begin{Bmatrix} b_1 \\ b_2 \end{Bmatrix} - R_Q \begin{bmatrix} 6 & 0 \\ 0 & 2 \end{bmatrix} \begin{Bmatrix} b_1 \\ b_2 \end{Bmatrix} = \{0\}$$

Notice that there are no off-diagonal terms in this set of equations. As we saw in Chapter 4, this means that the equations are completely uncoupled. Thus what we've got are essentially two individual equations, neither of which influences the other. Solving them gives us

$$R_{Q1} = .\bar{6} \ (\text{rad/s})^2, \quad B_1 = \begin{Bmatrix} 1 \\ 0 \end{Bmatrix} \quad \text{and} \quad R_{Q2} = 2.0 \ (\text{rad/s})^2, \quad B_2 = \begin{Bmatrix} 0 \\ 1 \end{Bmatrix}$$

Note that we are using B_i to denote the vector of b_i's associated with R_{Qi} and that the B_i's components are used to calculate our estimate of the system's eigenvectors. As we expected, the addition of $\hat{X}_2$ didn't alter Est$_1$ in the slightest, as indicated by the fact that the second component is equal to zero. Thus

$$\text{Est}_1 = b_1 \hat{X}_1 + b_2 \hat{X}_2 = 1 \begin{Bmatrix} 1 \\ 2 \\ 1 \end{Bmatrix} + 0 \begin{Bmatrix} 1 \\ 0 \\ -1 \end{Bmatrix} = \begin{Bmatrix} 1 \\ 2 \\ 1 \end{Bmatrix}$$

Similarly, the estimate for the second natural frequency and eigenvector is unaffected by $\hat{X}_1$.

Example 6.5

Problem   Repeat Example 6.4 with different trial vectors.

Solution   Having seen that the choice of trial vectors in Example 6.4 didn't induce any improvement in the eigenvalue estimate, let's see how a different choice fares. For this example we'll use two dissimilar vectors that are neither symmetric nor antisymmetric. Hopefully the Rayleigh-Ritz procedure will use them to create approximations to the actual eigenvalues and eigenvectors.
   The two new trial vectors are

$$\hat{X}_1 = \begin{Bmatrix} 1 \\ 3 \\ 2 \end{Bmatrix} \quad \text{and} \quad \hat{X}_2 = \begin{Bmatrix} 2 \\ 1 \\ -1 \end{Bmatrix}$$

We can use these new values for our trial vectors to calculate $[\mathcal{K}]$ and $[\mathcal{M}]$, obtaining

$$\begin{bmatrix} 10 & 0 \\ 0 & 10 \end{bmatrix} \begin{Bmatrix} b_1 \\ b_2 \end{Bmatrix} - R_Q \begin{bmatrix} 14 & 3 \\ 3 & 6 \end{bmatrix} \begin{Bmatrix} b_1 \\ b_2 \end{Bmatrix} = \{0\} \tag{6.4.23}$$

If we divide $\mathcal{K}_{11}$ by $\mathcal{M}_{11}$ we get .7143 (rad/s)^2 as our single vector estimate for the square of the first natural frequency. You'll note that this is worse than our previous single vector estimate of $.\bar{6}$ (rad/s)^2. This makes sense because this trial vector is even further from the first eigenvector than the original trial vector was. If we calculate the solution to (6.4.23) we'll obtain

$$R_{Q1} = .\bar{6} \text{ (rad/s)}^2, \quad B_1 = \begin{Bmatrix} 3 \\ 1 \end{Bmatrix} \quad \text{and} \quad R_{Q2} = 2.0 \text{ (rad/s)}^2, \quad B_2 = \begin{Bmatrix} -1 \\ 3 \end{Bmatrix}$$

Interesting. We've obtained the exact same estimates as last time for the natural frequencies. However the associated eigenvectors have changed. Let's take a look at what our new estimated eigenvectors are. The first is

$$\text{Est}_1 = b_1 \hat{X}_1 + b_2 \hat{X}_2 = 3 \begin{Bmatrix} 1 \\ 3 \\ 2 \end{Bmatrix} + 1 \begin{Bmatrix} 2 \\ 1 \\ -1 \end{Bmatrix} = \begin{Bmatrix} 5 \\ 10 \\ 5 \end{Bmatrix}$$

This is simply five times our old first trial vector! And since constant multiples make no difference to an eigenvector, we can just as well say it is our old one. The second is given by

$$\text{Est}_2 = -1 \begin{Bmatrix} 1 \\ 3 \\ 2 \end{Bmatrix} + 3 \begin{Bmatrix} 2 \\ 1 \\ -1 \end{Bmatrix} = \begin{Bmatrix} 5 \\ 0 \\ -5 \end{Bmatrix}$$

which is also five times our old second trial vector. So the new set of two trial vectors must actually be just a linear combination of the old set. This is worth remembering. If you've got a set of trial vectors and another set that are just combinations of the first, then your new frequency estimates will just be the same as the old ones. Along with this goes the fact that your estimated eigenvectors will also remain unchanged. The b_i values will change, of course, since the trial vectors have been altered. But when you examine the final estimated eigenvectors that are indicated by the b_i, you'll see that you get the same answers.

We can gain some more insight into this by examining the $[\mathcal{K}]$ and $[\mathcal{M}]$ matrices and recalling our knowledge of normal coordinates. From Chapter 4 we know the following:

- The normal coordinates are those that decouple our equations of motion.
- The decoupled mass and stiffness matrices are diagonal.
- The normal coordinates are always linear combinations of our original coordinates.

Well, we can view our new set of $\hat{X}_i$'s as our original coordinate set. The Rayleigh-Ritz method takes these vectors and combines them so that Rayleigh's quotient is stationary about them. There's only *one* set of vectors that will satisfy this property of stationarity, and we'll know we have this set because the $[\mathcal{K}]$ and $[\mathcal{M}]$ matrices associated with them will be diagonal. Thus the resultant eigenvectors B_i will be unitary (ones on the diagonal with everything else equal to zero), which will mean that we can't improve the trial vectors by combining them; they're optimal as they are.

With this in mind, take a look at the $[\mathcal{K}]$ and $[\mathcal{M}]$ matrices of our first trial set. They are already in diagonal form. Thus any linear combination of them, when run through the Rayleigh-Ritz method, will simply reproduce them.

It's important to note that even though these matrices are diagonal, they aren't *necessarily* associated with the best possible results. The best would be found if the vectors were exactly equal to the actual eigenvectors. In this case the $[\mathcal{K}]$ and $[\mathcal{M}]$ matrices would be diagonal *and* the eigenvalue estimates would be exact (which they aren't yet for our examples). These are the two ways by which we can obtain completely uncoupled $[\mathcal{K}]$ and $[\mathcal{M}]$ matrices—by using the actual eigenvectors or by choosing (or calculating) the combination of any original trial set that renders Rayleigh quotient stationary.

Having seen what happened with our first two trial sets, let's try a third.

Example 6.6

Problem   Repeat Example 6.4 with different trial vectors.

Solution   This time, we'll use

$$\hat{X}_1 = \begin{Bmatrix} 4 \\ 3 \\ 0 \end{Bmatrix} \quad \text{and} \quad \hat{X}_2 = \begin{Bmatrix} 0 \\ 2 \\ 4 \end{Bmatrix}$$

The associated $[\mathcal{K}]$ and $[\mathcal{M}]$ matrices are

$$[\mathcal{K}] = \begin{bmatrix} 26 & -8 \\ -8 & 24 \end{bmatrix} \quad \text{and} \quad [\mathcal{M}] = \begin{bmatrix} 25 & 6 \\ 6 & 20 \end{bmatrix}$$

Just using the first trial vector gives us

$$R_Q = \frac{\mathcal{K}_{11}}{\mathcal{M}_{11}} = 1.04 \ (\text{rad/s})^2$$

This is not terribly close to the actual value of .5858 rad/s^2 (having an error of 78%). Let's see how adding the second trial vector helps. If we go ahead and solve the 2×2 eigenvalue problem, we'll find

$$R_{Q1} = .5961 \text{ (rad/s)}^2, \quad B_1 = \left\{ \begin{array}{c} .2500 \\ .2396 \end{array} \right\}$$

$$R_{Q2} = 2.0246 \text{ (rad/s)}^2, \quad B_2 = \left\{ \begin{array}{c} .2500 \\ -.3054 \end{array} \right\} \tag{6.4.24}$$

These are more typical results. Our original $[\mathcal{K}]$ and $[\mathcal{M}]$ matrices were not decoupled; thus our trial vectors were improved by the Rayleigh-Ritz procedure. Unlike the last time, neither of the frequency estimates was exact, both were simply approximations to the actual eigenvalues. At .5961 (rad/s)^2, our estimate for ω_1^2 is greatly improved from what we had using just our first trial vector. Our original 78% error has been reduced to 2%. To find the associated eigenvector estimate, we need to add .2500 times the first trial vector to .2396 times the second, that is, $B_1(1, 1) * \hat{X}_1 + B_1(2, 1) * \hat{X}_2$. If we do this, we'll obtain

$$\text{Est}_1 = \left\{ \begin{array}{c} 1.000 \\ 1.229 \\ .959 \end{array} \right\}$$

Clearly, this is a lot closer to the actual eigenvector (shown in 6.4.22) than the initial trial vector was. Our first trial vector was $\{4 \quad 3 \quad 0\}^T$, a crummy guess. By appropriately adding a judicious amount of the second trial vector $(\{0 \quad 2 \quad 4\}^T)$, we came up with $\{1.000 \quad 1.229 \quad .959\}^T$, a really good estimate. Similarly, neither initial vector is close to the system's second eigenvector, but by subtracting .3054 times the second trial vector from .2500 times the first, we get $\{1.000 \quad .139 \quad -1.222\}^T$, which is quite close to the actual eigenvector. Furthermore, the error in the estimate of ω_2^2 is only 1%.

We've seen in the preceding example that the Rayleigh-Ritz approach has done an excellent job. We fed in some trial vectors and got out good estimates of the actual system solutions. Of course, the next step is to ask what happens if we add a third trial vector. Well, if we've got three independent, 3×1 vectors to work with, we know that we can create *any* 3×1 vector we might wish to set up; any three independent trial vectors will span the system's configuration space. Therefore, if you give the Rayleigh-Ritz formulation three independent vectors and ask it for the combination that best approximates the actual solution, it will give you the actual solution. That's only fair, since at this point you're doing just as much work as you would have had to do to solve the original problem. Remember, the point of the Rayleigh-Ritz approach is to save time and get good estimates of the first few natural frequencies and eigenfunctions of a continuous system, one that has an infinite number of modes. In this section we restricted the problem to a finite-dimensional one to better understand the method. If the original problem has order n, then to find the natural frequencies and eigenvectors, we need to solve an nth-order eigenvalue problem. By going ahead and using n trial functions, we'll again get an nth-order eigenvalue problem to solve—i.e., no savings. But that's okay because we'd never use the method in this way. The whole point is to use it when you want only a few of the modes, not all of them.

Example 6.7

Problem   Add another trial function to Example 6.6.

Solution   Just to demonstrate that we do indeed get the precise answer back if we expand our trial function set, we'll again do a Rayleigh-Ritz analysis, this time using

$$\hat{X}_1 = \begin{Bmatrix} 4 \\ 3 \\ 0 \end{Bmatrix}, \quad \hat{X}_2 = \begin{Bmatrix} 0 \\ 2 \\ 4 \end{Bmatrix}, \quad \text{and} \quad \hat{X}_3 = \begin{Bmatrix} 1 \\ 1 \\ 1 \end{Bmatrix}$$

For these three trial vectors we'll find that

$$[\mathcal{K}] = \begin{bmatrix} 26 & -8 & 4 \\ -8 & 24 & 4 \\ 4 & 4 & 2 \end{bmatrix} \quad \text{and} \quad [\mathcal{M}] = \begin{bmatrix} 25 & 6 & 7 \\ 6 & 20 & 6 \\ 7 & 6 & 3 \end{bmatrix}$$

The full solutions to the Rayleigh-Ritz problem using these matrices are

$$R_{Q1} = .5858 \ (\text{rad/s})^2$$
$$R_{Q2} = 2.0000 \ (\text{rad/s})^2$$
$$R_{Q3} = 3.4142 \ (\text{rad/s})^2 \tag{6.4.25}$$

with the associated B_i vectors are

$$B_1 = \begin{Bmatrix} .4142 \\ .4142 \\ -.6569 \end{Bmatrix}, \quad B_2 = \begin{Bmatrix} .0 \\ -.5 \\ 1.0 \end{Bmatrix}, \quad \text{and} \quad B_3 = \begin{Bmatrix} -2.4142 \\ -2.4142 \\ 10.6569 \end{Bmatrix}$$

As predicted, these natural frequency estimates are exact. Furthermore, if you construct the first estimated eigenvector you'll see that it's

$$\text{Est}_1 = .4142 \begin{Bmatrix} 4 \\ 3 \\ 0 \end{Bmatrix} + .4142 \begin{Bmatrix} 0 \\ 2 \\ 4 \end{Bmatrix} + -.6569 \begin{Bmatrix} 1 \\ 1 \\ 1 \end{Bmatrix} = \begin{Bmatrix} 1 \\ \sqrt{2} \\ 1 \end{Bmatrix}$$

i.e., the actual first eigenvector. Similarly, Est_2 and Est_3 also give us the exact second and third eigenvectors.

Example 6.8

Problem   Example 6.7 gave us a good insight into how the Rayleigh-Ritz method functions. Now we'll try a more complicated problem, the solutions of which will more closely resemble "real-world" results. The system to be analyzed is shown in Figure 6.14. You'll note that it is again a chain of springs and masses. This time, however, it isn't symmetric. All the springs are the same, but the first mass is half the mass of the others.

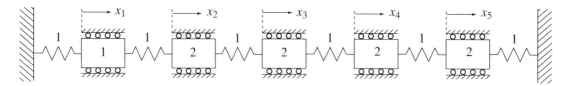

Figure 6.14 5 DOF spring-mass system, spatially nonsymmetric masses, and springs

Solution The $[K]$ and $[M]$ matrices are given by

$$[K] = \begin{bmatrix} 2 & -1 & 0 & 0 & 0 \\ -1 & 2 & -1 & 0 & 0 \\ 0 & -1 & 2 & -1 & 0 \\ 0 & 0 & -1 & 2 & -1 \\ 0 & 0 & 0 & -1 & 2 \end{bmatrix}$$

and

$$[M] = \begin{bmatrix} 1 & 0 & 0 & 0 & 0 \\ 0 & 2 & 0 & 0 & 0 \\ 0 & 0 & 2 & 0 & 0 \\ 0 & 0 & 0 & 2 & 0 \\ 0 & 0 & 0 & 0 & 2 \end{bmatrix}$$

Our trial functions will be

$$\hat{X}_1 = \begin{Bmatrix} 2 \\ 3 \\ 4 \\ 3 \\ 1 \end{Bmatrix}, \quad \hat{X}_2 = \begin{Bmatrix} 2 \\ 2 \\ 0 \\ -1 \\ -1 \end{Bmatrix}, \quad \hat{X}_3 = \begin{Bmatrix} 2 \\ 0 \\ -1 \\ 0 \\ 1 \end{Bmatrix}, \quad \hat{X}_4 = \begin{Bmatrix} 2 \\ -1 \\ 0 \\ 1 \\ -1 \end{Bmatrix},$$

and

$$\hat{X}_5 = \begin{Bmatrix} .5 \\ -.4 \\ 1. \\ -.3 \\ .4 \end{Bmatrix}$$

Based on these trial vectors and the given $[M]$ and $[K]$ matrices, the $[\mathcal{K}]$ and $[\mathcal{M}]$ matrices are found to be

$$[\mathcal{K}] = \begin{bmatrix} 12 & 2 & -1 & 4 & 1.8 \\ 2 & 10 & 4 & 2 & -.9 \\ -1 & 4 & 12 & 7 & 1.2 \\ 4 & 2 & 7 & 20 & 2 \\ 1.8 & -.9 & 1.2 & 2 & 5.36 \end{bmatrix}$$

and

$$[\mathcal{M}] = \begin{bmatrix} 74 & 8 & -2 & 2 & 5.6 \\ 8 & 16 & 2 & 0 & -.8 \\ -2 & 2 & 8 & 2 & -.2 \\ 2 & 0 & 2 & 10 & .4 \\ 5.6 & -.8 & -.2 & .4 & 3.07 \end{bmatrix}$$

If we now go through and solve all the possible Rayleigh-Ritz problems (one trial vector, two trial vectors, etc.) we'll finally obtain the estimates shown in Table 6.1.

Here we see that every time a new vector is added to our trial set, the estimates get better. We start with the Rayleigh's quotient answer of .1622. Upon adding a second vector, this estimate goes to .1612, and we also get an estimate of the square of the second natural frequency (.6423). Note that when a third trial vector is added, the first estimate drops by quite a bit more than it did when we went from one vector to two (.1612 down to .1546, vs. 1622 to .1612). The reason for this is that the second trial vector didn't really do much for us in terms of improving our approximation of the first eigenvector. However the first trial vector plus the third allow us to get a much better approximation, thus the bigger improvement in our estimate. The answers for five trial vectors are exact, since the original system only involves five eigenvectors, just as we saw in Example 6.7.

It's time for a brief recap before moving on to continuous systems. Let's say for the moment that the $[M]$ and $[K]$ matrices for our actual system are 10×10. The easiest attempt at finding the fundamental frequency involves a single equation. This is the Rayleigh's quotient approach. If we decide to use more than one trial vector, we've got to use the Rayleigh-Ritz approach. As we add more and more trial vectors, our estimates for the system natural frequencies and eigenvectors become better and better. Each time we're solving an eigenvalue problem whose size is equal to the number of trial vectors. If we go all the way up to 10 independent trial vectors, we're solving a problem that's every bit as hard as the actual problem we were trying to approximate.

We can also add a few more observations. First, you probably noticed that our Rayleigh-Ritz estimates always went down or remained unchanged as we added trial vectors. This isn't a coincidence. If you think about the problem physically, what the Rayleigh-Ritz method is doing is forcing the system to behave as if it can move only in certain vector directions. For instance, the actual system might involve five masses, as in the preceding example. But by using only three trial vectors, we're constraining the system. We're saying to it, "We don't care how the system

TABLE 6.1 ω_i^2 ESTIMATES FOR 1–5 TRIAL VECTOR $(\text{rad/s})^2$

	Number of Trial Vectors				
	1	2	3	4	5
R_{Q_1}	.1622	.1612	.1546	.1424	.1393
R_{Q_2}		.6423	.5705	.5688	.5558
R_{Q_3}			1.511	1.176	1.158
R_{Q_4}				2.153	1.733
R_{Q_5}					2.414

wants to move, it can move *only* in the ways allowed by these three vectors." Constraints invariably stiffen systems, and stiffer systems have higher natural frequencies. Thus our estimates are
generally greater than the actual values. If, by chance, we happen to pick the actual eigenvector, or if our trial set can be combined to give us an eigenvector, then the estimate will equal the
actual value. But the estimated value won't be lower than the true result. That's one of the nice
things about the Rayleigh-Ritz approach—it always gives results that are greater than or equal to
the actual values. Of course, if we use n independent trial vectors for an n-mass system, we're
still telling the system it has to move in the manner allowed by these vectors. However, any vector (including the eigenvectors) can be formed from the five independent vectors, and thus our
results will be precisely equal to the actual values because our constraints will then have gone
to zero.

Another item of concern is deciding how many trial vectors are "enough." Each time we added
a new trial vector, our estimates improved. However, we don't want to use too many or the method
won't be buying us anything in terms of reduced computation. The first thing to do is make sure
that our trial vectors are as good as we can make them. The convergence rate is certainly going
to be faster if the trial vectors are close to the actual eigenvectors. In fact, if they're equal to the
eigenvectors, the convergence is immediate. But our trial vector set has been established, we still
have the problem of deciding when our results are good enough. Although there's no absolute way
to determine how many are enough, a rough rule of thumb is to use twice as many vectors as the
number of eigenfrequencies we wish to estimate. Thus, if we want good estimates of the first three
eigenfrequencies, we should start with six trial vectors. To be really confident that we've got a good
handle on the actual eigenfrequencies, we would solve the problem again, this time with even more
trial vectors. We would then compare the results for the first three eigenfrequencies. If they didn't
change too much, we could be confident that we were close to the actual answer. Of course, now
you're going to ask how much is too much. Again, the answer will vary depending upon your own
application. Generally, however, if the result hasn't changed by more than 2%, you're safe in assuming
that you're close to the right result.

Keep in mind that this problem of convergence is usually encountered when one is dealing
with continuous problems, in which there are literally an infinite number of eigenfrequencies (at least
mathematically). In this case you know that the eigenfrequency estimates will continue to change as
long as you add new trial functions, and it therefore becomes important to know when to stop.

6.5 RAYLEIGH-RITZ METHOD: CONTINUOUS PROBLEMS

Having read the last section, you're now an expert on how the Rayleigh-Ritz method works for
finite-dimensional systems. Unfortunately, you'll rarely be able to apply this method to actual finite-
dimensional systems because most engineering structures are continuous in nature. Fortunately, the
differences between using the method for continuous systems and finite-dimensional ones are minimal.
As we've already seen, there are several qualitatively different kinds of continuous problem we can
examine: one-dimensional string-type problems, two-dimensional membrane problems, fourth-order

beam problems, etc. Whereas we easily showed what the Rayleigh-Ritz method was like for all finite-dimensional problems (they were all matrix problems), the continuous case as less straightforward. So we'll indicate with a couple of examples how the method works and then indicate how you can apply it to any other type of problem.

The first example we'll look at is the bar vibration problem from Chapter 5. You'll recall from (5.2.6) that the governing equation for free vibrations of a bar is

$$m(x)\ddot{\xi} = \frac{\partial}{\partial x}(EA(x)\xi_x) \tag{6.5.1}$$

We'll write m and EA from now on instead of $m(x)$ and $EA(x)$ to keep the notation clean, but don't forget that we're allowing both these to depend upon x. In Chapter 5 we were able to deal with only the case of constant mass and stiffness distributions, but much of the power of the Rayleigh-Ritz method lies in its capacity to handle nonconstant distributions of mass and stiffness and the constant cases with equal ease. If we assume sinusoidal motion at a frequency ω (i.e., $\xi = \bar{\xi}\cos(\omega t)$), then (6.5.1) becomes

$$\omega^2 m\bar{\xi} = -(EA\bar{\xi}_x)_x \tag{6.5.2}$$

and dividing by $m\bar{\xi}$ gives us

$$\omega^2 = \frac{-(EA\bar{\xi}_x)_x}{m\bar{\xi}} \tag{6.5.3}$$

At this point we'll introduce what will look like a trick. You can consider it one if you wish or simply view it as another mathematical tool in your black bag of mathematical techniques. What it consists of is multiplying the numerator and denominator of (6.5.3) by the dependent variable ($\bar{\xi}$) and then integrating each of them over the length of the bar. Since we're performing the same operation to both the numerator and denominator, the overall ratio is unaffected and we obtain

$$\omega^2 = \frac{-\displaystyle\int_0^l \bar{\xi}\left(EA\bar{\xi}_x\right)_x dx}{\displaystyle\int_0^l m\bar{\xi}^2 dx} \tag{6.5.4}$$

What we've got here is an expression for ω^2. If we use an eigenfunction for $\bar{\xi}$, then we come up with

$$\omega_i^2 = \frac{-\displaystyle\int_0^l \bar{\xi}_i\left(EA(\bar{\xi}_i)_x\right)_x dx}{\displaystyle\int_0^l m\bar{\xi}_i^2 dx} \tag{6.5.5}$$

This is a very close parallel to (6.3.16). We might even go so far as to think that we could use (6.5.4) to define a continuous form of Rayleigh's quotient:

$$R_Q(\bar{\xi}) \equiv \frac{-\int_0^l \bar{\xi}(EA\bar{\xi}_x)_x dx}{\int_0^l m\bar{\xi}^2 dx} \tag{6.5.6}$$

This would give us a scalar that equals ω_i^2 when the input is the ith eigenfunction and gives us some other answer is the input isn't exactly equal to an eigenfunction. And if we did think this, we'd be right. This is a perfectly acceptable form of Rayleigh's quotient for continuous problems. It's also true in this case (as it was for the finite-dimensional case) that the minimum value for Rayleigh's quotient is ω_1^2 and that Rayleigh's quotient is stationary about any eigenfunction. The only difference is that for the finite-dimensional case we knew that the highest obtainable value of Rayleigh's quotient was ω_n^2 if the problem had n natural frequencies. Since the natural frequencies go to infinity for a continuous problem, Rayleigh's quotient is unbounded for these systems.

Just as in the finite-dimensional case, we can use (6.5.6) to get a quick estimate of the first natural frequency of our system by using a trial function that we feel is close to the actual eigenfunction. And just as before, if the guess is pretty good, the estimate will be extremely good. In fact, the only other point in which our continuous Rayleigh's quotient seems to differ from that of the finite-dimensional case is in the form of the numerator. If you first compare the denominators of (6.5.6) and (6.3.17), you'll see that they're both quadratic in form and both represent the zero-frequency kinetic energy, $\overline{KE}^0$. This can be clearly shown for our continuous case. For a bar, the kinetic energy of a differential element is simply

$$dKE = \tfrac{1}{2}m\dot{\xi}^2 dx \tag{6.5.7}$$

The total energy is found by integrating over the length of the bar:

$$KE = \tfrac{1}{2}\int_0^l m\dot{\xi}^2 dx \tag{6.5.8}$$

Finally, using $\xi = \bar{\xi}\cos(\omega t)$, leads to

$$KE = \left(\tfrac{1}{2}\omega^2 \int_0^l m\bar{\xi}^2 dx\right)\sin^2(\omega t) \tag{6.5.9}$$

Since the maximum value of $\sin^2(\omega t)$ is 1.0, the maximum value of (6.5.9) is just $\tfrac{1}{2}\omega^2 \int_0^l m\bar{\xi}^2 dx$, and the zero-frequency kinetic energy is found by factoring out the ω^2 term. Again, just as in the finite-dimensional case, we're factoring out the common factor of $\tfrac{1}{2}$, which will appear in both numerator and denominator.

Since the denominator of (6.5.6) seems to be the zero-frequency kinetic energy, it seems reasonable that the numerator should represent the potential energy of the system. And yet it certainly doesn't look like a quadratic form. It is, however, in a form that's just made for integration by parts. Recall that integration by parts is given by

$$\int_0^l u \, dv = uv \Big|_0^l - \int_0^l v \, du \tag{6.5.10}$$

We can identify $(EA\bar{\xi}_x)_x dx$ as dv and $\bar{\xi}$ as u to find

$$-\int_0^l \bar{\xi}(EA\bar{\xi}_x)_x dx = -EA\bar{\xi}\bar{\xi}_x \Big|_0^l + \int_0^l EA\bar{\xi}_x^2 dx \tag{6.5.11}$$

The $EA\bar{\xi}\bar{\xi}_x \Big|_0^l$ terms are boundary terms. Notice that, for either a fixed or free boundary, they're equal to zero. Thus we'll pay no attention to them for the moment. Now look at the last term of the equation. It's quadratic. Not only that, but it's (minus the $\frac{1}{2}$) exactly equal to the potential energy of the bar. In a course on deformable bodies, you'd probably now be asked to find the internal energy of a bar, and it would turn out to be equal to this. So by integrating by parts we see that the numerator really *is* proportional to the system's potential energy, just as in the finite-dimensional case. Therefore we can just as easily write Rayleigh's quotient as

$$R_Q(\bar{\xi}) = \frac{-EA\bar{\xi}\bar{\xi}_x \Big|_0^l + \int_0^l EA\bar{\xi}_x^2 dx}{\int_0^l m\bar{\xi}^2 dx} \tag{6.5.12}$$

What we've got are two acceptable forms of Rayleigh's quotient, (6.5.6) and (6.5.12). All our continuous problems will support qualitatively similar forms, one in which part of the differential equation itself appears in the numerator and one in which the potential energy and boundary conditions appear. To distinguish them, we'll refer to the the first type, illustrated here by (6.5.6), as a *force formulation*, since the numerator contains a force term, in this case $(EA\bar{\xi}_x)_x$. We'll refer to the second type (6.5.12) as an *energy formulation*, since the primary term in the numerator is the system's potential energy.

Because we have two acceptable forms for Rayleigh's quotient, it follows that we'll have two versions of the Rayleigh-Ritz equations. We'll derive one version in detail and simply present the other, since its derivation exactly parallels that of the first.

Just as in the case of finite-dimensional systems, we'll need to pick a set of trial functions and then make Rayleigh's quotient stationary with respect to them. The resulting equations, which will be in exactly the same form as the ones for the discrete case, will tell us how much of each trial function to use to approximate the eigenfunctions and, of course, we'll also get the best approximation to the system natural frequencies. There's actually a bit more to choosing trial functions than you might think. However, now's not the time to explain it. We'll assume for the present that any set of independent functions is okay to use and go into more depth later in the chapter.

If we denote our trial functions by $\psi_i(x)$, then

$$\text{Est}(x) = \sum_{i=1}^{n} b_i \psi_i(x) \tag{6.5.13}$$

Note the similarity between (6.5.13) and (6.4.1). In both cases we're approximating our displacement function (or vector) by a finite summation of trial functions (or vectors). Therefore the only difference between the finite and the continuous formulations at this point (beyond the fact that the dependent variable in the first case is a vector whereas in the second it is a function) is that one involves matrix operations while the other requires integration. Once again, we're using Est to indicate that this is going to be our estimate of the eigenfunction. The (x) is included just to make it doubly clear that we're dealing with a function, not a vector.

Substituting (6.5.13) into (6.5.12) yields

$$R_Q\left(\text{Est}(x)\right) = \frac{-EA\left(\text{Est}(x)\right)\left(\text{Est}(x)\right)_x \Big|_0^l + \int_0^l EA\left(\frac{\partial \text{Est}(x)}{\partial x}\right)^2 dx}{\int_0^l m\left(\text{Est}(x)\right)^2 dx} \tag{6.5.14}$$

or, in terms of our trial functions,

$R_Q(\text{Est}(x))$

$$= \frac{-EA\left(\sum_{i=1}^{n} b_i \psi_i(x)\right)\left(\sum_{j=1}^{n} (b_j \psi_j(x))_x\right)\Big|_0^l + \int_0^l EA\left(\sum_{i=1}^{n} b_i \frac{\partial \psi_i(x)}{\partial x}\right)\left(\sum_{j=1}^{n} b_j \frac{\partial \psi_j(x)}{\partial x}\right)dx}{\int_0^l m\left(\sum_{i=1}^{n} b_i \psi_i(x)\right)\left(\sum_{j=1}^{n} b_j \psi_j(x)\right)dx} \tag{6.5.15}$$

Notice that the terms like $\left(\text{Est}(x)\right)^2$ were expanded out as the product of two summations. This is always necessary when one is dealing with products of summations and subsequent differentiations (as we'll be doing).

Now we can recognize that, as before, Rayleigh's quotient can be written as

$$R_Q = \frac{\text{Num}}{\text{Den}} \tag{6.5.16}$$

where

$$\text{Num} = -EA\left(\sum_{i=1}^{n} b_i \psi_i(x)\right)\left(\sum_{j=1}^{n} (b_j \psi_j(x))_x\right)\Big|_0^l$$

$$+ \int_0^l EA\left(\sum_{i=1}^{n} b_i \frac{\partial \psi_i(x)}{\partial x}\right)\left(\sum_{j=1}^{n} b_j \frac{\partial \psi_j(x)}{\partial x}\right)dx \tag{6.5.17}$$

and

$$\text{Den} = \int_0^l m\left(\sum_{i=1}^n b_i\psi_i(x)\right)\left(\sum_{j=1}^n b_j\psi_j(x)\right)dx \tag{6.5.18}$$

At this point we can start to tidy things up a bit. For instance, the summations can be pulled out of the various terms as follows:

$$\text{Num} = -\sum_{i=1}^n\sum_{j=1}^n b_ib_j EA\psi_i(x)\frac{\partial}{\partial x}\psi_j(x)\bigg|_0^l + \sum_{i=1}^n\sum_{j=1}^n b_ib_j\int_0^l EA\psi_i(x)\psi_j(x)dx \tag{6.5.19}$$

and

$$\text{Den} = \sum_{i=1}^n\sum_{j=1}^n b_ib_j\int_0^l m\psi_i(x)\psi_j(x)dx \tag{6.5.20}$$

Our expression for Num can be tightened up further by combining the two terms, giving us

$$\text{Num} = \sum_{i=1}^n\sum_{j=1}^n b_ib_j\left(-EA\psi_i(x)\frac{\partial\psi_j(x)}{\partial x}\bigg|_0^l + \int_0^l EA\frac{\partial\psi_i(x)}{\partial x}\frac{\partial\psi_j(x)}{\partial x}dx\right) \tag{6.5.21}$$

Notice that everything within the large parentheses is known and, once evaluated, will simply leave us with a constant. You've probably also figured out that this is shaping up into the exact same form we had in the finite-dimensional case. By differentiating with respect to the b_i and grouping terms, we're going to end up with equations that look just like (6.4.14) and (6.4.15). The only difference is the particular components of the $[\mathcal{K}]$ and $[\mathcal{M}]$ matrices. For the bar, after carrying out the appropriate differentiations, we'll obtain

$$\mathcal{K}_{ij} \equiv -\tfrac{1}{2}EA\left(\psi_i(x)\frac{\partial\psi_j(x)}{\partial x} + \psi_j(x)\frac{\partial\psi_i(x)}{\partial x}\right)\bigg|_0^l + \int_0^l EA\frac{\partial\psi_i(x)}{\partial x}\frac{\partial\psi_j(x)}{\partial x}dx \tag{6.5.22}$$

and

$$\mathcal{M}_{ij} \equiv \int_0^l m\psi_i(x)\psi_j(x)dx \tag{6.5.23}$$

In other words, the denominator and numerator of Rayleigh's quotient can be written as

$$\text{Num} = \sum_{i=1}^n\sum_{j=1}^n b_ib_j\mathcal{K}_{ij} \tag{6.5.24}$$

and

$$\text{Den} = \sum_{i=1}^n\sum_{j=1}^n b_ib_j\mathcal{M}_{ij} \tag{6.5.25}$$

Because the form of Rayleigh's quotient is the same at this point for both the finite- and infinite-dimensional cases, there is no need to perform any more derivations; we've already done them. We immediately can see that the final form of the Rayleigh-Ritz equations will be

$$[\mathcal{K}]B - R_Q[\mathcal{M}]B = 0 \tag{6.5.26}$$

just as it was in Section 6.4.

If we wanted to use the force formulation (6.5.6), our definition of $[\mathcal{K}]$ would change to

$$\mathcal{K}_{ij} \equiv \frac{1}{2}\int_0^l \left((\psi_i(x)\frac{\partial}{\partial x}\left(EA\frac{\partial \psi_j}{\partial x}\right) + \psi_j(x)\frac{\partial}{\partial x}\left(EA\frac{\partial \psi_i}{\partial x}\right)\right)dx \tag{6.5.27}$$

Here we've seen that the Rayleigh-Ritz approach for finite-dimensional problems is really just the same as it was for continuous problems. The only difference is that in one case we're adding trial *vectors*, whereas in the other we're adding trial *functions*. Similarly, the $\mathcal{K}_{ij}$ and $\mathcal{M}_{ij}$ values for the finite-dimensional case are found from matrix operations on vectors while for the continuous case they're made up of integral functions. However, the end result for each case is the same—a constant scalar quantity.

Having established all this, some of you might still be a trifle unsatisfied. After all, the continuous Rayleigh's quotient did have one little qualitative difference in the numerator, namely, those boundary condition terms that popped in the energy formulation. The exact analogue of the finite-dimensional case's $X^T[K]X$ term is the continuous case's $\int_0^l EA\xi^2 dx$ term. The $EA\xi\xi_x\big|_0^l$ term would seem to be unique to the continuous case. Well, in a way there really isn't a difference, since the boundary conditions for the finite-dimensional case were implicitly included by whether the last mass was free or spring restrained. However boundary conditions are going to pop up quite a bit in continuous problems, and they'll be more bothersome than in finite-dimensional problems. Thus we will be looking more closely at boundary conditions and how they affect our solution procedures. But first, an example of the Rayleigh-Ritz method for a continuous problem.

Example 6.9

> **Problem**   Apply the Rayleigh-Ritz method to a tensioned string with a spring at one end, shown in Figure 6.15.
>
> **Solution**   Before we even start, we can get a feel for what the natural frequencies and eigenfunctions will be by considering the simpler examples shown in Figure 6.16. Figure 6.16*a* shows is a fixed-fixed

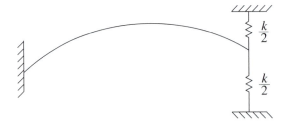

Figure 6.15   Fixed-spring restrained tensioned string

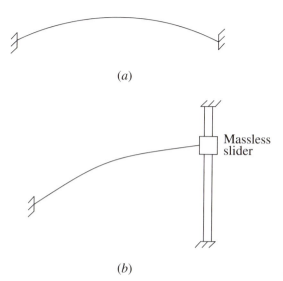

(a)

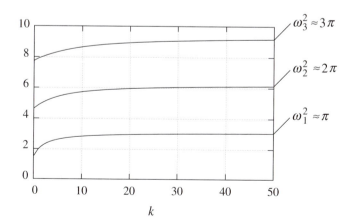

(b)

Figure 6.16   (a) Fixed-fixed string and (b) fixed-free string

string and Figure 6.16b a fixed-free string. The free condition simply means that the string is free to move vertically but not horizontally. These represent limiting cases for the more general problem of a string that is both fixed and spring restrained. If the spring constant goes to infinity, the the boundary condition becomes equivalent to a fixed one (Figure 6.16a), and if the stiffness goes to zero, we approach the fixed-free configuration (Figure 6.16b). Since we know that adding stiffness to a system will increase (or leave unchanged) the natural frequencies of the system, we know that our natural frequencies for a finite spring constant will lie between the natural frequencies associated with no spring and those corresponding to the infinite spring constant case. Figure 6.17 shows these limiting values of the natural frequencies and indicates the range within which the actual frequencies must lie for the finite stiffness case. The values at $k = 0$ correspond to the natural frequencies of a fixed-free system, and the limiting values shown at the rightmost end of the plot ($k = 50$) closely correspond to the fixed-fixed solutions.

$$\omega_3^2 \approx 3\pi$$

$$\omega_2^2 \approx 2\pi$$

$$\omega_1^2 \approx \pi$$

k

Figure 6.17   Variation of natural frequencies as a function of spring constant

We've already seen how to obtain these results in Chapter 5. The equation of motion for free vibrations of a string (Appendix A) is

$$\rho \ddot{y} = (T y_x)_x$$

For sinusoidal motion at a frequency ω, we therefore have

$$\rho \omega^2 \bar{y} + (T \bar{y}_x)_x = 0 \qquad (6.5.28)$$

If the mass density ρ and the tension T are constant, we can solve this problem in the same way we solved the bar problem of Chapter 5. We can rewrite (6.5.28) as

$$\bar{y}_{xx} + \beta^2 \bar{y} = 0 \qquad (6.5.29)$$

where $\beta^2 = \frac{\rho \omega^2}{T}$.

Equation (6.5.29) has solutions

$$\bar{y} = a_1 \sin(\beta x) + a_2 \cos(\beta x)$$

Applying the fixed boundary condition at $x = 0$ shows us that a_2 must equal zero, leaving us with

$$\bar{y} = a_1 \sin(\beta x)$$

If we're concerned with the fixed-free case, then $y_x = 0$ at $x = 1$, giving us

$$\frac{\partial \bar{y}(1)}{\partial x} = a_1 \beta \cos(\beta) = 0$$

which means that

$$\beta_n = \frac{(2n - 1)\pi}{2}$$

On the other hand, if we look at the fixed-fixed case, then $y = 0$ at $x = 1$, and so

$$\bar{y}(1) = a_1 \sin(\beta) = 0$$

In this case we see that

$$\beta_n = n\pi$$

These results are the values indicated in Figure 6.17. Since the finite spring stiffness case is only a bit more difficult than our limiting-case examples, we can go ahead and solve the problem in closed form, giving us an exact answer for comparison to our approximate results. Note that for this example we're using $T = 1$ N, $k = 8$ N/m, and $\rho = 1$ kg/m, not because these values have a particular physical meaning, but because they simplify the example. In addition, we'll take l to be 1 m.

An easy way to approach the problem of formulating the correct boundary conditions is to add a small mass to the end point you're concerned with. You can then apply Newton's law to this mass. For our problem, the appropriate force balance will be:

$$m\ddot{y}(1) = -ky(1) - T\frac{\partial y(1)}{\partial x}$$

The $ky(1)$ term is the restoring force due to the spring and the $-T\frac{\partial y(1)}{\partial x}$ term is the force due to the tension in the string. Since the angle of the string is approximated by $\frac{\partial y(1)}{\partial x}$, and when this is negative the tension is oriented upward, we need to put in a minus sign so that negative slopes of the string generate upward forces on the mass.

Once we have a force balance, all we need do is set m to zero (since in actuality there really isn't any mass at the end) to obtain

$$ky(1) = -T \frac{\partial y(1)}{\partial x}$$

This represents a combination of the two boundary condition types we just examined, both displacement and slope. Since $\bar{y} = a_1 \sin(\beta x)$, evaluating this at $x = 1$ gives us

$$ka_1 \sin(\beta) = -Ta_1 \beta \cos(\beta)$$

which can be rearranged to yield

$$\frac{-k}{T} \tan(\beta) = \beta$$

As we saw in Chapter 5, an easy way to visualize the answer to this kind of problem is to plot both the left- and right-hand sides versus β and see where they intersect. Such a plot is shown in Figure 6.18 and the solutions for the first six β_i are $\beta_1 = 2.8044$, $\beta_2 = 5.667$, $\beta_3 = 8.6031$, $\beta_4 = 11.5993$, $\beta_5 = 14.6374$, and $\beta_6 = 17.7032$.

For this problem, we'll choose the trial functions

$$\psi_n(x) = \sin\left(\frac{n\pi x}{2}\right)$$

As we recently saw, these include both the eigenfunctions for a fixed-free string, as well as those of a fixed-fixed string. Since the actual boundary condition lies somewhere between these, it seems reasonable that both sets will allow a convergence to the correct solution. Note that the actual solution will have finite displacement and finite slope at the spring end of the string. Thus none of our trial functions matches the boundary condition. The fixed-free solutions have finite displacement but zero slope, and the fixed-fixed ones exhibit finite slope with zero displacement at the spring end. Combining

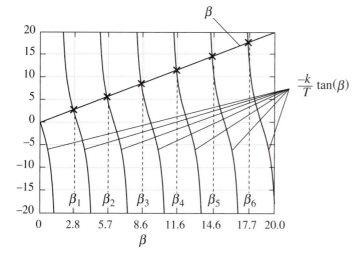

Figure 6.18 β values for a fixed-spring restrained string

sets of trial functions in this way can often improve the performance of the Rayleigh-Ritz method, and we'll examine this more closely in a few pages. For now, let's just see what sort of results we get when using these trial functions in the Rayleigh-Ritz equations.

Since the functions don't match the boundary conditions, we must use the energy formulation. We can obtain the correct from (6.5.22)–(6.5.26) simply by replacing the parameters of the bar with the appropriate string parameters. Doing so will leave us with

$$\mathcal{K}_{ij} = -\frac{1}{2}T\left(\psi_i(x)\frac{\partial\psi_j(x)}{\partial x} + \psi_j(x)\frac{\partial\psi_i(x)}{\partial x}\right)\Big|_0^l + \int_0^l T\frac{\partial\psi_i(x)}{\partial x}\frac{\partial\psi_j(x)}{\partial x}dx \qquad (6.5.30)$$

and

$$\mathcal{M}_{ij} = \int_0^l \rho\psi_i(x)\psi_j(x)dx \qquad (6.5.31)$$

Since our trial functions don't satisfy the boundary conditions, we have to get the boundary information into the problem. Luckily, the energy formulation lets us do that quite efficiently. The string is pinned at the left and so the boundary terms go to zero. We already know that the right boundary condition is

$$ky(1) = -T\frac{\partial y(1)}{\partial x}$$

We simply have to apply the same condition for each of our trial functions:

$$k\psi_i(1) = -T\frac{\partial\psi_i(1)}{\partial x}$$

Now we'll look at (6.5.30), and where we see terms like $T\frac{\partial\psi_i(1)}{\partial x}$ we simply replace them with $-k\psi_i(1)$. Doing so will change $[\mathcal{K}]$ to

$$\mathcal{K}_{ij} = k\psi_i(1)\psi_j(1) + \int_0^l T\frac{\partial\psi_i(x)}{\partial x}\frac{\partial\psi_j(x)}{\partial x}dx$$

Note that the boundary contribution appears as an energy. If we multiplied everything by .5, we'd see that the first term of $[\mathcal{K}]$ looks like a potential energy of the discrete spring while the second term looks like the potential energy of the string itself. Similarly, $[\mathcal{M}]$ will be the kinetic energy of the string (with ω^2 removed).

Since the trial functions are just sines and the tension and mass distribution are constant, we can evaluate $[\mathcal{K}]$ and $[\mathcal{M}]$ by hand. In more complex problems you'd obviously want to use the computer. Just so you can check yourself if you want, the upper 2×2 blocks of our matrices are given, respectively, as

$$[\mathcal{K}] = \begin{bmatrix} 9.2337 & 1.0472 \\ 1.0472 & 4.9348 \end{bmatrix}$$

and

$$[\mathcal{M}] = \begin{bmatrix} .50000 & .42441 \\ .42441 & .50000 \end{bmatrix}$$

Using these matrices will give you the β estimates for the first two modes. If we record these results, as well as those obtained for 1, 3, 4, 5, and 6 included trial functions, we'll obtain Table 6.2.

Table 6.2 β ESTIMATES FOR 1–6 TRIAL FUNCTIONS

	Number of trial functions					
Exact	6	5	4	3	2	1
2.80443	2.80443	2.80443	2.80443	2.80448	2.80675	4.29737
5.66687	5.66687	5.66687	5.66695	5.69515	8.98819	
8.60307	8.60308	8.60393	8.82614	14.9677		
11.5993	11.62067	12.3305	22.6011			
14.6374	16.3272	32.0714				
17.7032	43.4629					

Here we see similar behavior to that seen in the finite-dimensional example. As more trial functions are added, the approximation becomes better and better. Note that when the second trial function is added, the estimate for β_1 becomes much better, going from 4.30 to 2.81. This is quite close to the exact solution of 2.80. The reason for this abrupt improvement is that the first trial function correctly has a finite displacement at the spring end but doesn't have a finite slope. As we've already noted, the exact solution will have both a finite displacement and a finite slope at the end. We're clearly not matching this boundary condition with just a single $\sin(\frac{\pi x}{2})$ trial function. Adding the second trial function doesn't change the displacement (since $\sin(\pi x)$ vanishes at $x = 1$), but it *does* contribute a finite slope. Thus, by combining two functions, we're doing a very good job of matching both the boundary conditions as well as the overall shape of the eigenfunction. For reference, Figures 6.19 and 6.20 show the actual eigenfunction as well as what the one- and two-mode approximations actually look like. Note also that the rule of thumb is holding. Looking at the six-mode case, we see that the first three β estimates are excellent and also that the error increases markedly for higher β estimates.

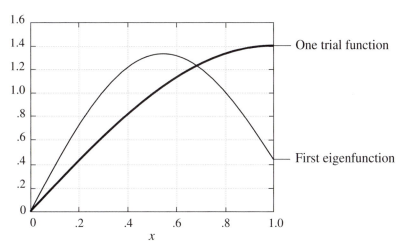

Figure 6.19   One-trial-function approximation to the first mode of a fixed-spring restrained string

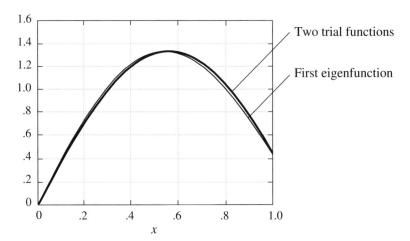

Figure 6.20   Two-trial-function approximation to the first mode of a fixed-spring restrained string

Now that we've seen a simple example of how one can use the Rayleigh-Ritz method, it's appropriate to look at a more complex problem—that of a beam undergoing transverse vibrations. At the same time, we'll get more specific about the ways in which we can formulate Rayleigh-Ritz problems, and we'll see how the boundary conditions affect our approach.

We've already seen how uniform beams behave and know that their eigenfunctions can be somewhat complicated, involving sine, cosine, sinh, and cosh terms. One of the advantages of the Rayleigh-Ritz approach is that we needn't concern ourselves with how complex the actual eigenfunctions are; we'll be approximating them by means of much simpler functions.

Example 6.10

> **Problem**   Determine how to apply the Rayleigh-Ritz approach to the problem of a vibrating beam.
>
> **Solution**   You'll recall from Chapter 5 that the equation of motion for transverse beam vibrations is
>
> $$\rho(x)\ddot{y} = -(EI(x)y_{xx})_{xx} + f(x,t)$$
>
> and setting the forcing to zero gives us
>
> $$\rho(x)\ddot{y} = -(EI(x)y_{xx})_{xx}$$
>
> We can now do what we did for the bar problem: assume sinusoidal motion ($y(x,t) = \bar{y}e^{i\omega t}$) and divide by $-\rho(x)\bar{y}$ to obtain
>
> $$\omega^2 = \frac{(EI(x)\bar{y}_{xx})_{xx}}{\rho(x)\bar{y}}$$

As before, we can multiply by the dependent variable and integrate over the domain (the bar's length) to define our Rayleigh's quotient

$$R_Q = \frac{\displaystyle\int_0^l \bar{y}(EI(x)\bar{y}_{xx})_{xx}dx}{\displaystyle\int_0^l \rho(x)\bar{y}^2 dx}$$

This is the first way in which we can express Rayleigh's quotient. In line with our bar example, we'll refer to this as the force formulation.

We can also use integration by parts to bring the boundary conditions out and put the numerator into an energy form, thus giving us our energy formulation. Our first integration by parts gives us

$$R_Q = \frac{\bar{y}\,(EI(x)\bar{y}_{xx})_x\,\Big|_0^l - \displaystyle\int_0^l \bar{y}_x(EI(x)\bar{y}_{xx})_x dx}{\displaystyle\int_0^l \rho(x)\bar{y}^2 dx}$$

and a second brings us to

$$R_Q = \frac{\bar{y}\,(EI(x)\bar{y}_{xx})_x\,\Big|_0^l - EI(x)\bar{y}_x\bar{y}_{xx}\,\Big|_0^l + \displaystyle\int_0^l EI(x)\bar{y}_{xx}^2 dx}{\displaystyle\int_0^l \rho(x)\bar{y}^2 dx}$$

What can we say about the differences between the energy formulation and the force formulation? Well, for one thing, the boundary conditions are explicitly included in the energy formulation. We saw in the Example 6.9 that this allowed us to include the energy of the boundary condition elements and not worry about having our trial functions satisfy all the problem's boundary conditions. Since the boundary conditions are not explicitly present in the force formulation, our trial functions must all match the appropriate boundary conditions. If they don't, we don't have a hope of getting a good answer, since there's no other way to introduce the boundary conditions into the problem. And it's not enough to match some of them; if you're planning to use the force formulation approach, the trial functions have to match all the boundary conditions of the problem. If you think about it, it's clearly easier to come up with boundary conditions that don't meet all the boundary conditions than to look for ones that do. Thus, using the force formulation implies we'll be doing more work in getting our trial functions.

The energy formulation, on the other hand, already has the boundary terms in the expression for Rayleigh's quotient. What this means is that our trial functions needn't satisfy all the boundary condition of the problem. If they don't, the boundary terms in the numerator of the Rayleigh's quotient will include them for us. This is generally good, since it means we have a wider choice of trial functions (there are more trial functions that don't satisfy all the boundary conditions than there are ones that do).

You should also notice that the dependent variable in the force formulation is differentiated four times, while in the energy formulation it's only differentiated twice. This means that we can get by with trial functions in the energy formulation that are only half as differentiable as in the force formulation. Although this isn't overly important from a theoretical standpoint (most of our trial functions will be infinitely differentiable), it is important in a practical sense. We may well be getting our trial functions from a finite element analysis or from experimental data, and therefore we will not have precise information about the functions at all points, and the information we have will contain some amount of noise contamination. Since differentiation amplifies the effects of noise, the less we differentiate, the better off we are. Thus the energy formulation again looks like a good choice.

Break for Boundary Conditions

We've sidestepped the issue of boundary conditions long enough; it's time to look carefully at them and talk about how they affect our solution procedures. We'll start by considering our trial functions. How can we decide if a given set is a good one or not? It helps to first ask what the best set would be. If we could find a set of functions that satisfied the differential equation and satisfied all the boundary conditions of the problem, then we'd have an excellent set of trial functions. In fact, we'd have the eigenfunctions themselves, since they're the only functions that do both these things. The next best set of functions wouldn't satisfy the differential equation but would satisfy *all* the boundary conditions of the problem. Except for the actual eigenfunctions, these are the best kind of function to have—obvious so,we can solve the Rayleigh-Ritz problem with these trial functions. After all, we know that the Rayleigh-Ritz method works to add together our trial functions to best approximate an eigenfunction. This is exactly analogous to a Fourier series analysis, in which we add together our Fourier series functions to approximate some other function. The only difficulty that can turn up in a Fourier series analysis is failure of the functions we've chosen to match the end conditions of the function we're trying to approximate. For example, say we're using sines to approximate one period of a cosine. Unfortunately, all the sines go to zero at the end points (the boundary conditions), whereas the cosine goes to 1. Therefore it's impossible to exactly match the cosine with a finite number of terms. As the number of included sine functions is increased, we'll come closer to approximating our function everywhere but at the end points. An unfortunate consequence of using sines, one that is not mitigated by adding more terms, is that the end points are always going to have zero deflections. Zero deflection at the ends is physically impossible for some situations (such as a free-free beam) and certainly suggests that our Rayleigh-Ritz approach will probably encounter some difficulties if we insist on using trial functions that violate physicality in such a way.

Clearly, if our trial functions had consisted of cosines, rather than sines, a single trial function would have been sufficient to exactly match the trial function. By choosing a set of trial functions that didn't match the proper boundary conditions (the sines), we doomed ourselves to needing a tremendous number of trial functions, hardly the thing for an efficient approximation technique. So if you can choose functions that match the system's boundary conditions, it behooves you to do so.

Functions that match all the boundary conditions are special enough to merit their own name, *comparison functions*. The easiest way to remember this is to say to yourself: "These are the ones we *compare* all the others against because they're the best." Comparison functions are the ones we have to use if we're going to be applying a force formulation form of the Rayleigh-Ritz equations, since there's no way to add any corrections for boundary condition mismatches. Clearly, the functions we choose must be $2p$ times differentiable, since they'll be differentiated this much in the numerator. (Recall from Chapter 5 that $2p$ refers to the number of differentiations occuring in the stiffness part of the differential equation.) If the trial functions weren't $2p$ times differentiable, then the numerator would be zero, hardly conducive to a good approximation.

If we now consider the energy formulation approach, we see that the bar problem has only one differentiation, and the beam problem two. This means that the trial functions need be only half as differentiable as in the force formulation approach. The trial functions also need satisfy only the geometric boundary conditions of the problem. We saw in Chapter 5 that we have two general kinds of boundary condition: geometric and natural. The geometric boundary conditions involved displacements and/or slopes at the end points, while the natural boundary conditions involved forces and/or moments. The functions that we can use in the energy formulation are called *admissible* functions. You can remember this by saying "We'll *admit* these function into our bag of trial functions." They need be only half as differentiable as comparison functions and they need satisfy only the geometric boundary conditions.

Now that we've discussed the different types of boundary condition and the differences between admissible and comparison functions, we'll present an in-depth look at a particular problem and see how the different types of trial function affect our results.

Example 6.11

Problem   Apply the Rayleigh-Ritz approach to a pinned-spring restrained beam (Figure 6.21).

Solution   We'll take the length of the beam to be 1.0 m and let the mass and stiffness distributions be constant. This example is an excellent one for us because we'll be able to see precisely what happens as we alter the degree to which we're matching the boundary conditions. Our trial functions will be chosen so that they always match the left boundary conditions (zero displacement and moment), but they'll differ in how well they match the conditions at the right end. To start, we'll let EI equal 1 N·m^2 and ρ equal 1 kg/m. We'll set k (the spring stiffness) at 100 N/m.

Recall from Chapter 5 that the general boundary condition at the right end is given by

$$u_{xx} = 0 \quad \text{at } l = 1 \tag{6.5.32}$$

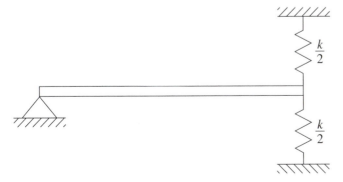

Figure 6.21   Pinned-spring restrained beam

and

$$EIu_{xxx} - ku = 0 \quad \text{at } l = 1 \tag{6.5.33}$$

We'll use

$$\psi_i(x) = \sin\left(\frac{i\pi x}{2}\right) \tag{6.5.34}$$

as our first set of trial functions, and we'll be using an energy formulation for the Rayleigh-Ritz equations. As in Example 6.10, this choice will highlight the power of the Rayleigh-Ritz method to get good answers from what seem like inappropriate choices for the trial functions. If we examine (6.5.32) we notice that u_{xx} is always zero at the right end of the beam while (6.5.33) tells us that there is a specific ratio between u and u_{xxx} at the right end: $\frac{u_{xxx}}{u} = \frac{k}{EI}$.

Do our trial functions all satisfy these boundary conditions? Certainly not! $\psi_1(1) = 1$ at the right end (which is good; the right end displacements will be nonzero) but $\frac{\partial^2 \psi_1(1)}{\partial x^2}$ is also going to be nonzero, whereas the boundary conditions call for it to be zero. In addition, $\frac{\partial^3 \psi_1(1)}{\partial x^3} = 0$, making it impossible to satisfy (6.5.33) when the actual displacement is nonzero. We've got similar problems with ψ_2. For example, $\psi_2(1)$ is zero, which doesn't match reality; the spring will definitely deflect. The first and third derivatives are nonzero, a good match to the physical behavior.

We see that we're not matching any of the natural boundary conditions exactly with any one trial function. However, by combining the trial functions, we should be able to get a good approximation. We're further helped out by the fact that we're using an energy formulation. Thus the spring energy will be accounted for, relieving us of the constraint of exactly matching $\frac{\partial^3 u}{\partial x}$ at $x = 1$. Since our trial functions do not all match the boundary conditions, they are admissible functions, not comparison functions.

We've already seen that the terms in the $[\mathcal{K}]$ and $[\mathcal{M}]$ matrices are simply the energies of the system, expressed in terms of the trial functions. Thus, since the energies of the beam are given by

$$KE = \tfrac{1}{2} \int_0^1 \rho \dot{u}^2 dx$$

and

$$PE = \tfrac{1}{2} \int_0^1 EIu_{xx}^2 dx$$

we can immediately identify our $[\mathcal{K}]$ and $[\mathcal{M}]$ elements as

$$\mathcal{K}_{ij} = \int_0^1 EI \frac{\partial^2 \psi_i(x)}{\partial x^2} \frac{\partial^2 \psi_j(x)}{\partial x^2} dx + k\psi_i(1)\psi_j(1) \tag{6.5.35}$$

and

$$\mathcal{M}_{ij} = \int_0^1 \rho \psi_i(x)\psi_j(x)dx \tag{6.5.36}$$

Just as we saw in Section 6.4, the discrete energy of the spring is added to the overall energy of the beam in the $[\mathcal{K}]$ entries.

Of course, it would be convenient if we could compare our approximate results with the exact answer. And luckily, since the problem is a relatively simple one, we can find the exact results ahead of time. From Section 5.3, we recall that our general solution for the eigenfunction of a beam is

$$b_1 \cos(\beta x) + b_2 \sin(\beta x) + b_3 \cosh(\beta x) + b_4 \sinh(\beta x).$$

Applying the boundary conditions for the left end of the beam gives us

$$b_1 + b_3 = 0$$

and

$$\beta^2(-b_1 + b_3) = 0$$

Together, these tell us that $b_1 = b_3 = 0$. Using this knowledge and applying the boundary conditions for the right end, we can write

$$\beta^2 (-b_2 \sin(\beta) + b_4 \cos(\beta)) = 0$$

and

$$EI\beta^3 (-b_2 \cos(\beta) + b_4 \cosh(\beta) - k(b_2 \sin(\beta) + b_4 \sinh(\beta))) = 0$$

As we saw in Chapter 5, this is simply an eigenvalue problem, which we can write out as

$$\begin{bmatrix} -\sin(\beta) & \sinh(\beta) \\ -EI\beta^3 \cos(\beta) - k \sin(\beta) & EI\beta^3 \cosh(\beta) - k \sinh(\beta) \end{bmatrix} \begin{Bmatrix} b_2 \\ b_4 \end{Bmatrix} = \begin{Bmatrix} 0 \\ 0 \end{Bmatrix} \tag{6.5.37}$$

The solutions of (6.5.37) give us the β_i for our beam vibration problem, and the associated values of b_2 and b_4 let us construct the eigenfunctions. The characteristic equation of (6.5.37) is

$$2k \tan(\beta) \tanh(\beta) + EI\beta^3 (\tanh(\beta) - \tan(\beta)) = 0$$

and the first five values of β are 2.9886, 5.1482, 7.3872, 10.310, and 13.395.

   Now that we know what our answers should be (if the approximation works), we can move ahead and see how well we do. Using the trial functions given by (6.5.34) and the $[\mathcal{K}]$ and $[\mathcal{M}]$ formulations given by (6.5.35) and (6.5.36), we find that our $[\mathcal{M}]$ and $[\mathcal{K}]$ matrices for a six-function approximation:

$$[\mathcal{M}] = \begin{bmatrix} .5 & .42441 & 0.0 & -.16977 & 0.0 & .10913 \\ .42441 & .5 & .25465 & 0.0 & -.06063 & 0.0 \\ 0.0 & .25465 & .5 & .36378 & 0.0 & -.14147 \\ -.16977 & 0.0 & .36378 & .5 & .28294 & 0.0 \\ 0.0 & -.06063 & 0.0 & .28294 & .5 & .34725 \\ .10913 & 0.0 & -.14147 & 0.0 & .34725 & .5 \end{bmatrix}$$

and

$$[\mathcal{K}] = \begin{bmatrix} 103.04 & 10.335 & -100.00 & -16.537 & 100.00 & 23.919 \\ 10.335 & 48.705 & 55.811 & 0.0 & -36.912 & 0.0 \\ -100.00 & 55.811 & 346.57 & 318.92 & -100.00 & -279.06 \\ -16.537 & 0.0 & 318.92 & 779.27 & 689.03 & 0.0 \\ 100.00 & -36.912 & -100.00 & 689.03 & 2002.5 & 1,902.7 \\ 23.919 & 0.0 & -279.06 & 0.0 & 1,902.7 & 3945.1 \end{bmatrix}$$

Solving the Rayleigh-Ritz equations associated with these matrices gives us β_i values of 2.9886, 5.1482, 7.3874, 10.310, 14.345, and 31.125. A comparison of these results with the actual β values shows that the first four are right on the money. Only with the fifth approximation do we see some significant error start to creep in.

Having seen how a "good" set of trial functions did in the foregoing example, we'll now look at the consequences of reducing the size of the trial set.

Example 6.12

Problem   Revisit Example 6.9 but drop all $\sin(n\pi x)$ terms.

Solution   Dropping these terms means that our reduced set will only include the $\sin\left(\frac{(2n-1)\pi x}{2}\right)$ terms as trial functions. If we restrict our trial function set in this way, then the β_i estimates will be 3.3772, 5.2136, 7.9638, 11.034, 14.155, and 17.289. Notice that the errors of the lower β_i estimates have gone up. This is because we've lost the ability to match the slope boundary conditions, thus hurting our approximation ability. Figure 6.22 compare the actual eigenfunction for the first mode and the approximations generated by the full and reduced sets of trial functions. Figure 6.22a compares the reduced set's approximation to the actual eigenfunction. Note that the approximation is quite good at the left end but becomes poorer as it approaches the right end. At $x = 1$, the slope of the trial function approximation is zero (unavoidable considering the type of trial functions used), while the actual solution has a finite, negative slope. Because of the mismatch, we generated a poorer natural frequency estimate. Now look at Figure 6.22b. The match is now excellent for all points along the beam. As a consequence, our natural frequency estimate is also excellent.

The same type of result can be seen in Figure 6.23, in which the two approximations to the second eigenfunction are compared with the exact result.

This illustrates the sort of problem that can occur with a boundary condition mismatch. If you've got a complete set of functions (such as those in Example 6.10), then convergence is quick and painless. Failure to match the physical constraints of the system, such as occurred in Example 6.11, will severely compromise the performance of the method. Thus it pays to think carefully about boundary conditions.

6.6 ASSUMED MODES METHOD

The Rayleigh-Ritz method has shown itself to be a powerful tool in calculating the modal responses of complicated systems. However another approach is perhaps even more widespread—the *assumed modes method*. This approach has the advantage of being readily applicable to both linear and nonlinear systems. Although the approach seems different from that of the Rayleigh-Ritz method, it turns out that the final equations are of the same form if the problem is linear and the energy formulation was used in obtaining the equations of motion. The name is actually quite descriptive. What we do is "assume" that a given set of trial functions will work for us. Then we use Lagrange's equations to find the equations of motion. To see how it works, we'll consider the same example we looked at in Section 6.5. We'll call our assumed modes $\psi_i(x)$. We'll assume that each trial function ψ_i is multiplied by a coordinate that depends on time and denote it by $a_i(t)$. This is analogous to what we know will result from a modal analysis, namely, a summation of the system's eigenfunctions, each

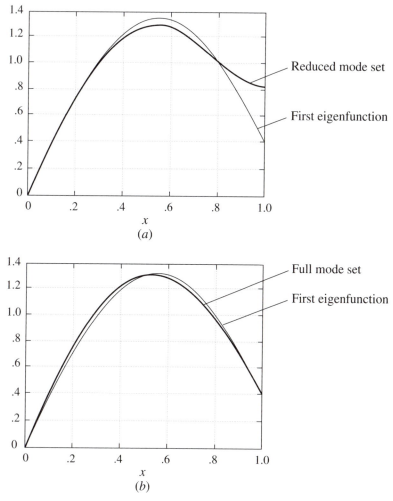

Figure 6.22   Comparison of reduced and full mode set approximation for a fixed-spring restrained beam, first eigenfunction

multiplied by a function of time. In the free vibration case this function of time is just sinusoidal motion at the system's natural frequency, and for the forced case it's a more involved function. Thus we're saying

$$u(x, t) = \sum_{i=1}^{n} a_i(t)\psi_i(x) \tag{6.6.1}$$

To illustrate the method, we'll examine the problem that we just looked at—a beam that is both pinned and spring restrained.

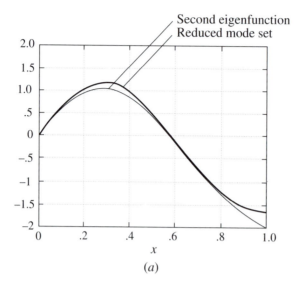

(a)

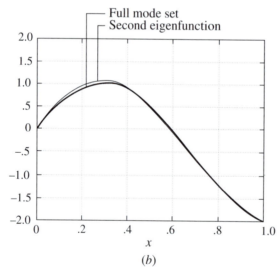

(b)

Figure 6.23 Comparison of reduced and full mode set approximation for a pinned-spring restrained beam, second eigenfunction

The potential energy of the beam is given by

$$PE = \tfrac{1}{2} \int_0^l EI(x)u_{xx}^2 \tag{6.6.2}$$

and the kinetic energy is given by

$$KE = \tfrac{1}{2} \int_0^l \rho(x)\dot{u}^2 \tag{6.6.3}$$

The potential energy of the spring is given by $\frac{1}{2}ku(1)^2$ Using (6.6.1)–(6.6.3) allows us to form the Lagrangian $(KE - PE)$,

$$L = \frac{1}{2}\int_0^l \rho(x)\left(\sum_{i=1}^n \dot{a}_i(t)\psi_i(x)\right)^2 dx - \frac{1}{2}\int_0^l EI(x)\left(\sum_{i=1}^n a_i(t)(\psi_i(x))_{xx}\right)^2 dx$$

$$- \frac{1}{2}k\left(\sum_{i=1}^n \dot{a}_i(t)\psi_i(1)\right)^2 \tag{6.6.4}$$

Now we must apply Lagrange's equations, which means we have to differentiate with respect to the system's generalized coordinates (in this case the a_i's and $\dot{a}_i$'s). The only tricky part is the squared summation. This means that we have to be careful to catch all the cross-terms during the differentiation. For example, consider the simple summation

$$y = \left(\sum_{i=1}^2 a_i\psi_i\right)^2 \tag{6.6.5}$$

Expanding this out we get

$$y = a_1^2\psi_1^2 + 2a_1a_2\psi_1\psi_2 + a_2^2\psi_2^2 \tag{6.6.6}$$

Differentiating with respect to a_1 gives us

$$\frac{dy}{da_1} = 2a_1\psi_1^2 + 2a_2\psi_1\psi_2 \tag{6.6.7}$$

You'll note the cross-term $2a_2\psi_1\psi_2$. This is the sort of term that must be correctly accounted for.

Applying Lagrange's equations to (6.6.4) (with respect to the ith generalized coordinate) results in

$$\int_0^l \rho(x)\left(\ddot{a}_1\psi_1(x)\psi_i(x) + \ddot{a}_2\psi_2(x)\psi_i(x) + \cdots\right)dx$$

$$+ \int_0^l EI\left(a_1\frac{\partial^2\psi_i(x)}{\partial x^2}\frac{\partial^2\psi_1(x)}{\partial x^2} + \frac{\partial^2\psi_i(x)}{\partial x^2}\frac{\partial^2\psi_2(x)}{\partial x^2} + \cdots\right)dx$$

$$+ k\left(a_1\psi_1(1)\psi_i(1) + a_2\psi_2(1)\psi_i(1) + \cdots\right) = 0 \tag{6.6.8}$$

Just as in the Rayleigh-Ritz development, this can be seen as the ith row of the matrix expression

$$[\mathcal{M}]\{\ddot{a}\} + [\mathcal{K}]\{a\} = 0 \tag{6.6.9}$$

where

$$\mathcal{K}_{ij} = \int_0^1 EI\frac{\partial^2\psi_i(x)}{\partial x^2}\frac{\partial^2\psi_j(x)}{\partial x^2}dx + k\psi_i(1)\psi_j(1) \tag{6.6.10}$$

and

$$M_{ij} = \int_0^1 \rho \psi_i(x) \psi_j(x) dx \tag{6.6.11}$$

Note that these matrices are identical to (6.5.35) and (6.5.36). These equations aren't in precisely the same form as the Rayleigh-Ritz equations because these are equations of motion, while the Rayleigh-Ritz equations are already in eigenvalue form. However if we were looking for oscillatory solutions, we'd assume that $\{a\} = \{\bar{A}\}e^{i\omega t}$, leading to

$$[\mathcal{K}]\bar{A} - \omega^2[\mathcal{M}]\bar{A} = 0 \tag{6.6.12}$$

This is exactly the form of the Rayleigh-Ritz equations. Thus either viewpoint, the Lagrangian approach of assumed modes or the stationarity approach of Rayleigh-Ritz, ultimately leads to the same system of equations.

6.7 HOMEWORK PROBLEMS

Section 6.2

6.1. In this problem we'll come up with a lumped discretization model for a uniform bar and compare its performance against the exact analytical results. Start by approximating the bar as a series of lumped spring and mass elements, as in Figure P6.1. A first cut might be to make all the springs and masses equal. The difficulty with this approach is that it causes a modeling problem. Putting two of our models together would produce a bar with a spring nonuniformity in the center, since the two end springs would result in an equivalent spring of only half the stiffness of the other springs. Thus you might want to have uniform interior springs and a spring with twice the interior stiffness at the ends. Compare the results generated from this model with those found from an exact analysis. Consider both free-free and fixed-fixed conditions. Use an eight-mass approximation.

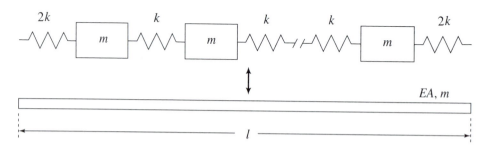

Figure P6.1

6.2. Repeat Problem 6.1 but this time use the approximation shown in Figure P6.2. In this case the interior stiffness will be uniform but the end masses will differ from the interior ones.

6.3. In this problem we'll consider the effect of adding a lumped mass to the end of a bar, using the approximation for the bar that was developed in Problem 6.1. Thus we're considering the system shown in

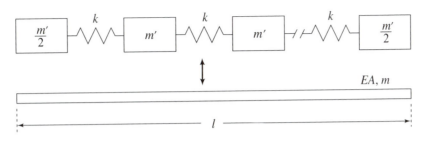

Figure P6.2

Figure P6.3. Let the lumped mass be 50 times greater than the bar mass and discretize the beam into eight mass elements. Compare the accuracies of

Figure P6.3

(a) an SDOF approximation that ignores the mass of the bar and treats it simply as a spring element

(b) a lumped-mass analysis

(c) an exact solution

6.4. For this problem we'll again consider the approximation of Problem 6.1 but include a spring constraint on the right end of the bar, as in Figure P6.4. How well does the approximation method do in solving this problem? Compare your approximate results with the exact solution. $k_5 = \frac{30EA}{l}$.

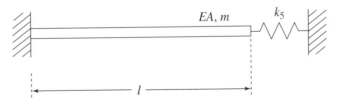

Figure P6.4

Section 6.3

6.5. Consider the system shown in Figure P6.5. (We considered this system in Chapter 4, Problem 4.106). $f_i = 0$. Calculate the exact eigenvectors and natural frequencies. Mass normalize your eigenvectors and determine how much contamination from the second-mode eigenvector is needed before the natural frequency estimate is off by 5%. $m_i = 3$ kg, $k_1 = 2000$ N/m, $k_2 = 3000$ N/m, $k_3 = 4000$ N/m, $k_4 = 1000$ N/m.

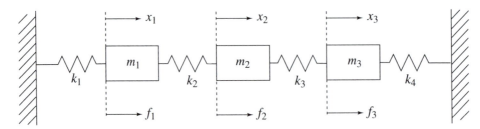

Figure P6.5

6.6. Consider the system of Problem 4.106, Chapter 4. Set the f_i and k_4 to zero. If k_1 were zero the system would have a rigid body mode in which all x_i were equal. How much error is there in assuming a rigid body mode for finite values of k_1? How large is k_1 when the error hits 5%? What does the actual first mode look like at this point? $k_2 = 2000$ N/m, $k_3 = 2000$ N/m, $m_1 = 4$ kg, $m_2 = 4$ kg, $m_3 = 4$ kg.

6.7. For the system in Figure P6.7 we'll ignore rotation of m_3. For m_3 going to zero we'd have a mode for which all x_i were equal. Use this as your assumed solution for a Rayleigh's quotient approximation for the case of finite m_3. How large can m_3 get before the error between the approximation and the exact result exceeds 5%? What does the actual first mode look like? $k_i = 10$ N/m, $m_1 = 10$ kg, $m_2 = 10$ kg.

6.8. The system shown in Figure P6.8 supports a mode for which $x_1 = a$, $x_2 = -a$, and $x_3 = 0$, for uniform mass and stiffness values. Let $k_1 = 400$ N/m, $k_2 = 400$ N/m, $k_3 = 800$ N/m, $m_3 = 15$ kg,

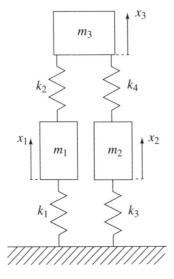

Figure P6.7

$m_1 = 8 + \alpha$ kg, and $m_2 = 8 - \alpha$ kg. How large can α get before the frequency error between the Rayleigh's quotient estimate found from using the mode just described and the actual natural frequency exceeds 2%? What do you notice in your Rayleigh's quotient results as you vary α? Do not allow rotation of m_3.

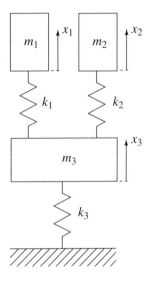

Figure P6.8

6.9. Look again at the system of Figure P6.5. Make an educated guess on the first mode's displacement and compare the results of using your guess in Rayleigh's quotient with the actual first natural frequency. What was your error? $k_1 = 100$ N/m, $k_2 = 100$ N/m, $k_3 = 100$ N/m, $k_4 = 1000$ N/m, $m_1 = 1$ kg, $m_2 = 1$ kg, $m_3 = 1$ kg.

Section 6.4

6.10. Say you are shown three vectors that satisfy orthogonality with respect to the mass and stiffness matrices of a discrete system. Explain how it would be possible for them to do this and yet *not* to be eigenvectors of the system.

6.11. Your assistant has given you the linear graph of Figure P6.11, the results of his Rayleigh-Ritz analysis of a 5 DOF structural problem. The actual system natural frequencies are indicated by the ω_i and his estimates are shown as dots. Are the results plausible? How many errors can you spot? What might have caused them? Four trial functions were used.

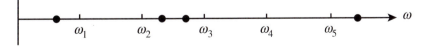

Figure P6.11

6.12. Consider the system shown in Figure P6.12 (k_i and m_i values shown on the figure). Without actually solving the problem, can you tell which of the following is the better set of trial vectors?

$$\text{Set A:} \quad \begin{Bmatrix} 1 \\ 1 \\ 1 \\ 1 \end{Bmatrix}, \quad \begin{Bmatrix} 1 \\ 1.1 \\ 1.1 \\ 1 \end{Bmatrix}, \quad \begin{Bmatrix} 1 \\ .9 \\ .8 \\ 1 \end{Bmatrix}$$

or

$$\text{Set B:} \quad \begin{Bmatrix} 1 \\ 2 \\ 2 \\ 1 \end{Bmatrix}, \quad \begin{Bmatrix} 1 \\ 1 \\ -1 \\ -1 \end{Bmatrix}, \quad \begin{Bmatrix} 1 \\ -.4 \\ -.4 \\ 1 \end{Bmatrix}$$

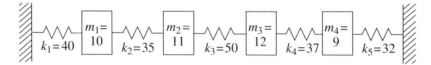

Figure P6.12

6.13. Explain why the Rayleigh-Ritz procedure will break down if you use a trial vector that is a linear combination of previous trial vectors. You can consider a 3 DOF system for simplicity. (*Hint*: Look at the $[\mathcal{M}]$ and $[\mathcal{K}]$ matrices and consider the consequences of the third trial vector being a linear combination of the first two.)

6.14. Are

$$\begin{Bmatrix} 1 \\ 2 \\ 1 \end{Bmatrix}, \quad \begin{Bmatrix} 1 \\ -1 \\ 1 \end{Bmatrix}, \quad \text{and} \quad \begin{Bmatrix} 1 \\ 0 \\ 2 \end{Bmatrix}$$

good trial vectors to use in a Rayleigh-Ritz analysis?

6.15. If two 10×1 vectors are orthogonal with respect to $[M]$ and $[K]$, are they necessarily eigenvectors? What if ten 10×1 vectors are considered?

Sections 6.5 and 6.6

6.16. Use a Rayleigh's quotient approach to estimate the first natural frequency of a fixed-free bar whose stiffness and mass varies linearly along the bar according to

$$m(x) = \overline{m} \left(1 - \frac{x}{l} \right)$$

$$EA(x) = \overline{EA} \left(1 - \frac{x}{l} \right)$$

6.17. Consider the set of trial functions

$$\psi_1 = x(1 - x)$$
$$\psi_2 = x(.5 - x)(1 - x)$$
$$\psi_3 = x(.\bar{3} - x)(.\bar{6} - x)(1 - x)$$
$$\vdots$$

Would these be reasonable trial functions to select if you were interested in using the Rayleigh-Ritz method for a fixed-fixed, tensioned string of length 1 m? Note that the tension and density of the string are not constant.

6.18. Repeat Problem 6.17 for the system shown in Figure P6.18. The tension and density are variable. $l = 1$ m.

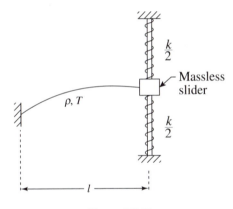

$$\frac{k}{2}$$

Massless slider

ρ, T

$$\frac{k}{2}$$

l

Figure P6.18

6.19. Consider again the system of Figure P6.18. Let the string properties be constant. For $k = 0$, the first four natural frequencies are 1, 2, 3, and 4 rad/s. You then calculate the natural frequencies for $k \neq 0$ and come up with a new set of natural frequencies: $\omega_1 = 2.5$ rad/s, $\omega_2 = 5$ rad/s, $\omega_3 = 7.5$ rad/s, and $\omega_4 = 10$ rad/s. Do you have confidence that the analysis for the $k \neq 0$ case was done correctly?

6.20. In this problem we'll revisit the first-order analysis of Problem 6.16, in which we saw that the first natural frequency of a bar whose mass and stiffness changed linearly along its length was $\frac{2.415}{l}\sqrt{\frac{EA}{m}}$. Rework the problem from a Rayleigh-Ritz approach. Include six trial functions. Use the uniform bar's eigenfunctions as your trial functions. How close was the one-trial-function approximation of Problem 6.16 to the more accurate Rayleigh-Ritz solution?

6.21. Set up the appropriate equations from which one can use a Rayleigh-Ritz approach to solve for the free vibration solution of the system illustrated in Figure P6.21; ρ and T are constant and m is a lumped mass attached to the string.

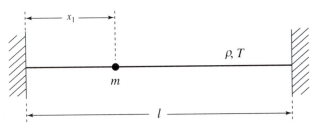

Figure P6.21

6.22. Use the Rayleigh-Ritz method employing a uniform, tensioned string's eigenfunctions as trial functions to calculate the first two natural frequencies of the tensioned string illustrated in Figure P6.22.

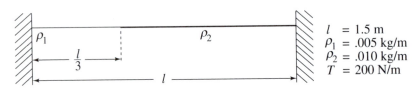

$$l = 1.5 \text{ m}$$
$$\rho_1 = .005 \text{ kg/m}$$
$$\rho_2 = .010 \text{ kg/m}$$
$$T = 200 \text{ N/m}$$

Figure P6.22

6.23. Consider the problem of a tensioned string with a uniform mass density over one part of the string and zero mass density over the rest of the string, as shown in Figure P6.23. Use the eigenfunctions of a uniform, tensioned string as trial functions to calculate the system's natural frequencies. Let $T = 1 \text{ N}$, $\rho_1 = 1 \text{ kg/m}$, and $l_1 = 1.0 \text{ m}$, $l = 1.125 \text{ m}$. You'll note that the results match those of Example 6.9. Why? How are the problems related? Why is this approach a bit more accurate than that in Example 6.9? (You can assess the accuracy by doing a four-mode approximation and comparing the results with those of a six-mode approximation matching the number of functions used in Example 6.9.)

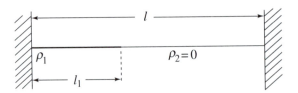

Figure P6.23

6.24. Set up the appropriate equations from which one can use a Rayleigh-Ritz approach to solve the free vibration of the system illustrated in Figure P6.24; ρ and T are constant. Two springs are placed at $x = x_1$ to restrain motions of the string.

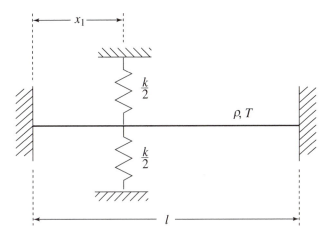

Figure P6.24

6.25. Consider again the system of Problem 6.21. Let the lumped mass be denoted as m_1. How much will the frequencies of the first four modes change if we look at two cases: $m_1 = 1$ kg and $m_1 = 50$ kg? Use $\psi_n(x) = \sin\left(\frac{n\pi x}{l}\right)$ as your trial functions—8 trial functions will be a good number to use. Do the results make sense? $\rho = 1$ kg/m, $T = 1000$ N, $l = 2$ m, $x_1 = .5l$.

6.26. Consider a tensioned string with constant tension but variable mass, $T = 1000$ N, and $l = 1$ m. Use trial functions $\psi_n(x) = \sin(\frac{n\pi x}{l})$ to determine the system's natural frequencies. Plot the approximation to the first eigenfunction. Use three mass distribution:

(a) $\rho(x) = \rho_0$; $\rho_0 = 1$ kg/m

(b) $\rho(x) = \rho_0(1 + .5\sin\frac{\pi x}{l})$; $\rho_0 = 1$ kg/m

(c) $\rho(x) = \rho_0(1 + 5\sin\frac{\pi x}{l})$; $\rho_0 = 1$ kg/m

   You should be able to verify analytically the correctness of the first case, since it is simply the case of uniform stiffness and density.

6.27. Consider again the system of Figure P6.18. Calculate the approximate natural frequencies and mode shapes for the system. Use $\psi_n(x) = \sin(\frac{n\pi x}{2l})$ as trial functions, $n = 1, 2, \ldots, 8$. Plot the first two mode shapes. The first two natural frequencies are 88.68 and 179.2 rad/s. $\rho = 1$ kg/m, $T = 1000$ N, $k = 8000$ N/m, $l = 1$ m.

6.28. In this problem, an extension of Problem 6.27, you'll be analyzing a tensioned string with constant tension and variable mass. The end points of the string are fixed. Use trial functions $\psi_n = \sin\left(\frac{n\pi x}{l}\right)$ in a Rayleigh-Ritz analysis to determine the system's natural frequencies. Plot the approximation to the first eigenfunction. Use the mass distribution

$$\rho = \left[1000\delta \left(x - \frac{l}{2} \right) + 1 \right] \text{kg/m}$$

This is equivalent to a lumped mass of 1000 kg placed at midspan along with a small con-stant density. Try using an eight-trial-function approximation. When you have your results, compare your estimate for ω_1 with the SDOF approximation (in which the string simply acts as an equivalent spring).

6.29. This problem builds upon the results of Problem 6.28. Instead of using $\psi_8(x) = \sin\left(\frac{8\pi x}{l}\right)$, use

$$\psi_8(x) = x; \quad 0 \le x < \frac{l}{2}$$

$$\psi_8(x) = l - x; \quad \frac{l}{2} \le x \le l$$

that is, a linear increase followed by a linear decrease (a *tent* mode). We know intuitively that this is very close to the actual response shape for the first mode when a large mass is placed in the center of the string. Compare the two Rayleigh-Ritz results from this and Problem 6.28, both for the natural frequencies and the model deflections. Go ahead and compare ω_1 and the actual model deflection with those found from an exact analysis as well. Explain why using the tent mode improves the estimate for ω_1 so dramatically. What's the problem with using the Rayleigh-Ritz approach and $\psi_n(x) = \sin(n\pi\frac{x}{l})$ as our only trial functions?

6.30. Repeat Problem 6.28 but with the following much more regular mass distributions:

$$\rho = \left[8 \left(x - \frac{l}{2} \right) + 1 \right] \text{kg/m}$$

Does the Rayleigh-Ritz method do better for this case when $\psi_n = \sin\left(\frac{n\pi x}{l}\right)$ than it did in Problem 6.28? Compare your results with the exact solution and also see how replacing ψ_8 with the tent mode of Problem 6.29 helps the analysis.

6.31. In Problem 6.27 we computed the natural frequencies and eigenmodes for the system shown in Figure P6.18, using $\psi_n = \sin\left(\frac{n\pi x}{2l}\right)$. This time use $\psi_n = \sin\left(\frac{(2n-1)\pi x}{2l}\right)$. Plot the first and second mode along with the exact results. Why aren't the approximate results as good as you might hope? $\rho = 1$ kg/m, $T = 1000$ N, $k = 8000$ N/m, $l = 1$ m. The first two natural frequencies are 88.68 and 179.2 rad/s.

6.32. In Problem 6.31 we solved a spring restrained, tensioned string problem using $\psi_n(x) = \sin\left(\frac{(2n-1)\pi x}{2l}\right)$ as trial functions, and in Problem 6.27 we used $\psi_n(x) = \sin\left(\frac{n\pi x}{2l}\right)$. These trial functions correspond to the eigenfunctions of the problem when the spring stiffness approaches zero and infinity, respectively.

Let's see what happens if we use trial functions that have nothing particular to do with either limiting case. Choose $\psi_n(x) = \sin(\sqrt{2}n\pi)$ and see how well you can approximate the solution. The first six natural frequencies are 88.68, 179.2, 272.05, 366.8, 262.9, and 559.8 rad/s. $\rho = 1$ kg/m, $T = 1000$ N, $k = 8000$ N/m, $l = 1$ m.

6.33. Show via a Rayleigh-Ritz analysis that the β values for the system illustrated in Figure P6.33 approach the pinned-pinned case for $k_\theta \rightarrow 0$ and the pinned-clamped case for $k_\theta \rightarrow \infty$. Show that the β_i lie between these two extremes when k_θ is finite. Let the overall numerical ratio of EI and k_θ be $\frac{k_\theta}{EI} = 8$ (all units used consistently). Use $\psi_n(x) = \sin\left(\frac{n\pi x}{l}\right)$. Eight trial functions will be sufficient. EI and ρ are constant, and k_θ is a torsional spring attached to the right end of the beam.

Figure P6.33

6.34. In Problem 5.53 you analyzed a pinned-pinned beam on a resilient foundation. Use a Rayleigh-Ritz approach (utilizing $\gamma_n(x) = \sin(\frac{n\pi x}{l})$ as the trial functions) to find the first natural frequency of the system (shown again in Figure P6.34). Compare your answer found with a four-mode trial set to the exact solution. Check how the answer varies when you approximate the foundation with 3 springs, 10 springs, and 40 springs. Also, comment on the modal participation reflected by the B vectors.

Figure P6.34

6.35. Show by a Rayleigh-Ritz analysis that the first three natural frequencies for the system illustrated in Figure 6.35 (top) match the 3 DOF lumped approximation shown in Figure P6.35 (bottom). (You'll have to determine $k_1 - k_4$ from a static analysis.) Use $\psi_n(x) = \sin\left(\frac{n\pi x}{l}\right)$, $n = 1, 2, \ldots, 8$. $m_1 = 10$ kg, $m_2 = 20$ kg, $m_3 = 30$ kg, $T = 1000$ N, $\rho = .3$ kg/m, $l = 1$ m.

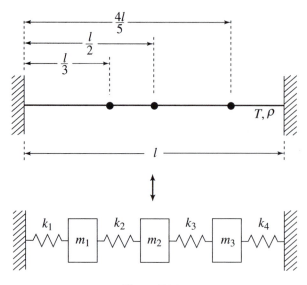

Figure P6.35

6.36. Assume that upon completing a Rayleigh-Ritz analysis of a continuous system using four trial functions, you obtained the following vectors as the resultant eigenvectors of the Rayleigh-Ritz formulation.

$$\left[[\mathcal{K}] - \omega_i^2 [\mathcal{M}] \right] A_i = 0$$

$$A_1 = \begin{Bmatrix} 0 \\ 1 \\ 0 \\ 0 \end{Bmatrix} ; \quad A_2 = \begin{Bmatrix} 1 \\ 0 \\ 0 \\ 0 \end{Bmatrix} ; \quad A_3 = \begin{Bmatrix} 0 \\ 0 \\ 0 \\ 1 \end{Bmatrix} ; \quad A_4 = \begin{Bmatrix} 1 \\ 1 \\ 1 \\ 1 \end{Bmatrix}$$

These aren't the displacement eigenvectors; they are the eigenvectors that indicate trial function participation as calculated from the Rayleigh-Ritz procedure. What do they tell you about your original trial functions?

7

Seat-of-the-Pants Engineering

7.1 INTRODUCTION

The next few pages will examine explicitly what we've already seen in some of the homework problems—the question of how to get a *good enough* answer instead of an exact one. Although this chapter and its problems could have been broken up and distributed among the earlier chapters, the approach seems to be distinctive enough to merit a more unified presentation. To a large extent the ability to do good seat-of-the-pants engineering is an acquired skill, one that comes after having gone through the tedium of producing exact results for a great many problems. The difficulty for the student is that the whole idea of seat-of-the-pants analysis is very seldom referred to in any text even though it underlies much of the successful engineering of this world. Really good engineers rarely sit down and precisely analyze a problem at the start. Rather, they consider many options, get quick approximations, redesign, get some more approximate solutions, and then finally zero in on a final solution. Only after a particular problem has been adequately pinned down do really serious analysis tools come into play. This chapter attempts to give the student some sense of what can be accomplished by stepping back and thinking about a particular problem in a general way before bringing out all the analytical tools the preceding chapters have given us.

7.2 GETTING APPROXIMATE RESULTS

Seat-of-the-pants engineering refers to the ability to analyze a system reasonably well, *but not exactly*, in a short time. Obviously, you can analyze a given problem in great depth and detail if you're given the necessary time and resources. Nevertheless, it's often the case that you'll have to produce an answer without having all the time you might like. In addition to situations like these, you will sometimes want to generate an approximate answer quickly because you're considering different design options and you don't want to spend too much development time on an idea that's not going to pan out. Thus, there's a good deal of value in having the ability to get a reasonably good answer fairly quickly.

As an example of how we can cut quickly to an answer, let's revisit our spring-mass problem of Section 1.2. You'll recall that in this problem we were trying to model the system as a lumped spring and mass and to determine the combined natural frequency of the system. What made the analysis approximate was the fact that the spring element had mass of its own, mass that wasn't accounted for in the analysis. To illustrate how we can reasonably account for this added mass, we'll consider longitudinal motions of a bar with a lumped mass at one end, as shown in Figure 7.1.

If the bar actually was a pure spring, then the deflected position of the bar under an applied load at the end would vary linearly, from zero at the fixed end to a maximum value at the free end. In Section 1.2 we found the relation between the deflection at the end and the applied force (1.2.47) to be

$$\xi(l) = \frac{fl}{EA} \tag{7.2.1}$$

If our simplifying assumptions hold, then a linearly varying displacement along the bar implies that the velocities will vary linearly also, since finding the velocities simply involves differentiation in time. No spatial operations are involved, so the spatial relations stay the same.

We've already seen that the natural frequencies of a system can be found by taking ratios of the system energies (Rayleigh's quotient). Since we know what the velocity field looks like within the bar (if it behaves according to our assumptions), we can modify the kinetic energy part of Rayleigh's quotient and solve for the modified natural frequency. Originally, the kinetic energy was given by

$$KE = \frac{1}{2}m_1\dot{\xi}^2(l) \tag{7.2.2}$$

That is, the only component of the kinetic energy was that due to the lumped mass at the right end. Now we're going to add the energy of the bar. If the velocity distribution were truly linear, we'd have

$$\dot{\xi}(x) = \frac{x}{l}\dot{\xi}(l) \tag{7.2.3}$$

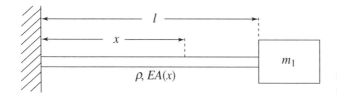

Figure 7.1 Uniform bar with a lumped mass at the right end

In reality, the actual distribution *won't* be precisely linear. For large end masses, the first mode will be very close to a linear distribution, and the closeness will degrade as the end mass is reduced with respect to the bar's mass.

Taking the bar's mass into account, our new kinetic energy becomes

$$KE' = \frac{1}{2}m_1\dot{\xi}^2(l) + \frac{1}{2}\int_0^l \rho\dot{\xi}^2(x)dx \qquad (7.2.4)$$

or, using (7.2.3)

$$KE' = \frac{1}{2}m_1\dot{\xi}^2(l) + \frac{1}{2}\int_0^l \rho\left(\frac{x}{l}\right)^2\dot{\xi}^2(l)dx \qquad (7.2.5)$$

We can factor out a couple of terms to leave us with

$$KE' = \frac{1}{2}\dot{\xi}^2(l)\left(m_1 + \int_0^l \rho\left(\frac{x}{l}\right)^2 dx\right) \qquad (7.2.6)$$

If we now evaluate the integral, we'll be left with

$$KE' = \frac{1}{2}\left(m_1 + \frac{\rho l}{3}\right)\dot{\xi}^2(l) \qquad (7.2.7)$$

Comparing (7.2.2) and (7.2.7), we see that what we've got is an equivalent mass. Whereas formerly our mass was just equal to m_1, now it's equal to $m_1 + \frac{\rho l}{3}$. Thus we can go ahead and use our old $\omega_n = \sqrt{\frac{k}{m}}$ formulation (with $k = \frac{EA}{l}$), but this time we'll use the modified mass rather than just the lumped mass. Therefore

$$\omega_n = \sqrt{\frac{EA}{l\left(m_1 + \frac{\rho l}{3}\right)}} \qquad (7.2.8)$$

If we calculate the natural frequency for the case of $EA = 10$ N, $l = 1$ m, $m_1 = 10$ kg, and $\rho = 1$ kg/m, we'll find that $\omega_n = .9837$ rad/s. Without the correction for ρ, our estimate for the natural frequency is 1.0 rad/s, giving us an error of about 2% from the exact result of .9836 rad/s (found from solving the actual partial differential equation). By modifying our mass term, we've reduced the error in the estimate to about one-hundredth of a percent.

Let's next go ahead and tackle the tougher problem of $\rho = 10$ kg/m. For this case we previously (Section 1.2) had a 16% error, reflecting the fact that since the bar's mass was equal to the lumped mass, the assumption of a massless spring element in place of the bar became less reasonable. Our modified estimate for the natural frequency (using the new modified mass) is .8660 rad/s. If we compare this with the exact value of .8603 rad/s, we see that we've cut the error from 16% to about .5%.

These are major improvements! Just by using a little insight, and realizing that the motions of the bar will contribute to the system's kinetic energy, we've improved our frequency estimates tremendously. Of course, if you really want the exact solution, you can always use the techniques of Chapter 5 to find it. But what we've gotten instead is a very accurate estimation with very little effort.

Okay, we've now seen how we can improve on our frequency estimate for an isolated SDOF problem. Let's move beyond that and into a MDOF estimation problem. For this example, we'll assume that you work for a major automobile manufacturer and your job involves suspension design. A colleague has come up with a very nice model of a vehicle, one that includes seven degrees of freedom. The precise model isn't important because we won't be doing an exact analysis, but Figure 7.2 shows the basic setup. The tires are modeled as spring elements, as are the suspension pieces between the axles and the body. The front and rear axles are included as rigid bars, and the body mass is supported by the suspension springs. Since the model has seven degrees of freedom, it also has seven eigenvectors, eigenvectors that can be associated with specific types of responses. For instance, the lowest natural frequency mode is associated with what automotive designers call *bounce*. This is the same mode that airplane designers call *heave*, and it corresponds to purely up-and-down motions of the body. Although the physical mode that's called bounce doesn't *exactly* correspond to pure up-and-down motions (there's some small amount of pitch, etc.), the approximation found by ignoring everything but the up-and-down motions is actually quite good. There's also a *roll* mode, which corresponds to rotations about the longitudinal axis of the car. *Pitching* refers to rotations about a lateral axis through the car (the sort of motion you'd get from strongly applying the brakes). These low frequency modes are *rigid body modes* because they reflect motions of the car body itself, oscillating on its suspension. The remaining modes are higher frequency modes that involve motions of smaller subsystems within the vehicle.

The data to be used in this example are taken from an actual car analysis discussed elsewhere [7]. In this model, there are two kinds of mass element. The car body is called a *sprung* mass because it's supported by the suspension. The mass elements that aren't directly attached to the body, like the rear axle, are called *unsprung* mass elements. Because these masses aren't directly connected to the body, they can experience their own, higher frequency modes. One of these is called *wheel hop*, which occurs when the front or back wheels bounce up and down in unison. If a set of wheels is moving in

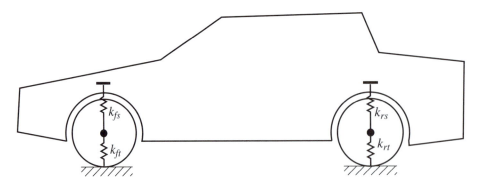

Figure 7.2 Simplified automobile vibration model

TABLE 7.1 PARAMETER VALUES FOR 1975
COMPACT TWO-DOOR

Parameters	Parametric values
Tire stiffness	184 kN/m
Front suspension stiffness	18.7 kN/m
Rear suspension stiffness	26.1 kN/m
Vehicle mass	1065 kg
Total unsprung mass	175 kg
Front unsprung mass	73 kg
Rear unsprung mass	102 kg
Rear axle	52 kg
Body roll inertia	190 kg·m^2
Body pitch inertia	1080 kg·m^2
Rear axle rotational inertia	8 kg·m^2
Wheelbase	2.42 m
Front wheels to CG	1.2 m
Rear wheels to CG	1.22 m
Ground to CG	.46 m
Front track	1.38 m
Rear track	1.35 m

an antisymmetric way (one up and the other down), the mode is called *tramp*. The various physical parameters of the vehicle model are given in Table 7.1.

To analyze the system and to determine what the frequencies and mode shapes are, you could construct the equations of motion for the complete system, code it up, and analyze it. But what we're going to show now is how you can get a quick and reasonable estimate of the responses without going to all this trouble. Just for comparison, Table 7.2 lists the exactly calculated natural frequencies for the modes we'll be looking at.

Let's start with the bounce mode. We know that this mode ideally consists of the body moving up and down. If you ignore the mass between the tire springs and the suspension springs, you can model them as two springs in series. Thus the equivalent spring at each front corner of the car is

TABLE 7.2 NATURAL FREQUENCIES FOR SOME
CAR VIBRATION MODES

Vibration mode	Frequency of oscillation (Hz)
Bounce (sprung mass)	1.33
Pitch (sprung mass)	1.71
Roll (sprung mass)	2.27
Front-wheel hop	11.87
Front-tramp	11.87
Rear-wheel hop	10.19
Rear-wheel tramp	12.63

equal to

$$k'_f = \frac{k_{ft}k_{fs}}{k_{ft} + k_{fs}} = 16{,}975 \text{ N/m} \qquad (7.2.9)$$

and

$$k'_r = \frac{k_{rt}k_{rs}}{k_{rt} + k_{rs}} = 22{,}858 \text{ N/m} \qquad (7.2.10)$$

The subscripts f and r indicate whether we're dealing with the front or rear springs and the subscripts t and s indicate whether we're looking at the tire or the suspension spring. Thus k_{ft} is referring to the stiffness due to the tire at the front of the car.

Remember that we're after a single mode, the one corresponding to bounce. If we call x_b the coordinate of the bounce mode, then we can write

$$1065\ddot{x}_b + 2(16{,}975 + 22{,}858)x_b = 0 \qquad (7.2.11)$$

Note that 1065 is the total mass (in kilograms) and 2(16,975+22,858) is the total spring constant. (We've got spring forces from both the left and the right, thus giving us the factor of 2.)

Solving for the natural frequency gives us

$$\omega_n = \sqrt{\frac{79{,}666}{1065}} = 8.65 \text{ rad/s} = 1.38 \text{ Hz} \qquad (7.2.12)$$

This can be compared to the exact value found in Table 7.2, which is 1.33 Hz. Thus our quick-and-dirty solution gave us an answer that is within 4% of the actual answer.

Given this success, let's go crazy and try to replicate another answer. The exact solution for the roll mode gives a natural frequency of 2.27 rad/s for the sprung mass. The moment of inertia for the sprung mass about the longitudinal axis is 190 kg·m^2. All we need do is determine the angular stiffness. Figure 7.3 shows what we've actually got to deal with. When the car rolls, the equivalent springs at the front and back are deformed. As Figure 7.3 shows, the equivalent torsional spring for each spring is equal to $k_{eq}l^2$, where l is the distance from the roll center to the spring and k_{eq} corresponds to the k'_f and k'_r defined. Note that the distances from the roll center to the equivalent spring is given by the front and rear track data in Table 7.1.

Our equation of motion for the roll mode is therefore given by

$$190\ddot{x}_r + 2\left(16{,}975\left(\frac{1.38}{2}\right)^2 + 22{,}858\left(\frac{1.35}{2}\right)^2\right)x_r = 0 \qquad (7.2.13)$$

For this case, the natural frequency turns out to be

$$\omega_n = \sqrt{\frac{34{,}598}{190}} = 14.0 \text{ rad/s} = 2.22 \text{ Hz} \qquad (7.2.14)$$

a result that's off only by 2% from the exact solution.

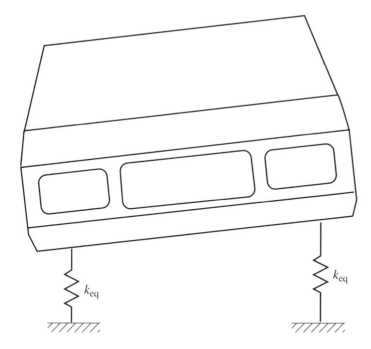

Figure 7.3 Roll mode

Finally, let's look at one more mode: that of front-wheel hop. This is an interesting one because we can choose from among a few different approximation levels. The most basic observation to make is that we've got a relatively large mass (890 kg) connected to a relatively small one (73 kg) by way of two suspension springs, each with stiffness equal to 18,700 N/m. The 890 kg mass corresponds to the sprung mass while the 73 kg mass represents the front unsprung mass (the mass between the tire spring and the suspension). The unsprung mass is sprung to ground by means of the two tire springs, each having stiffness equal to 184,000 N/m. If we completely ignore the sprung mass and suspension springs, we're left with the unsprung mass supported by two tire springs

$$73\ddot{x}_h + 2(184{,}000)x_h = 0 \tag{7.2.15}$$

This gives us a natural frequency estimate of 11.3 Hz., which is only 5% off from the exact value of 11.87 Hz. But wait, there's more! We observe that this frequency is much higher than the bounce frequency. Now, recall what happened in your physical experiment with the mass and spring when you moved your hand at different speeds. When you vibrated your hand quickly, the mass didn't move; it was isolated by means of the spring. Well, just think about our car. According to our estimates, if the rest of the car isn't moving, then the front-tire hop occurs at a frequency of 11.3 Hz. Let's make believe that the tires *are* bouncing at this frequency. Will the large mass of the car itself respond? No way! The total sprung vehicle mass is equal to 890 kg. The springs between this mass and the wheels have a combined spring constant of $2(18{,}700) = 37{,}400$ N/m. Thus the natural frequency of this mass on these springs is

$$\omega_n = \sqrt{\frac{37,400}{890}} = 6.48 \text{ rad/s} = 1.03 \text{ Hz} \tag{7.2.16}$$

This is more than 10 times lower than the excitation frequency. Thus we've got exactly the same situation that existed when we vibrated our hand at a much higher frequency than the system's natural frequency. End result? No motion of the big mass. Thus we can improve our approximation by saying that the sprung mass of the vehicle stays fixed, and we're left with the wheels being restrained by two sets of springs, namely, the suspension and the tire springs. Since these are acting in parallel on the sprung mass, we'll have

$$73\ddot{x}_h + 2(184,000 + 18,700)x_h = 0 \tag{7.2.17}$$

which gives us a natural frequency of

$$\omega_n = \sqrt{\frac{405,400}{73}} = 74.5 \text{ rad/s} = 11.86 \text{ Hz} \tag{7.2.18}$$

The result? An amazingly good approximation, one in which the error is less than a tenth of a percent.

   Now that you've seen the power of this kind of analysis, let's review more carefully what we're doing and see when it will work and when not. First of all, the approach depended upon the existence of distinct and *identifiable* modes of vibration in our system. For a general system, there might be no easily identified modes of the sort we have in the car problem. The first mode might involve both front wheels moving up and down in unison while the rear ones moved antisymmetrically. Or it might have involved equal amounts of bounce and roll happening simultaneously. If this had been the case, our assumptions of pure bounce or pure wheel hop would have been wrong and the frequency estimates associated with them would have been equally wrong. This is where engineering intuition first comes in. Whether from analyzing different cars in the past or from an examination of the model, you have to suspect that the bounce, roll, and hop modes will show up. Only then can you solve for their frequencies of vibration.

   After you've decided that certain physical modes should predominate in the response, you've still got to determine the essential factors that govern its motion. For the bounce mode, the key fact was in deciding that the two springs (tire and suspension) could be modeled as a series of springs, with a resultant equivalent single spring. For the wheel hop mode, we had to use a bit more insight and realize that certain elements of the vehicle wouldn't be involved in the motions (the sprung mass) because they were isolated from the high-frequency vibrations by their relatively soft springs. If the suspension had been a very hard one, one in which the suspension spring constants were as large as the tire spring constants, then the sprung mass *wouldn't* have been isolated and would have moved along with the wheels. This would have complicated our analysis.

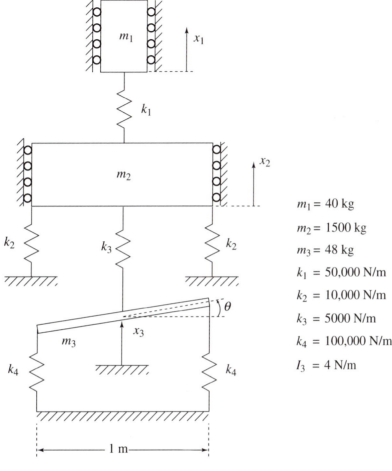

$m_1 = 40$ kg

$m_2 = 1500$ kg

$m_3 = 48$ kg

$k_1 = 50,000$ N/m

$k_2 = 10,000$ N/m

$k_3 = 5000$ N/m

$k_4 = 100,000$ N/m

$I_3 = 4$ N/m

Figure 7.4   4 DOF system

Example 7.1

Problem   To illustrate the concepts we've just learned, let's analyze the system shown in Figure 7.4. What we want to do is determine the modes of vibration and their natural frequencies without solving the entire coupled problem.

Solution   As you can see, we have three masses and six springs. The first item of business is to determine how many modes there will be. m_1 can only move vertically, as can m_2. However, m_3 can both translate vertically and rotate. Thus this mass supports two degrees of freedom. This means that our system has a total of four degrees of freedom, and we expect to find four modes of oscillation.

First we'll note the relative masses: m_1 and m_3 are tiny in comparison to m_2. It seems very likely that for the lowest frequency mode, we'll have m_2 ponderously going up and down, dragging everything else along with it. To find this mode, we need to determine the equivalent spring that's restraining m_2. In total, we have two k_2 springs plus an equivalent spring that's made up of the k_3

spring in series with two k_4 springs. Thus the overall spring constant is

$$k_{equiv} = 2k_2 + \frac{k_3(2k_4)}{k_3 + 2k_4} = 24{,}878 \text{ N/m}$$

using the values given in Figure 7.4. The simple answer for the oscillations of this mass is to set it equal to $1500 + 40$ which is the total of m_1 and m_2. The trickier move is to realize that m_3 is basically along for the ride, adding inertia but not doing much else in this low frequency mode. Since we know the spring constants surrounding this mass, we can determine how much it moves for a given motion of the main mass m_2. We'll just do a static analysis, like that shown in Figure 7.5. If we move the m_2 by x, then we need to do a force balance to decide how much m_3 moves. The force balance gives us

$$200{,}000y = 5000(x - y)$$

or

$$y = \frac{5000}{205{,}000}x$$

Thus the effect of m_3 will be to add $\left(\frac{5000}{205{,}000}\right)$ (48 kg) $= 1.17$ kg to the overall mass. This reflects the fact that m_3 doesn't move anywhere near as much as m_2, so we shouldn't include all its inertia.

Using our equivalent mass and spring, we can now solve for the natural frequency

$$\omega_n = \sqrt{\frac{24{,}878}{1541}} = 4.018 \text{ rad/s}$$

We can also find the approximate mode itself. If we use the coordinates

$$X^T = \{x_1 \ x_2 \ x_3 \ \theta\}$$

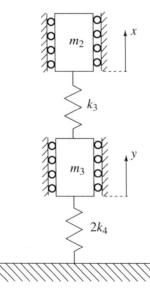

Figure 7.5   Static analysis to find equivalent mass contribution of m_3

(as shown in Figure 7.4), then we see that the mode is equal to

$$\tilde{X}_1^T = \{1.0 \ \ 1.0 \ \ .024 \ \ 0\}$$

where the tilde shows that this is just our estimate, not the actual eigenvector. This eigenvector was constructed by setting the response amplitude of x_2 to 1.0. Since m_1 is modeled as being rigidly attached to m_2, it also has amplitude 1.0. We just saw that the motion of m_3 is $\frac{5}{205}$ the size of m_2, which in this case is .024. Finally, the rotational mode isn't excited in this motion. Therefore it has an amplitude of zero.

Solving the fully coupled problem exactly results in a frequency of 4.019 rad/s and a mode shape of

$$X_1^T = \{1.000 \ \ .987 \ \ .024 \ \ 0\}$$

You can't get much better than this. There's hardly any difference at all between the approximate and exact results.

Now how about the other modes? One has already been alluded to. If m_3 is given an initial rotation, it will experience a rotary vibration. All the other motions in the system are purely translational, and there's no coupling between rotation and translation for this system. Thus we can analyze this mode immediately by setting all translations to zero and examining the rotational mode that's left. Note that this would definitely not be the case if the problem wasn't as symmetric as it is. For our system, m_3 is uniform, the springs from m_3 to ground are identical, and the attachments of the k_4 and k_3 springs are symmetric. These properties are what let us state that the rotations will not couple with the translational motion.

Taking the rotational part by itself, we have

$$4\ddot{\theta} + 2(100{,}000)\left(\frac{1}{2}\right)^2 \theta = 0$$

where the factor of 4 comes from the rotational inertia of m_3 and the rotational spring term comes from determining the restoring moment due to the k_4 springs. This supports a natural frequency of

$$\omega_n = 111.8 \ \text{rad/s}$$

and an eigenmode estimate of

$$\tilde{X}_2^T = \{0 \ \ 0 \ \ 0 \ \ 1\}$$

Since there really *are* no other interactions for this mode, the exact solution and the approximate one are identical.

For our next approximate solution, let's examine the small mass m_1 that's attached to the very large mass m_2. Common sense tells us that the vibrations of a very small mass aren't likely to do much to affect the motion of a huge one. Thus we can approximate our next mode as one for which m_1 oscillates and m_2 remains fixed. This leaves us with the following SDOF problem:

$$40\ddot{x}_1 + 50{,}000x_1 = 0$$

The associated natural frequency is equal to 35.36 rad/s, and the approximate mode is given by

$$\tilde{X}_3^T = \{1 \ \ 0 \ \ 0 \ \ 0\}$$

That was certainly quick. Does it agree with the exact answer? Well, the exact solution will produce a natural frequency of 35.83 rad/s and an eigenvector of

$$X_3^T = \{1.000 \quad -.027 \quad -.001 \quad 0\}$$

You'd have to be a very stubborn individual not to agree that these two results are just about identical.

Finally, we have to address what's happening with the last mode. Calling on the same insight as that used in the car problem, we can check to see if there's a mode that oscillates at a much higher frequency than the large mass would want to do, thus isolating it. The first choice is to look at the motion that hasn't gotten any play yet, namely, translational motions of m_3. If we rearrange our system as shown in Figure 7.6, we can view it as two SDOF systems joined by a central spring. (We've neglected the m_1-k_1 combination because it's too small to worry about.) As shown in Figure 7.6, m_2 is sprung to ground through the two k_2 springs and attached to m_3 through k_3. We've already seen that the mode involving m_2 occurs at 4.018 rad/s. If we examine the m_3-k_3-k_4 combination, we've got a natural frequency of $\omega_n = \sqrt{\frac{205,000}{48}} = 65.35$ rad/s. Thus m_3 would like to oscillate at a much higher frequency than the low frequency mode of m_2. Therefore m_2 is isolated from the vibrations, and we can approximate it as being stationary.

If m_2 is taken as stationary, then we know from the preceding paragraph that the natural frequency associated with m_3's motion is 65.35 rad/s. The mode shape would be

$$\tilde{X}_4^T = \{0 \quad 0 \quad 1 \quad 0\}$$

The actual eigenvector is

$$X_4^T = \{.000 \quad -0.001 \quad 1.000 \quad .000\}$$

and the natural frequency occurs at 65.35 rad/s. Not much difference, eh?

It wasn't just an accident that the approximation results were so close to the exact ones; the system was designed so that this would happen. The point is that if you hadn't taken the time to think

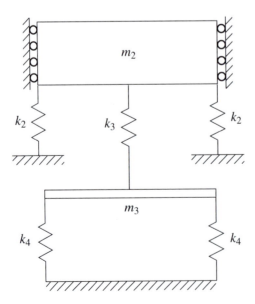

Figure 7.6   Reduced system ignoring m_1 and k_1

about what the system represented and considered ways to get approximate solutions, you'd have been forced to solve the full-blown, 4 DOF problem to get any results—and this isn't always the most intelligent approach. Real systems will invariably be complex, and solving them will require time and effort. If all you really want is an estimate of a couple of eigensolutions, it's worth your while to think a bit about intelligent approximations before jumping in.

7.3 LIMITING CASES

Up to now, our examples have had finite values for the mass and spring stiffness. Although this is very reasonable, it is also of use to consider what happens when one or both of these terms goes to zero or infinity. This is true for a variety of reasons. For one thing, we may wish to suggest design changes involving large reductions of mass or increases in spring stiffness in a given structure. In these cases, knowing the limiting value response for zero mass or infinite spring stiffness will let us know the greatest effect such an alteration could have on our system. In addition, an easy way to alter the system we're considering, if it's coded on the computer, is to have the computer make certain mass or stiffness elements very small or very large (approximating zero and infinity, respectively). This will serve to lock out or release degrees of freedom. Again, to correctly interpret the computer's output, we would want to know the analytical effect of this ahead of time.

This last technique, that of driving some terms to zero or infinity, is perhaps most useful in debugging programs. When the program is complex, it's hard to determine whether the output is really reasonable. You can often push the system into a simpler configuration, one that's easier to check, if you set certain parameters to their limiting values.

In Chapter 1 we considered the case of zero spring stiffness. Equation (1.2.2) showed the response of a free mass, which is the same as a mass-spring system for which the spring constant has been set to zero. As Equation (1.2.2) indicated, the mass can have an initial displacement and may also translate at a constant velocity if no forces are applied. This behavior is also indicated from our solution of a finite spring-mass system. Since the natural frequency of such a system is equal to the square root of $\frac{k}{m}$, we see that the consequence of k going to zero is that the natural frequency also goes to zero. And a zero frequency implies that the period of the oscillation is infinity. Thus the mass will move off and not come back. This is a very important point to remember: zero frequency implies translation (or rotation) with no oscillations. It is the free body motion referred to earlier. Note that increasing the mass a great deal while holding k fixed will give us the same result; it drives the natural frequency to zero.

This observation also has a more compelling intuitive feel. It is easy to see, given a huge mass and a tiny spring, that once the mass has started to move, the spring is going to have a very hard time reversing that motion. Since the force generated in the spring is directly proportional to its stretch, it's also clear that the little spring will have to stretch a great deal to generate sufficient force to bring the mass back; i.e., the mass will translate for a long time before the spring can affect it appreciably.

Next we'll look at the case of the mass going to zero (or equivalently, the stiffness going to infinity). In this case, the natural frequency will go to infinity, since either m going to zero or k going to infinity will cause $\frac{k}{m}$ to go to infinity. This makes sense if you consider a fixed mass attached to a spring that gets stiffer and stiffer. As the stiffness becomes huge, the amount of motion needed to counteract any inertial forces is quite small. The system looks more and more like a rigid body. This is exactly the opposite of the preceding case. For small stiffnesses, the system became free. Now we see that large stiffnesses lock the system up.

These SDOF examples also hold when we look at MDOF systems. For instance, let's look at the system shown in Figure 7.7. This system consists of three masses and four springs. We'll fix the values of m_1-m_3 and k_1-k_3 but let k_4 vary from zero to 12 N/m. As we'll see, this range will show all the effects that we'd have had if we'd varied the stiffness from zero to infinity.

Since there are three masses, the system will display three distinct natural frequencies for each value of k_4. The variation of ω_i^2 is shown in Figure 7.8 for all k_4 values between zero and 12 N/m.

We already know that because we're adding stiffness to the system, we should expect all the natural frequencies to increase as k_4 is increased. We can also predict what the final frequencies

$$
\begin{aligned}
m_1 &= 1 \text{ kg} \\
m_2 &= 1 \text{ kg} \\
m_3 &= 1 \text{ kg} \\
k_1 &= 1 \text{ N/m} \\
k_2 &= 1 \text{ N/m} \\
k_3 &= 1 \text{ N/m} \\
k_4 &= 0\text{–}12 \text{ N/m}
\end{aligned}
$$

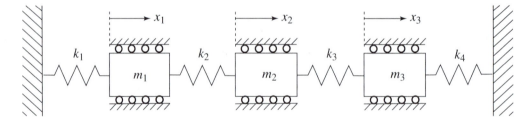

Figure 7.7   Spring-mass system with variable end spring

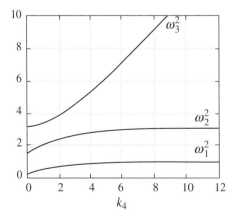

Figure 7.8   Variation of natural frequencies as a function of spring constant k_4

will be. When k_4 becomes huge, it locks down the motions of m_3. Thus we're left with a 2 DOF system. Analytically, we can solve for the natural frequencies, and find that they are 1 and $\sqrt{3}$ rad/s. Since Figure 7.8 plots the squares of the natural frequencies (the eigenvalues of the problem), we should see results of 1 and 3 for large k_4, which we do. In fact, you'll notice that essentially all the variation of the first two natural frequencies has occurred by the time k_4 has reached 8 N/m. This is quite typical. As soon as the effect of a particular element has become sufficiently large, the effect on the overall system response is *as if* it were equal to infinity.

   We can also see that the third natural frequency does not level off but in fact climbs quite steeply as k_4 is increased. This frequency is associated with the reduced system of m_3, k_3, and k_4. For high values of k_4, the other masses are unaffected; the vibration is too high for them to respond. Thus m_1 and m_2 are essentially stationary, leaving the mass m_3 to vibrate, restrained by k_3 and k_4. The natural frequency squared of this mode is given by

$$\omega_3^2 = \frac{1 + k_4}{1} \tag{7.3.1}$$

(i.e., $\frac{k_3 + k_4}{m_3}$ with the numerical values of m_3 and k_3 equal to 1). Therefore we would expect it to rise linearly as soon as k_4 becomes substantially greater than 1, which it does.

   Next, let's consider small values of k_4. Setting k_4 at zero still leaves us with a 3 DOF problem. If you go ahead and solve for the exact natural frequencies for this case, you'll see that they're equal to those given in Figure 7.8 for the case of $k_4 = 0$.

   You also will obtain similar results by letting damping and/or mass elements vary from small values to very high values. For instance, letting a mass element go to infinity will mean that you'll have a natural frequency that goes to zero (because of the huge mass) while the others go to the values they'd have if the mass were fixed in space. Damping can also lock out degrees of freedom. When the damping goes to infinity, it's as if the mass is moving in incredibly thick tar. Obviously, it won't move much in this case.

   These insights can and should be used in your engineering work. Let's say you want to increase a natural frequency in a system because one of the current natural frequencies is near an external excitation frequency. You may want to move it up by 20% and you may think you can do this by increasing a certain spring stiffness in the system. By analytically setting this stiffness to infinity, and analyzing the resulting vibrations, you can determine what the maximum frequency changes are going to be. If they're not sufficient, you know not to waste time trying to alter that particular spring—it's not going to be sufficient to do the job.

7.4 VERIFYING YOUR ANALYSES

One of the most important topics that you'll need to master is verifying your analyses. You may have developed a nice analytical or approximate model for a system and you're ready to produce some results. But how do you know if your model is any good? Or perhaps you've included some typos in your computer code, typos that don't keep the code from running but do give you incorrect answers. How can you tell?

The first secret is a simple one. Go slow. Go REALLY SLOOOOWWWW. This probably sounds silly, but it isn't. It's *incredibly* easy to make an error when flying through a derivation. You should try to develop the discipline to go a line at a time and *check each line*. You'll catch more than enough errors to justify the time spent in checking.

Okay, let's say you went slowly and did what you think is a good job. What now? Well, you can check that the limiting cases of your written equations work out. For instance, do your equations give you the correct static result? If you go through your equations and cross off all the time-dependent terms, you'll be left with a static equation, one that's easy to solve. It's usually pretty straightforward to determine what the physical response should be for a static situation. Thus you can check that at least parts of your equations are correct. The same reasoning also applies to the other extreme. If you kill all but the highest frequency terms, you'll be left with equations that describe the upper frequency limit for your problem, something you can then verify.

If you're attempting a modal approximation, you could include only a single mode to start. If your system is restrained to ground, this could be a rigid body mode. Your approximate analysis should then match the results you'd get from letting the distributed system behave rigidly. If it doesn't, you've got an error somewhere.

Most of the time, you'll end up with computer code that's supposed to accurately represent your equations, which in turn represents some physical reality. By looking at limiting cases for all of these, and cross-checking, you'll go a long way toward verifying that your work is correct.

7.5   HOMEWORK PROBLEMS

Section 7.2

7.1. Figure P7.1 shows two snapshots of a tensioned string undergoing free vibration. Explain how such a situation could arise if the system modes were proportionally damped. Explain how it could also occur under nonproportional damping.

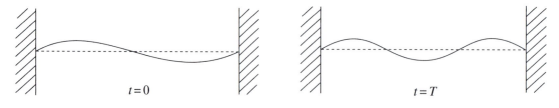

Figure P7.1

7.2. Your research assistant has told you that three given vectors represent the three eigenvectors of a particular system. You put these together into the model matrix $[U]$ and form the matrices $[U]^T[M][U]$ and

$[U]^T[K][U]$, which turn out to be

$$[U]^T[M][U] = \begin{bmatrix} 1 & 0 & 0 \\ 0 & 1 & 0 \\ 0 & 0 & 1 \end{bmatrix} \quad \text{and} \quad [U]^T[K][U] = \begin{bmatrix} 6 & 0 & 0 \\ 0 & 4 & 1 \\ 0 & 1 & 3 \end{bmatrix}$$

What can you say about the three vectors?

7.3. Figure P7.3 shows a seismically excited system. For an input displacement $y = \bar{y}(\cos \omega t)$, what will the phase lag of the output x to input y equal? An answer that's within 10% is good enough. $m = 22$ kg, $k = 20,000$ N/m, $c = 135$ N·s/m, $\bar{y} = .1$ cm, $\omega = 600$ rad/s.

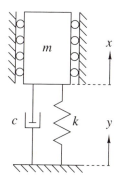

Figure P7.3

7.4. Consider the system shown in Figure P7.4: it's made up of a torsional rod with a heavy mass having a rotational inertia J_0 at one end. An external moment M is applied at $x = l$. For what frequency range (in terms of ρ, G, and J) would it be appropriate to model this system as an SDOF system (i.e., the lumped inertia of the mass at the end and the rod stiffness act together as a SDOF spring-mass system)?

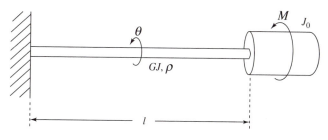

Figure P7.4

7.5. What response variable would produce the amplitude/phase outputs shown in Figure P7.5? (*Hint*: Consider the types of system we've seen—(mass excited, rotating imbalance, seismically excited—and what could be measured—$x, \dot{x}, \ddot{x}$, etc.). Note that the magnitude of the output goes to zero as the forcing frequency goes to infinity.

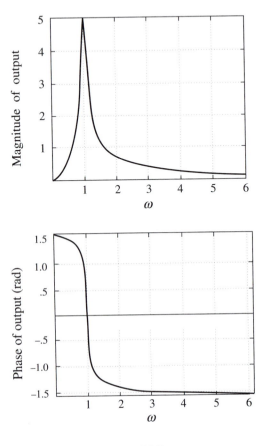

Figure P7.5

7.6. Consider the tensioned string with a string-mass system suspended from it, as shown in Figure P7.6. The string can be forced at a variety of frequencies. If the forcing is equal to the frequency of the first string mode (ignoring the spring-mass), what should k/m be to minimize the response? What if the forcing was at the second natural frequency? Can the spring-mass system be placed anywhere along the string in both these cases, or are there restrictions on its location? Note that this is NOT a Rayleigh-Ritz problem. $\rho = 1$ kg/m, $l = 1$ m, $T = 1$ N.

7.7. Find the eigenvectors and natural frequencies for the system illustrated in Figure P.7.7. Describe the modes that seem to involve only part of the structure and explain why the whole structure (all x_i's) isn't involved. $m_1 = 1$ kg, $m_2 = 10$ kg, $m_3 = 50$ kg, $m_4 = 1$ kg, $m_5 = 200$ kg, $k_0 = 1000$ N/m, $k_1 = 10$ N/m, $k_2 = 50$ N/m, $k_3 = 75$ N/m, $k_4 = 2000$ N/m, $k_5 = 1$ N/m, $k_6 = 2000$ N/m.

7.8. What's wrong with the amplitude-versus-frequency response illustrated in Figure P7.8?

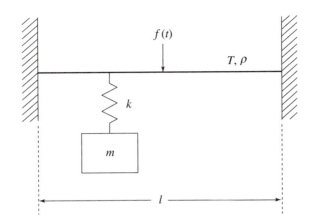

Figure P7.6

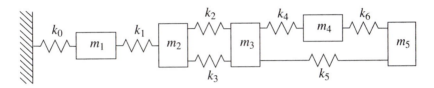

Figure P7.7

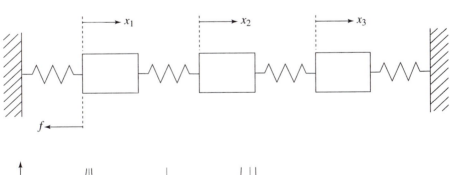

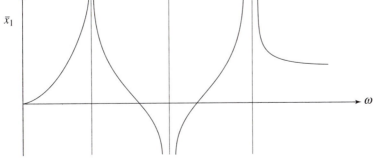

Figure P7.8

7.9. Show how you would find the *approximate* natural frequencies and eigenvectors for the system illustrated in Figure P7.9. Don't solve it exactly. The interest here is in how you'd approach it in an approximate sense. Then solve it exactly and compare the results. $k_i = 10$ N/m, $m_1 = 1$ kg, $m_2 = 100$ kg, $m_3 = 2$ kg.

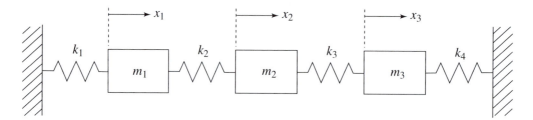

Figure P7.9

7.10. In this problem we'll try to gain some insight with respect the analyses of simpler subsystems, to determine when is a valid approach. We'll consider the four different configurations of the relative masses and stiffnesses of Figure P7.10. Solve the problems exactly and comment on which aspects of the solution are predictable from an analysis of a simpler subsystem. For all the problems, $f_i = 0$ and $k_3 = 0$, the base system is shown in part (a).

$$(a) \quad \begin{aligned} m_1 &= 100 \text{ kg} \\ m_2 &= 1 \text{ kg} \quad \text{base system} \\ k_1 &= 1000 \text{ N/m} \\ k_2 &= 10 \text{ N/m} \end{aligned}$$

$$(b) \quad \begin{aligned} m_1 &= 100 \text{ kg} & k_1 &= 1000 \text{ N/m} \\ m_2 &= 1 \text{ kg} & k_2 &= 1000 \text{ N/m} \end{aligned}$$

$$(c) \quad \begin{aligned} m_1 &= 100 \text{ kg} & k_1 &= 1000 \text{ N/m} \\ m_2 &= 100 \text{ kg} & k_2 &= 1000 \text{ N/m} \end{aligned}$$

$$(d) \quad \begin{aligned} m_1 &= 100 \text{ kg} & k_1 &= 1000 \text{ N/m} \\ m_2 &= 100 \text{ kg} & k_2 &= 10 \text{ N/m} \end{aligned}$$

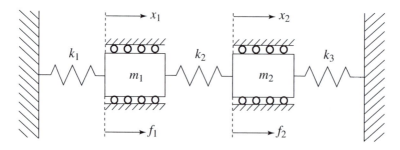

Figure P7.10

7.11. Let's spend some more time with Example 7.1 (Section 7.2). A revised version of the system is shown in Figure P7.11. The change is the fact that k_3 is no longer centered, but is offset by an amount e. Determine how much effect such a misalignment can have. Do the changes in the frequencies and eigenvectors make physical sense?

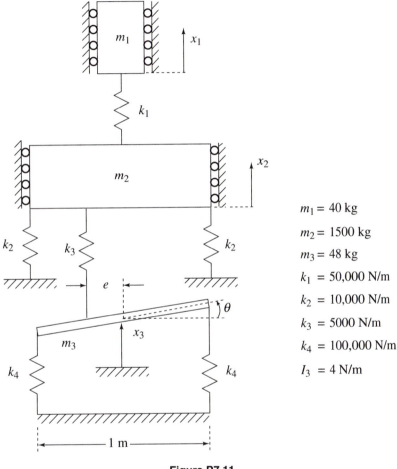

$m_1 = 40$ kg

$m_2 = 1500$ kg

$m_3 = 48$ kg

$k_1 = 50,000$ N/m

$k_2 = 10,000$ N/m

$k_3 = 5000$ N/m

$k_4 = 100,000$ N/m

$I_3 = 4$ N/m

Figure P7.11

7.12. In this problem we'll again consider Example 7.1. This time, you should vary the mass m_1. When it was small, it decoupled nicely from the rest of the system. As its mass increases, the assumption of decoupling will become invalid. At what mass do the errors in the frequency estimates of Example 7.1 exceed 10% in comparision to exact results?

7.13. The system illustrated in Figure P7.13 is a horizontal beam supported by two vertical beams. We want to determine the appropriate boundary conditions to use with the horizontal beam. One possibility is to view the ends as fixed. But perhaps this isn't reasonable. After all, we know that bars support longitudinal motions and the vertical vibrations of the horizontal beam might interact with the vertical compression modes of the two supporting beams. In this last case the supporting beams are actually acting like bars in longitudinal motion.

Look at the eigenfrequencies and deduce which boundary condition makes the most sense—fixed or spring restrained (the vertical supports providing the spring forces). All elements of the structure have the same parameter values. $E = 2.1 \times 10^{11}$ N/m, $\rho = 7.8 \times 10^3$ kg/m^3, $l = 1$ m, $A = .0001$ m^2, $\bar{I} = 8.\bar{3} \times 10^{-10}$ m^4.

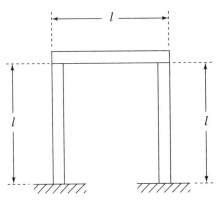

Figure P7.13

Section 7.3

7.14. The system illustrated in Figure P7.14 is composed of a lumped mass m_1 that's connected to ground by means of two tensioned strings. The length of each string is 1 m. The linear density of the strings is 2 kg/m, the lumped mass is variable, and the tension is equal to 1800 N. A lumped approximation (ignoring the string's mass) produces a natural frequency of $\sqrt{\frac{2T}{m_1}}$. How small can m_1 be before the actual first natural frequency of the string/mass system differs from this estimate by 10%?

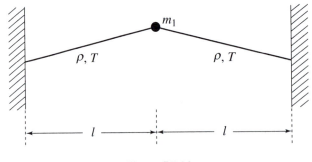

Figure P7.14

7.15. Figure P7.15 shows, a uniform, rigid bar, spring-restrained to ground, with a lumped mass attached to it. The lumped mass can be placed anywhere along the bar. Analyze the system for different placements of the mass (from midspan to the end) and comment on how badly the symmetry of the solution (when the mass is at midspan) is affected by its movement off from the center. $l = 2$ m, $m_2 = 3$ kg, $m_1 = 1$ kg, $k = 10,000$ N/m.

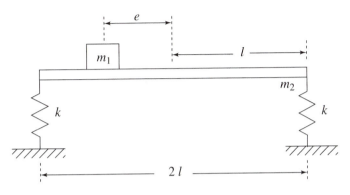

Figure P7.15

7.16. Figure P7.16 shows a uniform, rigid bar that's spring-restrained to ground. Analyze the system for the case of equal spring constants and then allow the right spring to become progressively stiffer. How do the solutions alter as the spring constant is increased? How much can it change and still leave the symmetric solutions (associated with equal spring constants) relatively unchanged? $m_1 = 1$ kg, $k_1 = 1000$ N/m, $l = 1$ m.

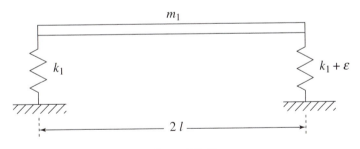

Figure P7.16

7.17. Figure P7.17 shows a tensioned string with an attached mass ε. All motion takes place in a horizontal plane, and the mass moves without friction along the guide. If the mass is large, the first vibratory mode can be analyzed as if the system is SDOF, with the string tension supplying the restoring force and the lumped mass acting as the dominant mass in the system. As the mass reduced, however, the other modes of the string problem become more important and, eventually, the SDOF approximation cases to be accurate. At what value of ε does the SDOF approximation for the lowest natural frequency exceed 5% error compared with the actual solution of the coupled string-mass problem? The tension in the string is 100,000 N, $\rho = 1$ kg/m, $l = 1$ m. Neglect gravity.

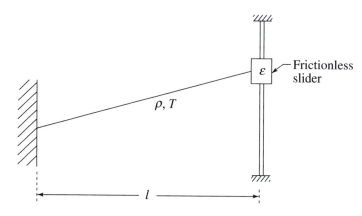

Figure P7.17

7.18. Consider again the 2 DOF system shown in Figure 7.10. Let $f_1 = f_2 = 0$ and let $k_3 = 0$. If k_1 is reduced sufficiently, the system can be viewed as supporting two decoupled modes. One involves motion of both m_1 and m_2, which move like a rigid body. The other involves the two masses moving in opposition to each other, with k_1 contributing little to the response (a *breathing* mode). Both these modes are shown in Figure P7.18. For what values of k_1 is the assumption of modes like those shown in the figure a good one? (*Note*: You'll have to define an error measure for when the modes look reasonable.) $m_1 = m_2 = 1$ kg, and $k_2 = 900$ N/m.

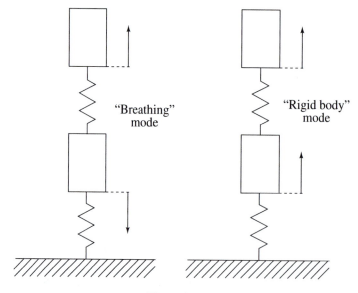

Figure P7.18

7.19. Consider a tensioned string with a lumped mass and spring at the right end, as shown in Figure P7.19. Solve for the actual natural frequencies. Next, use your insight into how the first mode will differ qualitatively from the higher ones and show how you can obtain an approximate first natural frequency without solving the entire continuous problem. Don't forget to include the equivalent lumped mass and stiffness provided by the string. $l = 1$ m, $m_1 = 10$ kg, $k = 9000$ N/m, $\rho = .1$ kg/m, $T = 50$ N.

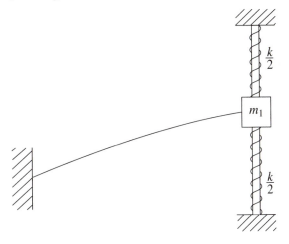

Figure P7.19

7.20. Assume that you've solved for the responses of the undamped vibration absorber problem shown in Figure 7.20. $m_1 = 1000$ kg, $k_1 = 36,000$ N/m, $m_2 = 2$ kg, $k_2 = 72$ N/m, $\omega = 6.03$ rad/s, and the forcing amplitude is .7 N. You determine that $x_1 = -.00783$ m and $x_2 = .781$ m. Verify that these results make sense by analyzing the SDOF system made up of m_2 and k_2, presuming that the system is seismically excited by m_1.

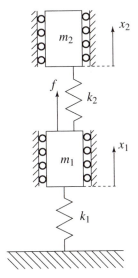

Figure P7.20

8

Experimental Methods and Real-World Behavior

8.1 INTRODUCTION

Up to this point we've been concerning ourselves with deterministic vibrations, i.e., periodically excited systems. The nice thing about periodic forcings is that once you know one period of the input, you know them all. Nothing else can happen—the input is completely determined. Thus we call the excitation deterministic. Although we could continue to restrict ourselves in this way, it turns out that expanding the scope of our excitations to include random vibrations will be very beneficial. For one thing, many systems are subjected to some form of random loading. The wind loads on an airliner, the forces acting on an offshore oil rig, and the accelerations felt by an object in a delivery truck are all well described by a random signal, not a periodically varying one. Furthermore, many of the algorithms used by modal analyzers to determine a system's natural frequencies and mode shapes are based on a random vibrations viewpoint. This chapter presents some of the important information you will need if you ever plan to carry out modal analyses, but please remember that it is just an introduction. The field of random vibrations is quite large, and there is a great deal of material that we won't be considering. Those interested in studying random vibration and modal analysis in more depth can consult other texts [3,9].

We'll start the chapter by laying out some fundamentals of random vibrations and finish by introducing some of the concepts of modal analysis.

8.2 SIGNAL DESCRIPTIONS

Let's begin this section by asking what a random signal is in the first place. Actually, you probably have a feel for what a random signal is, even if you haven't thought about it in a technical sense. Random signals look irregular, with no discernible pattern, like the example shown in Figure 8.1. At a simple level, we can think of random signals as waveforms that never repeat themselves. Thus a sinusoid isn't random because it repeats every T seconds, where T is the period of the oscillation. This is, however, not an adequate definition. Obviously, we wouldn't want to call $x(t) = at$ a random signal, even though it doesn't repeat itself. It's intuitively clear that $x(t) = at$ isn't random; it's just a constantly increasing signal. This is an example of a nonstationary situation. By "stationary," we mean that one or more averaged quantities remain fixed. For instance, in Figure 8.1 we see that the average value of x is zero; the signal seems to be centered about $x = 0$. Thus this signal exhibits stationarity. Signals that aren't stationary will drift, with time, as at does. Generally we will be concerned with stationary processes.

Even if the signal is stationary, nonperiodicity isn't enough. For example,

$$f(t) = \sin(t) + \sin(\sqrt{2}t)$$

isn't periodic but it certainly isn't random, as Figure 8.2 shows. Adding two sine waves whose frequencies aren't rationally related will always produce a nonperiodic result. We know that each individual sinusoidal signal isn't random, and thus adding two sinusoidal signals together shouldn't be considered to be random, since the function's value at a time t is completely predictable—we know it is simply composed of two sinusoids. This observation leads to the idea that random signals cannot

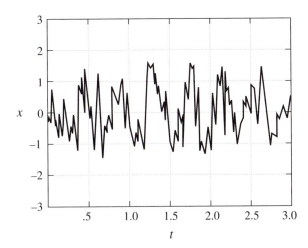

Figure 8.1 Random time response

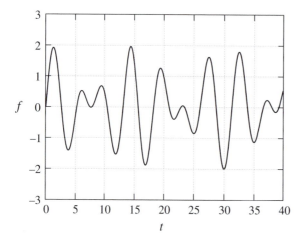

Figure 8.2   Response of two sinusoids with nonrationally related frequencies

be broken down into purely harmonic components. Thus we won't be able to find a Fourier series representation of a truly random signal.

Although people like to talk about random signals as if they truly exist, as far as our *computations* are concerned, they don't. Even if we had a steady state, truly random signal, one that continually and unpredictably changed over time, we'd have to cut off the data stream eventually. At this point our analyses can go one of two ways. We can view the actual signal as being zero from $t = -\infty$ to 0, equal to our data from $t = 0$ to $t = t_1$, and equal to zero for time greater than $t = t_1$, as shown in Figure 8.3. In this case the actual signal we're concerned with is a transient one, not a continuously varying, aperiodic one, and we obtain it by simply truncating the actual signal.

The other option is to impose a periodicity on the signal by assuming that the segment we captured repeats endlessly, with a period equal to t_1 (Figure 8.4). Just as in the preceding case,

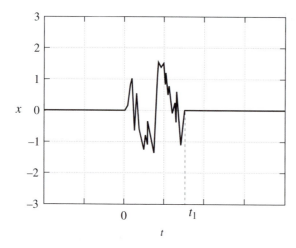

Figure 8.3   Truncated random time response

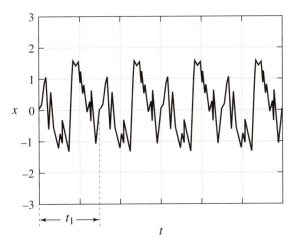

Figure 8.4 "Periodic" response

this viewpoint is not completely accurate. Whereas before we truncated the signal (making it into a transient), now we've made the signal periodic. Thus we can analyze either transient or periodic signals. The trick is to realize that as we increase the time t_1, we should be *approaching* the actual random signal. If we're lucky, these approximations will asymptotically approach the true random signal and we'll be left with only a small error for large enough t_1. It is important for the reader to keep these points in mind as we introduce the notions of random signals later in this chapter. Just as it was mathematically convenient to deal with viscous damping rather than the more realistic hysteretic or dry friction damping models, so it is more convenient to assume truly random signals, even when we know our analyses will actually be based on nonrandom ones.

You may have noted the use of the terms "random process" and "random signal" in the foregoing paragraphs. It is now time to ask what these words mean. Technically, there can be only random processes. It really makes no sense to call an acquired signal random since, once it has been acquired, it is completely determined. There isn't any randomness at all because the value of the signal for each point in time has been recorded, and there is no surprise about what value might occur for a given moment of time. Having said this, it's clear that this is a pretty pedantic viewpoint. Given a signal such as that shown in Figure 8.1, if you covered over the graph for t greater than 1 second, you would be hard pressed to reproduce the graph from a knowledge of the signal from zero to 1 second. That's what we really mean by random: that you can't predict what the signal will be at a later time, even if you know exactly what it was in the past. This is quite different from the case of a sinusoid, for which we know what all future values will be once we know the period and amplitude of the signal.

When we say random, what we're really concerned with are random *processes*, processes that generate an output that's unpredictable. Luckily, even if a random process produces an output that isn't predictable in the fine details, the outputs are nevertheless quite predictable in an averaged sense. For instance, if a skylight was opened during a rainstorm, you could pretty well predict that the floor beneath the skylight will be wet while the floor away from it will have stayed dry. Furthermore, you would expect the amount of water to be uniform, on average, over the wetted surface (this is ignoring drips from the sides of the skylight, of course). You won't be able to predict exactly where the next drop of water will fall, but you'll be able to predict the drop's general location, and you'll also be able to predict the *average* amount of rain per unit time, given your observations of how much has been falling over the preceding minutes.

If the average amount of water per unit time stays constant, then the process is stationary. If the average amount of water per unit time increases (the rainfall moves from a gentle sprinkle to a thunderstorm), then the process is nonstationary. Obviously the property of stationarity depends upon our viewpoint. The rain will eventually stop; therefore it can't be strictly stationary. Thus stationarity depends on the time span of interest. If the average rainfall stays constant over a 5-minute span and we're not interested in times greater than this, then for our purposes we can consider the process to be stationary. This kind of view, looking at averages rather than fine details, is the key to describing random processes.

The most common signal descriptor we'll deal with is the mean value of a signal, given by

$$E(y) = \frac{1}{n} \sum_{i=1}^{n} y_i \qquad (8.2.1)$$

Note that $E(y)$ is used to indicate the *expected value* (also called the *mean value*) of a variable. If we were concerned with a continuous process $y(t)$, we would have defined the mean as

$$E(y(t)) = \frac{1}{t_1} \int_0^{t_1} y(t)dt \qquad (8.2.2)$$

where our data sample has a duration of t_1 seconds. The only problem with such a representation is that we will almost never be dealing with continuous signals. All real signals must be sampled, and the samples must be processed. Granted, a continuous data stream is more familiar and is sometimes easier to work with mathematically (just as viscous damping is easier than hysteretic), but the sad fact is that all our processing will necessarily be digital. Thus we'll often be concerned with a digital outlook.

Certainly the mean of a signal gives us important information. If we know the dynamic pressure loading on an airplane wing, then the mean of this loading gives us the static pressure the wing must be designed to withstand. The mean, however, isn't usually enough. Consider Figure 8.5, in which two different force loadings are shown, each on having a mean value of 10. Clearly these loadings are very different, even though that difference doesn't show up in the mean value. The signal in Figure 8.5*a* has a much lower amplitude as well as a lower frequency than that of Figure 8.5*b*. Thus, although the mean value is important, we can see that more information is needed.

The next descriptor we need to construct is one that will capture how much the signal deviates from its mean value. One choice would be to calculate

$$\sum_{i=1}^{n} [y_i - E(y)] \qquad (8.2.3)$$

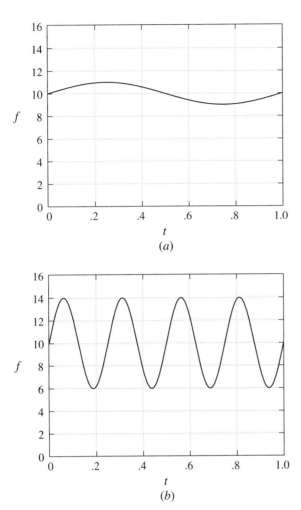

Figure 8.5 Same averaged behavior, but different dynamic behavior

Although this perhaps the most obvious choice, it's actually quite a poor one. For instance, this calculation would yield zero for both signals shown in Figure 8.5. This is because we've allowed both positive (when $y_i > E(y)$) and negative (when $y_i < E(y)$) values, which can act to cancel each other out. What we actually care about is how far the signal deviates from the mean value, and we aren't concerned about whether the deviation is positive or negative. So it is sensible to ensure that *all* deviations give a positive contribution to our descriptor. Although we could come up with a variety of different descriptors having this property, the one we'll introduce is the one used in common practice, called the *variance*:

$$\operatorname{var}(y) = \frac{1}{n} \sum_{i=1}^{n} [y_i - E(y)]^2 \tag{8.2.4}$$

You'll note that this could also be written as

$$\text{var}(y) = E[y - E(y)] \tag{8.2.5}$$

The variance is also called the *mean-square value*.

Often we'll talk about σ, the *standard deviation* of the signal

$$\sigma = \sqrt{\text{var}(y)} \tag{8.2.6}$$

which gives us a sense of exactly how far our signal is likely to be from the mean value (in an averaged way). This is also called the *root-mean-square* (RMS) *value* of the signal. If our signals are continuous in time rather than discrete, then the variance is found from

$$\text{var}(y) = \frac{1}{t_1} \int_0^{t_1} [y - E(y)]^2 \tag{8.2.7}$$

Although t_1 is always finite in any real analysis, we'll often set it to infinity for the purposes of our mathematical analyses.

Example 8.1

Problem   Consider two different discrete signals. The first (s_a) is given by

$$s_a = 1, 1, 1, 1, 1, 13$$

and the second (s_b) is given by

$$s_b = 1, 5, 1, 5, 1, 5$$

Determine the expected value and variance of these signals.

Solution   The expected value for both these discrete signals is 3. However the variances are quite different. For s_a we have

$$\text{var}(s_a) = \tfrac{1}{6} \left(2^2 + 2^2 + 2^2 + 2^2 + 2^2 + 10^2 \right) = 20$$

while the variance of s_b is

$$\text{var}(s_b) = \tfrac{1}{6} \left(2^2 + 2^2 + 2^2 + 2^2 + 2^2 + 2^2 \right) = 4$$

The variance for s_b is quite a bit smaller than that of s_a. This is because s_b varies only moderately from the mean value for all the entries, while s_a has a huge variation in its last entry. Thus we see that the variance allows us to distinguish between two signals as a function of how scattered the data are.

The foregoing descriptions have dealt with averaged quantities. It's possible to come up with many more averaged descriptors, but for our purposes mean and variance are sufficient. Next we'll want to look at a way of describing signals that depends on the particular value of the signal at particular times. Before doing this, however, we'll need to introduce the Fourier transform. The next section will describe this transform, and then we'll get back on track, discussing correlation functions and spectral density functions, both of which are important components of *spectral analyses*.

8.3 FOURIER TRANSFORM ANALYSIS

The basic building block of spectral analyses is Fourier transform analysis. As you'll recall from Section 3.2, we can take any signal of a fixed duration and determine the frequency components within it by calculating the Fourier series corresponding to the signal. The definition of a Fourier series was given by

$$x(t) = \sum_{n=0}^{\infty} a_n \cos(n\omega_0 t) + \sum_{n=1}^{\infty} b_n \sin(n\omega_0 t) \tag{8.3.1}$$

for which

$$a_0 = \frac{\omega_0}{2\pi} \int_0^{\frac{2\pi}{\omega_0}} x(t)dt \tag{8.3.2}$$

$$a_n = \frac{\omega_0}{\pi} \int_0^{\frac{2\pi}{\omega_0}} x(t) \cos(n\omega_0 t)dt \tag{8.3.3}$$

and

$$b_n = \frac{\omega_0}{\pi} \int_0^{\frac{2\pi}{\omega_0}} x(t) \sin(n\omega_0 t)dt \tag{8.3.4}$$

Although this is approach is fine for periodic signals, random signals are decidedly *not* periodic. Indeed, we've already given, a working definition of a random signal as a signal that is not periodic and yet doesn't die down to zero. (Obviously, a signal that goes to zero isn't periodic but isn't necessarily random.)

If the signal isn't periodic, then we can't identify a fundamental frequency ω_0, and we can't use (8.3.1)–(8.3.4). So what do we do? One approach is to view the signal as we did in Figure 8.4. By artificially making the signal look periodic, we can apply the Fourier series approach and find its spectral components. It's clear that this approach will give more accurate approximations as the period we're using increases. To see where this approach ultimately leads us, we'll mathematically let the period go to infinity. This means that ω_0 goes to zero and the difference between the various harmonics also goes to zero. As this happens, our summation in (8.3.1) mutates into an integration. This isn't surprising when you realize that to simulate an integration on a computer we represent it as a finite summation. For more accuracy, we use more terms in the summation. Conceptually, we're running this in reverse here, going from a finite summation to an integration. The difference between the harmonics (ω_0) will become $d\omega$ (the differential frequency) and our individual harmonics won't be indicated by $n\omega_0$ anymore, but by ω. If you go through the details of this process you'll obtain

$$x(t) = \int_{-\infty}^{\infty} a(\omega) \cos(\omega t)d\omega + \int_{-\infty}^{\infty} b(\omega) \sin(\omega t)d\omega \tag{8.3.5}$$

for which

$$a(\omega) = \frac{1}{2\pi} \int_{-\infty}^{\infty} x(t)\cos(\omega t)dt \qquad (8.3.6)$$

and

$$b(\omega) = \frac{1}{2\pi} \int_{-\infty}^{\infty} x(t)\sin(\omega t)dt \qquad (8.3.7)$$

Equations (8.3.5)–(8.3.7) are one way to define the *Fourier transform* of a signal $x(t)$.

The other way of representing a Fourier transform is to use complex notation, just as we did when dealing with deterministic forced vibrations. If we let $X(\omega)$ represent the Fourier transform of $x(t)$, we'll have

$$x(t) = \int_{-\infty}^{\infty} X(\omega)e^{i\omega t}d\omega \qquad (8.3.8)$$

and

$$X(\omega) = \frac{1}{2\pi} \int_{-\infty}^{\infty} x(t)e^{-i\omega t}dt \qquad (8.3.9)$$

Complex and real forms are equally valid. For the complex case, $X(\omega)$ is complex and $x(t)$ is real, while in the real form $x(t)$, $a(\omega)$, and $b(\omega)$ are real.

Example 8.2

Problem Determine the Fourier transform of the signal shown in Figure 8.6. Consider both real and complex forms of the transform equations.

Solution Starting with the real form, we have

$$a(\omega) = \frac{1}{2\pi} \int_{0}^{a} \cos(\omega t)dt = \frac{1}{2\pi\omega}\sin(\omega a)$$

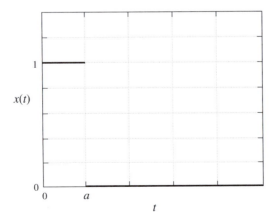

Figure 8.6 Rectangular pulse

and

$$b(\omega) = \frac{1}{2\pi} \int_0^a \sin(\omega t)dt = \frac{-1}{2\pi\omega}(\cos(\omega a) - 1)$$

Now let's consider the complex form. Equation (8.3.9) tells us that

$$X(\omega) = \frac{1}{2\pi} \int_0^a e^{-i\omega t}dt = \frac{-1}{2\pi i \omega}\left(e^{-i\omega a} - 1\right)$$

Expanding the complex exponential in terms of sine and cosine components gives us

$$X(\omega) = \frac{1}{2\pi\omega}(\sin(\omega a) + i(\cos(\omega a) - 1))$$

From this it seems like the real part of $X(\omega)$ is simply equal to the $a(\omega)$ we calculated from (8.3.6) and the imaginary part is equal to negative $b(\omega)$, found from (8.3.7).

To see that this wasn't just a coincidence, we can rewrite (8.3.9) as

$$X(\omega) = \frac{1}{2\pi} \int_{-\infty}^{\infty} x(t)(\cos(\omega t) - i\sin(\omega t))\,dt \qquad (8.3.10)$$

Comparing (8.3.10) with (8.3.6) and (8.3.7), we see that the real part of the integral gives us $a(\omega)$ while the imaginary part yields $-b(\omega)$, thus supporting the validity of our calculated results

When Fourier transforms are first encountered they often seem advanced and difficult to understand. We've now seen, however, that they're easily viewed as simply limiting cases of a Fourier series, something you're probably quite comfortable with. In addition, you'll rarely have to worry about actually computing the Fourier transform in any real-world application; such calculations are carried out as internal functions in the *spectrum analyzers* you'll be using.

8.4 SPECTRAL ANALYSES

Now that we've seen what Fourier transform analysis is all about, we can get back to finding useful descriptors of our signals. The next approach we'll examine is not one that looks at an overall average; rather, it asks if there's any correlation between the signal at two distinct points in time. Clearly, a signal is always perfectly correlated with itself. But what if we ask whether there's a correlation between a signal at time t and at $t + \tau$? If the signal is a sine wave with period τ, then we know there should be an excellent correlation, since the responses at these two times will be identical. A purely random signal should be uncorrelated for any τ's besides zero, since the response changes in a completely unpredictable way. The autocorrelation function, $R_{xx}(\tau)$, deals with just this sort of question and is defined by

$$R_{xx}(\tau) \equiv \lim_{T \to \infty} \frac{1}{T} \int_{-\frac{T}{2}}^{\frac{T}{2}} x(t)x(t + \tau)dt \qquad (8.4.1)$$

We've already mentioned that infinitely long time traces aren't really practical; thus the limit in (8.4.1) really applies to T getting large, not truly infinite. Also, if our signal $x(t)$ has been truncated, the integral will have finite limits defined by where $x(t)$ goes to zero.

If $\tau = 0$, then (8.4.1) becomes

$$R_{xx}(0) \equiv \lim_{T \to \infty} \frac{1}{T} \int_{-\frac{T}{2}}^{\frac{T}{2}} x^2(t) dt \tag{8.4.2}$$

This is simply the mean-square value of $x(t)$. Thus the autocorrelation of a signal is equal to the signal's mean-square value when the delay $\tau = 0$.

A similar correlation function can be found by asking what the correlation is between two *different* signals, again with varying time delay τ:

$$R_{xy}(\tau) \equiv \lim_{T \to \infty} \frac{1}{T} \int_{-\frac{T}{2}}^{\frac{T}{2}} x(t)y(t+\tau) dt \tag{8.4.3}$$

This form, called the *cross-correlation*, is more general because we can always let y equal x if we want the autocorrelation.

Cross-correlations would be useful if you were trying to determine whether a vibration at one point of a structure was being influenced by vibrations at some other point. If the cross-correlation is high for some value of τ, you can deduce that there is a relation between the two vibrations and that the transit time for the motions to be transmitted from one point to the other is equal to τ. Cross-correlations are also used in radar applications. By sending out a signal and looking for the same signal delayed in time (because of being bounced off a plane that's being tracked), a radar unit can use the detected time delay and knowledge of the signal's transmission speed to determine how far away an aircraft is.

Most of the usefulness of these correlation functions for our purposes doesn't actually come directly from them but rather from functions that are defined from them. The first of these functions is known as the *power spectral density*, $S_{xx}(\omega)$, and is found from

$$S_{xx}(\omega) = \frac{1}{2\pi} \int_{-\infty}^{\infty} R_{xx}(\tau)e^{-i\omega\tau} d\tau \tag{8.4.4}$$

If you look back at (8.3.9) you'll note that the form shown in (8.4.4) suggests that the power spectral density is the Fourier transform of the autocorrelation function. Logically, we can therefore deduce that R_{xx} can be written as the inverse transform of $S_{xx}(\omega)$:

$$R_{xx}(\tau) = \int_{-\infty}^{\infty} S_{xx}(\omega)e^{i\omega\tau} d\omega \tag{8.4.5}$$

The reason this is highly relevant is that the power spectral density does what its name implies. It tells us the density level of power that's present in a signal for any frequency of interest. If we want to know the total power between any two frequencies, then we need to integrate $S_{xx}(\omega)$ between these frequencies, taking into account both positive and negative ω's.

This particular descriptor is one of the cornerstones of spectral analysis. The outlook here is purely frequency based. Rather than worrying about the specific details of a signal in *time*, we're concerned with what's happening with respect to *frequency*. This can be made clear by an example.

Example 8.3

Problem   To see how the power spectral density gives us a frequency-based insight into our analyses, determine $S_{xx}(\omega)$ by means of $R_{xx}(t)$ for

$$x(t) = a \sin(\omega_0 t)$$

Solution   The autocorrelation is found from

$$R_{xx}(\tau) \equiv \lim_{T \to \infty} \frac{1}{T} \int_{-\frac{T}{2}}^{\frac{T}{2}} a \sin(\omega_0 t) a \sin(\omega_0(t + \tau)) \, dt$$

Expanding out the second sine term yields

$$R_{xx}(\tau) = \lim_{T \to \infty} \frac{a^2}{T} \int_{-\frac{T}{2}}^{\frac{T}{2}} \sin(\omega_0 t) \sin(\omega_0 t) \cos(\omega_0 \tau) \, dt$$

$$+ \lim_{T \to \infty} \frac{a^2}{T} \int_{-\frac{T}{2}}^{\frac{T}{2}} \sin(\omega_0 t) \cos(\omega_0 t) \sin(\omega_0 \tau) \, dt \qquad (8.4.6)$$

The second integral goes to zero as T goes to infinity, while the first will leave us with

$$R_{xx}(\tau) = \frac{a^2 \cos(\omega_0 \tau)}{2} \qquad (8.4.7)$$

This result makes sense because it tells us that the sine is maximally correlated with itself for delays of $\frac{2\pi}{\omega_0}$ (the times when $\cos(\omega_0 \tau)$ equals 1.0) and is not correlated at all for a time shift of $\frac{\pi}{2\omega_0}$. Since a time shift of $\frac{\pi}{2\omega_0}$ alters a sine into a cosine (which we know is an independent function), we'd expect zero correlation.

Equation (8.4.7) predicts that the mean-square value $R_{xx}(0)$ is equal to $\frac{a^2}{2}$. We can verify that this is so analytically. If $x(t) = a \sin(\omega_0 t)$, then $x^2 = a^2 \sin^2(\omega_0 t) = a^2 \left(\frac{1}{2} - \frac{\cos(2\omega_0 t)}{2} \right)$. Thus, since $\cos(2\omega_0 t)$ will average out to zero, we're left with the mean-square value being $\frac{a^2}{2}$, just as predicted.

Now that we've got the autocorrelation function, we can compute $S_{xx}(\omega)$:

$$S_{xx}(\omega) = \lim_{n \to \infty} \frac{1}{2\pi} \int_{-\frac{n\pi}{\omega_0}}^{\frac{n\pi}{\omega_0}} \frac{a^2 \cos(\omega_0 \tau)}{2} e^{-i\omega\tau} \, d\tau$$

You'll notice that this is a bit different from (8.4.4) in that it involves a limit. If we immediately tried to use the infinite limits of integration, we'd be unable to evaluate the integral. By taking the time integral to be integer multiples of the autocorrelation's period (using $\frac{n\pi}{\omega_0}$ instead of ∞), and letting the number of multiples go to infinity, we'll get the correct final solution and also be able to see the way in which we approach this solution.

Carrying out the foregoing integrations will leave us with

$$S_{xx}(\omega) = \lim_{n \to \infty} \frac{1}{4\pi} \left(\frac{a^2}{\omega_0 - \omega} \sin\left(n\pi - \frac{n\pi\omega}{\omega_0}\right) + \frac{a^2}{\omega_0 + \omega} \sin\left(n\pi + \frac{n\pi\omega}{\omega_0}\right) \right) \tag{8.4.8}$$

Figure 8.7 shows what this function looks like for different values of n. Only ω values greater than zero are shown. Since $S_{xx}(\omega)$ is an even function (which we haven't proved but is true nonetheless), there's no need to plot it for negative ω values—the behavior is identical to that for positive ω's. For $n = 4$ (Figure 8.7a) we see that a spike is developing at $\omega = 5$ rad/s (for this plot ω_0 was chosen to be 5). If we increase the integration duration by a factor of 20 (bringing n to 80) we get Figure 8.7b. What you can see from these two plots is that as n is increasing, the spike at $\omega = 5$ is also increasing. We can determine this analytically in a straightforward manner. Let $\omega = \omega_0 - \varepsilon$ in (8.4.8). The second term isn't going to be important because the first is going to be completely dominate the response. The first term becomes

$$S_{xx}(\omega) = \frac{a^2}{2\pi} \lim_{n \to \infty} \frac{\sin\left(\frac{n\pi\varepsilon}{\omega_0}\right)}{\varepsilon}$$

As we look at smaller and smaller values of ε, we obtain an indeterminant fraction. Expanding the sine in terms of its trigonometric representation (or using L'Hôspital's rule) gives us

$$S_{xx}(\omega_0) = \lim_{n \to \infty} \frac{na^2}{2\omega_0}$$

Clearly, as n goes to infinity, so does $S_{xx}(\omega)$.

Everything away from $\omega = \omega_0$ or $\omega = -\omega_0$ goes to zero, and we're left with two Dirac delta functions, one located at $\omega = \omega_0$ and the other at $\omega = -\omega_0$. The value of each Dirac delta function turns out to be .25, i.e., in the limit of $n \to \infty$ we have Figure 8.7c

$$S_{xx}(\omega) = \frac{a^2}{4}\delta(\omega + \omega_0) + \frac{a^2}{4}\delta(\omega - \omega_0) \tag{8.4.9}$$

This is a very relevant finding. It tells us that all the power in the sine wave is concentrated at one frequency. Thus when we run modal tests and find sharp power spectral density peaks, we'll know it means that there are distinct harmonics in the signal under analysis and that the frequencies of the harmonics correspond to the frequencies of the peaks. Note that we got this frequency-based information from an analysis of a time-based function (the autocorrelation function). The whole field of spectral analysis revolves around this viewpoint that's based in the frequency domain rather than in time.

A tremendously useful result, which we won't derive but shall simply present, is the following:

$$S_{xx}(\omega) = |g(\omega)|^2 S_{ff}(\omega) \tag{8.4.10}$$

where $g(\omega)$ is the transfer function of the system under consideration, $f(t)$ is the input to the system, and $x(t)$ is the system's output. $S_{xx}(\omega)$ is the power spectral density of the output, and $S_{ff}(\omega)$ is the power spectral density of the input.

For example, assume that our equation of motion was

$$m\ddot{x} + kx = f(t) \tag{8.4.11}$$

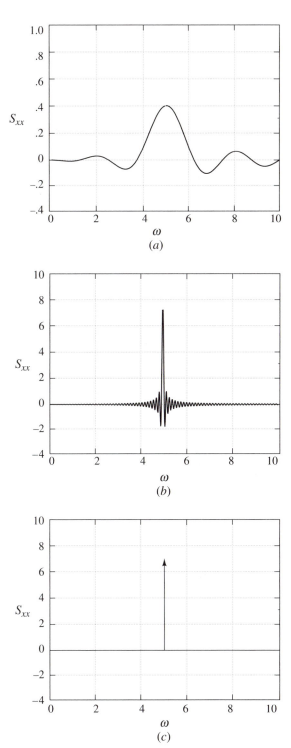

Figure 8.7 Power spectral density of a 5 rad/s sine wave with increasing numbers of cycles included in the calculation

in which case the transfer function is

$$g(t) = \frac{1}{k - m\omega^2} \tag{8.4.12}$$

If we know what the power spectral density of the forcing is, then we can immediately calculate the power spectral density of the output from (8.4.10) as

$$S_{xx}(\omega) = \left| \frac{1}{k - m\omega^2} \right|^2 S_{ff}(\omega) \tag{8.4.13}$$

Equation (8.4.10) can be used in a number of ways. The most obvious was just discussed, namely, finding the response power spectral density from knowing the system transfer function and the forcing power spectral density. But think about how else we can use it. One other use would be to determine the input excitation to a system by measuring the output power spectral density and dividing it by the square of the transfer function. This might be useful if we can't easily determine the input directly but have access to the system output and know the system characteristics.

This equation is most commonly used as a means of actually determining the system characteristics simply by measuring the input and output signals. Dividing (8.4.10) by $S_{ff}(\omega)$ gives us

$$|g(\omega)|^2 = \frac{S_{xx}(\omega)}{S_{ff}(\omega)} \tag{8.4.14}$$

Thus

$$|g(\omega)| = \sqrt{\frac{S_{xx}(\omega)}{S_{ff}(\omega)}} \tag{8.4.15}$$

Obtaining the power spectra of the input and output signals is usually easy. Common transducers for acquiring data of this sort would be an impact hammer to excite the system and produce an input signal and an accelerometer to produce output data. Both signals would be fed into a modal analyzer, which would calculate the system's transfer function in a manner similar to that just shown. Of course, in a real experimental setup there will be noise in the measurements, and the analyzer won't carry through exactly as shown in (8.4.15). The general approach, however, is valid and the differences between the analyzer's algorithms and (8.4.15) would simply be modifications introduced to minimize computational errors due to experimental error sources. Further information on this subject can be found elsewhere [3] and also (briefly) later in this section. One interesting point is that (8.4.15) doesn't include any phase information for the system's transfer function. This information can be found with a slightly different approach. Again, details are available [3].

Example 8.4

Problem   As an example of how the foregoing approach is used, we'll look at a problem we already know how to solve, namely, the response of a spring-mass under sinusoidal forcing. Consider the system shown earlier in Figure 2.10. Let the input force be $2\sin(\omega t)$ N, $m = 2$ kg, and $k = 100$ N/m. For this example we'll let $\omega = 10$ rad/s. Use two approaches to determine the mean-square value of the response, and compare the results. First, determine the power spectral density of the output, find the autocorrelation function from this, and evaluate the autocorrelation at $\tau = 0$ (giving the mean-square response). Next, solve the problem exactly and simply calculate the mean-square response from the known solution. Both methods should yield the same answer.

Solution   The transfer function for this problem is given by

$$g(\omega) = \frac{1}{k - m\omega^2}$$

The power spectral density of the input is given by (8.4.9)

$$S_{ff}(\omega) = \frac{2^2}{4}\,(\delta(\omega + 10) + \delta(\omega - 10))$$

Using (8.4.10) we find

$$S_{xx}(\omega) = \left|\frac{1}{k - m\omega^2}\right|^2 (\delta(\omega - 10) + \delta(\omega + 10))$$

Substituting the values for m, k, and ω yields

$$S_{xx}(\omega) = \frac{1}{10{,}000}\,(\delta(\omega - 10) + \delta(\omega + 10))$$

We can next take the inverse transform of this to find the autocorrelation $R_{xx}(\tau)$:

$$R_{xx}(\tau) = \frac{1}{10{,}000} \int_{-\infty}^{\infty} (\delta(\omega - 10) + \delta(\omega + 10))\, e^{i\omega\tau}\, d\omega$$

which yields

$$R_{xx}(\tau) = \frac{1}{5000}\cos(10\tau)$$

Evaluating this at $\tau = 0$ gives us

$$R_{xx}(0) = \frac{1}{5000} \tag{8.4.16}$$

Now let's see if this matches the result we'd get from calculating the actual forced response and evaluating the mean-square value from it. The amplitude response of the system is given by

$$\bar{x} = \frac{\bar{f}}{k - m\omega^2}$$

where $\bar{x}$ is the response magnitude and $\bar{f}$ is the forcing magnitude. Substituting the values for this example yields

$$\bar{x} = \frac{2}{100 - 2(100)} = -\frac{1}{50}$$

Thus the total response is given by

$$x(t) = -\frac{1}{50} \sin(10t)$$

The mean-square value of this is given by

$$E(x^2) = \left(\frac{1}{50^2}\right)\left(\frac{1}{2}\right) = \frac{1}{5000}$$

Note that this agrees with (8.4.16). We got the same answer by directly determining the mean-square value from a knowledge of the actual response, and by deducing it from a knowledge of the response's power spectral density.

We've now seen that the transfer function of a system relates the input power spectral density and the output power spectral density,

$$S_{xx}(\omega) = |g(\omega)|^2 S_{ff}(\omega) \qquad (8.4.17)$$

Two other formulas that are useful in modal analysis, which were alluded to a page or so ago, are

$$S_{fx}(\omega) = g(\omega)S_{ff}(\omega) \qquad (8.4.18)$$

and

$$S_{xx}(\omega) = g(\omega)S_{xf}(\omega) \qquad (8.4.19)$$

where the xf and fx indicate that these are power spectral densities found from the cross-correlation functions between the input and output. We can estimate the system transfer function in two more ways now. The first estimate, found from (8.4.18), is

$$g_1(\omega) = \frac{S_{fx}(\omega)}{S_{ff}(\omega)} \qquad (8.4.20)$$

and the second, found from (8.4.19), is

$$g_2(\omega) = \frac{S_{xx}(\omega)}{S_{xf}(\omega)} \qquad (8.4.21)$$

Noise in the measurements enters into these expressions in different ways, and they can be used together to find the best overall transfer function estimate. Normally, as a result of experimental problems, the estimates are not exactly the same. If we use the two estimates shown in (8.4.20) and (8.4.21), we can form their ratio:

$$\gamma^2 = \frac{g_1(\omega)}{g_2(\omega)} \qquad (8.4.22)$$

where γ^2 is called the *coherence function*. This frequency-dependent function is a widely used measure of how "believable" our measurements are. We won't derive it here, but it can be shown that γ^2 always lies between 0 and 1. If the coherence is near zero, our measurements should be viewed with great suspicion, whereas a coherence near 1.0 indicates that the $g_1(\omega)$ and $g_2(\omega)$ are in very close agreement, which in turn implies that the measurements are believable and that noise and nonlinear effects aren't significant. Most analyzers will output the coherence whenever they compute a system transfer function, allowing the user to decide whether to accept or reject the computed results.

8.5 NOISE

Now that we've seen how to analyze general random inputs, it is time to examine the most common random signal around, namely *white noise*. Returning to Figure 8.1, you can see what a white noise signal might look like when plotted versus time—clearly, it is a very jagged function, not at all regular. It is so irregular, in fact, that it is completely unpredictable from one instant to the next. This means that its autocorrelation function is simpy a Dirac delta function, $\delta t - \tau$. For $\tau = 0$ the signal is completely correlated, and for any other τ it is completely uncorrelated. If we denote the magnitude of the Dirac delta function by w_0, then our signal's autocorrelation is given by

$$R_{xx}(\tau) = w_0 \delta(\tau) \tag{8.5.1}$$

The power spectral density for this function is given by

$$S_{xx}(\omega) = \frac{w_0}{2\pi} \tag{8.5.2}$$

As you can see (Figure 8.8), the power spectral density is a constant, regardless of frequency. Thus the density of power is the same at all frequencies.

Does this make sense from a physical point of view? Well, if the output is a constant at all frequencies, then we can integrate over all frequencies to find the total power of the signal. Since

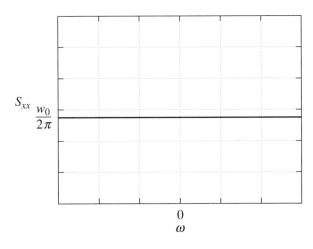

Figure 8.8 Constant power spectral density

the power spectral density has the same value everywhere and extends out to infinity, the integral is infinite. Thus the signal must have infinite energy. This is a problem. It's pretty obvious that we can't produce a signal that has an *infinite* amount of energy; all real signals must be finite.

What's the conclusion? Simply that our assumption of a constant power spectral density over all frequencies must be erroneous. At some point, the power spectral density has to drop off, as shown in Figure 8.9. This figure shows what all real-world, "white" signals really look like. They're reasonably constant over some range, and then the energy drops off to zero. An idealization of this situation is shown in Figure 8.10, in which we assume a constant level to ω_1 and then have an immediate cutoff to zero. Such signals are called *band-limited* noise.

Luckily, the difference between ideal white noise and actual white noise doesn't really affect our analyses. As we've already seen, the transfer function of our mechanical system will drop to zero as we pass its highest natural frequency. Thus, even if we had an ideal white noise input, the response

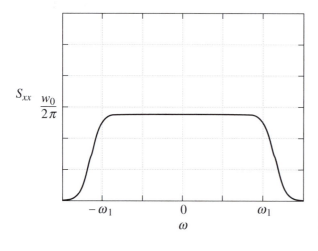

Figure 8.9 Power spectral density of band-limited white noise (gradual frequency cutoff)

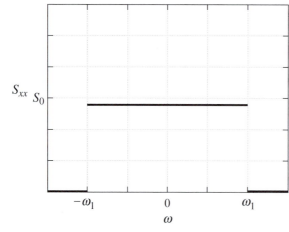

Figure 8.10 Power spectral density of band-limited white noise (sharp frequency cutoff)

will drop to zero at high frequencies because the transfer function itself is going to zero. We just have to make sure that the white noise signal is pretty much constant up to and a little beyond the highest natural frequency of the system of concern.

Another way of seeing how ideal white noise differs from what we find in real life is to realize that the autocorrelation won't really be a delta function. If the output in time was in fact changing in a completely random manner, by implication the acceleration experienced in the system would be infinite, since only an infinite acceleration could move the output from one state to another in an infinitesimal time span. Since we know that real systems are to some extent limited regarding the speed with which they can change, we know that an actual signal at $t = \tau$ will be close to its value at $t = 0$, for small τ. An "infinitely jagged" function (such as ideal white noise will have) requires energy at ultrahigh frequencies. This high frequency content is what lets the signal "wiggle" so quickly.

We can compute the autocorrelation corresponding to a more realistic, band-limited signal to see how cutting off the spectral energy at some finite frequency affects matters. Figure 8.10 shows our power spectral density. It is constant at a level S_0 from $-\omega_1$ to ω_1 and zero everywhere else. Calculating the autocorrelation for this case is straightforward. Recalling that for the general problem we have

$$R_{xx}(\tau) = \int_{-\infty}^{\infty} S_{xx}(\omega)e^{i\omega\tau}\,d\omega \tag{8.5.3}$$

for our particular problem we write

$$R_{xx}(\tau) = S_0 \int_{-\omega_1}^{\omega_1} e^{i\omega\tau}\,d\omega = 2S_0\frac{\sin(\omega_1\tau)}{\tau} \tag{8.5.4}$$

This autocorrelation function is plotted in Figure 8.11. For this plot the value of S_0 was set equal to .5 and ω_1 was 50 rad/s. As you can see, the correlation drops off quickly, but not instantly. As τ increases, the autocorrelation goes to zero. Finally, the magnitude at $\tau = 0$ is finite, not infinite as would be the case for a Dirac delta function.

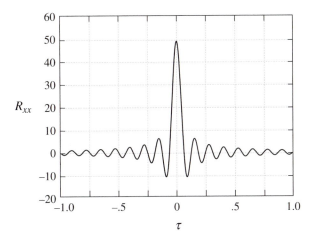

Figure 8.11 Autocorrelation of band-limited white noise

Figures 8.12–8.15 give an overview of the most basic responses we might expect to see. Figure 8.12 shows the relationship between white noise (a function of time), its autocorrelation (a function of time delay τ), and its power spectral density (a function of frequency). For white noise, the autocorrelation $R_{xx}(\tau)$ is a pure delta function and the power spectral density $S_{xx}(\omega)$ is a constant.

Real-world effects mean that the closest we'll ever get to Figure 8.12 is the situation shown in Figure 8.13. The time function looks similar to that of the preceding case. The only difference is that if we examined it in fine detail we'd see that the response of Figure 8.13a is continuous, while that of Figure 8.12a is discontinuous (implying finite changes of output for infinitesimal changes in time). The autocorrelation is therefore not a pure Dirac delta function, and the power spectral density drops off at high frequencies.

Figure 8.14 shows what we'll see if the signal is a pure sinusoid. The autocorrelation for this case is also sinusoidal (telling us that the signal is periodic), and the power spectral density is composed of Dirac deltas at the frequency of the input signal. As we've already found (Figure 8.7), this precise response won't ever be seen because we'd have to observe the input sinusoid for an infinitely long time to get a power spectral density that's in the form of a delta function. Furthermore, most real signals will have some noise superimposed on them, as shown in Figure 8.15a. The autocorrelation of this signal is shown in Figure 8.15b. The effect of the noise is to create a large response at $\tau = 0$, whereas the rest of the response is dominated by the sinusoidal signal (and thus gives us largely periodic behavior). Finally, Figure 8.15c shows the power spectral density. Since the noise is independent of the sinusoidal signal, we end up with the delta functions associated with sinusoids, along with a low level of constant energy density (associated with the noise).

8.6 SENSORS AND ACTUATORS

If you're going to run a modal test, a few basic ingredients must be present. You need some way of exciting the system, you need a way to sense the system's response, and you need a device that can use the input and output information to deduce some of the system properties. Let's start with the question of excitation. The two most widely used excitations are the *impact hammer* and the *shaker* (electrohydraulic or electromechanical).

The *impact hammer* (Figure 8.16) looks very much like an ordinary hammer and is used in a similar way. One holds the impulse hammer in much the same way as one holds a regular hammer and quickly strikes the structure being tested. Some care is necessary to make sure that the hammer hits the structure only once. (If the hammer isn't held correctly, the head may contact the structure two or more times in quick succession, ruining the data acquisition.) We saw in Section 3.4 that loadings that occur quickly look and act like delta functions. We've also found that delta functions have a constant power spectral energy. This characteristic is great for our analyses. Recall that the absolute value of the transfer function can be found from (8.4.15)

$$|g(\omega)| = \sqrt{\frac{S_{xx}(\omega)}{S_{ff}(\omega)}} \tag{8.6.1}$$

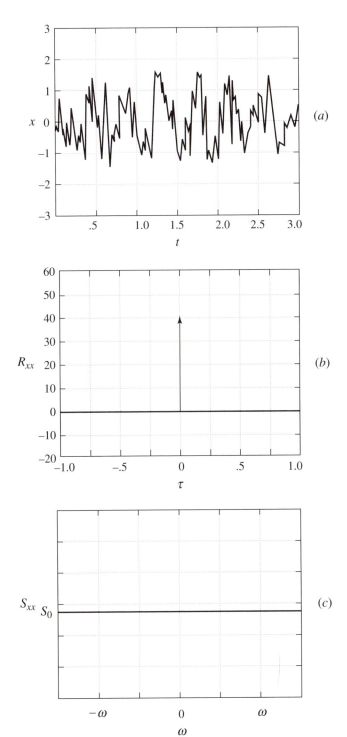

(a)

(b)

(c)

Figure 8.12 (a) Time trace, (b) autocorrelation, and (c) power spectral density for white noise

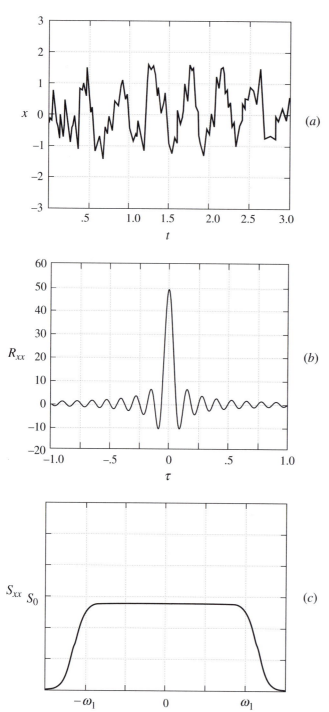

Figure 8.13   (*a*) Time trace, (*b*) autocorrelation, and (*c*) power spectral density for band-limited white noise

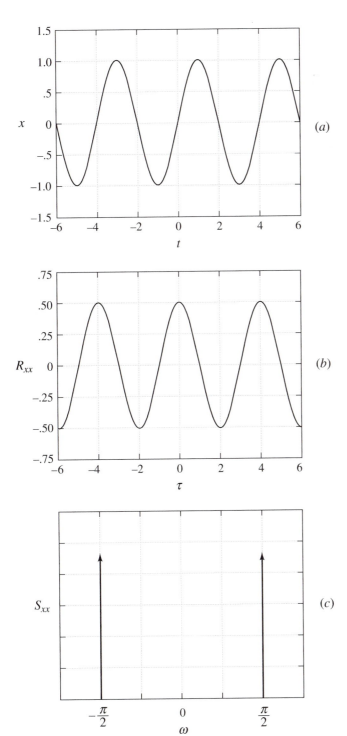

Figure 8.14 (a) Time trace, (b) autocorrelation, and (c) power spectral density for $\sin\left(\frac{\pi}{2}t\right)$

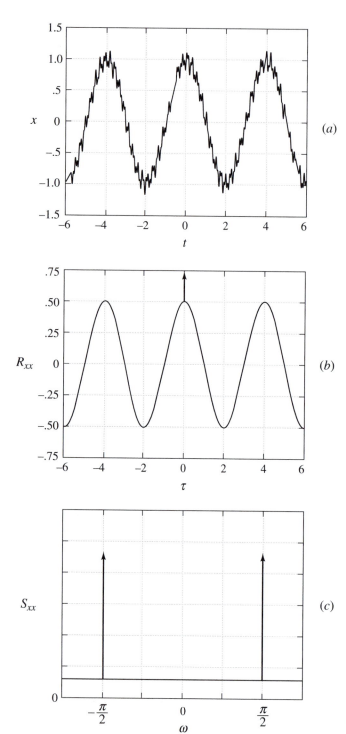

Figure 8.15 (*a*) Time trace, (*b*) autocorrelation, and (*c*) power spectral density for a combination of white noise and a sinusoidal signal

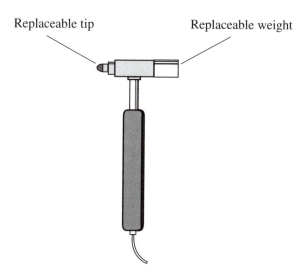

Replaceable tip   Replaceable weight

Figure 8.16 Impact hammer

If the input power spectral density is a constant (call it S_0), then we have

$$|g(\omega)| = \frac{\sqrt{S_{xx}(\omega)}}{\sqrt{S_0}} \tag{8.6.2}$$

Thus instead of having to take the ratio of two frequency-dependent functions, we need only divide the output's power spectral density by a constant. Actually, our analyzers will still calculate the ratio of the two power spectral densities, but since the input is essentially flat, calculation errors are minimized.

Just behind the tip of the head (the part striking the structure) is a piezoelectric sensor that generates a signal proportion to the impact force. By itself, this element is known as a *force transducer*. A cable runs from the hammer to a modal analyzer, supplying the output of the force transducer as an input to the analyzer. The analyzer can then determine what the precise force profile versus time was during the strike and from that calculate the power spectral density of the input force.

Impact hammers come with weights that can be added to the head, and different tips. The weights allow different force levels to be input to the structure, and the tip hardness determines how long the tip stays in contact with the structure (which affects the spectral energy input). Using a hard tip gives a wider frequency input but lowers the overall energy that's transferred to the structure. Thus the hard tips allow you to analyze higher frequency responses but might not supply enough excitation energy for accurate calculations. Of coure, softening the tip lets you put more energy in at the expense of a restricted excitation bandwidth.

The advantages of impact hammers are that they are fast to use and relatively inexpensive (on the order of $1000). In addition, they aren't attached to the structure and therefore don't alter its dynamic characteristics. They also have the advantage of being able to input energy at any point of the structure you can get a clean hit on. One disadvantage is that you don't know *precisely* where you're hitting the structure, and thus you're bringing some uncertainty into your measurements. In addition, you might hit the structure too hard and damage it, but if you don't hit it hard enough, you may not be able to input enough energy to adequately excite the system.

The second type of excitation device is known as a *shaker*. These devices move the test structures mechanically (electromechanical shaker) or hydraulically (electrohydraulic shaker). The more common type is the electromechanical shaker (Figure 8.17). If you're familiar with a loudspeaker then you're also familiar with electromechanical shakers—there's really little fundamental difference between the two. The speaker cone moves because a voice coil is caused to vibrate owing to an alternating current. This is precisely what happens in an electromechanical shaker. The only difference is in the power and accuracy involved. Shakers must be capable of producing a significant time-varying force with which to excite the test structure (unlike loudspeakers, which merely have to move the speaker cone and a bit of air).

The shaker is physically connected to the test structure by means of a thin rod of plastic or metal, known as a stinger. The stinger is much stiffer along its length than laterally. Thus if it is placed in line with the shaker's axis of movement, it effectively transmits the shaker forces but doesn't cause unwanted twisting of the structure.

Because we are able to control the electrical signal reaching the shaker, we can produce any input waveform we like. The most obvious signal would be a sinusoidal one, allowing us to determine the structure's response at a single frequency. We could also input a random signal

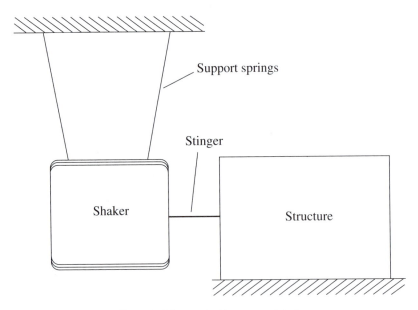

Figure 8.17   Elastically supported shaker

(band-limited white noise), which would let us calculate the entire frequency response of the system. Another common input is called a swept sine. This consists of a sinusoidal excitation that starts at low frequencies and ramps upward. This signal also allows us to determine the structure's complete frequency response. Transient inputs are also possible and are used to mimic shock loadings.

The advantages of a shaker are the variety of excitations that can be used, the range of force levels, and the accuracy with which the inputs can be applied. The disadvantages are cost (small shakers and associated power amps can run $2,000, with larger ones hitting $10,000–$20,000 without difficulty) and the fact that shakers modify the structures they are attached to. The stinger and the moving piece of the shaker to which the stinger is attached have their own mass and stiffness and therefore affect the structure. Shakers are also heavy, and therefore it is often difficult to attach them to the desired place on a structure. Often they are suspended near the structure by elastic cords, which serve to dynamically isolate them as well as to position them. Shakers come in a wide range of sizes. The smallest ones in general use are about the size of a baseball. The largest are called shaker tables, and the intent is to place the entire structure *on* the shaker to seismically excite it. Custom-made shaker assemblies (used in earthquake studies, not surprisingly) can be big enough to shake a small building.

Of course, it's not enough to just excite a structure—we also have to measure the response. The most common device used to measure a structure's response are *accelerometers* (discussed earlier in Section 2.12). These devices are usually piezoelectric in nature and allow us to determine the structure's acceleration as long as we make sure that the natural frequency of the accelerometer is greater than the highest frequency we're interested in measuring. Accelerometers come in a variety of masses, from .5 to 500 g. The mass is clearly of concern if the test structure is light. Since the accelerometer is physically attached to the structure, the mass is increased and must be analytically accounted for if the test results are to be accurate.

It always seems strange to someone new to modal testing, but the most common means of attaching an accelerometer is to simply stick it to the structure with a thin layer of bees wax. Alternatively, one can mount it semipermanently by screwing it to the structure.

A fairly new class of measurement devices is laser based. A laser beam is aimed at a structure, reflected back to the apparatus (called a *laser vibrometer*), and compared with the output beam. The laser vibrometer uses this information to determine what the structure's vibrational motions are like at the target point. Since there is no loading on the structure, this method is an excellent way of examining very lightweight test objects (for which an accelerometer would represent so much additional mass loading that accurate results would be impossible). The major drawbacks to this approach are cost (the instruments start at $15,000–$20,000) and the difficulty of getting the laser signal reflected back to the vibrometer.

An old sensor that is still widely used is the *strain gauge*, a device composed of a material that experiences a resistance change when subjected to strain. By measuring the resistance variation, one can deduce the local strain of the test specimen. The sensors are very cheap but are also relatively noisy and have a limited dynamic range.

Another new way of sensing and actuating a structure is by means of a distributed piezoelectric material. In addition to piezoelectric crystals (used in force transducers and accelerometers), large flat pieces of flexible or rigid piezoactive materials are available. Since the material is distributed over a wide area, one can use it to directly sense particular modes of the structure, or conversely, to excite particular modes. Another variant on this approach is to combine the sensing and actuation into a single piece of piezoelectric material, resulting in a single device that both excites a structure and measures the structure's vibrations.

After a signal has been sensed, it is amplified through an appropriate *signal conditioner* (charge amplifier or voltage amplifier) and then fed to an *analyzer*. The range of analyzers (known as vibration analyzers, modal analyzers, spectrum analyzers, etc.) is very wide nowadays. One can purchase simple two-channel units (one input and one output) as well as sophisticated multichannel ones capable of handling 64 channels or more of data. A few years ago all the analyzers were stand-alone pieces of equipment, but it is becoming more common to see analyzers that work along with a dedicated computer. In this way you can mix and match the amount of analyzer function and computational speed you desire. If you've already got the computer, you can add the analyzer section for a relatively low cost and buy only as many channels of input and output capability as you wish.

All these devices attempt to determine the transfer functions between the input and output signals. For instance, the output might be a sinusoidal signal that is fed to a shaker. The input would be a signal proportional to the structure's acceleration, as measured by an accelerometer. If the test was done at a single frequency, the analyzer would determine the amplitude and phase response of the structure, for this frequency of excitation. If the input was a random signal, the analyzer would use the spectral analysis we've gone over to determine the complete transfer function over a range of frequencies. By analyzing the peak responses, the half-power points, the zeros, etc., the analyzer could also construct estimates of the natural frequencies and damping of the system.

By getting responses from multiple points along the structure, we can build up a database of response information that will allow the analyzer to compute the actual mode shapes of the structure. Analyzers that do this are called *modal analyzers*. Sophisticated units will allow the user to create a model of the system from experimental data and then modify the system to see how changes will affect the overall vibrational response.

8.7 NONLINEAR EFFECTS

A central assumption in all the work so far has been linearity. At the risk of being obvious, this means that all the physical quantities we measure vary in a linear manner. Thus the force in a spring is linearly proportional to the stretch—double the extension and the force doubles. The same is true for the force necessary to accelerate a mass—double the force and mass's acceleration doubles (if there are no additional forces acting on it). Thus a nonlinearity in the system implies that something isn't behaving in a linear manner. One obvious example is a coil spring, like those found

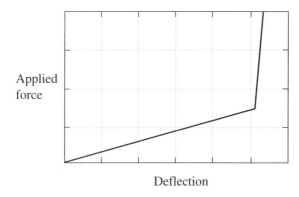

Applied
force

Deflection

Figure 8.18   Force versus deflection for
a coil spring

in automobile suspensions. For small oscillations this spring will behave linearly. What happens, however, if the load in the car becomes so great that the spring is completely compressed and all the coils touch each other? In this case it won't act anything like a linear spring. Additional forces now serve to compress the actual metal of the spring, rather than simply bending a coil. Thus the effective spring constant will be immense. This behavior is shown in Figure 8.18. The deflection versus force shows a normal linear relationship up until the critical deflection, at which point the coils become fully compressed. After this the force versus deflection curve still behaves linearly, but at a much increased spring constant. Physical elements that display such relationships are called piecewise linear. The overall behavior is nonlinear, but in fixed regimes the behavior can be well approximated as linear.

Sources of nonlinearity are common. One example occurs whenever two surfaces rub against each other. The friction between a sliding surface and a fixed surface is given by μN, where μ is the coefficient of friction and N is the normal force. Certainly this isn't linear—it shows no variation at all with velocity, let alone a linearly proportional one. This kind of friction (called Coulomb or dry friction) will always be present in the joints and connections of any structure.

What happens to our linear analyses when an unmodeled nonlinearity is present? If the non-linearities are small, there's no great effect. However as they become more significant they can substantially alter the modal analyzer's calculations (not unexpectedly). One of the difficulties facing the analyst is trying to determine whether problems with the analyses are due to non-linearities in the system or to noise. To illustrate how nonlinearities can alter a system's response, look at Figures 8.19–8.23. The system is a simple vibrating beam—with a twist (Figure 8.19), a metal block placed near the beam's midspan. When the block is far from the beam, the beam never touches it, resulting in a linear system. As the restraint is moved closer, it becomes more and more common for the beam to strike the restraint as it is vibrating—a strongly nonlinear effect. Four cases are examined, corresponding to linear, weakly nonlinear, moderately nonlinear, and strongly nonlinear.

Figure 8.20 shows the fully linear response (case 1). Two peaks are present, at approximately 30 and 175 rad/s. A zero is also indicated at about 105 rad/s. This simply means that the output is

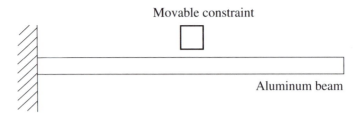

Movable constraint

Aluminum beam

Figure 8.19 Clamped-free beam with rigid constraint

close to zero when the system is excited at that frequency. The phase response is similarly clear. The phase angle versus frequency is very well defined and shows that the system is purely in phase or out of phase, depending upon the frequency. The phase switches around each resonance peak as well as at the zero. Finally, the coherence is excellent. Everywhere away from the two peaks and the zero we see a coherence of 1.0. This result is typical of a good modal test. You shouldn't expect a good coherence around resonances or zeros. At resonance, the amplification effect magnifies any noise in the excitation device, affecting $S_{ff}(\omega)$. When at a zero, the actual response is very small. In this case the measured response will be dominated by sensor noise and $S_{xx}(\omega)$ will be adversely affected. The fact that the coherence is excellent everywhere else is a good assurance that the test results are believable.

In Figure 8.21 (case 2) we show the results of having just a little bit of contact with the block. Because of this contact, we've got a bit of nonlinear contamination. Of course the linear algorithms in the analyzer don't know this, and so they go ahead as if the system were linear. The phase and amplitude responses have changed a bit. The most obvious change is the apparent reduction in damping (shown by the flatter resonance peak) and the "noise" in the phase response. The real indicator that we've got a problem is the coherence results. As you can see, the coherence has degraded quite a bit, especially around the zero (105 rad/s).

A moderate nonlinearity (Figure 8.22, case 3) causes further degradation. The amplitude response is now very irregular, as is the phase response. Hardly any frequencies show a good coherence response. Finally, Figure 8.23 (case 4) shows the results for a severe nonlinearity. For this case the block is being struck for almost every oscillation of the beam. The results are clearly a mess. The phase response is completely garbled, and we've lost any sense of separate resonance peaks in the amplitude response. As expected, the coherence is terrible, indicating that we shouldn't have any faith at all in these responses.

Table 8.1 shows the first two calculated natural frequencies for the tests in Figure 8.20–8.23, as well as the calculated linear damping approximation. Since the block provides a constraint, and constraints typically stiffen systems, we would expect it to cause the calculated first natural frequency to increase, as indeed happens. Note also the wide variation in the damping estimates. This simply reflects the fact that the measurements are contaminated by the nonlinearity and thus the damping algorithms work badly, trying to find the linear damping response for a system that isn't truly linear.

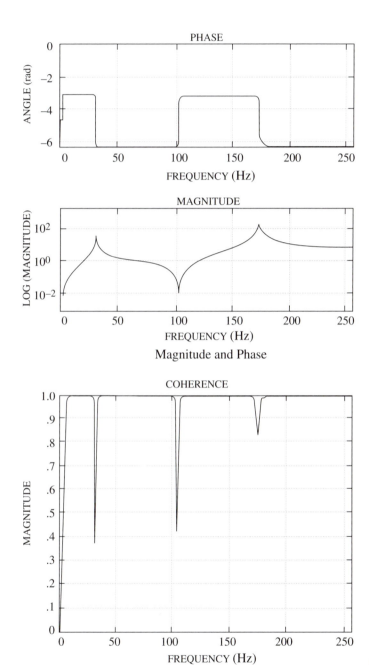

Magnitude and Phase

Coherence

Figure 8.20 Linear beam response (case 1)

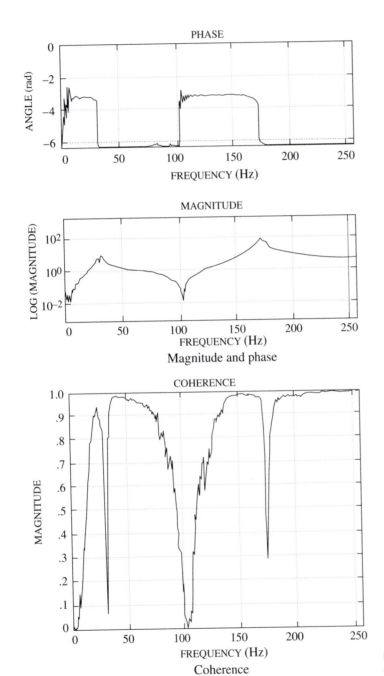

Magnitude and phase

Coherence

Figure 8.21 Nonlinear beam response (case 2)

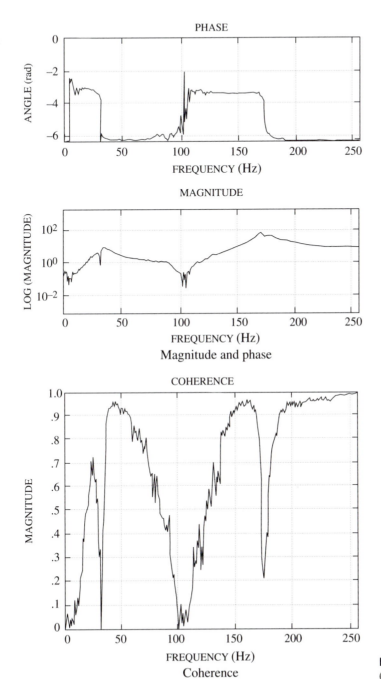

Magnitude and phase

Coherence

Figure 8.22   Nonlinear beam response (case 3)

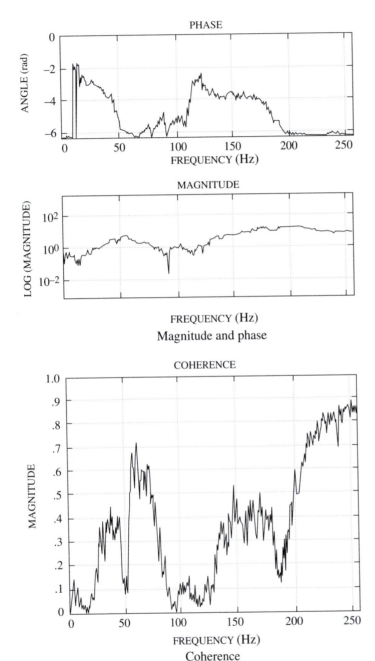

Magnitude and phase

Coherence

Figure 8.23 Nonlinear beam response (case 4)

TABLE 8.1 FREQUENCY/DAMPING ESTIMATES FOR NONLINEAR BEAM VIBRATION

	Mode 1		Mode 2	
	Frequency	Damping	Frequency	Damping
Case 1	31.2	0.009	174	0.005
Case 2	32.2	0.011	174	0.005
Case 3	34.8	0.003	173	0.020
Case 4	56.0	0.110	174	0.051

8.8 HOMEWORK PROBLEMS

Section 8.2

8.1. Assume that

$$x(t) = \sin^3(t)$$

is the signal of concern. Generate a discrete time series by recording n values of $x(t)$ for

$$t = \frac{2\pi i}{n} \quad i = 1, \dots, n$$

Let $n = 10, 20$, and 100. Does the average of these values equal 0 (which is the correct average value for the continuous signal $x(t)$)?

8.2. Repeat Problem 8.1 but let

$$t = \frac{\pi}{4} + \frac{2\pi i}{n} \quad i = 1, \dots, n$$

That is, include a phase shift. Does the average come out to zero? Why?

8.3. Repeat Problem 8.1 but take the average of a 20-second interval, i.e., average

$$x(t) = \sin^3(t)$$

over $0 \le t \le 20$. Use a 20-point discretization. Does the average equal zero? Why? How does this impact a real-world data analysis?

8.4. Repeat Problem 8.1 but vary the averaging period. Take the average over 10, 20, and 100 seconds. To keep the resolution constant, use 10 data points for the 10-second average, 20 for the 20-second average, and 100 for the 100-second average. Does the average for each trial equal zero? How do longer averaging times affect the average? Comment on the implications for data analysis.

8.5. Can you find a finite averaging time for which a discrete sampling of

$$x(t) = \cos(t) + \sin(2t)$$

will average to zero? What about

$$x(t) = \cos(t) + \sin(\sqrt{3}t)$$

8.6. Calculate the variance of

$$x(t) = 2\sin(4\pi t) - 5\cos(2\pi t)$$

over the time interval $0 \leq t \leq 1$. Demonstrate numerically that the variance of the sum of the two signals is equal to the sum of the variance of the individual components.

8.7. Problem 8.6 showed that the sum of the variance is equal to the variance of the sum when several signals are being added together. Examine the same function:

$$x(t) = 2\sin(4\pi t) - 5\cos(2\pi t)$$

and compute the variance for 20, 200, and 600 data points. Use a sampling interval of .0415 second. Explain the results.

Section 8.3

8.8. Compute the complex Fourier transform of the $x(t)$ shown in Figure P8.8.

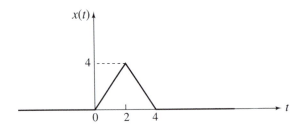

Figure P8.8

8.9. Compute the complex Fourier transform of the $y(t)$ shown in Figure P8.9. How does the answer differ from that of Problem 8.8? Explain how you can obtain the answer to Problem 8.9 if you already know the answer to Problem 8.8 (i.e., without explicitly calculating the Fourier transform).

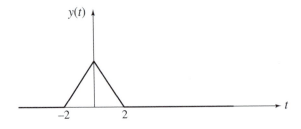

Figure P8.9

8.10. Show that the Fourier transform of a signal $y(t)$, where $y(t)$ is simply a delayed version of a signal $x(t)$, is equal to the Fourier transform of $x(t)$ multiplied by $e^{-i\tau\omega}$, where τ is the time delay.

8.11. Show that the complex Fourier transform is purely imaginary if the time function is odd and the transform is purely real for even functions. (*Hint*: Consider what the Fourier series representation of odd and even functions looks like.)

8.12. Calculate the Fourier transform of $x(t)$ and $y(t)$ shown in Figure P8.12. How does the delay in time appear in the Fourier transform?

$$x(t) = \delta(t)$$

$$y(t) = \delta(t - t_0)$$

Figure P8.12

Section 8.4

8.13. Find the solution to $\ddot{x} + .2\dot{x} + 4x = \delta(t - 1)$; $x(0) = 0$, $\dot{x}(0) = 0$.

8.14. Find the solution to $\ddot{x} + .2\dot{x} + 4x = \delta(t - 1)$; $x(0) = 1$ cm, $\dot{x}(0) = 0$.

8.15. Find the solution to $\ddot{x} + .2\dot{x} + 4x = \delta(t)$; $x(0) = 0$, $\dot{x}(0) = 2$ m/s.

8.16. What does the impulse response look like for an overdamped system?

8.17. Calculate and plot $R_{xx}(\tau)$ for the power spectral density illustrated in Figure P8.17. Let $\omega_1 = 1.6$ rad/s and $\omega_2 = 2$ rad/s.

8.18. How does the answer to Problem 8.17 change as ω_1 goes toward ω_2? (Keep the total energy constant.)

8.19. Consider the system shown in Figure P8.19. Let the forcing be described by band-limited white noise. Such an excitation has a power spectral density like that shown in the Figure P8.19. The input is unpredictable and has energy from $\omega = 10$ to 100 rad/s. Determine the mean-square value of the system's response. $m = 1$ kg and $k = 90$ N/m.

8.20. Verify your answer to Problem 8.19. Break the forcing spectra into segments of width $\Delta\omega$ as shown in Figure P8.20. Using the information in the text, determine what a delta function of magnitude

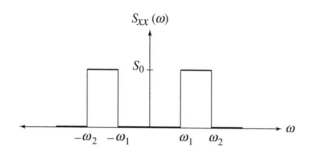

Figure P8.17

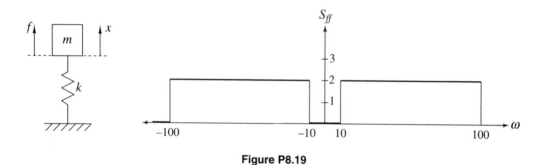

Figure P8.19

equal to $2\Delta\omega$ and centered at $10 + \frac{\Delta\omega}{2} + i\Delta\omega$ (i.e., the center points of s_1, s_2, s_3, etc.) corresponds to in the time domain (*Hint*: It corresponds to a sine wave of frequency $10 + \frac{\Delta\omega}{2} + i\Delta\omega$ with a magnitude that remains for you to determine.) Calculate the mean-square response of the system to each sine and sum the responses. This should give you the same result you found from Problem 8.19, since we're just approaching the problem from the time domain rather than the frequency domain.

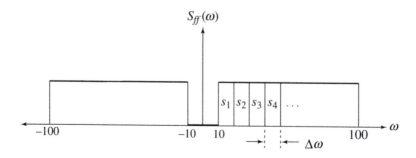

Figure P8.20

8.21. In this problem we'll show an alternative way to numerically evaluate the answer to Problem 8.19. As in Problem 8.20, break S_{ff} into segments and approximate the input by a string of δ functions (Figure P8.21):

$$\omega_1 = 10 + \frac{\Delta\omega}{2}$$

$$\omega_2 = 10 + \frac{3\Delta\omega}{2}$$

$$\omega_3 = 10 + \frac{5\Delta\omega}{2}$$

$$\vdots$$

Evaluate the spectral integral relating $S_{xx}(\omega)$, $\frac{1}{k-m\omega^2}$, and $S_{ff}(\omega)$ by utilizing the δ approximation and show that you obtain the same answer as in Problems 8.19 and 8.20.

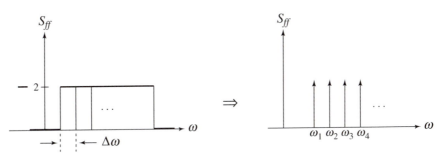

Figure P8.21

8.22. Determine the autocorrelation function for the $x(t)$ shown in Figure P8.22.

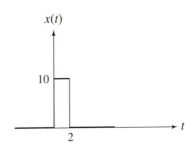

Figure P8.22

8.23. How does the autocorrelation change from that of Problem 8.22 if the rectangular function is repeated as shown in Figure P8.23?

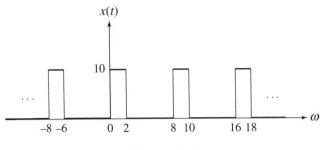

Figure P8.23

8.24. Determine the autocorrelation function for the function having the power spectral density function shown in Figure P8.24. The magnitude of each delta function is 20. What kind of signal does this correspond to?

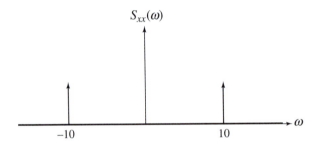

Figure P8.24

8.25. Use Figure P8.25 to determine the autocorrelation function for the function having the following power spectral density function:

$$S_{xx}(\omega) = 20\delta(\omega - 10) + 20\delta(\omega + 10) + 40\delta(\omega)$$

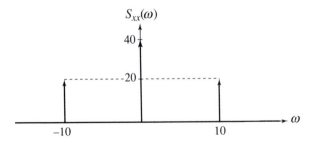

Figure P8.25

8.26. Find the mean-square response of a system having a transfer function

$$g(\omega) = \frac{2}{8 + 2i\omega}$$

for an ideal white noise input; i.e., the input has a power spectral density equal to

$$S_{ff}(\omega) = S_0$$

8.27. How does the answer to Problem 8.26 change if the input is band limited so that

$$S_{ff}(\omega) = S_0; \quad -7 \leq \omega \leq 7$$
$$S_{ff}(\omega) = 0; \quad |\omega| > 7$$

Section 8.7

8.28. If you're concerned about nonlinearities in your system, you might use an impact hammer and tap the structure lightly, identify the system parameters by means of a modal analyzer, and then repeat the procedure with a more forceful hammer strike. How could you use the resulting information to determine whether the structure is acting nonlinearly?

8.29. Numerically integrate

$$\ddot{x} + x + x^3 = 0; \quad \dot{x}(0) = 0$$

and compare the frequency of the resultant oscillations for $x(0) = .1$, $x(0) = 0.2$ and $x(0) = 0.6$. Does the nonlinearity act to "stiffen" the system (raise the natural frequency or "soften" it (lower the natural frequency)? Plot the force versus deflection curve for the equivalent spring

$$F = x + \varepsilon x^3$$

and try to explain why the system behaves as it does.

Four Continuous Systems

Figures A.1–A.4 present four simple continuous systems that have qualitatively identical equations of motion, Figure A.5 gives the equation of motion for a uniform beam.

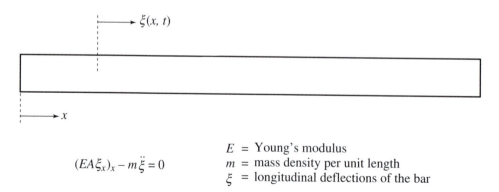

$$(EA\xi_x)_x - m\ddot{\xi} = 0$$

E = Young's modulus
m = mass density per unit length
ξ = longitudinal deflections of the bar

Figure A.1 Equation of motion for a bar

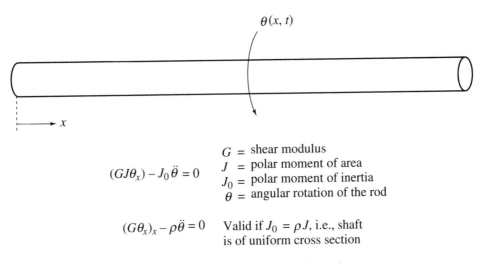

$$(GJ\theta_x) - J_0\ddot{\theta} = 0$$

G = shear modulus
J = polar moment of area
J_0 = polar moment of inertia
θ = angular rotation of the rod

$$(G\theta_x)_x - \rho\ddot{\theta} = 0$$

Valid if $J_0 = \rho J$, i.e., shaft
is of uniform cross section

Figure A.2   Equation of motion for a rod

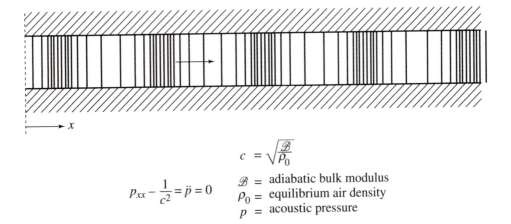

$$c = \sqrt{\frac{\mathscr{B}}{\rho_0}}$$

$$p_{xx} - \frac{1}{c^2} = \ddot{p} = 0$$

$\mathscr{B}$ = adiabatic bulk modulus
ρ_0 = equilibrium air density
p = acoustic pressure

Figure A.3   Equation of motion for one-dimensional duct acoustics

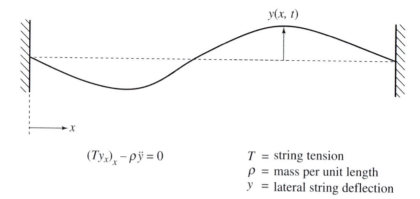

$$(Ty_x)_x - \rho \ddot{y} = 0$$

T = string tension
ρ = mass per unit length
y = lateral string deflection

Figure A.4 Equation of motion for a tensioned spring

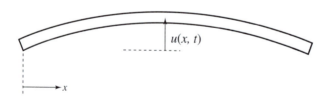

$$(EIu_{xx})_{xx} + m\ddot{u} = 0$$

E = Young's modulus
m = mass density per unit length
I = area moment of inertia
A = cross-sectional area
u = beam deflection

Figure A.5 Equation of motion for an Euler-Bernoulli beam

B

Lumped Spring Constants

Figures B.1–B.7 illustrate single DOF, lumped spring constants for several common situations.

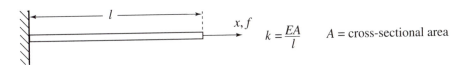

$$k = \frac{EA}{l} \qquad A = \text{cross-sectional area}$$

Figure B.1 Longitudinal bar

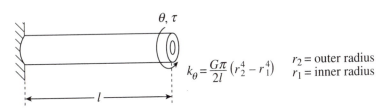

$$k_\theta = \frac{G\pi}{2l}\left(r_2^4 - r_1^4\right) \qquad \begin{array}{l} r_2 = \text{outer radius} \\ r_1 = \text{inner radius} \end{array}$$

Figure B.2 Torsional hollow rod

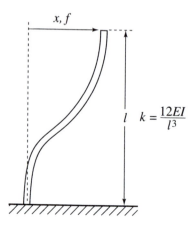

$$k = \frac{12EI}{l^3}$$

Figure B.3 Lateral motion of cantilevered beam (both ends constrained to remain vertical)

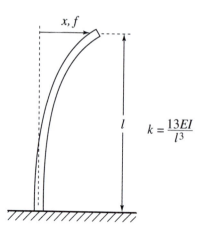

$$k = \frac{13EI}{l^3}$$

Figure B.4 Lateral motion of cantilevered beam (free end allowed to rotate)

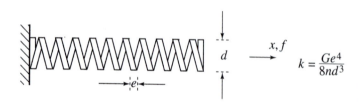

$$k = \frac{Ge^4}{8nd^3}$$

Figure B.5 Extensional motion of a coil spring ($n =$ number of turns in coil)

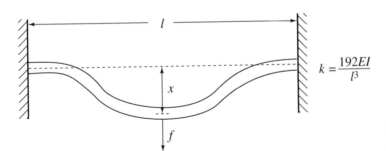

$$k = \frac{192EI}{l^3}$$

Figure B.6 Lateral motion of clamped-clamped beam

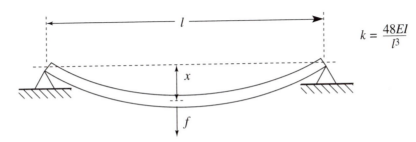

$$k = \frac{48EI}{l^3}$$

Figure B.7 Lateral motion of pinned-pinned beam

C

Assorted Material Constants

Tables C.1 and C.2 list some material constants for several common materials.

TABLE C.1 DAMPING FACTORS OF COMMON MATERIALS

Material	Damping factor
Steel	$(5-8) \times 10^{-4}$
Aluminum	$(2-3) \times 10^{-4}$
Concrete	$(2-3) \times 10^{-2}$
Wood	$(2-4) \times 10^{-3}$
Riveted steel truss	$(2-4) \times 10^{-2}$
Soft rubber	$(3-5) \times 10^{-2}$

TABLE C.2 MATERIAL CONSTANTS

Material	Density ρ (kg/m^3)	Young's modulus E (N/m^2)	Shear modulus G (N/m^2)
Steel	7.8×10^3	2.1×10^{11}	7.9×10^{10}
Aluminum	2.7×10^3	7.1×10^{10}	2.6×10^{10}
Concrete	1.1×10^3	4.2×10^9	—
Beryllium copper	8.2×10^3	1.2×10^{11}	4.8×10^{10}
Soft rubber	$.95 \times 10^3$	2.0×10^9	7.8×10^8

D

Elementary Matrix Relations

This appendix is by no means meant to be a substitute for a good course in linear algebra. All we will be doing here is reminding you of a few matrix facts and operations that you should have already seen in a previous course.

To start, a matrix is an ordered rectangular array of numbers, arranged in rows and columns. The simplest matrix we can have is a 1×1 matrix, that is 1 row and 1 column. Of course, this is just a scalar. Matrices that have only one row or one column of numbers are the next jump in complexity. We usually call such a one-dimensional array a *vector*. Obviously, a vector in the form of a column is called a column vector and a vector in the form of a row is termed a row vector. If X is a 4×1 column vector made up of 1's, for example, then

$$X = \begin{bmatrix} 1 \\ 1 \\ 1 \\ 1 \end{bmatrix} \tag{D.0.1}$$

Finally, an honest-to-goodness matrix consists of a number of columns and rows, such as the matrix illustrated below:

$$[P] \equiv \begin{bmatrix} p_{1,1} & p_{1,2} & \cdots & & p_{1,n} \\ p_{2,1} & p_{2,2} & \cdots & & \vdots \\ \vdots & \cdots & p_{m-1,n-1} & p_{m-1,n} \\ p_{m,1} & \cdots & p_{m,n-1} & p_{m,n} \end{bmatrix} \tag{D.0.2}$$

Note that we've used the same notation here as in the main text. A matrix is always indicated by a capital letter, surrounded by square brackets. A vector is indicated by a capital letter without any brackets and a scalar is indicated by a lowercase letter. If we're considering a particular matrix, say $[H]$, then we'll always indicate the entries of the matrix by a lowercase h, for example $h_{2,3}$. Writing $[h_{i,j}]$ is equivalent to writing $[H]$; they both represent the matrix composed of the elements $h_{i,j}$. We indicate the dimension of a matrix by the number of rows and columns it has. Thus the matrix $[P]$ illustrated above is an $m \times n$ matrix.

Taking the transpose of a matrix simply means interchanging the rows and columns. The transpose of an $m \times n$ matrix is therefore an $n \times m$ matrix. We indicate the transpose by the superscript T. Thus if $[B]$ is the transpose of $[A]$, we'd write

$$[B] = [A]^T \tag{D.0.3}$$

In this case we'd have

$$b_{i,j} = a_{j,i} \tag{D.0.4}$$

Matrices $[P]$ for which $p_{i,j} = p_{j,i}$ are called symmetric. Such matrices must be square (that is, the same number of rows as columns). The elements $p_{i,i}$ of a square matrix are known as the main diagonal elements (and lie on the main diagonal), while the other elements ($p_{i,j}$ with $i \neq j$) are known as the off-diagonal elements of the matrix.

For two matrices to be equal, all elements must be equal and in the same relative position, that is, for $[P]$ to equal $[Q]$ we'd require $p_{i,j}$ to equal $q_{i,j}$ for all i and j. Only if two matrices are the same size can they be added or subtracted. If $[C] = [A] \pm [B]$, then $c_{i,j} = a_{i,j} \pm b_{i,j}$.

Multiplication is a bit more tricky. If we're multiplying a matrix by a scalar, then we simply need multiply each element in the matrix by the scalar. Thus if $[B] = \alpha[A]$, where α is some scalar constant, then $b_{i,j} = \alpha a_{i,j}$. This isn't too hard. It's when we're multiplying two matrices together that we'll have difficulties. If we want to multiply two matrices together, such as $[P] = [Q][R]$, then

$$p_{i,j} = q_{i,1}r_{1,j} + q_{i,2}r_{2,j} + q_{i,3}r_{3,j} + \cdots = \sum_{k=1}^{o} q_{i,k}r_{k,j} \tag{D.0.5}$$

In the above operation, $[Q]$ is $m \times o$, $[R]$ is $o \times n$, and the resultant matrix $[P]$ is $m \times n$. It's pretty clear that, in general,

$$[Q][R] \neq [R][Q] \tag{D.0.6}$$

In fact, only if $[Q]$ and $[R]$ are square matrices can we form both $[Q][R]$ and $[R][Q]$ and, even then, it's usually not the case that both multiplications lead to the same result.

Division by matrices will be defined soon. First, let's see what is meant by matrix division. With scalars, division is straightforward. If

$$c = \frac{a}{b} \tag{D.0.7}$$

then we know that $a = cb$ and also $a = bc$.

When we talk about matrix division we don't write $\frac{[Q]}{[R]}$. Rather we'd write $[Q][R]^{-1}$ or $[Q]^{-1}[R]$. $[R]^{-1}$ is the inverse of $[R]$ and is the matrix equivalent of $\frac{1}{r}$. Note that the meanings of the two "divisions" just mentioned are quite different. If $[P] = [Q][R]^{-1}$ then $[P][R] = [Q]$. If $[P] = [Q]^{-1}[R]$ then $[Q][P] = [R]$. You can see that the important fact is what side the inverse is on. Why? If you look at the intermediate steps it'll be clear. We started with

$$[P] = [Q][R]^{-1} \tag{D.0.8}$$

If we postmultiply both sides by $[R]$, we'll have

$$[P][R] = [Q][R]^{-1}[R] \tag{D.0.9}$$

One of the facts about matrix inverses is, if a matrix has an inverse, then the matrix times its inverse is the identity matrix, that is,

$$[R][R]^{-1} = [R]^{-1}[R] = [I] \tag{D.0.10}$$

Thus we can simplify (D.0.9) to

$$[P][R] = [Q] \tag{D.0.11}$$

which is what we wanted to show. The same holds for $[P] = [Q]^{-1}[R]$, except in this case you'd premultiply both sides by $[Q]$. Note that a matrix can't have an inverse unless it is square. Furthermore, its determinant must be nonzero. And now, having mentioned and used both determinants and inverses, let's define them.

The determinant of a matrix $[P]$ is represented by $\det[P]$ and by $\|[P]\|$ and is defined by

$$\|[P]\| = \sum_{j=1}^{n} p_{i,j}\pi_{i,j} \tag{D.0.12}$$

where

$$\pi_{i,j} = (-1)^{i+j} \left| [\tilde{P}_{i,j}] \right| \tag{D.0.13}$$

$[\tilde{P}_{i,j}]$ stands for the matrix (not matrix entry) that's composed of what's left over when the ith row and jth column of $[P]$ have been removed. This smaller matrix is called a *minor* of $[P]$. $\pi_{i,j}$ is called the *cofactor* of $[P]$.

It's easy to calculate low-order determinants. The determinant of a scalar is just the scalar itself. If the matrix $[P]$ is 2×2, then its determinant is given by $p_{1,1}p_{2,2} - p_{1,2}p_{2,1}$. For higher order determinants you must rely on (D.0.12).

Using minors and cofactors, we can define a matrix inverse. First, we need to construct the adjoint of $[P]$. The adjoint is simply the transpose of the matrix composed of the cofactors of $[P]$,

$$\text{adj}[P] = [\pi_{i,j}]^{T} \tag{D.0.14}$$

One nice fact about the adjoint is that

$$[P]\text{adj}[P] = \|[P]\| [I] = \|[P]\| \tag{D.0.15}$$

Since

$$[P]^{-1}[P] = [P][P]^{-1} = [I] \tag{D.0.16}$$

we can use the foregoing two relations to find

$$[P]^{-1} = \frac{\text{adj}[P]}{\|[P]\|} \tag{D.0.17}$$

Notice that if the determinant of $[P]$ is equal to zero, then we won't get an inverse. In this case we'd say that $[P]$ is *singular*. Clearly, matrices that have nonzero determinants are nonsingular.

You should realize that you'll rarely be performing these operations by hand; it's far more efficient to use software such as MATLAB. It is, however, nice to know where the algorithms come from, especially if you run into some tricky matrices that give the software problems.

TABLE D.1 ASSORTED MATRIX RELATIONS

$$\begin{array}{l}
\text{If } [P] = [M] + [N] \text{ then } p_{i,j} = m_{i,j} + n_{i,j} \\
[M] + [N] = [N] + [M] \\
[M][N] \neq [N][M] \quad \text{in general} \\
[M]\,([N][P]) = ([M][N])\,[P] \\
([M] + [N])^T = [M]^T + [N]^T \\
([M][N])^T = [N]^T[M]^T \\
([M][N][P])^T = [P]^T[N]^T[M]^T \\
([M][N][P])^{-1} = [P]^{-1}[N]^{-1}[M]^{-1}
\end{array}$$

Bibliography

[1] Boyce, W. E., and DiPrima, R. C. (1977). *Elementary Differential Equations and Boundary Conditions* (3rd ed.). John Wiley & Sons, New York.

[2] Dowell, E. H. (1984). "On Asymptotic Approximations to Beam Mode Shapes." *Journal of Applied Mechanics*, **51**, pp. 439.

[3] Ewins, D. J. (1984). *Modal Testing: Theory and Practice*. Research Studies Press, Letchworth, England.

[4] Farley, C., Glasheen, J., and McMahon, T. (1993). "Running Springs: Speed and Animal Size." *Journal of Experimental Biology*, **185**, pp. 71–86.

[5] Inaudi, J. (1991). Active Isolation: Analytical and Experimental Studies, Ph.D. dissertation, Department of Civil Engineering, University of California, Berkeley.

[6] Inman, D. J. (1994). *Engineering Vibration*. Prentice Hall, Englewood Cliffs, NJ.

[7] Kamal, M. M., and Wolf, J. A., Jr. (eds.). (1982). *Modern Automotive Structural Analysis*. Van Nostrand Reinhold, New York.

[8] Karnopp, D., and Rosenberg, R. (1975). *System Dynamics: A Unified Approach*. John Wiley & Sons, New York.

[9] Newland, D. E. (1984). *An Introduction to Random Vibrations and Spectral Analysis* (2nd ed.). Longman Scientific and Technical, New York.

[10] Rayleigh, J. W. S. (1945). *The Theory of Sound*, Volume Two. Dover Publications, New York.

[11] Synge, J. L., and Griffith, B. A. (1959). *Principles of Mechanics*. McGraw-Hill, New York.

[12] Wahl, A. M. (1949). *Mechanical Springs*. Penton Publishing, Cleveland.

Selected Readings

There are many vibrations text out there that can and should be perused by the interested student. Some are listed below.

Structural Dynamics: An Introduction to Computer Methods, R. R. Craig, Jr., John Wiley and Sons, New York, 1981.

Mechanical Vibration (4th ed.), J. P. Den Hartog, McGraw-Hill, New York, 1956.

Vibration for Engineers, A. D. Dimarogonas and S. Haddad, Prentice Hall, Englewood Cliffs, New Jersey, 1992.

Engineering Vibration, D. J. Inman, Prentice Hall, Englewood Cliffs, New Jersey, 1994.

Elements of Vibration Analysis (2nd ed.), L. Meirovitch, McGraw-Hill, New York, 1986.

Analytical Methods in Vibration, L. Meirovitch, Macmillan, New York, 1967.

Fundamentals of Vibration Analysis, N. O. Myklestad, The Maple Press Co., York, Pennsylvania, 1956.

Mechanical Vibrations (2nd ed.), S. S. Rao, Addison-Wesley, Reading, Massachusetts, 1990.

The Theory of Sound, Volume Two, J. W. S. Rayleigh, Dover Publications, Inc., New York, 1945.

An Introduction to Mechanical Vibrations (3rd ed.), R. F. Steidel, John Wiley and Sons, New York, 1989.

Theory of Vibration with Applications (4th ed.), W. T. Thomson, Prentice Hall, Englewood Cliffs, New Jersey, 1993.

Just the Facts

$$mx\ddot{} + kx = 0$$

$$\ddot{x} + \omega_n^2 x = 0$$

Equations of motion Spring-Mass system (1) (2)

$$x(t) = a_1 e^{it\sqrt{\frac{k}{m}}} + a_2 e^{-it\sqrt{\frac{k}{m}}}$$

$$x(t) = b_1 \cos\left(\sqrt{\frac{k}{m}}t\right) + b_2 \sin\left(\sqrt{\frac{k}{m}}t\right)$$

Free vibration solution (3) (4)

$$k_{eq} = k_1 + k_2 \quad \} \text{ Springs in parallel}$$

$$k_{eq} = \left(\frac{k_1 k_2}{k_1 + k_2}\right) \quad \} \text{ Springs in series}$$

$$\ddot{\theta} + \frac{g}{l}\theta = 0 \quad \} \text{ Pendulum equation of motion}$$

$$m\ddot{x} + c\dot{x} + kx = 0$$

$$\ddot{x} + 2\zeta\omega_n\dot{x} + \omega_n^2 x = 0$$

Equations of motion-spring-mass-damper system (5) (6)

$$x(t) = e^{-\zeta\omega_n t}[b_1 \cos(\omega_d t) + b_2 \sin(\omega_d t)] \quad \} \text{ Free vibration solution} \quad (7)$$

$$\frac{d}{dt}\frac{\partial L}{\partial \dot{q}_i} + \frac{\partial RD}{\partial \dot{q}_i} - \frac{\partial L}{\partial q_i} = Q_i, \quad i = 1\cdots n \quad \} \text{ Lagrange's equations} \quad (8)$$

Chapter 2

$$\ddot{x} + \omega_n^2 x = \omega_n^2 y \quad \left.\right\} \begin{array}{l} \text{Equation of motion-seismically} \\ \text{excited spring-mass system} \end{array} \qquad (9)$$

$$x(t) = \frac{\omega_n^2 \bar{y}}{(\omega_n^2 - \omega^2)} \sin(\omega t) \quad \left.\right\} \text{Solution} \qquad (10)$$

$$\ddot{x} + \omega_n^2 x = \frac{f(t)}{m} \quad \left.\right\} \begin{array}{l} \text{Equation of motion-direct force} \\ \text{excited spring-mass system} \end{array} \qquad (11)$$

$$\bar{x} = \frac{\bar{f}}{m(\omega_n^2 - \omega^2)} \quad \left.\right\} \text{Solution} \qquad (12)$$

$$\ddot{x} + 2\zeta \omega_n \dot{x} + \omega_n^2 x = \frac{\bar{f}}{m} \sin(\omega t) \quad \left.\right\} \begin{array}{l} \text{Equation of motion-direct force} \\ \text{excited spring-mass-damper} \end{array} \qquad (13)$$

$$x(t) = -\frac{(2\zeta \omega \omega_n)\frac{\bar{f}}{m}}{(\omega_n^2 - \omega^2)^2 + (2\zeta \omega \omega_n)^2} \cos(\omega t) + \frac{(\omega_n^2 - \omega^2)\frac{\bar{f}}{m}}{(\omega_n^2 - \omega^2)^2 + (2\zeta \omega \omega_n)^2} \sin(\omega t) \quad \left.\right\} \text{solution} \qquad (14)$$

$$\phi = \tan^{-1}\left(\frac{2\zeta \omega \omega_n}{\omega_n^2 - \omega^2}\right) \quad \left.\right\} \text{Phase lag-output to input} \qquad (15)$$

and

$$|\bar{x}| = \frac{\bar{f}}{m} \frac{1}{\sqrt{(\omega_n^2 - \omega^2)^2 + (2\zeta \omega_n \omega)^2}} \quad \left.\right\} \begin{array}{l} \text{Magnitude of displacement} \\ \text{response} \end{array} \qquad (16)$$

$$\ddot{x} + 2\zeta \omega_n \dot{x} + \omega_n^2 x = \omega_n^2 y + 2\zeta \omega_n \dot{y} \quad \left.\right\} \begin{array}{l} \text{Equation of motion-seismically} \\ \text{excited spring-mass-damper} \end{array} \qquad (17)$$

$$g(\omega) = \sqrt{\frac{\omega_n^4 + (2\zeta \omega \omega_n)^2}{(\omega_n^2 - \omega^2)^2 + (2\zeta \omega \omega_n)^2}} e^{i\phi}, \quad \phi = \phi_1 - \phi_2 \quad \left.\right\} \begin{array}{l} \text{Transfer function and} \\ \text{phase lag} \end{array} \qquad (18)$$

$$\phi_1 = \tan^{-1}\left(\frac{2\zeta \omega}{\omega_n}\right) \text{ and } \phi_2 = \tan^{-1}\left(\frac{2\zeta \omega \omega_n}{\omega_n^2 - \omega^2}\right)$$

$$\ddot{x} + 2\zeta \omega_n \dot{x} + \omega_n^2 x = e\beta \omega^2 \cos(\omega t) \quad \left.\right\} \begin{array}{l} \text{Equation of motion-rotating} \\ \text{imbalance} \end{array} \qquad (19)$$

where $\beta = \frac{m}{m_1+m}$, $\omega_n^2 = \frac{k}{m_1+m}$ and $2\zeta\,\omega_n = \frac{c}{m_1+m}$

$$x(t) = \frac{e\beta\omega^2\cos(\omega t - \phi)}{\sqrt{(\omega_n^2 - \omega^2)^2 + (2\zeta\,\omega\omega_n)^2}}, \qquad \phi = \tan^{-1}\left(\frac{2\zeta\,\omega\omega_n}{\omega_n^2 - \omega^2}\right) \left.\right\} \text{Solution} \qquad (20)$$

$$\sigma \equiv \ln\left(\frac{x(0)}{x(t_p)}\right) \left.\right\} \text{Log decrement} \qquad (21)$$

$$\zeta \approx \frac{\sigma}{2\pi} \qquad (22)$$

$$\omega_p = \omega_n\sqrt{1 - 2\zeta^2} \left.\right\} \text{Approximations to } \zeta \text{ and } \omega_p \qquad (23)$$

$$\zeta \approx \frac{\delta_h}{2\omega_n} \qquad (24)$$

$$c_{eq} = \frac{\gamma k}{\omega} \left.\right\} \begin{array}{l}\text{Equivalent linear damping for}\\ \text{structural damping}\end{array} \qquad (25)$$

$$x(t) = -\frac{1}{\omega_n^2}\ddot{y}(t) \left.\right\} \text{Accelerometer response} \qquad (26)$$

Chapter 3

$$f(t) = \sum_{n=0}^{\infty} a_n\cos(n\omega_0 t) + \sum_{n=1}^{\infty} b_n\sin(n\omega_0 t) \qquad (27)$$

$$a_0 = \frac{\omega_0}{2\pi}\int_0^{\frac{2\pi}{\omega_0}} f(t)\,dt \qquad (28)$$

Fourier series representation

$$a_n = \frac{\omega_0}{\pi}\int_0^{\frac{2\pi}{\omega_0}} f(t)\cos(n\omega_0 t)\,dt \qquad (29)$$

$$b_n = \frac{\omega_0}{\pi}\int_0^{\frac{2\pi}{\omega_0}} f(t)\sin(n\omega_0 t)\,dt \qquad (30)$$

$$x(t) = \int_0^t f(\tau)h(t-\tau)\,d\tau \left.\right\} \text{Convolution integral} \qquad (31)$$

$$\left|\frac{\ddot{x}}{g}\right|_{max} = \sqrt{1 + \frac{2x_h}{x_{eq}}} \left.\right\} \begin{array}{l} \text{Acceleration response due to a drop} \\ \text{impact} \end{array} \tag{32}$$

Chapter 4

$$[M]\ddot{X} + [K]X = O \left.\right\} \text{n DOF - Unforced} \tag{33}$$

$$X = \bar{X}e^{i\omega t} \left.\right\} \text{Assumed eigensolution}$$

$$\left[[K] - \omega^2[M]\right]\bar{X} = O \left.\right\} \text{n DOF eigenvalue problem} \tag{34}$$

$$\det\left([K] - \omega^2[M]\right) = 0 \left.\right\} \text{Characteristic equation} \tag{35}$$

$$[M]\ddot{X} + [K]X = \bar{F}\cos(\omega t) \left.\right\} \text{n DOF - forced} \tag{36}$$

$$\left[[K] - \omega^2[M]\right]\bar{X} = \bar{F} \left.\right\} \text{n DOF - response forcing relation} \tag{37}$$

$$[M']\ddot{H} + [K']H = F'\cos(\omega t) \left.\right\} \text{n DOF - normal form} \tag{38}$$

$$[M'] \equiv [U]^T[M][U], \quad [K'] \equiv [U]^T[K][U], \quad F' \equiv [U]^T F \tag{39}$$

$$[M]\ddot{X} + [C]\dot{X} + [K]X = \bar{F}\cos(\omega t) \left.\right\} \text{n DOF - forced with damping} \tag{40}$$

$$\bar{X} = \left[-\omega^2[M] + i\omega[C] + [K]\right]^{-1}\bar{F} \left.\right\} \begin{array}{l} \text{n DOF - forced response with} \\ \text{damping} \end{array} \tag{41}$$

$$X = [A]F \left.\right\} \begin{array}{l} \text{Flexibility influence coefficient} \\ \text{matrix} \end{array} \tag{42}$$

$$F = [K]X \left.\right\} \begin{array}{l} \text{Stiffness influence coefficient} \\ \text{matrix} \end{array} \tag{43}$$

Chapter 5

$$m(x)\ddot{\xi} = \frac{\partial}{\partial x}(EA(x)\xi_x) \quad \left.\right\} \text{ Equation of motion for a bar} \tag{44}$$

$$\xi(x, t) = \mathbf{x}(x)\mathbf{t}(t) \quad \left.\right\} \text{ Separated variable form} \tag{45}$$

$$\ddot{\mathbf{t}} + \omega^2 \mathbf{t} = 0 \quad \left.\right\} \text{ Time equation of motion} \tag{46}$$

$$\mathbf{x}_{xx} + \frac{m\omega^2}{EA}\mathbf{x} = 0 \quad \left.\right\} \text{ Deflection equation of motion} \tag{47}$$

$$\beta = \omega\sqrt{\frac{m}{EA}} \quad \left.\right\} \beta \text{ definition}$$

$$\xi(x, t) = \sum_{n=1}^{\infty} a_n \sin\left(\frac{n\pi x}{l}\right) \cos(\omega_n t + \phi_n) \quad \left.\right\} \begin{array}{l}\text{Full solution to particular end}\\\text{conditions}\end{array} \tag{48}$$

$$\rho(x)\ddot{u} + (EI(x)u_{xx})_{xx} = f(x, t) \quad \left.\right\} \text{ Equation of motion for a beam} \tag{49}$$

$$\mathbf{x}_{xxxx} - \frac{\omega^2 \rho}{EI}\mathbf{x} = 0 \quad \left.\right\} \text{ Deflection equation of motion} \tag{50}$$

$$m_m \ddot{a}_m(t) + k_m a_m(t) = c_m \bar{f}_m \cos(\omega t) \quad m = 1 \cdots, \infty \quad \left.\right\} \text{ Generalized forced response} \tag{51}$$

Chapter 6

$$\dot{y} \approx \frac{y_{i+1} - y_i}{\Delta} \quad \left.\right\} \text{ Discretization of velocity}$$

$$R_Q(X) = \frac{X^T[K]X}{X^T[M]X} \quad \left.\right\} \begin{array}{l}\text{Rayleigh's quotient for a discrete}\\\text{system}\end{array} \tag{52}$$

$$\overline{PE} = \tfrac{1}{2}X_i^T[K]X_i \quad \left.\right\} \begin{array}{l}\text{Maximal potential}\\\text{energy—discrete sytsem}\end{array} \tag{53}$$

$$\overline{KE}^0 = \tfrac{1}{2} X_i^T [M] X_i \quad \left\} \begin{array}{l} \text{Zero-frequency kinetic} \\ \text{energy—discrete system} \end{array} \right. \tag{54}$$

$$\begin{aligned} \mathcal{K}_{ij} &\equiv \hat{X}_i^T [K] \hat{X}_j \\ \mathcal{M}_{ij} &\equiv \hat{X}_i^T [M] \hat{X}_j \\ [\mathcal{K}]B - R_Q [\mathcal{M}]B &= 0 \end{aligned} \quad \left\} \begin{array}{l} \text{Rayleigh-Ritz formulation—} \\ \text{discrete system} \end{array} \right. \tag{55} \tag{56}$$

$$R_Q(\bar{\xi}) \equiv \frac{-\displaystyle\int_0^l \bar{\xi}(EA\bar{\xi}_x)_x \, dx}{\displaystyle\int_0^l m\bar{\xi}^2 dx} \tag{57}$$

$$\left. \begin{array}{l} \\ \\ \end{array} \right\} \begin{array}{l} \text{Rayleigh-Ritz formulation—} \\ \text{continuous system} \end{array}$$

$$R_Q(\bar{\xi}) = \frac{-EA\bar{\xi}\bar{\xi}_x \big|_0^l + \displaystyle\int_0^l EA\bar{\xi}_x^2 dx}{\displaystyle\int_0^l m\bar{\xi}^2 dx} \tag{58}$$

$$u(x,t) = \sum_{i=1}^n a_i(t)\psi_i(x)$$

$$[\mathcal{K}]\bar{A} - \omega^2[\mathcal{M}]\bar{A} = 0 \quad \left\} \text{Assumed modes formulation}\right.$$

Chapter 8

$$E(y) = \frac{1}{n} \sum_{i=1}^n y_i \tag{59}$$

$$\left. \begin{array}{l} \\ \\ \end{array} \right\} \text{Mean value of a signal}$$

$$E(y(t)) = \frac{1}{t_1} \int_0^{t_1} y(t) \, dt \tag{60}$$

$$\text{var}(y) = \frac{1}{n} \sum_{i=1}^n [y_i - E(y)]^2 \tag{61}$$

$$\left. \begin{array}{l} \\ \\ \end{array} \right\} \text{Variance and standard deviation}$$

$$\text{var}(y) = E[y - E(y)]^2 \tag{62}$$

$$\sigma = \sqrt{\text{var}(y)}$$

$$x(t) = \int_{-\infty}^{\infty} a(\omega) \cos(\omega t) \, d\omega + \int_{-\infty}^{\infty} b(\omega) \sin(\omega t) \, d\omega \quad \left.\right\} \tag{63}$$

$$a(\omega) = \frac{1}{2\pi} \int_{-\infty}^{\infty} x(t) \cos(\omega t) \, dt \qquad\qquad \text{Real form of Fourier transform} \tag{64}$$

$$b(\omega) = \frac{1}{2\pi} \int_{-\infty}^{\infty} x(t) \sin(\omega t) \, dt. \tag{65}$$

$$x(t) = \int_{-\infty}^{\infty} X(\omega) e^{i\omega t} \, d\omega \tag{66}$$

Complex form of Fourier transform

$$X(\omega) = \frac{1}{2\pi} \int_{-\infty}^{\infty} x(t) e^{-i\omega t} \, dt \tag{67}$$

$$R_{xx}(\tau) \equiv \lim_{T \to \infty} \frac{1}{T} \int_{-\frac{T}{2}}^{\frac{T}{2}} x(t) x(t + \tau) \, dt \quad \left.\right\} \text{Auto/correlation} \tag{68}$$

$$R_{xy}(\tau) \equiv \lim_{T \to \infty} \frac{1}{T} \int_{-\frac{T}{2}}^{\frac{T}{2}} x(t) y(t + \tau) \, dt \quad \left.\right\} \text{Cross-correlation} \tag{69}$$

$$S_{xx}(\omega) = \frac{1}{2\pi} \int_{-\infty}^{\infty} R_{xx}(\tau) e^{-i\omega \tau} \, d\tau \quad \left.\right\} \begin{array}{l}\text{Power Spectral density from auto-}\\\text{correlation}\end{array} \tag{70}$$

$$R_{xx}(\tau) = \int_{-\infty}^{\infty} S_{xx}(\omega) e^{i\omega \tau} \, d\omega \quad \left.\right\} \begin{array}{l}\text{Auto-correlation from power}\\\text{spectral density}\end{array} \tag{71}$$

$$|g(\omega)| = \sqrt{\frac{S_{xx}(\omega)}{S_{ff}(\omega)}} \quad \left.\right\} \begin{array}{l}\text{Transfer function from power}\\\text{spectral densities}\end{array} \tag{72}$$

$$\gamma^2 = \frac{g_1(\omega)}{g_2(\omega)} \quad \left.\right\} \text{Coherence function} \tag{73}$$

Index